Advances in Intelligent Systems and Computing

Volume 250

Series editor

Janusz Kacprzyk, Warsaw, Poland

For further volumes:
http://www.springer.com/series/11156

About this Series

The series "Advances in Intelligent Systems and Computing" contains publications on theory, applications, and design methods of Intelligent Systems and Intelligent Computing. Virtually all disciplines such as engineering, natural sciences, computer and information science, ICT, economics, business, e-commerce, environment, healthcare, life science are covered. The list of topics spans all the areas of modern intelligent systems and computing.

The publications within "Advances in Intelligent Systems and Computing" are primarily textbooks and proceedings of important conferences, symposia and congresses. They cover significant recent developments in the field, both of a foundational and applicable character. An important characteristic feature of the series is the short publication time and world-wide distribution. This permits a rapid and broad dissemination of research results.

Advisory Board

Srikanta Patnaik · Xiaolong Li
Editors

Proceedings of International Conference on Soft Computing Techniques and Engineering Application

ICSCTEA 2013, September 25–27, 2013, Kunming, China

 Springer

Editors
Srikanta Patnaik
Department of Computer Science
    and Engineering
SOA University
Bhubaneswar, Orissa
India

Xiaolong Li
Electronics and Computer
    Engineering Technology
Indiana State University
Indiana, IN
USA

ISSN 2194-5357   ISSN 2194-5365   (electronic)
ISBN 978-81-322-1694-0 ISBN 978-81-322-1695-7   (eBook)
DOI 10.1007/978-81-322-1695-7
Springer New Delhi Heidelberg New York Dordrecht London

Library of Congress Control Number: 2013956601

Preface

On behalf of the Programme Committee, we welcome you to *International Conference on Soft Computing Techniques and Engineering Application (SCTEA 2013)* held during September 25–27, 2013 in Kunming China. The main objective of SCTEA 2013 is to provide a platform for researchers, engineers, and academicians from all over the world to present their research results and recent developments in soft computing techniques and engineering application. This conference provides opportunities for them to exchange new ideas and application experiences face to face, to establish business or research relations, and to find global partners for future collaboration.

During the last decade, soft computing techniques have been recognized as one and most attractive alternatives to the standard, well-established "hard computing" paradigms. Traditional computing methods are often too cumbersome for the present day multiobjective problems, which requires a precisely stated analytical model and often consumes large computational time. Soft computing techniques, on the other hand, emphasize gains in understanding system behavior in exchange for unnecessary precision have proved to be important practical tools for many contemporary problems. There are various soft computing techniques available and popular for the engineering and other applications. They are artificial neural networks (NN), genetic algorithms (GA), fuzzy logic models (FLM), and particle swarm techniques. NNs and FLMs are universal approximators of any multivariate function because they can be used for modeling highly nonlinear, unknown, or partially known complex systems, plants, or processes. GA and particle swarm optimization techniques have emerged as potential and robust optimization tools in recent years.

The International Conference on Soft Computing Techniques and Engineering Application (SCTEA 2013) covers all areas of soft computing and its applications to engineering domain. We have received 232 numbers of papers through "Call for Paper," out of which 67 numbers of papers were accepted for publication in the conference proceedings through double blind review process. The broad areas, which are covered in this Conference, are: theoretical computer science, computer systems organization, soft computing, computing methodologies, computer applications, and software engineering application.

We are sure this proceeding will contribute to the knowledge of soft computing and particularly its application to various engineering domains. We are sure the participants will share their innovations in this conference.

Srikanta Patnaik
Xiaolong Li

Organizing Committee

General Chairs

Prof. Srikanta Patnaik, Computer Science and Engineering in SOA University, India

General Co-Chair

Dr. Xiaolong Li, Indiana State University, USA

Technical Program Committee

Prof. Ilona Weinreich, Koblenz University of Applied Sciences, Germany
Prof. Flaminia L. Luccio, Università Ca' Foscari Venezia, Italy
Dr. Rajarshi Gupta, University of Calcutta, India
Prof. Yi-Hsing Chang, Southern Taiwan University of Science and Technology, Taiwan
Prof. Min-Shiang Hwang, Asia University, Taiwan
Prof. Chien-Lung Hsu Chang-Gung University, Taiwan
Prof. Chien-Jen Wang, National University of Tainan, Taiwan
Prof. W. David Pan, University of Alabama in Huntsville, USA
Prof. Sotirios G. Ziavras, New Jersey Institute of Technology, USA
Dr. Afolabi Babajide Samuel, The Computer Professionals Registration Council of Nigeria, Nigeria
Prof. Mauricio Papa, The University of Tulsa, USA
Prof. Tianshi Lu, Wichita State University, USA
Prof. Rita Moura Fortes, Mackenzie Presbyterian University, Brazil

Contents

Research on Hierarchical Clustering Algorithm Based on Cluster Outline..................................... 1
Hai-Dong Meng, Jing-Pei Ren and Yu-Chen Song

Exploration of Fault Diagnosis Technology for Air Compressor Based on Internet of Things.............................. 11
Zheng Yue-zhai and Chen Xiao-ying

An Enhancement Algorithm Based on Fuzzy Sets Algorithm Using Computer Vision System for Chip Image Processing......... 17
Chengxiang Tan, Lina Yang and Xichun Li

Development of DDoS Attack Defense System Based on IKEv2 Protocol 25
Qing Tan and Xiaojing Yue

Power Data Network Dynamic Simulation Platform............. 33
Qian Guo, Xingchuan Bao and Gu Feng

An Emotional Model Based on Multiple Factors................ 43
Qiong Xiao, Gangyi Ding and Yongkang Liu

Multi-Source and Heterogeneous Knowledge Organization and Representation for Knowledge Fusion in Cloud Manufacturing 55
Jihong Liu, Wenting Xu and Hongfei Zhan

Research of Dynamics and Deploying Control Method on Tethered Satellite 63
Lei Gang, Xian Yong, Feng Jie and Wang Kui

Using Auto-Associative Neural Networks for Signal Recognition Technology on Sky Screen   71
Yan Lou, Zhipeng Ren, Yiwu Zhao and Yugui Song

Research on the Hotspot Information Push System for the Online Journal Based on Open-Source Framework   81
Jiya Jiang, Tong Liu, Yanqing Shi and Changhua Lu

Development of Control System of Wheel Type Backhoe Loader   87
Luwei Yang

Some Results on Fuzzy Weak Boolean Filters of Non-commutative Residuated Lattice   97
Wei Wang, Yang Xu, Dan Tong, Xiao-yan Cheng and Yong-fei Li

CGPS: A Collaborative Game in Parking-Lot Search   105
Peng Li, Demin Li and Xiaolu Zhang

A Fuzzy-Based Context-Aware Privacy Preserving Scheme for Mobile Computing Services   115
Eric Ke Wang and Yunming Ye

Research and Application of Trust Management System   123
Fengyin Li and Peiyu Liu

Ranaad-Xek: A Prototype Design of Traditional Thai Musical Instrument Application for Android Tablet PC   131
Kasikrit Damkliang, Chawee Kaeoaiad and Sulkiplee Chehmasong

Identifying Accurate Refactoring Opportunities Using Metrics   141
Yixin Bian, Xiaohong Su and Peijun Ma

Research on Neural Network Predictive Control of Induction Motor Servo System for Robot   147
Chaofa Yu, Zelong Zhou, Zhiyong Chen and Xiangyong Su

Research on Scale-Out Workloads and Optimal Design of Multicore Processors   157
Qiong Wang, Li Shen and Zhiying Wang

Study of Modified Montgomery's Algorithm and Its Application to 1,024-bit RSA   167
Yulin Zhang and Xinggang Wang

A MVS-Based Object Relational Model of the Internet of Things 177
Huijuan Zhang and Ran Xu

**Rateless Code-Based Unequal Loss Protection for Layer-Coded
Media Delivery** 185
Xuan Dong, Shaohe Lv, Hu Shen, Junquan Deng, Xiaodong Wang
and Xingming Zhou

Construction of the Grade-3 System for GJB5000A-2008 195
Yonggang Li, Jinbiao Zhou, Jianwei He, Xiangming Li and Libing Guo

**Virtual Training System of Assembly and Disassembly
Based on Petri Net** 205
Xiaoqiang Yang, Jinhua Han and Yi Pan

**The Design of Visual RBAC Model Based on UML
and XACML Integrating** 213
Baode Fan and Mengmeng Li

**A New Approach to Reproduce Traffic Accident Based on the Data
of Vehicle Video Recorders** 223
Hong Li, Qing Kang and Jing He

**Improved RNS Montgomery Modular Multiplication
with Residue Recovery** 233
Tao Wu, Shuguo Li and Litian Liu

**Functionally Equivalent C Code Clone Refactoring by Combining
Static Analysis with Dynamic Testing** 247
Xiaohong Su, Fanlong Zhang, Xia Li, Peijun Ma and Tiantian Wang

**Architecture Designing of Astronaut Onboard Training System
Based on AR Technology** 257
He Ning, Hou Quanchao and Hu Fuchao

Design and Implementation of Bibliometrics System Based on RIA . . . 263
Geyang Han and Bing Sun

**SAR Image Filtering Based on Quantum-Inspired Estimation
of Speckle Variance** 273
Xiaowei Fu, Li Chen, Jing Tian, Xin Xu and Yi Wang

**Image Semantic Annotation Approach Based
on the Feature Matching** 281
Cong Jin and Jinglei Guo

**Research on Transmission and Transformation Land Reclamation
Based on BP Neural Network** 289
Xi Wu, Hai-Ting Ming, Xue-Huan Qin and Wen-Jing Zhu

**TTP-ACE: A Trusted Third Party for Auditing
in Cloud Environment** 299
Songzhu Mei, Haihe Ba, Fang Tu, Jiangchun Ren and Zhiying Wang

**Pattern Recognition Based on the Nonparametric Kernel
Regression Method in A-share Market** 309
Huaiyu Sun, Mi Zhu and Feng He

**The Research on the Detection and Defense Method
of the Smurf-Type DDos Attack** 315
Wantian Cao and Xingchuan Bao

**A Preliminary Analysis of Web Usage Behaviors from Web
Access Log Files** 325
Thakerng Wongsirichot, Sukgamon Sukpisit and Warakorn Hanghu

**Assessment of BER Performance of a Power Line Communication
System in the Presence of Transformer and Performance
Improvement Using Diversity Reception** 333
Munshi Mahbubur Rahman and S. P. Majumder

**A Multi-Constraint Anonymous Parameter Design Method
Based on the Attribute Significance of Rough Set** 345
Taorong Qiu, Lu Liu, Wenying Duan, Xiaoming Bai and Zhongda Lin

**Design and Implementation of a Middleware for Service-Oriented
Distributed Systems** 351
Hong Xie, Donglin Su, Yijia Pan and Zhongfu Xu

**Refactoring Structure Semantics Similar Clones Combining
Standardization with Metrics** 361
Xia Li, Xiaohong Su, Peijun Ma and Tiantian Wang

A Retail Outlet Classification Model Based on AdaBoost 369
Kai Liu, Bing Wang, Xinshi Lin, Yeyun Ma and Jianqiang Xing

Extensions of Statecharts with Time of Transition, Time Delay of Message Transmitting, and Arrival Probability of Message 381
Junqiao Li, Jun Tang and Shuang Wan

The Optimization of Hadoop Scheduling Algorithms on Distributed System for Processing Traffic Information 389
Weizhen Sun and Xiujin Wang

Understanding the Capacity Scaling of Personal Communications Services 397
Zheng Wang

VHDL Implementation of Complex Number Multiplier Using Vedic Mathematics 403
Laxman P. Thakare, A. Y. Deshmukh and Gopichand D. Khandale

Key Security Technologies of Cloud Computing Platforms 411
Liang Junjie

The Measurement and Analysis of Software Engineering Risk Based on Information Entropy 419
Ming Yang, Hongzhi Liao, Rong Jiang and Junhui Liu

GRACE: A Gradient Distance-Based Peer-to-Peer Network Supporting Efficient Content-Based Retrieval 427
Jianming Lv, Can Yang and Kaidong Liang

Image Denoising Using Discrete Orthonormal S-Transform 435
Feng-rong Sun, Paul Babyn, Yu-huan Luan, Shang-ling Song and Gui-hua Yao

Analyzing Services Composition Using Petri Nets 443
Jiajun Xu and Shuzhen Yao

FPGA-Based Image Processing for Seamless Tiled Display System . . . 451
Mingyu Wang, Yan Han, Rui Wang, Xiaopeng Liu and Yuji Qian

2D Simulation of Static Interface States in GaN HEMT with AlN/GaN Super-Lattice as Barrier Layer 457
Imtiaz Alamgir and Aminur Rahman

Study on Model and Platform Architecture of Cloud Manufacturing for Aerospace Conglomerate 467
Jihong Liu, Hongfei Zhan and Wenting Xu

Service Composition Algorithm for Vehicle Network Based on Multiple Ontology 475
Yamei Xia and Chen Liu

Design and Application of Virtual Laboratory for Photography 483
Xuefei Shi

The Research of Travel-Time Tomography Based on Forward Calculation and Inversion 491
Yaping Li and Suping Yu

Protein Secondary Structure Prediction Based on Improved C-SVM for Unbalanced Datasets 499
Ao Pei

An Algorithm for Speckle Noise Based on SVD and QSF 507
Weizhou Zhao, Hui Zhang, Baozhen Yang and Huili Jing

A Generation Model of Function Call Based on the Control Flow Graph 513
Weizhen Sun and Xiangyan Du

A Novel Community-Based Trust Model for P2P Networks 521
Songxin Wang

The Algorithm of Mining Frequent Itemsets Based on MapReduce... 529
Bo He

Multi-Feature Metric-Guided Mesh Simplification 535
Hailing Wang, Fu Qiao and Bo Zhou

Research on Medical Image Fusion Algorithms Based on Nonsubsampled Contourlet 543
Junyu Long, Hong Yu and Aiming Yu

Influence of Previous Cueing Validity on Gaze-Evoked Attention Orienting 553
Qian Qian, Yong Feng, Lin Shi and Feng Wang

Application of RBF Neural Network in Intelligent Fault Diagnosis System 561
Yingying Wang, Ming Chang, Hongwei Chen and Ming Qian Wang

An Analysis of the Keys to the Executable Domain-Specific Model . . .   567
Qing Duan, Junhui Liu, Zhihong Liang, Hongwei Kang
and Xingping Sun

About the Editors   575

Author Index   577

Research on Hierarchical Clustering Algorithm Based on Cluster Outline

Hai-Dong Meng, Jing-Pei Ren and Yu-Chen Song

Abstract The traditional hierarchical methods always fail to take both the features of connectivity and proximity of clusters into consideration at the same time. This paper presents a hierarchical clustering algorithm based on cluster outline, which effectively addresses clusters of arbitrary shapes and sizes, and is relatively resistant to noise and easily detects outliers. The definition of the boundary point and cluster outline is firstly given, and the standard and approach of measuring similarity between clusters is then taken with the feature of connectivity and proximity of clusters. The experiments on the Iris and image data sets confirm the feasibility and validity of the algorithm.

Keywords Connectivity · Proximity · Cluster outline · Hierarchical clustering

1 Introduction

Hierarchical algorithm in cluster analysis as a main method in data mining, no matter what the field of research and application is, has been developed largely and promoted greatly. In recent years, there are various hierarchical clustering algorithm put forward, which are partitioned into two categories in terms of the similarity standard: one for connectivity-based similarity and the other for proximity-based similarity. The first one is free to find clusters of arbitrary shapes, but

H.-D. Meng (✉) · J.-P. Ren · Y.-C. Song
Inner Mongolia University of Science and Technology, Baotou 014010, China
e-mail: haidongm@imust.edu.cn

J.-P. Ren
e-mail: fm190135@126.com

Y.-C. Song
e-mail: songyuchen@imust.edu.cn

S. Patnaik and X. Li (eds.), *Proceedings of International Conference on Soft Computing Techniques and Engineering Application*, Advances in Intelligent Systems and Computing 250, DOI: 10.1007/978-81-322-1695-7_1, © Springer India 2014

lack of the consideration for density of certain cluster; the second is only limited in discovering spherical-shaped clusters, but the best consideration for density is given in clustering. At present, there are some representative hierarchical clustering algorithms such as BIRCH, CURE, ROCK, and Chameleon algorithm, and all of them have both of merits and inadequacies. The BIRCH [1] algorithm cannot effectively discover non-spherical-shaped clusters because of the method of dealing with borders of clusters with the radius or diameter of clusters, and it is not feasible in addressing the high-dimensional data sets because the exact number of clusters and diameters can be hardly obtained. The CURE [2] algorithm can handle various shaped clusters, but the efficiency of the algorithm is low, and it is subjective to choosing the number of typical points, regardless of consideration for the connectivity between clusters. In 1999, Sudipno Guha put forward the ROCK [3] algorithm that adopts a approach about the global information based on the number of public linking points to measure the correlation of data points, this means the feature of connectivity is considered, but the proximity is ignored. In 1999, Karypis proposes the Chameleon [4] hierarchical clustering algorithm based on the dynamic modeling technology, though the Chameleon algorithm takes both connectivity and proximity into consideration, but there are still some shortcomings as follows: (1) The value of K in the K-nearest figure must be given before the experiment; (2) there is difficulty in the Min-Bisection. (3) the threshold of similarity must be given before the experiment [5], and in the literature [6], the probability of the limitation of dealing with high-dimensional data sets is pointed out in the Chameleon algorithm; In the stage of partition, an object may be assigned to a cluster, which doesn't actually belong to the cluster.

Considering the above clustering methods, the hierarchical clustering algorithm based on cluster outline is proposed in this paper, which reduces the influences of the existence of outliers, considers reasonably both the connectivity and proximity between clusters with an automatic choice of cluster centers, and optimizes the clustering result effectively.

2 Definition and Solution

2.1 The Definition of Boundary Point and Cluster Outline

Cluster boundary points are defined as follows: In a cluster with n-dimensional data set, all data objects are regarded as spatial points distributed in n-dimensional data space. If a spatial point is discussed and considered in certain scope (such as the similarity-measuring standards based on distance and density), and a spatial point does not belong to all of 2^n quadrants in n-dimensional data space at the same time, the point can be defined as a boundary point (as Figs. 1 and 2 shown in two-dimensional data space). The outline of a cluster consists of all boundary points belonging to the cluster (as Figs. 3 and 4 shown in two-dimensional data space).

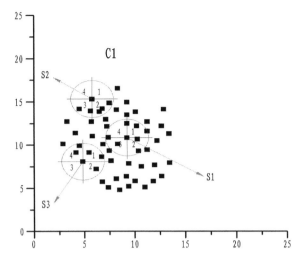

Fig. 1 Boundary points of cluster C_1

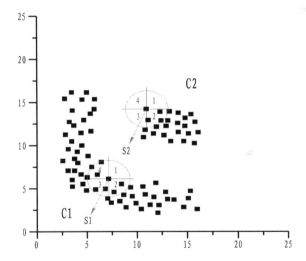

Fig. 2 Boundary points of cluster C_2

In Fig. 1, the point S_1 of the cluster C_1 is not an boundary point, because in its scope based on distance or density, there are always one or more points that can be found in any of its own quadrants of 1, 2, 3, and 4; the scope of the point S_2 does not touch any point in its own quadrants of 1 and 4, so the point S_2 is an boundary point; the third quadrant of the S_3 can cover no point, so the S_3 is an boundary point.

In Fig. 2, the points S_1 and S_2, respectively, from C_1 and C_2 are boundary points with the same as shown in Fig. 1.

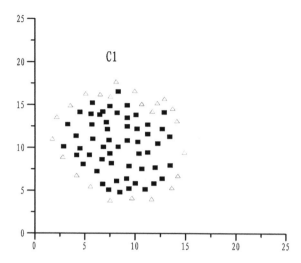

Fig. 3 Cluster outline C_1

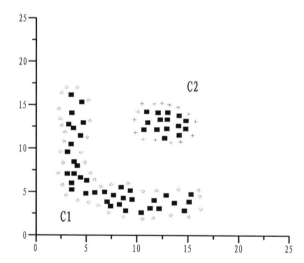

Fig. 4 Outlines of clusters C_1 and C_2

Cluster outline is defined as follows: All boundary points consist of the cluster outline; that is also the cluster outline is a collection of all boundary points.

In Fig. 3, the cluster outline C_1 consists of all red triangle symbols.

In Fig. 4, the cluster outline C_1 consists of all red diamond symbols, and C_2 all blue cross symbols.

2.2 The Solution of Cluster Outline

The solution of finding cluster outline aims at searching for all boundary points of cluster. Boundary points are the space within the cluster of a relative positional relationship of the points, so the cluster constructs the relative position of the point and point matrix. A complete solution of cluster outline includes three steps.

(1) Construction of the relative matrix P

Suppose a cluster C, $C = \{S_1, S_2, S_3, \ldots S_n\}$, and the cluster C is discussed and considered in two-dimensional data space consisting of the x axis and the y axis. The symbol D_0 represents the threshold of distance or density. The expression $D(S_i, S_j)$ represents the relative distance or density between the points S_i and S_j. The relative distance or density between the points S_i and S_j is defined as follows:

$$D(S_i, S_j) \leq D_0 \tag{1}$$

If the formula $D(S_i, S_j) \leq D_0$ is true, and the expression $S_{jX} > S_{iX}$ & $S_{jY} > S_{iY}$ is true, this means that S_j is in the first quadrant of S_i and $P_r = 1$. So we can get the following expressions:

$$\begin{aligned}
S_{jX} > S_{iX} \;\&\; S_{jY} > S_{iY} &\Rightarrow P_r = 1 \\
S_{jX} > S_{iX} \;\&\; S_{jY} < S_{iY} &\Rightarrow P_r = 2 \\
S_{jX} < S_{iX} \;\&\; S_{jY} < S_{iY} &\Rightarrow P_r = 3 \\
S_{jX} < S_{iX} \;\&\; S_{jY} > S_{iY} &\Rightarrow P_r = 4
\end{aligned} \tag{2}$$

Note: for the point S_i, the value of X is the value of S_{iX}, and the value of Y is the value of S_{iY}.

The relative matrix P is constructed as follows:

$$P = \begin{array}{c} \\ S_1 \\ S_2 \\ S_i \\ S_n \end{array} \begin{array}{cccc} S_1 & S_2 & S_i & S_n \\ \begin{pmatrix} 0 & P_r & P_r & P_r \\ P_r & 0 & P_r & P_r \\ P_r & P_r & 0 & P_r \\ P_r & P_r & P_r & 0 \end{pmatrix} \end{array} \quad \left(i < n; \text{ if } D(S_i, S_j) > D_0, \text{ then } P_r = 0 \right) \tag{3}$$

(2) Search for boundary point

To figure out whether the point S_i is an boundary point or not, we can check the i column of S_i to find all different values of P_r. If the set composed of all values of P_r in the i column of S_i is not $\{1, 2, 3, 4\}$, the point S_i is an boundary point.

(3) Search for cluster outline

The cluster outline is a definite set of all boundary points belonging to cluster in the step (2).

3 The Design of Hierarchical Clustering Algorithm Based on Cluster Outline

3.1 The Clustering Between Clusters

(1) Standard of connectivity

Suppose that the outline of the cluster C_1 is P_{r1} and C_2 is P_{r2}. The symbol N shows the number of all shared boundary points between clusters. The calculation of N is as follows:

$$N_1 = \text{count}(P_{r1} \cap P_{r2}) \tag{4}$$

$$N_2 = \text{count}(P_{r1} \cap C_2) \tag{5}$$

$$N_3 = \text{count}(P_{r2} \cap C_1) \tag{6}$$

$$N_4 = N_2 + N_3 - N_1 \tag{7}$$

Note: N_1 represents the number of the same points of P_{r1} and P_{r2}; N_2 represents the number of the same points of P_{r1} and C_2; N_3 represents the number of the same points of P_{r2} and C_1. Figure 5 shows the connectivity of the cluster outlines.

In Fig. 5, the clusters C_1 and C_2 share with the points of the same shape and color. The number of all boundary points of C_1 and C_2 is 17, which contains 10 yellow diamonds, 4 dark red stars, and 3 light red sunshine-shaped points. That also means the connectivity of the outline of the clusters C_1 and C_2 is 17.

(2) Standard of proximity

Suppose that the number of clusters is N with its corresponding outline of each cluster: $S_1, S_2, \ldots, S_n, S_i = \{B_{i1}, B_{i2}, \ldots, B_{ij}\}$, where S_i represents the outline of the cluster i and B_{ij} represents the boundary point j of the cluster i, then the average distance as the measurement of proximity of clusters is defined. The following formula shows the average distance between the cluster outlines:

$$\overline{D}(S_i, S_j) = \frac{1}{n_{S_i} n_{S_j}} \sum_{B_i \in S_i} \sum_{B_j \in S_j} \text{dist}\|B_i - B_j\|. \tag{8}$$

3.2 The Clustering Between Clusters and Outliers

Considering a ratio r, the symbol n_r represents the number of all boundary points of one cluster, and the symbol n represents all points of the cluster. The ratio r can be set up as follows:

Fig. 5 Connectivity of the outlines of clusters C_1 and C_2

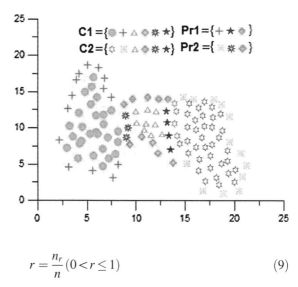

$$r = \frac{n_r}{n} \quad (0 < r \leq 1) \tag{9}$$

The smaller the value of r, the denser the cluster, and vice versa.

When an outlier is assigned to a cluster to form a new cluster, the ratio of the new cluster is r'. If r' is less than or equal to r (the ratio of the original cluster), the outlier belongs to the new cluster. If the condition just mentioned is not true, we should keep the original situation not to be changed. The snake-shaped and linear clusters can be easily detected in this way to some degree.

3.3 The Process of the Algorithm

For a data set D_0, the hierarchical clustering algorithm based on cluster outline is as follows:

1. Input the value of threshold of distance or density $(d_{min} < r_0 < d_{max})$, the number of data points: n, and the number of clusters: K.
2. Construct distance or density matrix d_{matrix}.
3. Take a data point X from the data set D_0, search out all points based on the threshold of distance of X in d_{matrix}, and combine X and all the points into a cluster in the condition of the number of all data points more than n, or else set these points as temporary outliers.
4. Construct the relative matrix P and form outlines of all clusters with step 3.
5. Calculate the number of the shared boundary points between clusters (the connectivity matrix), begin to cluster from the biggest number, and readjust the cluster outline until there is K clusters or the number of the shared boundary points is up to n.
6. Repeat the step 4 for the new outline.

Fig. 6 Original image
(500 × 375)

7. If the number of clusters is still more than K, then the proximity matrix can be calculated and the min distance can be combined into a cluster. Renew the cluster outlines until there are K clusters.
8. Handle the outliers according to equation (9).

4 The Experimental Analysis

Experiment 1: In order to facilitate the realization of the data sets and visualization of the experiment results, the experiment takes an image data as an experimental object, which consists of 500 pixels of width and 375 pixels of height. Each data point is represented by all RGB color values of the pixel. The following figures show the result of the experiment by contrast between the original image in Fig. 6 and the processed image in Fig. 7.

In Fig. 7, there are 10 clusters and an outlier data set. The experiment shows the advantage effectiveness in clustering with this algorithm.

Experiment 2: In order to validate further the feasibility and effectiveness of the hierarchical clustering algorithm based on cluster outline, the experiment 2 is given with the Iris data set from the UCI. The Iris data set is a kind of biometric data set of three various flowers, which includes 150 records divided into three clusters, each cluster has 50 records, and each record consists of four attributes.

The results of the clustering in the Iris data set are shown in Fig. 8 by contrast between the clustering algorithm based on the cluster outline, the CURE algorithm, and the various algorithms provided by SPSS statistical analysis software.

As shown in Fig. 8, the algorithm accuracy of the cluster outline and the centroid clustering in Iris data set obtained is more than 90 % (there are 14 records in error). They show a relatively good clustering effect.

Fig. 7 Simulated image of experimental result

Fig. 8 Accuracy comparison between clustering algorithms

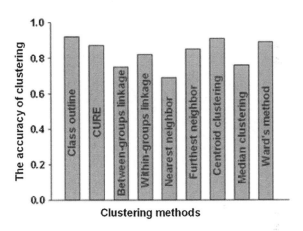

The values of threshold of the initial distance and number of points can be hardly determined in clustering. In order to address this, the initial value of threshold can be small, and then, the processed points are compared with the rest of the outliers; if the rest of the outliers are more than 10 % of the number of the total data points, the value of threshold of the outliers should be set bigger to cluster again. It can guarantee that the number of new clusters combining the subclusters in terms of connectivity is more than K, and then, subclusters are combined again in terms of proximity; if the value of threshold of the number of the initial data points is given too small, the number of clusters can be less than K, and it can also give rise to the worst situation that there is only one cluster including all data points.

There is almost no change in the time complexity of the cluster-outline-based algorithm, in contrast to other traditional clustering algorithms, and in the worst situation, the time complexity is $O(n^2)$, where the time is mainly consumed in dealing with distance matrixes and relative matrixes. However, the points of outline are processed instead of all points of whole clusters, so the space complexity is relatively low, especially for densely based clusters.

5 Conclusions

The feasibility and effectiveness of the hierarchical clustering algorithm based on cluster outline are proven in the experiments. Because of the standard of measuring the similarity between clusters based on both connectivity and proximity, the clusters with arbitrary shape, noise, or outliers can be easily discovered and detected. And the cluster outline is chosen to replace the whole cluster, so the task of dealing with huge data points can be avoided in the algorithm.

The concept of cluster outline has much practical utility in application, which can also be applied to other traditional clustering methods, such as discovering shapes of clusters by the visualization of cluster outline; measuring density of clusters by the proportion between the number of boundary points and all points of one cluster; and detecting noise or outliers.

Acknowledgments The research is founded in part by The Natural Science Foundation of Inner Mongolia under Grant No. 2012MS0611, Chunhui Project of Ministry of Education under Grant No. Z2009-1-01041, and Higher School Science Research Project of Inner Mongolia under Grant No. NJZZ11140.

References

1. Zhang, T., Ramakrishnan, R., Livny, M.: BIRCH: An efficient data clustering method for very large databases. Technical Report, Computer Sciences Department, University of Wisconsin–Madison (1995)
2. Guha, S., Rastogi, R., Shim, K.: Cure: An efficient clustering algorithm for large database. In: Proceedings of the 1996 ACM SIGMOD International Conference on Management of Data, pp. 73–84, Seattle, Washington (1998)
3. Guha, S., Rastogi, R., Shim, K.: ROCK: A robust clustering algorithm for categorical attributes. In: Proceedings of the 15th ICDE, pp. 512–521, Sydney (1999)
4. Karypis, G., Han, E.-H., Kumar, V.: Chameleon: hierarchical clustering using dynamic modeling. Computer **32**(8), 68–75 (1999)
5. Long, Zhen-Zhen, Zhang, C., Liu, F.-Y., Zhang, Z.-W.: An improved chameleon algorithm. Comput. Eng. **20**(35), 189–191 (2009)
6. Ma, X.-Y., Tang, Y.: Research on hierarchical clustering algorithm. Comput. Sci. **34**(7), 34–36 (2008)

Exploration of Fault Diagnosis Technology for Air Compressor Based on Internet of Things

Zheng Yue-zhai and Chen Xiao-ying

Abstract With the development of network and communication technology, this article puts forward to new methods and ideas on the fault diagnosis technology of air compressor based on Internet of Things, ensure the safe, stable, and reliable running of air compressors, and carry out some beneficial exploration for remote diagnosis technology of mechanical equipment.

Keywords Internet of Things · Air compressor · Fault diagnosis technology · Remote monitoring

1 Introduction

Compressed air is the second largest power after the electric power energy, air compressor (hereinafter referred to as the compressor) is the main part of manufacturing the compressed air, and it is the original motivation (typically a motor) that convert the mechanical energy into a air pressure energy. Air compressors are widely used in industrial production, and most of the work environments are harsh [1]. When a machine fails, it cannot be diagnosed and be treated immediately, this must cause the increase of the loss. So the establishment of remote intelligent control for monitoring the response of the system has become a pressing problem of air compressor technology for the early prediction of failure.

In this paper, the use of Internet of Things technology carries out a useful exploration of the air compressor fault diagnosis technology.

Z. Yue-zhai (✉) · C. Xiao-ying
Quzhou College, Quzhou, Zhejiang 324000, People's Republic of China
e-mail: zyz8602@126.com

C. Xiao-ying
e-mail: chxy1@163.com

S. Patnaik and X. Li (eds.), *Proceedings of International Conference on Soft Computing Techniques and Engineering Application*, Advances in Intelligent Systems and Computing 250, DOI: 10.1007/978-81-322-1695-7_2, © Springer India 2014

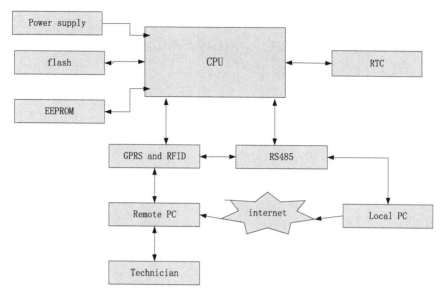

Fig. 1 Basic structural diagram of the system

2 The Remote Diagnosis System Designment of the Air Compressor Using Internet of Things

With the attention of the state to the Internet of Things technology, the Internet of Things technology has been more widely used in China. The application of the Internet of Things technology in air compressor for remote monitoring and diagnosis function conforms to the needs of society, and it will show considerable application prospects for the future development.

Basic structural diagram of the system, see Fig. 1

This system includes CPU with external GPRS system, flash memory, RS485 communication port; by the RS485 port, the real-time state of the compressor is transferred to a host computer and realizes the remote communication of the screw air compressor. The operator can start and stop the screw air compressor through the computer. The GPRS module and radio frequency identification (RFID) of the screw air compressor control system can fix the position of the air compressor, get to know where the parameters come, and ensure to control the machine correctly. The flash memory is used to store the historical running state, and when a mistake happens, the parameters can be drawn to help diagnose the problem and solve it. In addition, the acquisition of parameters can be transferred to the server through the Internet, there are many data from the working compressor in the server, and there is an expert diagnostic system according to the experience, the server can deal with collected data and make the plan of the equipment maintenance [2].

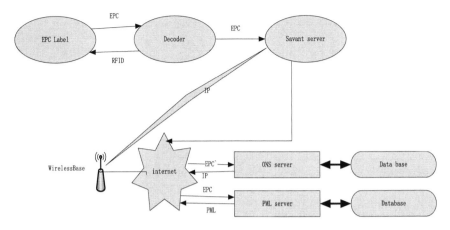

Fig. 2 Technical principle logic diagram

3 Field Information Collection with the Internet of Things Technology

In this paper, the technology of the Internet of Things is mainly used to locate the compressor, as well as its relevant technical indicators, operation parameters, etc.

Internet of Things is a kind of network that according to the agreed protocol, through the RFID, infrared sensors, global positioning system, laser scanners, and other information sensing devices, connects any objects with the Internet, and exchanges information and communication, and realizes the object's intelligent identification, location, tracking and monitoring and management. With the Use of the Internet of Things technology, people can implement intelligent and accurate regulation and operation of the devices in the network [3].

Internet of Things' key technology includes five segments: electronic product code (EPC), the information identification system (ID system), the EPC middleware (implementation information filtering and sampling), discovery service (information discovery service), the EPC information service (EPCIS). Its technical principle logic diagram is shown in Fig. 2:

There is a quite good coverage of wireless communication network in China, and it is a good infrastructure of the Internet of Things. The M2M business support platform launched by China mobile can provide the M2M business users of the data acquisition, transmission, processing, and business management, and other functions.

4 To Realize Remote Monitoring Using Communication Technology

This article adopts the type of site monitoring and remote monitoring, and through the GPRS network or the Internet, information is transmitted to the specified server, carries on the analysis processing, and achieves remote monitoring of equipment through the communication interface link.

Remote monitoring refers to the use of a computer to monitor and control of the remote industrial production process control system via a wired or wireless network system. A remote monitoring system is a computer hardware and software system that can achieve remote monitoring.

There are two types of remote monitoring system, one is no on-site monitoring system, but collects the data directly to a remote computer for processing, this remote monitoring has no difference with the general site monitoring except there is a long distance to transmit; the other is site monitoring and remote monitoring coexist. Generally, the remote monitoring adopts fieldbus technology connecting distributed sensors of monitoring equipment and develops the discrete cell to integrate unit, and at last, Intranet is built.

Remote monitoring function to achieve

1. Acquisition and process function: detect sample and pretreat all kinds of analog or digital signals, and output them in a certain form, such as printing form, display, or database server to help technician to analysis and know the situation of the compressor.
2. Supervision function: analyze, summarize, finish, and compute the detected real-time data, inputted parameters by the operators and so on, store them for the real-time data and historical data.
3. Management function: with the help of the existent valid data, images and reports related with the working condition, analysis, fault diagnosis, risk predict the working state, alarm with noise, light and electrical form.
4. Control function: Based on the detected processing information and predetermined control strategy, control output is formed and works on the compressor directly.

Because the remote monitoring system can realize real-time data collection of the site operation data and rapid concentration, the base for remote fault diagnosis technology is formed. The remote monitoring system connects with the enterprise Intranet; this makes it possible for manager to grasp the working states, and it works with the management strategy system, a much more advanced applying system will be built. It provides the possibilities of no-one on-duty at the site and achieves more profit ultimately.

5 To Build Fault Model

Due to the mechatronics degree is low and lack of the original fault data gathered, it will take a long time to accumulate the practice experiment and much fund to build the fault model, and it may be difficult to achieve in practice. In order to build up the fault document quickly, the most economic way is to combine experimental and computer simulation, namely to establish mathematical model of the compressor and verify the correctness of the model through the experiment, if there are some difference, change some parameters in the model to simulate the machine fault. With the deeply study of the compressor working process, mathematical model research can be achieved and working more accurately and credible. The specific steps are as follows: (1) establish a mathematical model of the compressor; (2) establish valve cavity fluctuating pressure calculation model; (3) through the "pressure arouse," combined the cylinder pressure with the body cavity of fluctuating pressure calculation; (4) verify the correctness of the mathematical model and calculating program; (5) change the related parameters, simulate the fault state, and set up a corresponding fault document.

To establish a standard machine fault-state model, you should take full account of their volatility. This is because, the first, the boundaries of the machine fault status and fault conditions are not clear, and no fault condition contains a certain state change; the second, with kinds of fault's severity difference, its characteristic parameters necessarily change; the third, parameter measurement on the control deviation, the changes of environmental conditions, etc., will make the characteristic parameter fluctuate. In order to improve the accuracy of fault diagnosis, the characteristic parameters (parameter fluctuation range) of the fluctuating nature is considered to construct standard mode.

6 To Establish Fault Diagnosis Expert System

The content of this paper is to build expert database in accordance with the public theoretical knowledge and long-term day-to-day empirical data, store them into SQL SERVER database, with a certain algorithm analysis, compare analysis result with knowledge database, plan the equipment's maintenance and repair, and the results are displayed in Web way, if possible, give some advice to improve the equipment designment [4].

The expert system consists of the knowledge base, database, data interface, and reasoning machine.

1. Knowledge base

Knowledge base is the core of the diagnosis expert system, its main function is to store and manage the knowledge of expert system. There are two main types of knowledge in the knowledge base, the first is the related theory in the field; the second is gained practical day-to-day maintenance data by the experts.

2. Database

Database is a data storage area that stores the real-time state in the expert system and the intermediate results that from the reasoning process of the device, and these values are changing in the running system.

3. Data interface

The data interface is responsible for transferring data in real-time database into the reasoning machine, and at the same time, the information inputted by the user converts the normalized expression within the system, output fault-type and disposal programs.

4. Reasoning machine

Reasoning machine is a group of program that compare the data from real-time database and single analysis system with the framework of the knowledge according some rules, and finally, fault conditions are obtained.

7 Summary

Remote monitoring and fault diagnosis system of air compressor based on Internet of Things and the mobile business of things adopts the advanced data communications, database and Web technologies; its application in high-voltage inverter air compressor products can achieve, transmit, process, and manage the run-time data, dynamic location information and user information collection of the air compressor, help to complete the fault diagnosis quickly, realize the product's maintenance services online, and quickly response after-sales. It makes it possible plan the overhaul and real-time development and help technicians exclude faults or hidden faults timely, improve efficiency, and reduce operating costs. It is a great significant thing for the long-term development for the air power industry, and it is a useful exploration of remote fault diagnosis in mechanical equipment.

Acknowledgments Project: Study & Develop of Remote Monitoring and Fault Diagnosis System for Air Compressor, Quzhou Science and Technology Bureau, Zhejiang Province (20121053).

References

1. Wang, X.: Screw air compressor's application and overtemperature fault cause analysis. Development **2**, 124–125 (2011)
2. Zhu, W.,Tao, B., He, L.: Monitoring and controlling system on internet. Netw. Intrument China Measur. **7**, 53–54 (2004)
3. Liu, Q.: Key technology and application of internet of things. Computer Science **6**, 1–4, 10 (2010)
4. Liu, X.: Fault feature extraction and identification of reciprocating compressor based on multifractal. Thesis for Master Degree, Daqing Petroleum Institute

An Enhancement Algorithm Based on Fuzzy Sets Algorithm Using Computer Vision System for Chip Image Processing

Chengxiang Tan, Lina Yang and Xichun Li

Abstract In industry, chip vision-based detection system cannot detect the dots and shatter within 2 pixels. In the process of chip image detection, image processing algorithm has great influence on the effectiveness and accuracy of detection and recognition. Among them, the image enhancement and edge extraction are the primary characteristics. The classical edge extraction methods mainly include Prewitt operator, Sobel operator, and traditional canny operator. By using these, the processing speed is fast and simple, but to shatter edge extraction is not efficient. In this paper, an enhancement algorithm based on fussy sets algorithm for the chip image processing is presented. We expect that the proposed algorithm can improve the detection accuracy within 2 pixels and improve the processing efficiency.

Keywords Image processing algorithm · Image enhancement · Edge detection · Computer vision detection · Fuzzy sets

1 Introduction

Chip makes the human society to enter the digital era, such as computer, mobile phone, digital camera, automotive electronic control, navigation system, medical equipment, and so on. Obviously, chip is everywhere. Meanwhile, by the fast development of high integration (LSI—large-scale integrated, VLSI—very large-scale integrated circuit LSI), high speed, and high performance, people desire to purchase the perfect chip badly [1, 2].

C. Tan (✉) · L. Yang · X. Li
Guangxi Normal University for Nationalities, Chongzuo 532200 Guangxi, China
e-mail: tcx6223@126.com

S. Patnaik and X. Li (eds.), *Proceedings of International Conference on Soft Computing Techniques and Engineering Application*, Advances in Intelligent Systems and Computing 250, DOI: 10.1007/978-81-322-1695-7_3, © Springer India 2014

China is one of the largest chip producing countries in the world. However, the global computer technology of chip design is controlled by the American companies: INTEL and AMD. INTEL company's production process is characterized by automatic control detection system, which is fixed in the production line. The chips are tested following the flow line one by one. AMD company's production technology is characterized by an increased detection system, which is for improving the detection of mobility, but the automatic control and detection system is more complex.

However, in the real world, it is difficult to avoid errors for the integrated circuit design, processing, manufacturing, and the production process [2]. The enormous price caused by the errors is too difficult to estimate in terms of the waste of resources, the risk of accidents, and casualties. In order to obtain the high quality and high reliability of the chip, detection is extremely important.

Machine vision detection is a technology which uses machines instead of human eyes to measure and make judgments [3]. In assembly line of chips, the method checking up the quality by human eyes is not only inefficient and has low precision but also easily makes eyes fatigue [2]. Machine vision can greatly improve production efficiency and degree of automation. For this case, we can design a system that can accurately find out spots, holes, threads, and other defects.

Section 1 of this paper is a background introduction of the edge detection algorithms based on different operators and the existing problems. Section 2 discusses related works for the edge detection methods. Section 3 proposes a new methodology of detecting edges of images with a detailed description of each step. Concluding remarks and a scope for further research are given in Sect. 5.

2 Related Works

2.1 Image Enhancement

The methods so far developed for image enhancement may be categorized into two broad classes, namely frequency-domain methods and spatial-domain methods [4]. The technique in the first category depends on modifying the Fourier transform, whereas in spatial-domain methods, the direct manipulation of the pixel is adopted. When an image is processed for visual "interpretation," it is ultimately up to the viewers to judge its quality for a specific application. The process of evaluation of image quality therefore becomes a subjective one.

Some differences between the image with the original scene or with the original image should be produced because of many factors in the course of image generation, transmission, and transformation. The differences are called deterioration or degeneration. Image deterioration should cause difficulties and inconvenience to get information from the image. The characteristics of image deterioration include the fall of image contrast, the image blur, the noise, and the image distortion in the

Fig. 1 Computer vision system [2]

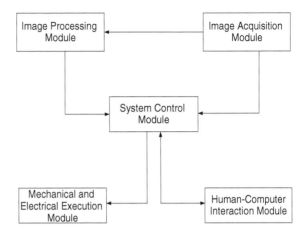

course of image acquisition. Image enhancement is a technique that enhances the interesting characteristics selectively in an image, weakens secondary information, and expands the characteristic differences between different objects, without considering the reasons for the image quality degraded [4].

2.2 Vision-Based Detection System

Research and application of vision-based detection system has become worldwide hot topic. Because the machine vision-based detection system can quickly obtain large quantity of information, it is easy to design and process the controlled information. In the production process, vision-based detection systems have been widely used in monitoring, product inspection, and quality control. Machine vision detection system is characterized by flexibility. The machine vision is used to replace the human vision in some dangerous work environment. In addition, in the process of industry production, it is obviously better using machine vision-based detection system than the artificial in terms of high speed and high precision, which can greatly improve the production efficiency and production automation degree.

Machine vision detection system can quickly display the image, output the data, and order the instructions according to the detection results. The actuator can match its completion position adjustment, quality screening, and other automated processes. Compared with the artificial vision, machine vision has the advantage of being accurate, fast, and reliable. Computer vision detection system mainly includes the following several parts: image acquisition module, image processing module, a system control module, mechanical and electrical system execution module, and a human–computer interaction module (Fig. 1).

2.3 Image Edge

Edge is often used in image analysis for finding region boundaries. Provided that the region has homogeneous brightness, its boundary is at the pixels where the image function varies and so in the ideal case without noise consists of pixels with high edge magnitude. It can be seen that the boundary and its parts (edges) are perpendicular to the direction of the gradient.

Image edge is one of the important characteristics of images, which contains rich and useful image information. The correct edge extraction can not only reduce the information content of the image processing greatly, but also can describe the shape characteristics of objects, which provides the foundation for further analysis [5]. Therefore, the edge detection is often the basis of other image processing, and it is also an important branch in the field of image processing. On the other hand, it is a big challenge to preserve the image edges for image enhancement during the course of the image filtering operation to remove or compress the noise pixels in the images. The traditional methods in the spatial-domain smoothing filters, such as mean filter and median filter, only possess good denoising performance, which weaken the image edges simultaneously.

The theory of fuzzy set [2, 6] provides a suitable algorithm in analyzing complex systems and decision processes when the pattern indeterminacy is due to inherent variability and/or vagueness (fuzziness) rather than randomness. Since a gray-tone picture possesses some ambiguity within pixels due to the possible multivalued levels of brightness, it is justified to apply the concept and logic of fuzzy set rather than ordinary set theory to an image processing problem. Keeping this in mind, an image can be considered as an array of fuzzy singletons [2, 6] each with a membership function denoting the degree of having some brightness level.

3 Methodology

3.1 Introduction of Fuzzy Sets

According to the theory of Pal and King algorithm [7, 8], the image X, having the size of $P * Q$ and the grayscale of L, can be expressed in fuzzy matrix of $P * Q$ size, which is achieved by the Eq. (1):

$$X = \bigcup_{i=1}^{P} \bigcup_{j=1}^{Q} \mu_{ij}/x_{ij} \qquad (1)$$

In the above formula, the element μ_{ij}/x_{ij} represents the membership grade μ_{ij} of the pixel grayscale value χ_{ij} relative to a special grayscale L, which represents the grade of possessing some property μ_{ij} by the (p, q)th pixel χ_{ij}. The fuzzy property

μ_{ij} may be defined in a number of ways with respect to any experiment. We defined it with respect to the level $L - 1$.

In general, each μ_{ij} in (1) may be modified to μ'_{ij} to enhance the image X in the property domain by a transformation function α_r where $r = 1, 2, \ldots$

$$\mu'_{ij} = \alpha_r(\mu_{ij}) = \alpha_1\left(\alpha_{r-1}(\mu_{ij})\right), \quad 0 \le \mu_{ij} \le 0.5 \tag{2}$$

$$\mu'_{ij} = \alpha_r(\mu_{ij}) = \alpha''(\mu_{ij}), \quad 0.5 \le \mu_{ij} \le 1 \tag{3}$$

The transformation function α_r is defined as successive application of T_1 by the recursive relationship

$$T_u(\mu_{ij}) = T_1\{T_{u-1}(\mu_{ij})\}, \quad u = 1, 2, \ldots \tag{4}$$

In order to enhance the image contrast, a nonlinear transformation is taken with μ_{ij}. The nonlinear transformation function defined by the previous work is shown as follows:

$$\alpha_1(\mu_{ij}) = 2\mu_{ij}^2 \quad 0 \le \mu_{ij} \le 0.5 \tag{5}$$

$$\alpha_1(\mu_{ij}) = 1 - 2(1 - \mu_{ij})^2 \quad 0.5 \le \mu_{ij} \le 1 \tag{6}$$

All the operations described above are restricted to the fuzzy property plane. The definition of membership function transformed from image sets to fuzzy sets is shown as follows:

$$\mu_{ij} = T(\chi_{ij}) = \left[1 + \frac{(L-1) - \chi_{ij}}{F_d}\right]^{-F_e} \quad i = 1, 2, \ldots, I; \, j = 1, 2, \ldots, J \tag{7}$$

where $(L - 1)$ denotes the maximum gray level. The parameters F_d, F_e depend on the shape of μ_{ij} and denote the exponential and denominational fuzzifiers. Generally, F_e is taken as 2. When $\mu_{ij} = P_C = F(X_C) = 0.5$, X_C is called crossover point. Therefore, we have

$$F_d = \frac{[(L-1) - X_C]}{2^{1/F_e} - 1} \tag{8}$$

Equation (5) shows that $\lim (\mu_{ij}) = 1$, whereas $\lim [(L-1) - \chi_{ij}] = 0$. It is to be noted that the Eq. (5) for $\chi_{ij} = 0$, μ_{ij} has a finite positive value δ, where

$$\delta = \left(1 + \frac{L-1}{F_d}\right)^{-F_e} \tag{9}$$

$L - 1$ is the maximum level in the image we use and χ_{max} replaces the $L - 1$. μ_{ij} plane becomes restricted in the interval $[\delta, 1]$ instead of $[0, 1]$. After enhancement, the enhanced μ'_{ij} contains the region where $\mu'_{ij} < \delta$ due to the transformation function.

Fig. 2 Chip image

4 Results and Discussion

This paper uses the grayscale image that is filmed by industrial CCD as the research object, as is shown in Fig. 2. Firstly, we preprocess the chip image, to eliminate noise and realize chip image edge accurate testing.

In this research, low-cost CCD cameras were used to capture the images. Image processing and analysis algorithms were developed to acquire the images to a computer, preprocess the images, and detect the edge for chip. A chip image acquisition device was developed, which consisted of one CCD sensor camera, two Light Source Card, Camera lens, Move Card, and a PC Computer, as shown in Fig. 3.

This new vision inspection system is to integrate the light, camera, movement and visual system. The system is based on the current visual detection system on the basis of the development of new algorithms to achieve the best detection system.

The camera had 1,024 × 1,024 pixel resolution and could record a video at about 40 frames per second. The color of two light sources will be white. A DELL computer is used. Software which captures the real-time images from the CCD camera will be developed in Visual C++ environment. Visual Studio 6.0 is one of the most rapid studies of visual system platform. We mainly concern its encapsulation, inheritance, fast calculation speed, high precision , and ease of transplantation.

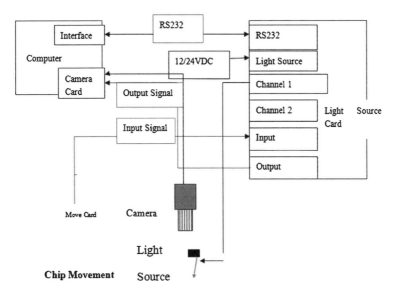

Fig. 3 Vision-based detection system

5 Conclusions

This paper presents a new algorithm based on fuzzy sets techniques to enhance the image edge detection algorithm. The theoretical analyses are expecting to show that the proposed algorithm improves the chip edge localization accuracy and then yields better edge detection results. The experiments on synthetic and natural images will be made to test the proposed method.

References

1. Zhao, Y., Wang, D.: Research on lower-layer computer vision processing algorithms in the detection of pear size. International Conference on Computer Application and System Modeling (ICCASM), pp. V10-557–V10-561 (2010)
2. Zadeh, L.A.: Outline of a new approach to the analysis of complex systems and decision processes. IEEE Trans. Syst. Man. Cybern. **SMC-3**, 28–44 (Jan 1973)
3. Shiwei, T., Guofeng, Z., Mingming, N.: An improved image enhancement algorithm based on fuzzy sets. International Forum on Information Technology and Applications (IFITA), pp. 197–199 (2010)
4. Farrell, P.G.: Influence of LSI and VLSI technology on the design of error-correction coding systems. IEE Proc. Commun. Radar Signal Proc. **129**(5), 323–326 (1982)
5. Pal, S.K., King, R.A.: Image enhancement using fuzzy sets. J. Electron. Lett. **16**(9), 376–378 (1980)

6. Zadeh, L.A., Fu, K.S., Tanaka, K., Shimura, M. (eds.): Fuzzy Sets and Their Applications to Cognitive and Decision Processes. Academic, London (1975)
7. Pal, S.K., King, R.A.: One edge detection of X-ray images using fuzzy sets. J. IEEE Trans. PAM I 5(1), 69–77 (1983)
8. Pal, N.R., Pal, S.K.: A review on image segmentation techniques. Pattern Recognit. 26(9), 1277–1294 (1993)

Development of DDoS Attack Defense System Based on IKEv2 Protocol

Qing Tan and Xiaojing Yue

Abstract IKE version second (IKEv2) simplifies the redundant function of IKEv1 and enhances the security of internet. This paper uses a DDoS attack detection technology, if the detection detected the DDoS attacks, with the establishment of good faith database TD filtering the network data flow, defense DDoS attack. The paper presents development of DDoS attack defense system based on IKEv2 protocol. This paper designs a secure and practical IKEv2 system implementation scheme.

Keywords IKEv2 · DDoS · Key distribution

1 Introduction

Flooding DDoS attacks, RoQ attack has several features: distributed, collaboratively large-scale attack (RoQ attack in pulse attacks stages) and a single-packet undetectable [1]. Exhausted the resources of the target machine, DDoS attacks target a port or multiple ports to send a large number of service requests or take advantage of the reflection server network implementation of DDoS attacks. A large number of attack packets bound to make some of the network traffic characteristics can be changed; through the study of changes in these characteristics, it can effectively detect DDoS attacks.

As a security protocol, the important point is the ability to resist attacks. Although the IKEv1 protocol developed a number of mechanisms to enhance its

Q. Tan (✉)
College of Information Technology, Luoyang Normal University, Luoyang 471022, China
e-mail: edutanqing@163.com

X. Yue
The High School Affiliated to Luoyang Normal University, Luoyang 471022, China

S. Patnaik and X. Li (eds.), *Proceedings of International Conference on Soft Computing Techniques and Engineering Application*, Advances in Intelligent Systems and Computing 250, DOI: 10.1007/978-81-322-1695-7_4, © Springer India 2014

security, there are still some security vulnerabilities, making it easy to attack, of which the most important is the Distributed Denial of Service (DDoS). So, here, we will mainly describe the improved IKE version second (IKEv2) attack that is against DDoS. The paper presents development of DDoS attack defense system based on IKEv2 protocol.

2 IKEv2 Protocol Analysis and Design

The IKE protocol plays a vital role in the whole IPsec system. It is a hybrid protocol, based on the Internet security association and key management protocol (ISAKMP) defined frame. IKE uses two stages of the ISAKMP, defines four kinds of mode [2]. In the first stage, it is regardless of the use of main mode or savage mode and it can use four kinds of authentication methods: pre-shared key, digital signature, public key encryption, and improved public key encryption.

As the certificate cannot be forged, the certificate can be placed in a directory for participants to access; the user can also directly send the certificate to other users. Certification Center (CA) plays a key role in the public key system, responsible for the management of the system, all users' (including people, all kinds of applications, host) certificate. We describe Certification Center (CA) that PKI's main purpose is to automatically manage keys and certificates for users to establish a secure network operating environment in a variety of application environments, allowing the users to easily use data encryption and signature technology to ensure the network data confidentiality, integrity, and effectiveness.

Key material IKEv1 in the calculation, due to the presence of significant flaw in the design of the protocol itself, can cause security vulnerabilities. The IETF working group took note of this point, and in IKEv2, using the new design avoids these defects, as is shown by Eq. 1.

$$X_{i+1,V}(m) = \bar{H}_i^* X_{i,V}(m) + \sum_{r=1}^{q-1} \bar{G}_{r,i}^* X_{i,r,D}(m) \qquad (1)$$

Between IKEv2 and IKEv1, a major difference is that the new agreement is a new definition of the independent encryption loads. For the IKE_AUTH stage message, the CREATE_CHILD_SA, and exchange of information messages, the message removed from the head outside portion is encrypted, and the encrypted portions on the encrypted IKE payload field are the encrypted payloads. Message encryption load must be placed in the tail of the message. From the defined encrypted payload, opinion from outside intuitive, the entire message contains only a payload (encrypted payload).

Exchange of IKEv2 corresponding to the first stage of IKEv1 is called the IKE_SA_INIT exchange, and in the basic exchange, the number of packets correspondingly reduced to two. IKEv2 from the IKE_SA_INIT exchange nonce's and the establishment of Diffie-Hellman is sharing to calculate a key seed secret

(SKEYSEED). The SKEYSEED is further used to calculate the seven key for other materials: SK_d is used to generate CHILD_SA key, only if he does not divide the direction; SK_ai and SK_ar were used as the integrity verification, and IKE_SA_INIT subsequent exchange of key materials: SK_ei and SK_er are used to encrypt and decrypt subsequent exchange; SK_pi and SK_pr is used to generate the AUTH load. The 6 keys for both sides of communication is different, SK_ai, SK_ei and SK_pi used to protect the originator of the message, SK_ar, SK_er and SK_pr used to protect the response message, as is shown by Eq. 2:

$$\begin{aligned} \mathbf{M} = \sigma_{s|a}^2 &= \mathrm{cov}(s(k) \mid a(k)) \\ &= E\{[s(k) - \mu_{s|a}]^2 \mid a(k)\} \\ &= \beta(k)^{\mathrm{T}} \Sigma_{\varepsilon(k)}^{-1} \beta(k) + \sigma_{s(k)}^{-2} \end{aligned} \qquad (2)$$

The kernel of IKEv2 system and IPsec protocol are used to provide VPN security gateway together. The system is the realization of the IKEv2 protocol; its ultimate aim is to peer entities in the system that provides the IPsec security protocol for secure network connections needed to establish security alliance. The IKEv2 system is responsible for handling user management configuration commands, the physical interaction, negotiation of packet encryption authentication, and the same kernel SAD interaction.

The IKEv2 message negotiation process IKEv1, a great improvement. IKEv1 negotiation process is very complex, with two stages and four modes. The main mode or aggressive mode is selected in the first stage, in which in the main mode, there are six exchange messages, and the second phase of the fast mode requires three exchange messages. And the messages exchanged are also different depending on the method of authentication barrier, making the entire negotiation process more complicated.

IPsec security protocol has two modes of operation: transport mode and tunnel mode (Transport Mode) (Tunnel Mode). Transport mode is used in the upper-layer protocol to protect IP packet; it is inserted into a special IPsec header between IP and upper-layer protocol header. Tunnel mode is used to protect the whole IP packet; in tunnel mode, to transmit, the IP packets are encapsulated and inserted into an IPsec head between the external and internal IP head [3]. Two kinds of IPsec security protocol (AH and ESP) can also be used in transport mode and tunnel mode, as is shown by Eq. 3.

$$\begin{bmatrix} X_{M-1}^T & X_{M-2}^T & \cdots & X_1^T \end{bmatrix} \begin{bmatrix} w_{M-1} \\ w_{M-2} \\ \vdots \\ w_1 \end{bmatrix} = X_M^T \qquad (3)$$

Information exchange is mainly used for error handling in IKE negotiation and solving the communication configuration; the messages carrying one or more deleted loads will be used for determining deleted SA SPI value. The best exchange of information is only after the completion of the initial exchange, so that it is protected by the negotiated IKE_SA. An IKE_SA control message must be sent under the protection of the IKE_SA, and part of a CHILD_SA control message must be produced under the protection of the IKE_SA CHILD_SA sent.

IKE provides key information to generate the encryption key and the authentication key for IPsec communication. Similarly, IKE using ISAKMP is used for other IPsec protocols' (AH and ESP) negotiation of SA. Later, with the higher requirements to the performance, security, there is a need to increase the NAT genetic authentication, remote address acquisition, and other contents, so the content of the agreement is getting more and more complex, with the lack of consistency in it [4]. IKE is of vital importance to the structure in the IPsec position and is used very frequently, so the performance inefficiency has become the bottleneck to the system. Therefore, the urgent need to reduce the complexity is performed by IKEv1. Therefore, IETF since the beginning of February 2002, IKE version second (IKEv2) of the drafting work, was released in October 2005, official version of IKEv2, as is shown by Eq. 4.

$$Wf(j, k) = 2^{-\frac{j}{2}} \sum_{n=0}^{N-1} f(n)\phi(2^{-j}n, n - k) \tag{4}$$

IKEv2 initial exchange in the first stage is equivalent to the IKEv1 exchange, usually consisting four new, in response to that attack by DoS circumstances, will add two more. All exchange messages in the IKEv2 appear in pairs. Before the initial exchange of messages called IKE_SA_INIT exchange, negotiation encryption algorithm, nonce value, Diffie-Hellman exchange, and various keys are needed after the value calculation. Second in the message is called the IKE_AUTH exchange, which is the authenticated message exchange front, has identity and certificate (optional), determines the traffic selectors, and establishes the first CHILD_SA. The two IKE_AUTH message exchanges are the encryption and authentication; encryption and authentication using key is generated in IKE_-SA_INIT exchange, to protect the identities of both sides of the play.

Corresponding the first version of the second stage in IKEv2 is the IKE_AUTH exchange. CHILD_SAs can be established through exchange of IKE_AUTH, also may establish the exchange by CREATE_CHILD_SA. The following key material generated in the way: KEYMAT = prf + (SK_d, Ni | Nr) if the establishment of the first CHILD_SA, then Ni, and Nr, is done by IKE_SA_INIT exchange in the nonces value; if not, then Ni and Nr are derived from CREATE_CHILD_SA exchange, as is shown by Eq. 5.

$$\sum h(n)h(n + 2k) = \delta_{k,0} \tag{5}$$

In order to be more flexible during forward compatibility, IKEv2 defines "1" and "0" for critical marks "C". The critical mark for "0" implies : the sender wants response but could not identify the next receiving load types of the load, and skips the load. The critical mark for "1": the sender wants response but could not identify the next receiving load types of the load, and refused to accept the message. If the receiver of the message is to identify the types of loads then it is neglecting the domain.

In the second stage, IKEv2 function and IKEv1 CREATE_CHILD_SA exchange in the exchange of similar are negotiating a new CHILD_SA. This exchange is from the two message composition. If the initial exchange is completed, it may be initiated by the exchange of any party, so the CREATE_CHILD_SA exchange in the initiator and the initial exchange of sponsors may be different. The use of exchanging the initial exchange of negotiating encryption and authentication algorithms is to protect the message. Sponsors include our proposal in SA load, nonce load in nonce value will be used to calculate the CHILD_SA key generation, KE load is optional, and the use of KE load can guarantee CHILD_SA a perfect forward protection. If Diffie-Hellman group is used in SA, then the message must contain a KE load, as is shown by Fig. 1. The response contains the accepted recommendations in SA load; if KEi load is the originator of the message, the responder of the message must have KEr load.

In the process of message exchange, a party may wish to send control messages; notification occurs with some errors or events [5]. In order to accomplish these operations, IKEv2 defines information exchange. If message exchange of information contains zero or more notice, delete, configuration load, an INFORMATIONAL message is received, and a party must response, and response message may not contain any load. Origination message can also do not contain any load; the sponsors may determine whether the other is survived by this method.

Each IKE message starts at IKE message header, followed by one or more IKE payload. IKE message header with biggest change is the initiator and responder of SPI IKEv2 message header instead of IKEv1 header initiator and responder cookie value.

A CREATE_CHILD_SA exchange can generate more than one SA, so according to the method, if IKEv2 protocol extends the key definition extends KEYMAT to less than the required length, and each key sequence capture is certainly required. Rules for determining key sequence are as follows: First of all, key all SA from the initiator to the response direction, all the keys of all SA direction are received from the direction response to initiator. The consultation of multiple IPsec protocol is keys in order to intercept data packet protocol head appeared in the package. If there is a single agreement with the encryption key and the authentication key, then take the encryption key and then the authentication key.

The main version and the major version are designed for compatibility with IKEv1. The main version of IKEv2 must be 2, while the main version of IKEv1 is 1, which is greater than the major version number 2 that was not accepted. Must be the version of IKEv2 implementation based on the minor version number that is set to 0. IKEv2 redefines the exchange type, number of 34–37, IKE_AUTH, representing the IKE_SA_INIT, CREATE_CHILD_SA, and INFORMATIONAL exchange. While retaining, the 0–33 number is compatible with IKEv1, as shown by Eq. 6:

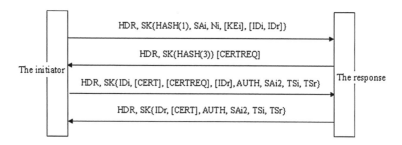

Fig. 1 The IKEv2 information exchange figure

$$P^{(\eta)}(m, s) = W_X^* \bar{P}^{(\eta)}(m, s) W_X \tag{6}$$

IKEv2 has canceled the three loads, the different structure security alliance load (substructure) according to the level relation together. IKEv2 SA load allows the inclusion of more suggested substructure, they must be in accordance with the arrangement from the best to the suboptimal order; each substructure can contain multiple IPsec protocol (e.g., IKE, esp., AH); each IPsec protocol can contain multiple transform substructure (e.g., encryption algorithm, integrity check algorithm, D-H group); transform structure can contain multiple attribute structure (e.g., key length).

The number of transformation of substructure is decided by both parties to choose the agreement, AH must implement a transform through integrity verification algorithm; ESP usually needs two transformations: encryption and integrity check algorithm. IKE needs four transformations: D-H group, integrity verification algorithm, PRF algorithm, and encryption algorithm. According to the different types of transformation and that defines the transformation of ID different, it appears in each transform head this definition and IKEv1 have great difference. In IKEv1 protocol, the protocol requires multiple algorithms, there is usually an algorithm in transform reflected, while the rest of the algorithm on the property is in contrast to the IKEv2 definition more clear.

IKEv2 do not record the state cookie (stateless cookie). In IKEv2, if a party is suspected of legitimacy of the request, it may request the other party with a cookie request, and it does not record any state or do any complex operation before such a request is received. So, when the receiver receives a large number of half open (half-open) of the IKE_SA request, it will reject them, unless the receipt notices of a load with cookie. This will enable the ability to resist DDoS attack greatly enhanced.

Network communication subsystem is responsible for the network interface. To provide communication, IKEv2 application and network complete the receiving and sending of IKEv2 messages. The module will do local message negotiation of IKEv2 consultation message encapsulation system generated into UDP packets and is sent out; UDP data packet stripping IKEv2 message is received and passed to the upper hand by message negotiation processing subsystem. The programming interface module of the full realization of the network socket and monitor is performed in the 500 and 4,500 ports, respectively.

3 Development of DDoS Attack Defense System Based on IKEv2 Protocol

Reflector DDoS attacks can be described as follows: First, the attacker host controls a large number of vulnerabilities puppet machine, in these puppets on placement control procedure and attack procedures. Then, attack the host attack instructions sent to the host computer; the host computer received attack instructions for the source IP address, the IP address of the target machine sends service requests (or packets) to the reflector server. Finally, the reflection to the server upon receiving a service request (or packets) of the source IP address in response to a large number of data packets, thus, forming the reflector DDoS attacks.

DDoS attacks in order to resist the IP source tracking IP attack packets generally use IP spoofing technology. In this way, even if detected, DDoS attacks cannot trace the source of the attack according to the source IP address in the IP packet items. The length of the IPv4 IP address is 32, the IP address entry in the database of traditional IP packet filtering technology. If the database stores all IP address, probably occupied a space of $232 = 4G$, then it is a very large number.

For stationary time, series can be analyzed by AR model and MA model. The PDD time series is nonstationary time series; DDoS attack detection and online analysis, therefore, used the online analysis ability and strong adaptive autoregressive (AAR) model to describe the PDD time series. Parameter sequence is fitting as a stationary current state (or generalized stationary time series) [6]. The PDD time series for arbitrary is the definition of P order AAR model.

IKEv2 for all messages appear in pairs; in each of the message, the initiator is responsible for retransmission event, response without retransmission of the response message, unless requested a retransmission for each other. Avoid both initiation and retransmission, resulting in a waste of resources, but also can prevent the attacker intercepting the message, disguised as a negotiator who is constantly sending out the retransmission of the message, the two parties of the resource consumption. The IKEv2 protocol is a very complex security protocols, engineering complex and difficult. The IKEv2 system for Fedora Core 4 operating system, and it is Linux 2.6.11 kernel. The use of the programming language is as the standard C++ language, as shown by Fig. 2.

At present, the distributed DDoS attack defense system structure is the trend of development, dynamic IP packet filtering technology, if the nodes in a distributed manner used in the network (such as servers, switches, and boundary router), each node to a certain extent filtering part of a DDoS attack packets, the target hit by

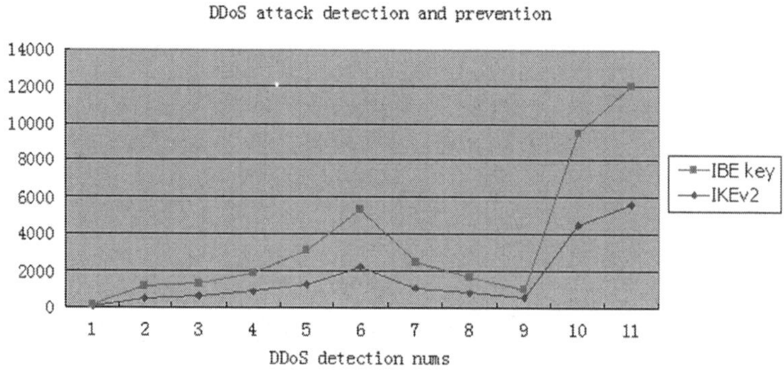

DDoS attack detection and prevention

Fig. 2 Comparison of DDoS attack detection and prevention based on the IKEv2 protocol with IBE key algorithm

DDoS attacks will be greatly diminished, conductive to the realization of defense strategy of DDoS attack on the target machine. The paper proposes the strategies of DDoS attack detection and prevention based on the IKEv2 protocol and IBE key distribution.

4 Conclusions

IKEv2 in agreement with SA is divided into two steps: first, the establishment of a certified safety channel in the communication between the two sides, namely the establishment of IKE SA and then under the security channel protection. Second, IPsec security service negotiating SA, namely the establishment of IPsec SA. In IKEv2, the IKE SA is still called IKE SA, and IPsec SA is called CHILD SA.

References

1. Soussi, H., Hussain, M., Afifi, H., Seret, D.: IKEv1 and IKEv2: A quantitative analyses. The 4th World Enformatika Conference (2005)
2. Zhao, L., Li, X., Dong, Q., Shi, L.: An IKEv2 based security authentication scheme for mobile network. AISS **3**(9), 191–198 (2011)
3. Kwok, Y.K., Tripathi, R., Chen, Y., Hwang, K.: HAWK: Halting anomalies with weighted choking to rescue well-behaved TCP sessions from shrew DDoS attacks. ICCNMC, pp. 423–432 (2005)
4. Yan, H., Wang, F., Cao, Z., Lin, F., Chen, C.: A novel method to defense against web DDoS. JDCTA **6**(19), 162–170 (2012)
5. Fan, L., Zeng, W., Jiang, Y., Li, J., Liang, Q.: A group tracing and filtering tree for REST DDos in cloud. JDCTA **4**(9), 212–224 (2010)
6. Perlman, R., Kaufman, C.: Key exchange in IPsec: Analysis of IKE. IEEE Internet Comput. **4**(6), 50–56 (2000)

Power Data Network Dynamic Simulation Platform

Qian Guo, Xingchuan Bao and Gu Feng

Abstract Simulation is a powerful tool for research. In many cases, it is needed for realistic network simulation and the power of data prediction, modeling a large number of simple interactions electricity Statute, without attention to details of a particular Statute, to identify a key component or the whole network's support of power protocol and the ability of stable work under critical conditions such as data stream change, long-term transmission and critical states. This paper proposes and implements a model of power data network dynamic simulation platform to solve the dynamic and power protocol simulation problem in large-scale network simulation, through multi-threaded and multi-agent dynamic changing, the layered implementation of OSI model, and simplified simulation methods of the power protocol.

Keywords Power data network · Power protocol · Dynamic · Simulation

1 Introduction

With the rapid development of intelligent power system, power applications, and power data network environment is increasingly complex, power data applications increasingly high demand for the transfer of existing devices on the power of the Statute, in particular based on the multi-cast the lack of real-time transmission

Q. Guo (✉) · X. Bao · G. Feng
Chinese Electric Power Research Institute, Nanjing, Jiangsu, China
e-mail: guoqian@epri.sgcc.com.cn

X. Bao
e-mail: baoxingchuan@epri.sgcc.com.cn

G. Feng
e-mail: fenggu@epri.sgcc.com.cn

S. Patnaik and X. Li (eds.), *Proceedings of International Conference on Soft Computing Techniques and Engineering Application*, Advances in Intelligent Systems and Computing 250, DOI: 10.1007/978-81-322-1695-7_5, © Springer India 2014

performance of the Statute, are directly related to the stability of power control [1]. Therefore, how to simulate changes in the real network environment to determine the carrying capacity of the power application, clearly the real power data network the ability to work in the case of data flow changes, long-term transmission, critical state stability to predict the types of events that may occur, related to the long-term stability of the grid. Data networks and related communication member simulation testing process often require the interaction of the high-load application data to support capacity and transmission performance for a specific application, without the need for strict Statute data details. To this end, a power data network simulation platform model, this paper proposes a power data network simulation platform model OSI model-layered, multi-threaded, and multi-agent dynamic changes, and power Statute-simplified simulation to solve large-scale network simulation of dynamic change and power Statute simulation.

2 Introduction of Existing Simulation Technology

The existing network protocol simulation technology application layer pressure simulation, flow simulation, the packet simulation and the Statute of simulation to simulate different levels of network data contents, these means of simulation in a certain extent, to meet certain needs of the network testing requirements. The simulation and detecting unity are applicable to the detection of the experimental environment, and there are some defects in the large-scale simulation.

The pressure simulation application layer and application layer interaction, simulation a large number of users to access. This type of simulation is mostly used in application systems and Web performance testing of the system, for example Load Runnerm QA Load. The test and simulation process is totally dependent on a common protocol and unable to provide effective support for custom protocols.

Flow simulation simulates frame structure and data stream of OSI 2–3 layers as switchers, routers et. Most of this type of simulation for performance testing of communications equipment and safety equipment, typical flow simulation tools the Spirent TestCenter series, Xixia Ethernet test kits, test and simulation emphasis on test accuracy, simulation process to determine the content, flow and times for the center, the support capacity of the specific content of the data packet and dynamic movements [2].

Packet simulation is used mainly for packet interception and playback mode; the simulation is repeated reality by real packet network data traffic, and this type of simulation accuracy and data packet structure is complicated; usually rare in the general, communication device testing in the security equipment testing and on-site environmental analysis are more common [3]. Test and simulation process flow structure is usually a certain period of time to do the template, and the dynamic changes of weak cannot simulate the dynamic process, and also the lack of support does not support the application of static resolution.

The Statute of the simulation for the professional protocol Statute simulation and analog initiates and responds to the Statute of the standard process; specifically for the class of devices, Statute standardized the typical power Statute simulation. The Netherlands Kema Statute simulation series is the most common [4] and IEC101, IEC104, and IEC61850 international standard protocol support are better. Traditional test and simulation for a single device or a single protocol simulation is more focused on the analysis of the protocol conformance, not large-scale simulation [5].

On the other hand, the simulation for real-world applications of electricity or power data network environment, especially for the power protocols passing through industrial switchers, is only need to focus on real-world multi-application high loads and complex environment, but do not focus on specific details of the power within the statute. This requires a large flow simulation to simulate the power data network simulation platform.

3 Dynamic Simulation Platform for the Design of the Electric Power Data Network

3.1 Logical Structure

In this paper, through data exchange between the various distributed components, the Power Data Network Dynamic Simulation Platform constitutes a logical network application environment, which can simulate various types of electrical applications, in the physical network environment to simulate the real environment of each application and the power of the Statute of the interaction data streams, and to achieve more accurate and credible simulation environment providing strong support for predicting changes in network and application specific circumstances visible data.

The logical structure shown in Fig. 1 consists of three basic elements constitute: Controller, Agent, and DUT.

1. Controller: mainly responsible for the graphics display, data analysis, policy formation, coordination during the simulation of load sharing between the agent terminal thread dynamic control and the coordination of agent and terminal emulation thread.
2. Agent: analog equipment and related agreements impersonated thread simulation object is responsible for loading. The section is responsible for carrying internal analog thread and, according to the instructions of the Controller, to its corresponding dynamic control.
3. DUT: measured or need simulation test object. It may be provided for a physical network and may also be a business application, or more complex multi-application environment.

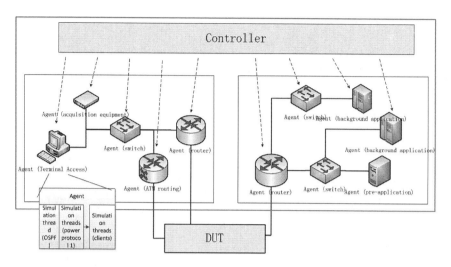

Fig. 1 Logical structure

In this way, the network model of the basic elements of the interaction together constitutes a realistic simulation of the application of the network environment.

3.2 Hierarchy

The power data Network dynamic simulation platform performs as four layers, the hierarchy shown in Fig. 2. Level functions are as follows:

1. Presentation layer: responsible for the performance of the test data shows the relevant data, including dynamic charts, bringing dynamic response test objects such as flow rate and changing data, simulation objects and the corresponding test object log data and a variety of error correction nuclear data.
2. Abstraction layer: responsible for all kinds of policy definition and summary of events functional layer reanalysis. Interaction is defined as each object layer mock objects carry data, logical relationships policy definitions, such as operating cycle and a special response status. Agent is responsible for the definition of the deployment of the Agent, type definitions, and management. Topology defined for each Agent object is the logical topology to achieve mutual relationship and form definition. Event analysis is a summary of events in the functional layer reanalysis.
3. Functional layers: the lower functional layer is mainly responsible for the management, including policy distribution, topology to achieve, time synchronization, event aggregation, and other functions.

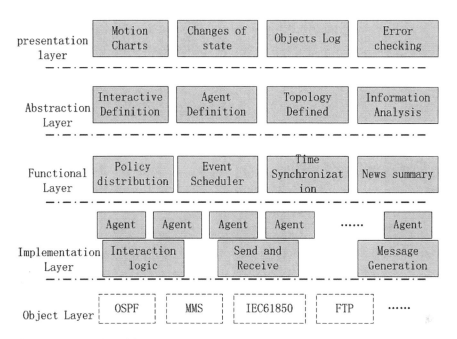

Fig. 2 Hierarchy of model

4. Implementation layer: agent through the achievement of specific interaction logic, data sent and received, as well as events related to the generation of test data for transmission to the upper layer.
5. Object layer: responsible for the Agent to specific protocols and applications simulation.

4 Working Mechanism of Power Data Network Dynamic Simulation Platform

According to this model, the power data network dynamic simulation platform should be dynamic while being able to meet the high-performance requirements.

4.1 Simulate the Dynamic Interaction Objects

For data network in the power dynamic simulation platform, each agent is independent of the individual objects are shown in Fig. 1. When an agent needs to communicate with another agent, its threads will interact within the simulation

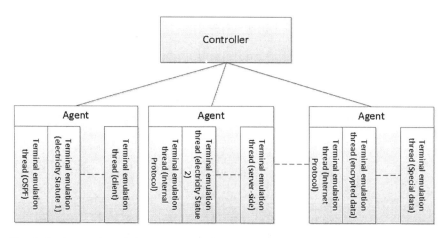

Fig. 3 Implementation of simulation object's dynamic changes

data in the form of event–event pushed to the functional layer for event scheduling, and different events will be forwarded to the target object and analog objects for dynamic interaction, to achieve dynamic changes such as OSPF, FEP acceptance, and forwarding dynamic response action.

4.2 Dynamic Mock Object

The platform is based on Remote Controller, Proxy Agent and Terminal Emulation Client Thread, shown in Fig. 3, making mock objects which can simulate a variety of dynamic changes typical situations.

1. Dynamic change proxy client Agent. Proxy Agent can simulate a number of interactive processes, and also can simulate a large network. Changing the amount of Agents, is actually increase or decrease a lot of simulated environments.
2. Change the dynamic simulation of thread. Each analog thread is designed to simulate the actual interaction of an application, the dynamic simulation of the thread change, changing the number of operational simulation, emulation communications network operator log, exit, operation, and other basic business operations.
3. Dynamically changing simulation threads. Threaded applications by changing the simulation process, changing the interactive process simulation, creating misuse, abnormal operation, etc.

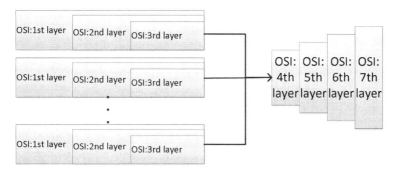

Fig. 4 Level constitutes of simulation protocol

4.3 Different Levels of Dynamic Change Agreement

With digital substation and smart grid in use the power of data transmitted over the Internet will not only include current common electricity protocols as IEC61850, but also include more complex protocols and applications, whcih will be more and more. The current mainstream flow simulation is mainly for OSI2–3 layer custom protocol emulation, usually by way of a data frame simulation, which uses OSI1–3 layer packet stepwise manner to construct. Such emulation mode cannot simulate OSI 4-7 layer protocols, such as IEC61850 mainly due OSI4–7 layer joined the interactive process, so that a complete OSI4–7 layer transmission requires multiple OSI1–3 layer packets combination is completed.

For data network in the power dynamic simulation platform model, OSI4–7 layer and 1–3 layers separated analog mode, see Fig. 4.

In the process of constructing the simulation protocol, when simulating IEC61850 class Statute [6], terminal emulation thread will call a type of OSI1–3-layer protocol packets matching a single-application process in order to provide support for such an agreement; And when simulating complex applications, then OSI1–3 layer protocol packet analysis section from recombination, while an application process is responsible for receiving data packets on the other hand reorganization, the application process for the corresponding packet exchange provides the underlying protocol support. As a result of such a separate analog mode, all 1–3 and 4–7 layer protocol interaction data can be run on the same analog thread; when you need dynamic switching, it does not need to be as similar to other simulation platform stop switch, only need to dynamically add, delete impersonated thread; you can achieve different levels of agreement dynamics.

4.4 Simulation Performance Improvements

Separating the application layer and the bottom, to solve the OSI layer protocol simulation problem, but it is complicated, and bulky Statute response protocol

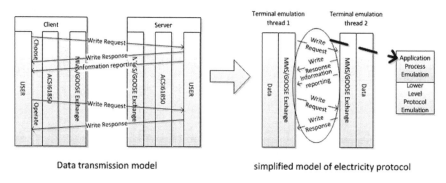

Fig. 5 Simplify the simulation of GOOSE

defined data definition and precision will inevitably lead to poor performance, and these parts will not be transmitted and the transmission network detailed device identification. In fact, the data network and transmission equipment only for electricity Statute underlying transport (i.e., OSI4 layer and below) [7] provide accurate identification of parts of the transmission path. Even the particular application layer filtering equipment, but also only the Statute otherwise specifically interested in the interactive process, those definitions specifically state the amount of data that only the final application will be specifically identified, and the entire transfer process will not be particularly concerned about. In view of this, we need to make the appropriate Statute of power cut, select the contents of concern, and discard irrelevant data in order to maximize performance and accuracy.

In case GOOSE message. According to IEC61850 standard, GOOSE messages directly mapped to the Ethernet protocol stack; but in a standard Ethernet header added a Tag, Tag contains a 12-bit virtual LAN identification code (IEEE802.1q) and 3 bit of packet priority code (IEEE802.1p), on priority tagging and virtual LANs details encoded in IEEE Std.802.1Q: virtual bridged Local Area Networks in detail [8].

Since GOOSE message is mapped directly from the application layer to the data link layer, message content just application protocol data unit (APDU) does not include other protocols such as TCP or UDP packet header. Format of the packet is based on the manufacturing message specification (MMS application layer protocol specification) and ASN.1 data to the correspondence between bits of code coding standards, so the Statute complies with the interactive part of the MMS interactive process. For simulation of the GOOSE message data transfer protocol based on GOOSE model, data can be simplified and application section to fill the data instead of, the individual packets, can be simplified as part of the underlying transport protocol OSI, GOOSE interaction protocols, and padding data part of the simulation, Client and Server, respectively, by a different terminal emulation thread simulation in Fig. 5.

5 Conclusions

With the gradual deepening of the development of the smart grid, smart meters and smart application of a large number of applications. The power data network and back-end applications, which support these smart devices running, will also play an increasingly important role in the stable operation of the power grid. Dynamic simulation predicted power data network will play an extremely important role. Power Data Network Dynamic simulation platform model proposes a clear and detailed platform solution. It makes electricity more operational data network simulation and related software for the development and application of reference.

Acknowledgments Thanks to Dr. Yu Yong. He spends a lot of time for this article and effort to provide guidance and assistance. Thanks to Dr. Yu Yuehai on the power of the Statute to provide technical support, and finally, thanks to Zhang Tao, who provided selfless help.

References

1. IEC 61850-8-1.: Communication networks and systems in substations-Part 8-1: Specific communication service mapping (SCSM)-Mapping to MMS (ISO 9506-1 and ISO 9506-2) and to ISO/IEC 8802-3 (2004)
2. Deng, Y.M., Xiong H.G., et al.: The computer simulation of switching ethernet. Comput. Eng. Appl. **04**, 167–169, 223 (2004)
3. De Lu, C. , L.V., Sui, Z, Jiang X.Z.: Development and application of simulation technologies in the power industry of China. J. Syst. Simul. **04** (1999)
4. Miao, W.G., Liu, F., Tang, K., Zhang, X.W.: Study on protocol testing technology in substation based on IEC61850. Jiangsu Electr. Eng. **08**(75–76), 82 (2007)
5. Zuo, Q.Y., Zhang, Y.L.: Study and application of IEC61850 protocol test technology. Relay **02**, 68–71 (2007)
6. Lin Z.M., Yang, F.P., Yu, Y.: Analyzing and implementation of mapping between IEC61850 and MMS. Relay **02**, 64–67 (2007)
7. Ren, Y.M., Qin, L.J., Yang, Q.X.: Study on IEC 61850 communication protocol architecture. Autom. Electric Power Syst. **8**, 62–64 (2000)
8. Ke, S.W., Liu, S.G., He N., Liu, Y.: Study of GOOSE messages on the substation transmission. Relay **35**, 308–310 (2007)

An Emotional Model Based on Multiple Factors

Qiong Xiao, Gangyi Ding and Yongkang Liu

Abstract In the research of emotional simulation, the existing emotional models, such as the emotional model based on cognition, the emotional model based on probability, multilayer emotional model and so on, always put particular emphasis on one or some factors and cannot describe complicated emotions; in addition, they do not consider the subconsciousness' impact on emotion. This paper comes from emotional space which based on dimensions, integrates heterogeneous personality factor, introduces the cognitive and non-cognitive factors of emotion's generation, and combines the factors such as character's influence components of emotion, decay of emotion, mutual influence between emotions, outside stimulus, subconsciousness, builds a personalized emotional model. At last, this model is used to simulate a crowd scene, which indicates that this emotion model can reflect different emotions generally.

Keywords Emotional model · Subconsciousness · Personality · Complicated emotions

Q. Xiao (✉)
School of Computer Science, Beijing Institute of Technology, Beijing, China
e-mail: qunxiao@foxmail.com

G. Ding
School of Software, Beijing Institute of Technology, Beijing, China
e-mail: dgy@bit.com

Y. Liu
Space Star Technology Co., Ltd, Beijing, China
e-mail: liuyk@spacestar.com.cn

S. Patnaik and X. Li (eds.), *Proceedings of International Conference on Soft Computing Techniques and Engineering Application*, Advances in Intelligent Systems and Computing 250, DOI: 10.1007/978-81-322-1695-7_6, © Springer India 2014

1 Introduction

In the virtual humans' simulation, emotion computation is a highly integrated research topic, the main research content includes emotion mechanism's researchment, emotion signal's acquisition, emotion signal's analysis, modeling and recognition, emotion understanding, emotion expression, and emotion computation's research focus is on acquisition of physiology and behavioral traits signals caused by emotion from many sensors, building the emotion model, generating the personal computer system which can react intelligently, sensitively, friendly to the users' emotion.

Emotion has complexity and fuzzy, in order to implement the intelligent system, an emotion model should be built first, building a reasonable math model to describe the emotion is the critical problem of emotion modeling's research, several models have been proposed for emotion, such as the OCC model, HMM model and so on.

2 Problem Description and Approach

2.1 Problem Description

The existing emotion models have many advantages and disadvantages, some of which will be introduced in detail. First, the emotional model based on cognition will be introduced, its most typical model is the OCC model [1], which only considered the emotion's cognition factor and its generating mechanism, but did not consider emotion's non-cognitive factor, such as the character's influence. Second, the emotional model based on probability's most typical model is the HMM model [2], and it simulated the emotion only from probability, while did not consider emotion's cognitive and non-cognitive factors, leading to the result that for the same stimulus, HMM model's perceptive situations are the same. The emotional model based on dimensions described the emotion by presuming a small number of discrete emotion and small-scale range of emotion variation, while did not consider emotion's non-cognitive factor. At last, the multilayer emotional model of "character-mood-emotion-expression" presented by Kshirsagar is the emotion model which related the character and emotion firstly and applied itself to virtual human's face expression successfully [3], but it deals with the mood obscurely and cannot describe the complex emotion.

In order to resolve the problems of these models, a new emotion model is introduced here, which considered many crucial factors. This emotion model includes perception module, inside variable module, character module, and mood module, and every module will be introduced in detail in the following part.

2.2 Perception Module

The perception module's function is perceiving the outside environment and obtaining its information; in here, the virtual human does not need to pay close attention to all the objects, suppose that the main concerned objects include the other virtual humans, static obstacles. The perception of the outside environment can be classified to five kinds: visual perception, auditory perception, smelling perception, tasting perception, and tactile perception, which can be defined as follows:

$$\text{Set}i = \{aj \mid aj \in \text{ElemSet}, \quad j = 1, 2, \ldots, n, \quad n \geq 0\} \tag{1}$$

Set$i$ denotes five kinds of perception, $i \in [1, 5]$, aj denotes a perception event, and ElemSet denotes the set of event element.

2.3 Subconscious Module

According to biological studies, emotion is formed before cognition in subconscious state, and some of the emotional reactions which are under the conscious control are the results of cognitive activities, but some of the emotions are stored in the brain by the way of conditioned responses. Once the conditioned stimulus is touched, these emotions will be realized automatically, which usually cannot be controlled or noticed by oneself. The latter one is called conditioned emotional response [4], some of which are related to the virtual human, after the perceived objects' visibility testing in perception module, their emotional effects will be proceed firstly in here. Suppose at time $t0$, a conditioned stimulus of ai is perceived; if ai is visual stimulus, $ai \in \text{Set}1$, the emotion intensity caused by it can be formulated as follows:

$$\text{ConFo(dis)} = \begin{cases} 0, \text{dis} > \text{MAXR} \\ k * \text{Cp}, \text{dis} \leq \text{MAXR} \end{cases} \tag{2}$$

ConFo(dis) denotes the emotion intensity caused by ai, dis denotes the distance between visual-conditioned stimulus and virtual human, and k is the environmental impact factor. Cp is the former emotion intensity which has formed when virtual human got the stimulus similar to ai; MAXR is the maximum distance in which the conditioned stimulus can take effect. If ai is the other stimulus, $ai \in \text{Set}2 \cup \text{Set}3 \cup \text{Set}4 \cup \text{Set}5$, its duration is Δt, the emotion intensity caused by it at time t is computed as follows:

$$ConFo(t) = \begin{cases} 0, t \leq t0 \\ k * Cp, t0 < t < t0 + \Delta t \\ k * Cp * e^{\wedge}(-b(t - t0)), t \geq t0 + \Delta t \end{cases} \tag{3}$$

ConFo(t) denotes the emotion intensity caused by ai, k is the environmental impact factor, Cp is identical with the one in formula (2), and b is the coefficient of mood attenuation.

2.4 Inside Variable Module

Inside variable module mainly includes some physiological variable and social variable, based on the Maslow's theory of a hierarchy of needs to be satisfied, and physiological variable includes the basic physiological needs; we consider three kinds of physiological demands, including fatigue, hunger, and thirsty. Social variable includes security requirement, respect demands, self-fulfillment; in here, we only consider the safe demand. Values of inside variables' demand intensity range from 0 to 1. 0 denote that demand intensity is very low, 1 denotes it is very high.

We presume that the distance between virtual human and outside stimulus is dis; in here, outside stimulus are the ones which can affect its physiological demands, the maximum distance in which these outside stimulus can affect virtual human is MAXR. The effect strength function to the virtual human is as follows:

$$Ivo(dis) = \begin{cases} 0, dis \geq MAXR \\ \frac{Ivi(t) \times k}{dis}, MINR \leq dis \leq MAXR \\ Ivi(t) \times k, dis \leq MINR \end{cases} \tag{4}$$

K is a constant, which denotes the effect strength of the outside stimulus to the virtual role. Ivi(t) denotes inside physiological demand intensity of the virtual human at the moment of t, Ivi(t) $\in$ [0, 1]. Ivi(t) varies with time; it can be denoted as follows:

$$Ivi(t) = \min\{Ivi(t - 1) + \Delta M, 1\} \tag{5}$$

ΔM denotes inside physiological demand's increment within the time span from $t - 1$ to t. The overall inside physiological demand intensity is the sum of its own physiological demand intensity and the effect strength of outside stimulus to it. It can be denoted as follows:

$$Ivs(t) = \min\{Ivi(t) + Ivo(t), 1\} \tag{6}$$

Ivs(t) denotes the overall inside physiological demand intensity of the virtual human, Ivs(t) $\in$ [0, 1].The thresholds of fatigue, hunger, and thirsty are set, when one or some of them are exceeded, the virtual human's emotion will be affected.

Table 1 Effective values range of behavior parameters

Parameter	Min	Max	Unit
Max.neighborsdist	3	30	m
Max.num.neighbors	1	100	n/a
Planning horizon	1	30	s
Agent radius	0.3	2.0	m
Preferred speed	1.2	2.2	m/s

Table 2 Mapping between adjectives and PEN factors

Trait	Adjectives
Psychoticism	Aggressive, impulsive
Extraversion	Assertive, active
Neuroticism	Shy, tense

Although virtual human may have many kinds of motives at the same time, when in the face of danger, it will consider the safety as the most important things, security needs will play a leading role; when it has no security needs and the needs intensity value of fatigue, hunger, or thirsty is bigger than the threshold, needs of fatigue, hunger or thirsty will play a leading role.

2.5 Character Module

Character module mainly includes the virtual human's character; different characteristics of the virtual persons are created according to three factors model's mapping between the mood and parameter. PEN model is used as the virtual human's character data. PEN model's three factors are psychoticism, extraversion, and neuroticism.

Guy et al. research explored the effect of the crowd's behavior parameters to the perceived behaviors, five behavior parameters in Table 1 were used. The effective values range of behavior parameters was obtained by data analysis, and it is shown in Table 1, which will be used as empirical data in this paper [5].

Perceived behaviors are described by six adjectives: aggressive, assertive, shy, active, tense, and impulsive. Virtual humans of different personalities are generated as follows: First, behavior parameters and perceived behavior's mapping are generated by linear regression model. Second, behavior parameters and PEN model's mapping are generated so that the virtual humans have different personalities. This is based on the research finding of Pervin, and six adjectives of behaviors are corresponding to three characteristic factors in Table 2.

2.6 Mood Module

The mood has an important influence on the emotion, good mood can repress negative emotions, and bad mood can repress positive emotions. Mood intensity expressed as Im, Im∈ [−1, 1]; positive value of Im denotes that virtual human is in a good state of mind, and negative value of Im represents the virtual human is in a bad mood. The greater the absolute value of Mood intensity, the stronger the influence to the emotion, conversely, the weaker the absolute value of Mood intensity, the stronger the influence to the emotion.

Mood space's selection. In our research, the mood model uses "pleasure," "arousal," and "dominance" to constitute a three-dimensional mood space, which is also called PAD space. The initial mood intensity values are inputted by the user to guide and control virtual human's emotion tendency to some extent.

Mood intensity can be represented by PAD space as: $Im(t) = [Ip(t), Ia(t), Id(t)]$. When mood intensity is equal to [0, 0, 0], it denotes the mood in peace.

Mood attenuation function. As time went on, the mood will gradually decay, this means mood intensity will tend to [0, 0, 0], which represents the state of calm. Suppose at time $t0$ user input the mood intensity of initial value, the following is mood intensity at time t:

$$Im(t) = \varphi(Im(t0)) = Im(t0) \times e^{-b(t-t0)} \qquad (7)$$

b is the coefficient of mood attenuation, which controls the mood attenuation speed, and different character has different value of b.

Mood strength update equation.

$$Im(t) = \varphi(Im(t0)) + R(\mu) \qquad (8)$$

$\varphi(Im(t0))$ denotes the mood intensity after the mood attenuation, $R(\mu)$ denotes the influence function of outside stimulus to mood. It can be represented as follows:

$$R(\mu) = \omega m \times PAD^* \times \rho^T \qquad (9)$$

In the above formulate, PAD* denotes the mapping matrix between emotion and mood, ρ^T is the outside stimulus of emotion, and ωm is the influence coefficient of outside stimulus to the fluctuation of mood. It is decided by character.

2.7 Emotion Module

Emotion space's selection. In this study, the Ekman's six basic emotions are as follows: happy, sad, anger, surprise, fear, and disgust constitute six-dimensional emotion space, so emotion vector is represented as follows:

$$E(t) = (Ehap(t), Esad(t), Eang(t), Esur(t), Efear(t), Edis(t)) \qquad (10)$$

Emotion intensity can be represented as follows:

$$Ie(t) = (I1(t), I2(t), I3(t), I4(t), I5(t), I6(t)) \tag{11}$$

Basic emotions' representation. Each component of emotion vector is computed by the following expression:

$$Ei(t) = f(Iei(t), \omega 0) = \begin{cases} 1, & Iei(t) \geq \omega 0 \\ 0, & Iei(t) < \omega 0 \end{cases} \tag{12}$$

$Iei(t)$ denotes the individual's value of emotion intensity, $\omega 0$ denotes the active threshold, $i \in [1, 6]$.

Emotion's outside external stimulus. Because the outside stimulation directly leads to generation and change of mood, it is represented as follows:

$$\rho = (eo1, eo2, \ldots, eo6) \tag{13}$$

$eoi \in [0, 1]$, eoi denotes the outside stimulation intensity of the number i basic emotion, and $i \in [1, 6]$, i denotes the basic emotion's number.

Interactive influence factor matrix between emotions [6]. Human's emotions often influence each other, suppose emotional interactive influence factor matrix is

$$\lambda, \ \lambda = \begin{bmatrix} \lambda 11 & \cdots & \lambda n1 \\ \vdots & \ddots & \vdots \\ \lambda n1 & \cdots & \lambda nn \end{bmatrix}, \ n \text{ is the basic emotion's number, } \lambda ij \in [-1, 1], \text{ it}$$

represents the factor matrix of emotion i to emotion j, $i \in [1, n]$, $j \in [1, n]$ [6].

Mood space and emotional space. According to Gebhard's research, mood space and emotional space have mapping matrix and the computing method of Gebhard is used in our work [7].

Emotion's self-attenuation and update. Emotion's self-attenuation and update equations can be expressed as follows:

$$Ie(t) = Ie(t0) \times e^{-ai(t-t0)} \tag{14}$$

$Ie(t)$ is the emotion's intensity at time t, ai is emotion's attenuation coefficient, which is different according to different characters.

Emotion's intensity and emotion status's update. Emotion's intensity and emotion status' update can be computed by Eqs. (15) and (16).

$$Ie(t) = \alpha(\rho) + \beta(\rho) + \phi(Ie(t-1)) + Iepad + \psi(Iez) \tag{15}$$

$$E'(t) = (f(Ie1, \gamma), f(Ie2, \gamma), f(Ie3, \gamma), f(Ie4, \gamma), f(Ie5, \gamma), f(Ie6, \gamma)) \tag{16}$$

$\alpha(\rho)$ is the emotion value caused by conditioned emotional response, $\beta(\rho)$ is the emotion value caused by consciousness, $(Ie(t-1))$is the emotion value after emotion's attenuation at previous moment, Iepad is the influence of current mood on emotional fluctuations, and $\psi(Iez)$ is the interaction between different emotions. In formula (16), $E'(t)$ is the emotion status at time t and γ is the activation threshold.

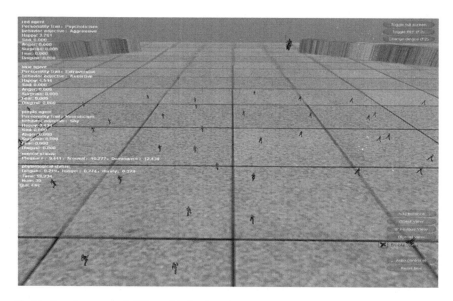

Fig. 1 Crossing scene of different kinds of agents

2.8 Behavior Module

When meeting the stimulus, different virtual human may have different behavior, which can be displayed by velocity, path selection, expression, language, etc. In here, the emotions' changes are mainly represented by path selection.

3 Experiments and Results

3.1 Experiments

In the experiment, there are 136 agents crossing a narrow passage, including virtual humans of three different personality traits, which are psychoticism, extraversion, and neuroticism according to the Eysnek 3-factor personality model. A dangerous source which is a knight with sword is at the end of the narrow passage, a red agent, a blue agent, and a purple agent is tested from the start position to the end position, the red agent is high psychoticism in personality trait and it is aggressive, the blue agent is high extraversion in personality trait and it is assertive, and the purple agent is high neuroticism in personality trait and it is shy, but it has ever been injured by a knight in the battle. They are probably in the second row of the crowd and must walk through the dangerous source. According to this emotion model, the six values of happy, sad, anger, surprise, fear, and disgust are recorded and displayed in the user interface of the simulation system. The crossing scene is displayed as Fig. 1.

Table 3 Agent of high psychoticism's status values

Time	Recorded variables				
	Happy	Fear	Pleasure	Arousal	Dominance
$t = 0.15$	0.65	0	13.54	14.73	17.81
$t = 61.35$	0.00	0.80	2.16	2.35	2.84
$t = 82.57$	0.335	0.20	1.14	1.24	1.50

Table 4 Agent of high extraversion's status values

Time	Recorded variables				
	Happy	Fear	Pleasure	Arousal	Dominance
$t = 0.14$	0.648	0	13.56	14.76	17.85
$t = 125.84$	0.035	0.171	1.10	1.19	1.44
$t = 145.0$	0.020	0.162	0.74	0.81	0.98

Table 5 Agent of high neuroticism's status values

Time	Recorded variables				
	Happy	Fear	Pleasure	Arousal	Dominance
$t = 0.13$	0.647	0	13.54	14.73	17.82
$t = 20.39$	0.00	0.90	5.59	6.09	7.36
$t = 70.42$	0.041	0.139	1.16	1.26	1.52

3.2 Results

When crossing the dangerous source, three recorded agents' emotion values are displayed in Tables 3, 4 and 5, and their paths are displayed as Fig. 2. Because the testing time is very short, physiological status' impact has been overlooked.

From the above three tables, we can see during the first part of time, all of the agents' fear values are increased when they come closer to the dangerous source, at the second record time, fear values reach their maximum values, conversely, during the latter part of time, fear values are decreased when they get farther to the dangerous source, although the purple agent is the furthest one from the dangerous source, but it has the biggest fear value at the shortest time, because it has ever been injured by a knight and has the conditioned emotional response to the dangerous source. The purple agent's status value has the biggest percentage change, next is the agent of high psychoticism, and last is the agent of high extraversion; the values of these agents' mental status also have the same trend.

In Fig. 2, three recorded agents' paths are shown: black lines on two sides represent the narrow passage, red line at the far right is the red agent's path, it is high psychoticism and aggressive, green line in the middle is the blue agent's path, it is high extraversion and assertive, purple line at the far left is the purple agent's

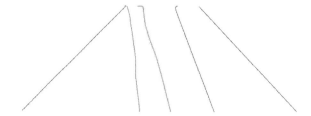

Fig. 2 Three recorded agents' paths

path, it is high neuroticism and shy. When meeting the dangerous resource, aggressive agent move toward the dangerous resource, assertive agent move away the dangerous resource, affected by conditioned emotional response, shy agent move away the dangerous resource along the edge of the narrow passage in the shortest time.

4 Conclusions

This emotion model considers various factors affecting the emotion, including the decay of emotion, mutual influence between emotions, outside stimulus, subconsciousness; the final three components of emotion are reflected by computed values, and the complex mood can be reflected to a certain extent, but this emotion model also has many limitations, for example, there may be more of the emotion components; furthermore, the emotions of human have uncertainty, and facial expression is not included, etc.; in the future work, these aspects should also be considered.

References

1. Wang, L., Wang, L.: Research of affective model of agent based on OCC. Microcomput. Inf. **23**, 256–258 (2007)
2. Gu, X., Wang, Z., Liu, J., Liu, S.: Research on modeling artificial psychology based on HMM. Appl. Res. Comput. **23**, 30–32 (2006)
3. Kshirsagar, S., Magnenat-Thalmann, N.: A multilayer personality model. In: Proceedings of the 2nd International Symposium on Smart Graphics, pp. 107–115. ACM Press, New York (2002)
4. Zhu, H.: Unconsciousness conditioned emotional response behavior effect and clinical applications. Master Dissertation of Liaoning Normal University, pp. 1–2 (2005)
5. Guy, S.J., Kim, S., Lin, M.C., Manocha, D.: Simulating heterogeneous crowd behaviors using personality trait theory. In: Proceedings of the 2011 ACM SIGGRAPH/Eurographics Symposium on Computer Animation (SCA'11). 2011, pp. 43–52. ACM Press, New York (2011)

6. He, H.: Research on a layered model of affect. Master Dissertation of Taiyuan University of Technology, pp. 14 (2010)
7. Gebhard, P.: ALMA: A layered model of affect. In: Proceedings of AAMAS 05, pp. 29–36. ACM Press, New York (2005)

Multi-Source and Heterogeneous Knowledge Organization and Representation for Knowledge Fusion in Cloud Manufacturing

Jihong Liu, Wenting Xu and Hongfei Zhan

Abstract In this knowledge-intensive world, knowledge contributes to the access and utilization of manufacturing resources as the most important intelligent resource. With the development of network technology, cloud manufacturing (CMfg) is proposed to meet the emerging requirement for high-efficiency, energy-saving, and service-orientated manufacturing. Focusing on the situation of the distributed and heterogeneous knowledge resources in Group Corporation, this paper presents a knowledge organization and representation model to support knowledge fusion (KF) and service. Meanwhile, a framework of KF and service is constructed to improve the efficiency of knowledge resource usage and the quality of knowledge services (KSs) in CMfg.

Keywords Cloud manufacturing · Knowledge organization · Knowledge representation · Knowledge fusion

1 Introduction

1.1 Cloud Manufacturing

In recent years, the development of cloud computing [1], Internet of things [2], and some other information technologies has laid a solid foundation for the improvement of advanced manufacturing models such as networked manufacturing, grid

J. Liu (✉) · W. Xu · H. Zhan
School of Mechanical Engineering and Automation, Beihang University, Beijing, China
e-mail: ryukeiko@buaa.edu.cn

W. Xu
e-mail: xuwenting8911@126.com

H. Zhan
e-mail: feihong1225@126.com

S. Patnaik and X. Li (eds.), *Proceedings of International Conference on Soft Computing Techniques and Engineering Application*, Advances in Intelligent Systems and Computing 250, DOI: 10.1007/978-81-322-1695-7_7, © Springer India 2014

manufacturing, and service manufacturing. According to the conceptions of cloud computing and "Manufacture as a Service," cloud manufacturing (CMfg) [3] is proposed to make manufacturing resources virtualized and manufacturing capacity service-oriented in order to provide active, agile, aggregative and all-aspects manufacturing resources and services for manufacturing enterprises.

1.2 Knowledge Fusion and Service

In CMfg, manufacturing resources contain not only all kinds of manufacturing equipments, but also the knowledge resources (models, data, software, etc.) in manufacturing process. As the core support of CMfg, knowledge provides technical services for the business processes (such as design, simulation, manufacture, and test). However, there are a large number of distributed and heterogeneous knowledge resources in Group Corporation, which cannot support the business processes effectively due to the lack of unified organization and management.

Knowledge fusion (KF) is proposed to combine this knowledge and exploit them to the fullest extent in a dynamic way [4]. As an important part of knowledge application, knowledge service (KS) aims to construct and provide effective knowledge resources according to the user requirements. Knowledge engineers fuse knowledge with different types, different contents, different features, and different positions to give full play to the integrated benefit of on-demand knowledge supply and thus achieve the goal of knowledge innovation. And complete solutions are provided to the owners and operators in CMfg platform to keep the advancements of knowledge as well as the high quality and efficiency of the services.

There have been many different approaches to achieve KF, such as agent-based KF architecture [4], fuzzy label fusion framework [5], and Horn fragment KF [6]. However, there has been little research reporting on service-orientated KF in CMfg. Therefore, a framework of KF and service is presented in this paper to get adapted to the new environment.

This paper is organized as follows. Section 2 analyzes the knowledge resources and the requirements for knowledge organization and representation of Group Corporation in CMfg. Section 3 proposes knowledge organization and representation model to support KF and service. Section 4 constructs a framework of KF and service. Section 5 draws the conclusion.

2 Analysis of Knowledge Resources and Requirements for Knowledge Organization and Representation of Group Corporation in Cloud Manufacturing

2.1 Features of Group Corporation Knowledge Management and Knowledge Resource Classification

The product development of Group Corporation has the characteristics of widely geographical distribution, great capacity of research data, frequent data exchange, and short development cycle. So there are three features of the knowledge management to consider in Group Corporation.

1. Knowledge Acquisition: It is difficult to discover, integrate, and accumulate the knowledge resources distributed in different kinds of systems in subunits and subdepartments as well as in minds of professional technicians, as the lack of effective knowledge acquisition.
2. Knowledge Organization: As the lack of unified knowledge organization, the knowledge resources scattered around the inside of Group Corporation cannot give full play to the integrated benefit of knowledge resources.
3. Knowledge Fusion and Service: There is no effective method for KF and service to support the sharing and reuse of knowledge resources.

Combining with the researches of knowledge engineering and knowledge management, this paper divides the knowledge resources in Group Corporation into the following classifications:

1. Framework knowledge: Different organizations are more or less suffering from knowledge barriers so that the accumulated knowledge resources cannot be integrated at the semantic level. Framework knowledge is know-what knowledge that can integrate and disintegrate the existing business resources. This kind of knowledge resources mainly contains the terminology standards and the terminology networks.
2. Tacit knowledge: Tacit knowledge exists in the minds of a great many of professional technicians, so it is difficult to deliver or transform to others. This kind of knowledge is know-why and know-how knowledge. It mainly consists of thinking processes and experience skills.
3. Explicit knowledge: Explicit knowledge exists in the existing resources and can be completely represented by symbols (mathematical formulas, mathematical languages, etc.). This kind of knowledge resources consists of resultant resources (documents, models) and procedural resources (processes, tools).

As shown in Fig. 1, the knowledge resources expressed by ontology and text mainly provide references for the business processes instead of solving business problems directly. The latter expressed by KBE with clearly defined input and

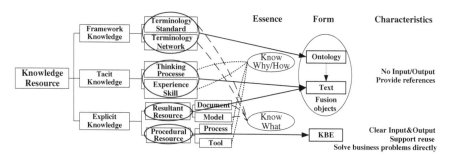

Fig. 1 Knowledge resource classifications in Group Corporation

output can support knowledge reuse and solve business problems directly, but could not be processed internally. In this paper, KF focuses on the former one without input and output that could be fused directly by the algorithms of text mining.

2.2 Requirement Analysis of Knowledge Organization and Representation in Group Corporation

In CMfg service platform, on the one hand, knowledge resources and manufacturing resources should be managed and serviced uniformly, and on the other hand, distinctions between them should be addressed a teach platform layer. In general, KF and service should meet the following three requirements:

1. As there are numerous knowledge resources, the scattered knowledge resources should be organized and the multi-source heterogeneous knowledge resources should be modeled uniformly.
2. After being packed and indexed, the knowledge resources should be accessed to system platform by virtualization.
3. Based on the analysis of knowledge demand, KSs should be provided by KF.

3 Knowledge Organization and Representation Model for Knowledge Fusion

Liu [7] has explored a knowledge organization model in CMfg. Based on the model, this paper proposes a knowledge organization and representation model which includes three layers: knowledge resource layer, knowledge unit layer, and fusion knowledge layer, as shown in Fig. 2. Through analyzing the knowledge demand, ontology is used to recognize knowledge segments for the text resources. Ontology tools are used to describe KBE tool resources to package the tools. Combining the knowledge segments and tools, the fusion knowledge is formed by using dynamic ontology as the framework of fusion knowledge.

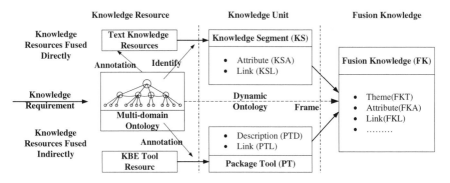

Fig. 2 Knowledge organization and representation model

3.1 Knowledge Unit

Knowledge unit (KU), as the meta-knowledge of KF, is the description of independent text resources or KBE tool resources to identify and represent the distributed knowledge resources.

KU represented by text is the related knowledge segment (KS) which is extracted from the text directly. It can be represented by the following tuple:

$$KS = \langle KSA, KSL \rangle.$$

Knowledge segment attribute (KSA), as knowledge property and description, includes {Ontology Terms, Knowledge Property, and Description}. Knowledge segment link (KSL), as the link of knowledge resources, includes {Inherited Property and Document Location}.

KBE tool resources should be added with the related descriptive information to get package tool (PT). It can be represented by the following tuple:

$$TP = \langle PTD, PTL \rangle.$$

Package tool description (PTD), as the basic properties of tools, includes {Ontology terms, General Description, Input, and Output}. Package tool link (PTL), as the link of tools, includes {Enterprise, Department, Creator, Security Level, Field, Product, and Node}.

3.2 Fusion Knowledge

Fusion knowledge (FK) is the knowledge with a certain structure that is fused from KU. The target is to reintegrate the scattered knowledge resources to form more targeted, more completed knowledge resources.

FK representation contains fusion knowledge theme (FKT), fusion knowledge framework (FKF), fusion knowledge attribute (FKA), and fusion knowledge link (FKL). It can be represented by the following triple:

$$FK = \langle FKT, FKF \ (FKA, FKL) \rangle.$$

FKT is the representation of theme of fusion knowledge resources, which is fused from integrated KUs.

FKF is the representation of whole structure of FK resources, which is obtained by forming the dynamic ontology. Both FKA and FKL inherit the knowledge fragments of KSA and KSL. Based on the FK, the scattered knowledge resources in Group Corporation can be organized to increase the efficiency of knowledge resources.

4 Framework of Knowledge Fusion and Service in Cloud Manufacturing

Based on the knowledge organization and representation model, this paper proposes a framework of KF and service in CMfg which includes four layers: knowledge resource layer (KRL), knowledge virtualization layer (KVL), knowledge fusion layer (KFL), and application service layer (ASL).

1. Knowledge Resource Layer

KRL describes the distributed knowledge resources of the subunits and departments, which mainly consists of documents (standards, research reports, and quality reports), models (product models, structure models, and simulation models), tacit experiences (thinking processes, experience summaries, and expert resources), process tools (process templates and computational tools), and ontologies constructed in various fields.

2. Knowledge Virtualization Layer

As the basis of KF, KVL is a unified access for distributed knowledge resources by dealing with heterogeneity of KU. KVL obtains critical knowledge fragments by analyzing ontologies for documents, gets knowledge resources by knowledge annotation for models, records design thinking processes by modeling for thinking processes, and achieves tool package by adding semantic description for computational tools. In this way, the distributed heterogeneous knowledge resources are acquired by knowledge virtualization. In the meantime, a networked organization system of knowledge resources can be established based on ontology annotation [8].

3. Knowledge Fusion Layer

Focused on the demands of knowledge, KFL provides supportable services for knowledge retrieval and push. Based on the analysis of the requirements for knowledge, the dynamic ontologies are constructed (including the structure and fragments of the ontology). KF focuses on constructing the dynamic ontologies on demand to form the specific KF and service.

4. Application Service Layer

ASL provides various types of KSs to the users such as business-oriented knowledge push and the solar enterprise-wide knowledge retrieval. It also provides entrance for the distributed users to achieve distributed knowledge resources.

In this framework, the Group Corporation Ontology formed by multi-domain ontology mapping can support enterprise knowledge management to organize the scattered knowledge resources.

5 Conclusions

Knowledge resource is an important kind of manufacturing resources in CMfg. This paper analyzes three features of knowledge in Group Corporation and divides the knowledge resources into 3 classifications. Based on the requirements, a knowledge organization and representation model is proposed to support KF and service. At last, this paper also presents a framework of KF and service, which includes four layers to improve the quality of KSs.

References

1. Rings, T., Caryer, G., Gallop, J., Grabowski, J., Kovacikova, T., Schulz, S., Stokes, R.: Grid and cloud computing: opportunities for Integration with the next generation network. J. Grid Comput. **7**(3), 375–393 (2009)
2. Bandyopadhyay, D., Sen, J.: Internet of things: applications and challenges in technology and standardization. Wirel. Pers. Commun. **58**(1), 46–69 (2011)
3. Li, B.H., Zhang, L., Zhang, S.L.: Cloud manufacturing: a new service-oriented networked manufacturing model. Comput. Integr. Manuf. Syst. **16**(1), 1–7 (2010). (in Chinese)
4. Preece, A., Hui, K., Gray, A., Marti, P., Bench-Capon, T., Jones, D., Cui, Z.: The KRAFT architecture for knowledge fusion and transformation. Knowl. Based Syst. **13**(2), 113–120 (2000)
5. Lawry, J., Hall, J.W., Bovey, R.: Fusion of expert and learnt knowledge in a framework of fuzzy labels. Int. J. Approximate Reasoning **36**(2), 151–198 (2004)
6. Dunin-Ke Plicz, B., Nguyen, L.A., Szałas, A.: Tractable approximate knowledge fusion using the Horn fragment of serial propositional dynamic logic. Int. J. Approximate Reasoning **51**(3), 346–362 (2010)
7. Liu, J., Li, B.: An Ontology-Based Architecture for Service-Orientated Design Knowledge Fusion in Group Corporation Cloud Manufacturing, pp. 811–816. IEEE Computer Society, Wuhan, China (2012)
8. Yu, X., Liu, J.H., He, M.: Design knowledge retrieval technology based on domain ontology for complex products. Comput. Integr. Manuf. Syst. **17**(2), 225–231 (2011). (in Chinese)

Research of Dynamics and Deploying Control Method on Tethered Satellite

Lei Gang, Xian Yong, Feng Jie and Wang Kui

Abstract Deployment of a tethered satellite system is the process of separating the two end bodies by spooling out the tether connecting them. In this paper, a 3D dynamic model of a tethered satellite was established. Three typical deploying parameters were chosen and discussed according to some correlative references, thus obtaining the changing law for the percentage of deployment. The numerical results show the process of deployment is much sensitive with the in-plane angle, and the stable points are prior to others in long-distance deployment considering energy needed. These results could provide ideas about systemic design and engineering practice for tethered satellite.

Keywords Tethered satellite · Dynamic model · Deploying parameter · Deployment

1 Introduction

The concept of tethered satellite system was first proposed by Tsiolkovsky [1]—the "Father of Space" of Russian—in 1895 in his book "Day-Dreams of Earth and Heaven". It has great prospect in the momentum exchange, conductive tether, artificial gravity generation, cargo space transportation, deep space exploration, spacecraft rendezvous, and capture, etc., [2, 3].

Deploying and reclaiming have been based on the issues of tethered satellites' dynamics. People has carried out a large research on the complex dynamics model, the programs of deploying and reclaiming control, rendezvous, precise capture, dynamic stability control, etc., [4–7]. In the release areas of the tethered satellite,

L. Gang (✉) · X. Yong · F. Jie · W. Kui
Xi'an Hongqing Research Institute of Hi-Tech, Xi'an 710025 ShanXi,
People's Republic of China
e-mail: leig603@163.com

S. Patnaik and X. Li (eds.), *Proceedings of International Conference on Soft Computing Techniques and Engineering Application*, Advances in Intelligent Systems and Computing 250, DOI: 10.1007/978-81-322-1695-7_8, © Springer India 2014

Williams and Trivailo [7] studied the problems of rendezvous and capture in 3D tether system and got the rendezvous where both accordant to position coincidence and tether release percentage is zero in radial within 30 days. Bibliography [8] studied the problem about sub-star's quick release in the close distance, analyzed the motion characteristics of tether and sub-star attitude. To the question of tethered satellite parameters for selection, a system dynamics model is established in this paper. Considering relevant literature, the typical values of deploying parameters were analyzed. By numerical simulation, it analyzed the effect of the initial parameters on the length of the deployed tether. The conclusion shows that stable points are prior to others in long-distance deployment.

2 A System Dynamics Model

In Fig. 1, $EX_EY_EZ_E$, $CX_oY_oZ_o$ and $CX_sY_sZ_s$ separately represent the geocentric inertial coordinate system, orbit coordinate system, and satellite coordinates. Of which, E is the center of mass for the Earth; C is the tethered system mass; A and B are the main star and sub-star, respectively; θ is the in-plane angles (pitch angle); and φ is the out-of-angle (roll angle). Both ends of the tethered satellite are regarded as particle, ignoring tether's mass. Therefore, system dynamics equation is:

$$\frac{F_A}{m_A} - \frac{F_B}{m_B} = a_{A/E} - a_{B/E} \tag{1}$$

Assuming the main star and sub-star are only affected by the force of gravity and the tension of tether, Eq. (1) can be written as:

$$\frac{F_A}{m_A} - \frac{F_B}{m_B} = \left(\frac{-GM}{|r_{A/E}|^3} r_{A/E} + \frac{T}{m_A} \frac{r_{C/A}}{|r_{C/A}|} \right) \\ - \left(\frac{-GM}{|r_{B/E}|^3} r_{B/E} + \frac{T}{m_B} \frac{r_{C/B}}{|r_{C/B}|} \right). \tag{2}$$

where $\dfrac{r_{C/A}}{|r_{C/A}|} = \dfrac{r_{C/B}}{|r_{C/B}|} = \begin{bmatrix} -1 \\ 0 \\ 0 \end{bmatrix}$.

In the satellite system:

$$\{r_{A/E}\}_s = \{r_{A/C}\}_s + \{r_{C/E}\}_s = \begin{bmatrix} L_a + R\cos\theta\cos\varphi \\ -R\sin\theta \\ -R\cos\theta\sin\varphi \end{bmatrix} \tag{3}$$

Fig. 1 The coordinate system of TSS

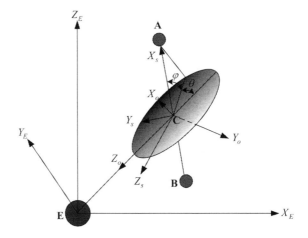

Then,

$$|r_{A/E}| = R\left(1 + \frac{L_a^2}{R^2} + \frac{2L_a}{R}\cos\theta\cos\varphi\right)^{1/2} \tag{4}$$

$L_a^2 \ll R^2$ when the tether length is much smaller than the orbital radius. Ignoring higher-order items, second item launched of Eq. (4) to get:

$$|r_{A/E}|^{-3} \approx \frac{1}{R^3}\left(1 - 3\frac{L_a}{R}\cos\theta\cos\varphi\right) \tag{5}$$

Similar to get:

$$\{r_{B/E}\}_s = \begin{bmatrix} -L_b + R\cos\theta\cos\varphi \\ -R\sin\theta \\ -R\cos\theta\sin\varphi \end{bmatrix} \tag{6}$$

$$|r_{B/E}|^{-3} \approx \frac{1}{R^3}\left(1 + 3\frac{L_b}{R}\cos\theta\cos\varphi\right) \tag{7}$$

Setting $GM/R^3 = \Omega^2$, by (3)–(7), Eq. (2) can be written as:

$$\left\{\frac{F_A}{m_A} - \frac{F_B}{m_B}\right\}_s = \begin{bmatrix} -\Omega^2 L + 3\Omega^2 L\cos^2\theta\cos^2\varphi - \frac{T}{m} \\ -3\Omega^2 L\cos\theta\sin\theta\cos\varphi \\ -3\Omega^2 L\cos^2\theta\cos\varphi\sin\varphi \end{bmatrix} \tag{8}$$

m is the effective mass of the system, and $m = m_A m_B/(m_A + m_B)$.

Relatively inertial coordinate system, the angular percentage, and angular of acceleration satellite coordinate system is:

$$\omega_s^E = \begin{bmatrix} (\dot{\theta} + \Omega)\sin\varphi \\ -\dot{\varphi} \\ (\dot{\theta} + \Omega)\cos\varphi \end{bmatrix} \qquad (9)$$

$$\dot{\omega}_s^E = \begin{bmatrix} \ddot{\theta}\cos\varphi + (\dot{\theta} + \Omega)\cos\varphi \\ -\ddot{\varphi} \\ \ddot{\theta}\cos\varphi - (\dot{\theta} + \Omega)\dot{\varphi}\sin\varphi \end{bmatrix} \qquad (10)$$

And the relationship of acceleration between inertial and satellite coordinate system is:

$$\{\ddot{r}_{A/B}\}_E = \left\{ \ddot{r}_{A/B} \right\}_s + \dot{\omega}_s^E \times r_{A/B}$$
$$+ 2\omega_s^E \times \{\dot{r}_{A/B}\}_s + \omega_s^E \times (\omega_s^E \times r_{A/B}) \qquad (11)$$

where

$$\{r_{A/B}\}_s = \begin{bmatrix} L \\ 0 \\ 0 \end{bmatrix}, \ \{\dot{r}_{A/B}\}_s = \begin{bmatrix} \dot{L} \\ 0 \\ 0 \end{bmatrix}, \ \{\ddot{r}_{A/B}\} = \begin{bmatrix} \ddot{L} \\ 0 \\ 0 \end{bmatrix}.$$

By (9)–(11), it can be simplified:

$$\{\ddot{r}_{A/B}\}_s = \begin{bmatrix} \ddot{L} - \dot{\varphi}^2 L - (\dot{\theta} + \Omega)^2\cos^2\varphi L \\ \ddot{\theta}\cos\varphi L - 2(\dot{\theta} + \Omega)\dot{\varphi}\sin\varphi + 2(\dot{\theta} + \Omega)\dot{L}\cos\varphi \\ \ddot{\varphi}L + 2\ddot{\varphi}\dot{L} + (\dot{\theta} + \Omega)^2\cos\varphi\sin\varphi L \end{bmatrix} \qquad (12)$$

Contrast to Eq. (8), tethered system dynamics equation is:

$$\begin{cases} \ddot{L} = L\left[(\dot{\theta} + \Omega)^2\cos^2\varphi + \dot{\varphi}^2 + 3\Omega^2\cos^2\theta\cos^2\varphi - \Omega^2 \right] - \dfrac{T}{m} \\[2mm] \ddot{\theta} = 2(\dot{\theta} + \Omega)\dot{\varphi}\tan\varphi - 2\dfrac{\dot{L}}{L}(\dot{\theta} + \Omega) - 3\Omega^2\cos\theta\sin\theta \\[2mm] \ddot{\varphi} = -2\dfrac{\dot{L}}{L}\dot{\varphi} - \left[\dfrac{(\dot{\theta} + \Omega)^2 + 3}{\Omega^2\cos^2\theta} \right]\cos\varphi\sin\varphi \end{cases} \qquad (13)$$

3 Analyze the Values of System Parameters

This paper focuses on the released process of the sub-star in the different initial conditions. For this, it can select the typical system parameters as follows: the initial tethered tension T_0, the system effective mass m, the initial separated velocity $\dot{L}_0$, the initial tether angle θ_0 and φ_0, track elevation H, and tether length

L_f. Among them, H and L_f may be regarded as objective parameters, the rest as the designed parameters. Values of the main parameters as follows:

1. The initial separated velocity $\dot{L}_0$

The initial velocity of tether can be gotten through ejection from the main star [8]. $\dot{L}_0$ is decided by the elasticity and compression degree of spring in the ejection institution. In 1993, SEDS-1 developed by NASA Marshall Space Flight Center in the USA, which was a small one-time-tethered expansion system, successfully carried out the long tether orbital flight tests, and the initial ejection speed was 1.64 m/s[9]. It is obvious that the capacity of catapult is limited by the device's narrow space. $\dot{L}_0$ was selected [3, 6] when simulated in Ref. [9].

2. The initial tension of tether T_0

By the release of the tethered reel device, as rope length is short, the gravity gradient force is negligible at the beginning. Therefore, the T_0 can be approximately equaled to the friction between the tether and reel device. The study about the reclaimed/deployed device showed that the friction is 50 mN approximately in room temperature [9]. Once deployed, the friction is 0, then the tension is related to the vibration cycle and the length of the tether. Taking into account the initial velocity of the tether is slow, focusing on the released process sensitivity to initial parameters, this article assumes that tension is constant in the process of entire release.

3. The effective mass of the system m

By Eq. (13), it shows that system dynamic model is directly related to the effective mass m. If the total mass of main star and the sub-star is constant, the m depends on the distribution of mass. This article assumes that total mass is 80 kg, the minimum mass of sub-star is 12 kg, and the m range 10.2–20 kg.

4 Simulation and Analysis

Simulation is based on assumptions that the tethered system in a circular orbit, with the conditions of $\dot{L} = 0$ or $L = L_f$, where L_f is scheduled deployed length of the tether. Table 1 shows the initial value of parameters.

The simulated results shown in Figs. 2 and 3.

Shown in Figs. 2 and 3, the extremum of tether's deployed percentage concentratedly present near the $\theta = 0$ resulted from the gravity gradient force, and the degree of increase gradually decreases with distance lengthen, which indicates that, compared to out-of-angle, the process of deployment is more sensitive to in-plane angles. When other conditions remain unchanged, the deployed percentage

Table 1 Initial value of parameters

T_0 (mN)	$\dot{L}_0$(m/s)	m (kg)	θ_0 (deg)	φ_0 (deg)	H (km)	L_f(km)
50, 100	3, 4.5, 6	10.2, 15.1, 20	$-90 \sim 90$	$-90 \sim 90$	1,000	3

Fig. 2 Deployed percentage of tether when $T_0 = 50$ mN, $m = 10.2$ kg, $\dot{L}_0 = 3$ m/s

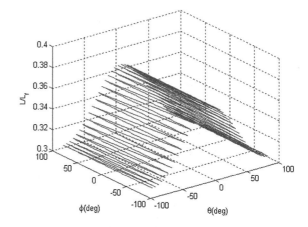

Fig. 3 Deployed percentage of tether when $T_0 = 100$ mN, $m = 10.2$ kg, $\dot{L}_0 = 3$ m/s

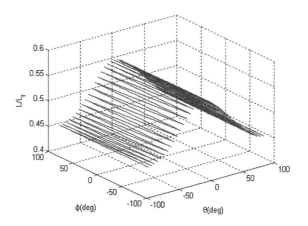

of tether is inversely proportional to T_0 and directly proportional to $\dot{L}_0$ and m. When sub-star is releasing, the stable equilibratory position is in radial direction. Therefore, the stable points are prior to others in long-distance deployment considering energy needed.

5 Conclusions

Based on the establishment of space tethered system dynamics model and analysis of typically released parameters, the relationship between the released parameters and the length of deployed tether was researched in this paper. The numerical results indicates that, compared to out-of-angle, the process of deployment is much sensitive to the in-plane angle and the stable points are prior to others in long-distance deployment considering energy needed. These results could provide ideas about systemic design and engineering practice for tethered satellite. With the in-depth study of tethered satellite complex models, deployment and recovery control, captured control, etc., the development prospects of tethered satellite systems will be broad.

References

1. Pearson, J.: Konstantin Tsiolkovski and the origin of the space elevator. In: 48th IAF, International Astronautical Congress, Turin, Italy, **10**, 6–10 (1997)
2. Cosmo, M.L., Lorenzini, E.C.: Tethers in Space Handbook, 3rd edn. Smithsonian Astrophysical Observatory, Cambridge (1997)
3. Carroll, J.A.: Tether applications in space transportation. IAF84-438, April 1984
4. Williams, P.: Nonlinear control and applications of tethered space systems. PhD Dissertation, School of Aerospace, Mechanical, and Manufacturing Engineering, Royal Melbourne Institute of Technology, Melbourne, Australia (2004)
5. Ohkami, Y., Yoshimura, S., Okamoto, O.: Evaluation of microgravity level fluctuation due to attitude/orbital motion in a tethered satellite system. Acta Astronaut. **35**(2–3), 187–191 (1995)
6. Cartmell, M., McKenzie, D.: A review of space tether research. Prog. Aerosp. Sci. **7**, 1–22 (2007)
7. Williams, P., Trivailo, P.: On the optimal deployment and retrieval of tethered satellites. In: 41st AIAA/ASME/SAE/ASEE Joint Propulsion Conference and Exhibit, Tucson, Arizona, USA: AIAA, 10–13 July 2005
8. Williams, P., Trivailo, P.: On the optimal deployment and retrieval of tethered satellites. AIAA Paper 2005-4291, July 2005
9. Mazzoleni, A.: Nonplanar deployment dynamics of the small expendable-tether deployment system. Acta Astronaut. **36**(3), 141–148 (1995)

Using Auto-Associative Neural Networks for Signal Recognition Technology on Sky Screen

Yan Lou, Zhipeng Ren, Yiwu Zhao and Yugui Song

Abstract This paper proposes an auto-associative Neural Network pattern that has been utilized for sky-screen signal recognition. The AANN mode has been prepared in VC++ platform. Compared to the level signal recognition, the recognition rate can be increased by 3 % using auto-associative neural network with 30 mm caliber projectiles at a frequency of 7,500 rounds/min for 10 s duration in the sky-screen-repeating projectile test, the accuracy and reliability of the system was fully verified.

Keywords Sky screen · Signal recognition · Hopfield auto-associative neural network · Recognition rate

1 Introduction

The sky screen is a kind of system that was widely used to measure the velocity of flying projectile in ranger. The flying projectile that goes through the screen outputs a trigger digital pulse signal where its front edge indicates the instant of projectile nose trigging and back edge indicates the instant of projectile base trigging, the duration of pulse indicates the time of flying through the screen [1, 2]. Some kinds of methods suppressed non-projectile signals that were given in Refs. [3–5]. However, those methods can only work at the case that shock wave and flying insects exist independently. Based on the above analysis, Hopfield neural

Y. Lou (✉) · Z. Ren · Y. Zhao
Institute of Space and Photoelectric Technology, Chang Chun University of Science and Technology, 7186 Weixing Road, Changchun 130022, China
e-mail: Louyan2008@126.com

Y. Song
School of Optoelectronic Engineering, Xi'an Technological University,
4 Jinhua Road, Xi'an 710032, China

S. Patnaik and X. Li (eds.), *Proceedings of International Conference on Soft Computing Techniques and Engineering Application*, Advances in Intelligent Systems and Computing 250, DOI: 10.1007/978-81-322-1695-7_9, © Springer India 2014

network pattern recognition methods have been proposed in this paper. A pattern decision by the memory of their closest signal samples of typical learning established memory. Many advantages are available to deal with some very complex environmental information, blurred background knowledge, unclear rules of inference problems, and allow a greater sample defects and aberrations.

In this paper, several typical projectile signals were identified from sky-screen acquisition at the shooting range scene. Recognition rate and the error rate are calculated as follows [6]:

Recognition rate = (number of identified signal/Theoretical valid signal number) × 100 %

Error rate = (number of misidentified signals/Theoretical valid signal number) × 100 %

2 Hopfield Associative Neural Network

2.1 Discrete Hopfield Network Theory

Hopfield network is divided into two types: discrete and continuous Hopfield network [7]. In this paper, the application of discrete Hopfield neural network (DHNN) is a discrete-time system. The basic structure of discrete Hopfield network is given in Fig. 1, the network consists of n units, $N_1, N_2, \ldots, N_{n-1}, N_n$ denotes n units, this neural network is basically fed forward a layered neural network that has same number of nodes in the input layer and the output layer, the transfer characteristic function is $f_1, f_2, \ldots, f_n$ and the threshold is $\theta_1, \theta_2, \ldots, \theta_n$. For discrete Hopfield network, generally, all nodes select the same transfer function, which is the sign function, that is [8]:

$$f_1(x) = f_2(x) = \cdots = f_n(x) = \text{sgn}(x) \tag{1}$$

all transfer function is equal to 0 as:

$$\theta_1 = \theta_2 = \cdots = \theta_n = 0 \tag{2}$$

meanwhile, $x = (x_1, x_2, \ldots, x_n), x \in \{-1, +1\}^n$, x is actually the computing inputs layer; $y = (y_1, y_2, \ldots, y_n)$, $y \in \{-1, +1\}^n$, y is actually the computing output layer; $V(t) = (V_1(t), V_2(t), \ldots, V_n(t)), V(t) \in \{-1, +1\}^n$ is the status of network at time t, wherein $t \in \{0, 1, 2, \ldots\}$ is variable of discrete time; where w_{ij} is connection weights from N_i to N_j, since Hopfield network is symmetrical that: $w_{ij} = w_{ji}$, $i, j \in \{1, 2, \ldots, n\}$.

All n nodes-associated connection strength in the network expressed by matrix W, and W is $n \times n$ matrix.

The figure shows the structure of discrete Hopfield network in layer feedback network that can handle bipolar discrete data (input $x \in \{-1, +1\}$). When the network is trained, the whole operation is a process of repeated feedback. If the

Fig. 1 Neural network
program flowchart

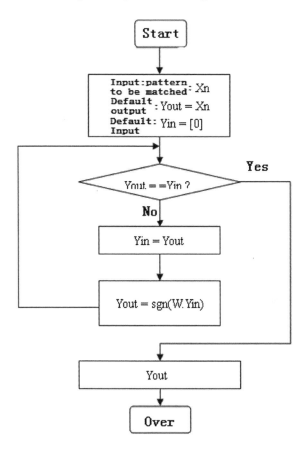

network is stable, then, with the number of times the feedback operation, the network status changes are reduced until the reach of steady state or no longer changes. In this case, the output of the network stable output can be obtained. The formula is expressed as follows:

$$
\begin{cases}
v_j(0) = x_j \\
v_j(t+1) = f_j\left(\sum_{i=1}^{n} w_{ij}v_i(t) - \theta_j\right)
\end{cases}
\tag{3}
$$

f_j was determined by formula (1). After a certain time t from the state network with no change, that is, $V(t+1) = V(t)$, then the output is:

$$
y = V(t)
\tag{4}
$$

Asynchronous (serial) pattern. Status update based on formula (3) of a neuron N_j at a time t, the state of $j - 1$ remaining neurons keeping unchanged,

$$V_j(t+1) = \text{sgn}\left(\sum_{i=1}^{n} w_{ij}v_j(t)\right). \tag{5}$$

To other neurons:

$$V_i(t+1) = V_i(t) \quad i \in \{1, 2, \ldots n\}, \; i \neq j. \tag{6}$$

Order update is defined by the change in accordance with formula (5) if the order selected according to a deterministic, random update is called the selected-based neurons according to the preset probability.

Synchronous (parallel) pattern. The state update of some neurons according to formula (3) at any time t, in which an important special case is at certain time, while the state of all the neurons changing in accordance with formula (3), as $V_j(t+1) = \text{sgn}\left(\sum_{i=1}^{n} w_{ij}v_i(t)\right)$, $j = \{1, 2, \ldots n\}$ can be written:

$$V(t+1) = \text{sgn}(V(t) \times W). \tag{7}$$

If the network has a limited period of time at any initial state $x(0)$ from $t = 0$, the network is called stable after the nerve network status with no change from the moment,

$$V(t + \Delta t) = V(t), \quad \Delta t > 0 \tag{8}$$

3 Application of Hopfield Network for Typical Projectile Signal

3.1 Extraction of Pattern Feature

The main purpose of feature extraction is characterized as centralized pattern information with differences of significant classification. Another purpose is to minimize the data sets, to improve the recognition efficiency, and to reduce the amount of calculation. Feature extraction and selection is very important.

According to the characteristics of the collected signal sky target, we use one-dimensional moment feature extraction method. One-dimensional moment feature is dominated and generated by one-dimensional pattern sequence. We define finite pattern sequence $\{x_j^{(i)}, i = 1, 2, \ldots, c, j = 1, 2, \ldots N^{(i)}\}$ r-order moments and central moments, respectively [9]

$$C_r^i = E[x^{(i)r}] \approx \frac{1}{N^{(i)}} \sum_{j=1}^{N^{(i)}} x_j^{(i)r} \quad i = 1, 2, \ldots, c \tag{9}$$

$$D_r^i = E\left[\left(x^{(i)} - \bar{x}^{(i)}\right)^r\right] \approx \frac{1}{N^{(i)}} \sum_{j=1}^{N^{(i)}} \left(x_j^{(i)} - \bar{x}_j^{(i)}\right)^r \quad i = 1, 2, \ldots, c \quad (10)$$

where $\bar{x}^{(i)}$ is the mean vector pattern sequence. Therefore, the usual moment feature may be:

1. Mean-variance:

$$m^{(i)} = E\left[\left(x^{(i)}\right)\right] \approx \frac{1}{N^{(i)}} \sum_{j=1}^{N^{(i)}} x_j^{(i)} \quad i = 1, 2, \ldots, c \quad (11)$$

2. Variance:

$$\sigma^{(i)} = E\left[\left(x^{(i)} - \bar{x}^{(i)}\right)^2\right] \approx \frac{1}{N^{(i)}} \sum_{j=1}^{N^{(i)}} (x_j^{(i)} - \bar{x}_j^{(i)})^2 \quad i = 1, 2, \ldots, c \quad (12)$$

3. Partial odd:

$$\text{sk}^{(i)} = \frac{E[(x^{(i)} - \bar{x}^{(i)})^3]}{\sigma^{(i)3}} \approx \frac{1}{N^{(i)}} \sum_{j=1}^{N^{(i)}} \left(\frac{x_j^{(i)} - \bar{x}_j^{(i)}}{\sigma^{(i)}}\right)^3 \quad i = 1, 2, \ldots, c \quad (13)$$

4. Kurtosis:

$$\text{ku}^{(i)} = \frac{E[(x^{(i)} - \bar{x}^{(i)})^4]}{\sigma^{(i)4}} - 3 \approx \frac{1}{N^{(i)}} \sum_{j=1}^{N^{(i)}} \left(\frac{x_j^{(i)} - \bar{x}_j^{(i)}}{\sigma^{(i)}}\right)^7 - 3 \quad i = 1, 2, \ldots, c. \quad (14)$$

In general, the pattern reflects the pattern cluster centers by the mean characteristics; patterns around the degree of deviation by the mean vector variance; and sample distribution shape information was given by partial odd and kurtosis. Partial odd is portrayed as the asymmetric degree on the sample mean vector pattern. sk > 0 indicates pattern-right, sk < 0 indicates pattern-left; while kurtosis reflects the peak flatness of pattern sample.

In normal subject, ku > 0 indicates peak distribution than the Gaussian distribution pattern, ku < 0 indicates peak distribution pattern below the Gaussian distribution.

3.2 Associative Memory Algorithm for Discrete Hopfield Neural Networks

The following is a use of a Hebb rule, according to the discrete Hopfield asynchronous update algorithm steps:

1. Initialize the weights, set $w = [0]$.
2. Input p sample pattern of $x^1, x^2, \ldots, x^p$ to network, determine the network weights.
3. Initialize unknown input pattern x^l, $x_j(0) = x_j^l$, $1 \leq j \leq n$, where x_j is the number of j in the input pattern of the $x_j \in \{-1, +1\}$.
4. Iteration until convergence $x_j(t+1) = \text{sgn}[\sum\limits_{i=1}^{n} W_{ij}x_i(t)]\ 1 \leq j \leq n$, where $x_j(t)$ is the output state at time t and neuron j.
5. When $x_j(t+1) = x_j(t)$ is steady-state output, $1 \leq j \leq n$, the steady-state output indicates its network best match with unknown input pattern.

3.3 Simulation and Application for Typical Signal of Sky Screens

Pattern neurons determine and feature extraction. Hopfield network memory capacity is limited according to the training by Hebb rule, when the pattern vector is orthogonal vectors, the network pattern is equal to the number of stable storage of the number of neurons in the network are all n. Based on sky screens test firing the collected data, select the penetrator signal, using 16 neurons nodes. Ten training samples and the test samples were selected.

Sky-screen data analyzed in this paper was collected through experiment of rifle in field, the sampling frequency (1 MHz), SNR $\geq$ 6 dB. All signals data saved in computer with text pattern. Using Matlab to obtain one-dimensional feature, vector extracted moments was shown in Table 1.

Pattern features corresponding binary vector. Using cluster encoding

$$x^1 = (1, -1, -1, -1,\quad 1, -1, -1, -1,\quad 1, -1, -1, -1,\quad 1, -1, -1, -1,)^{\text{T}}$$
$$x^2 = (-1, 1, -1, -1,\quad -1, 1, -1, -1,\quad -1, 1, -1, -1,\quad -1, 1, -1, -1)^{\text{T}}$$
$$x^3 = (-1, -1, 1, -1,\quad -1, -1, 1, -1,\quad -1, -1, 1, -1,\quad -1, -1, 1, -1)^{\text{T}}$$

As the network composed of a layer of saturated linear neurons, neurons with an output connected to the input through the weight matrix, the neuron output is specified as the initial output vector, requiring a discrete value, and value is a binary function, where to take 1 and -1, design of a steady-state value:

Five bursts signal can be memorized by x^1.
A single-slit sky target projectile signal can be memorized x^2.
Penetrators signal characteristics can be memorized x^3.

Update the network as many times, when the network has reached steady state at some point, the output of a vector with a value will be equal to the initial output and stabilize at a certain point on the output of the initial setting, the final output vector is the classification of the initial vector.

Table 1 Typical signal characteristics

Characteristic	Pattern		
	Five bursts of projectile signal of sky screens	Penetrators signal	Projectile signal bursts
Signal pulse width μs	100	460	220
The mean amplitude m/v	1.2899 to 6.6534	1.3446 to 5.5002	1.3520 to 4.3571
Amplitude variance d/v	0.7780 to 10.1460	1.941 to 7.6231	3.0229 to 10.0180
Amplitude of the odd side sk/v	−0.7363 to 0.0256	−2.9045 to −2.3589	−0.0051 to 0.1587
Amplitude of the odd side ku/v	−2.9796 to −0.0729	−0.2013 to −0.0932	−2.9913 to −2.8058

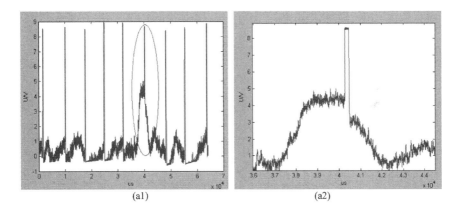

(a1)   (a2)

Fig. 2 Multiple bursts signal recognition of sky screens

The result of Matlab simulation. Multiple bursts of signal recognition of sky screens

Bullet: 30 mm diameter; Frequency: 7,500 rounds/min; Time: 10 s

1. Matlab simulation interface (Fig. 2):
2. Analysis

From Table 2, compared to the level signal recognition, the recognition rate can be increased by 3 % from auto-associative neural network with 30 mm caliber projectiles at a frequency of 7,500 rounds/min for 10 s duration in the sky-screen-repeating projectile test, the accuracy and reliability of the system was fully verified. In order to test the performance of the method, to promote, and to expand the sample set with 50 training and test samples, the objective-recognition effectiveness is proved achieving the approach by 93 % at signal noise ratio greater than 6 dB. Under the circumstances of high signal to noise ratio, especially for plus, a typical noise such as shock, fly insects, and other confounding factors in the case of low SNR, the neural network identification method is far superior level of recognition.

Table 2 Analysis of Hopfield network identification result

Test content	Penetrators signal									
	1	2	3	4	5	6	7	8	9	10
Network identification number	3	3	2	3	3	2	3	3	3	3
Recognition results	Correct	Correct	add1	Correct	Correct	add 2	Correct	Correct	Correct	Correct
Recognition rate (%)	100	100	93	100	100	86	100	100	100	100
The average recognition rate (%)	97.6									

4 Results

In this paper, the rule of the test accuracy and reliability of the sky-screen system was analyzed based on the principle and a variety of the interference noise such as warhead shock, shock projectile bottom, mosquito birds, vibration noise signal, and etc. Then, using approach of auto-associative neural network to identify and eliminate typical factors interference, through live fire test and simulation, the accuracy and reliability of the system was fully verified and also proved prototype testing system to achieve the technical specifications.

References

1. Lu, S.T., Yu, A.T., Chou, C.: Electro-optics target for position and speed measurement. In: IEE Proceedings-A, vol. 140(4), July 1993
2. McCarthy, B.D, Regan, B.J.: Position measuring apparatus and method. USA, US4885725, Dec 1989
3. Crittenden, E.C., et al.: Target measurement system for precise projectile location. USA, 3727069, April 1973
4. Li, H.-S., Yuan, Z.-S., Lei, Z.-Y.: The study of recognizing and eliminating fake projectile signals in the sky screen. In: 2008 Intelligent Information Technology Application Workshops, pp. 427–430, Dec 2008
5. Hartati, R.S., EI-Hawary, M.E.: A summary of application of Hopfield neural economic work to economic load dispatch. In: 2000 Canadian Conference on Electrical and Computer Engineering, vol. 2, pp. 707–711, 7–10 March 2000
6. Chiou, G.-J., Lee, C.-S.: New approach for solving optimization problems in economic load dispatch using Hopfield neural networks. In: 2000 Canadian Conference, vol. 2, pp. 722–725 (2000)
7. Yalcinoz, T., Altum, H., Hasan, U.: Constrained economic dispatch with prohibited operating zones a Hopfield neural network approach. In: 10th Mediterranean Electro Technical Conference. vol. 2, pp. 570–573, 29–31 May 2000
8. Barron, A.: Universal approximation bounds for superposition of a sigmoidal function. IEEE Trans. Inf. Theory **39**(1), 930–945 (1993)
9. Dreyfus, G.: Neural Networks Methodology and Applications. Springer, Berlin (2000)

Research on the Hotspot Information Push System for the Online Journal Based on Open-Source Framework

Jiya Jiang, Tong Liu, Yanqing Shi and Changhua Lu

Abstract This paper analyzes the technology of the open-source framework as HttpClient, HTMLParser, and IKAnalyzer, and then gives a system for the individual needs of researchers. The system can collect the online journal information automatically, analyze the hotspots and then push the hotspots to the researchers.

Keywords Online journal · Hotspot analysis · Information push · Open-source framework

1 Introduction

At present, all kinds of scientific papers increase at a rate of more than two million articles each year [1]. Finding and utilization of the massive data become the common concern of the researchers. There are three questions in the use of the journal articles: Firstly, for the copyright reasons, most of the journals that appear in the digital publisher's Web site have a few months lag, but these journals can update the information of the latest articles on their own official Web site usually; Secondly, the digital publishers have a large scale of the digital publication and offer a variety of convenient query for the researchers, but require the user to take the initiative to search, and lack of personalized hot push function. Thirdly, some research institutes do not buy data resources, which bring more inconvenience to the journal articles query and utilization.

J. Jiang (✉) · T. Liu · Y. Shi · C. Lu
Beijing Science and Technology Information Institute, Beijing 100044, China
e-mail: jiya_jiang@sina.com

T. Liu
e-mail: Liu_tongmiss@163.net

T. Liu · Y. Shi · C. Lu
Beijing Academy of Science and Technology, Beijing 100089, China

S. Patnaik and X. Li (eds.), *Proceedings of International Conference on Soft Computing Techniques and Engineering Application*, Advances in Intelligent Systems and Computing 250, DOI: 10.1007/978-81-322-1695-7_10, © Springer India 2014

For the above phenomenon, this paper proposes a tracking and hot push system based on open-source framework. According to the individual needs of researchers, the system crawls the latest online journal in a targeted manner regularly and automatically, and pushes the hot topics of concern to the user automatically. Thus, the researchers can use the information conveniently.

2 Key Technologies

In this paper, the main idea is as following: firstly, determine the collection Web site; secondly, crawl the latest information from concern online journals; thirdly, generate the knowledge database and analyze the hot spot; finally, push the analysis for the scientific and technological workers. All of the technology in this paper utilizes the java-based open-source framework, just call the simple interface to complete the complex data acquisition and data analysis for researchers. The open-source frameworks used in this paper include HttpClient [2], HTMLParser [3], and IKAnalyzer [4].

2.1 HttpClient

The JDK java net package provides HttpURL Connection technology, many of the early applications adopt the jar package for data acquisition, but for most applications, the functions provided by the JDK library itself are not enough rich and flexible. In recent years, the developers are keen to the HttpClient technology to achieve data acquisition. HttpClient is a subproject under the Apache Jakarta Common, can be used to provide efficient, latest, feature-rich support for the HTTP protocol client programming toolkit, and support for the latest version of the HTTP protocol and recommendations.

The HTTP protocol is the most used in the Internet latest years, the most important protocol, more and more Java applications need to be directly through the HTTP protocol to access the network resource. HttpClient has been applied in many projects, and two other open-source projects, such as Apache Jakarta famous Cactus and HTMLUnit, use the HttpClient. The latest version of HttpClient is HttpClient 4.2 (GA).

2.2 HTMLParser

HTMLParser is a pure java html-parsing library, and it does not depend on other java library files, mainly for the modification or withdrawal of HTML, and is currently the most widely used html-parsing and analysis tools, and its latest version is HTMLParser 2.0. The HTMLParser has two main functions for information extraction and conversion. Information extraction features include five subfunctions: text information extraction, such as HTML, effective information search; link extraction, used to link text with the link to the page automatically label; resource extraction, such as some of the pictures, the sound of the resources at its disposal; link Checker is used to examine the HTML link is valid; and monitor the content of the page. Information conversion function consists of five subfunctions: the link rewrite, used to modify all hyperlinks in the page; web content copy for the web content saved to the local; contents of the test can be used to filter some words on the page; HTML information cleaning, HTML format; and into XML format data.

2.3 IKAnalyzer

Most open-source software is from abroad, so the Chinese word is segmented as single word, this way is ineffective. IKAnalyzer is an open source, light weight java language development-based Chinese word segmentation tool kit. This open-source project is developed by Lin Liangyi et al. all of whom are Chinese. It is widely used as the Lucene Word Breaker for Chinese word. With Lucene version updates and constantly updated, it has been updated to IKAnalyzer 2012 version. Initially, it is based on the open-source project Lucene of the main application, combined with a dictionary of words and grammar analysis algorithm of Chinese word segmentation component. From version 3.0, IK develops from the common word components for Java, independent of the Lucene project, while providing a Lucene default optimization to achieve.

IKAnalyzer provides a unique forward iteration, the most fine-grained segmentation algorithm, with 600,000 words/s high-speed processing capability. And the use of multiprocessor analyzes the submode support: English alphabet, digital, Chinese vocabulary, etc.

3 The Realization of the System Framework

The system is mainly composed of two parts: One is the information crawl, and another is hot spot analysis. According to the interest of the researchers, information crawl set up the source sites, analyze of Web page structure, and design crawl mode; Then, use the HTMLParser and HttpClient to get the information from those Web sites and store these information into the database; By the analyses of these papers in the database, the hot spot is analyzed; At last, the hot spots are pushed to the researchers. The information system workflow is shown in Fig. 1.

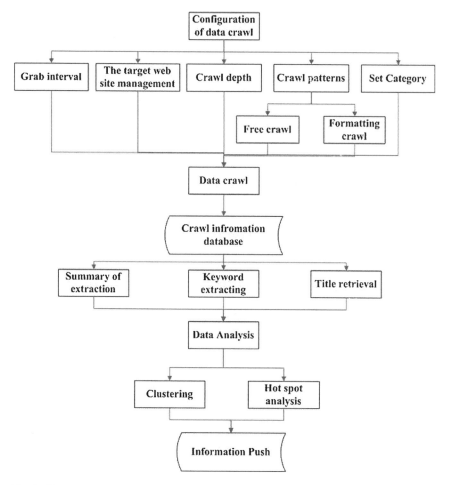

Fig. 1 The program of the system

4 Examples

For example, a research worker sets a Web site as his focus journal and selects concern "semantic" hot, and then, the system will be collected regularly for catalog of periodicals published in the journal's Web site. In the online journal, the system acquisition in recent years' paper information 442, the data are stored in the database. After the hot spot analysis, the high-frequency words are gotten as follows: semantic, cloud, mining, extraction, patterns, knowledge base, identity, search engine, OPAC, labels, a retrieval system, certification, CSSCI, acquisition, folksonomy, biomedical, k-means, public opinion, theme indexing, and crawling. The system will push the paper list about "semantic" to the researcher. Online publication in Springer Link.

5 Summary

This paper completed automatic acquisition and hot papers analyses, based entirely on open-source architecture to achieve the secondary development just need to make appropriate adjustments to the open source code. The system can realize information collection, information filtering, analysis of hot words, and information push.

Acknowledgments This work was financially supported by the program of Large-scale Network Authentication Center affiliated to Beijing Municipal Institute of Science and Technology Information (No. PXM2012_ 178214_000005), the program of Innovation Group for Internet Real-name System (No. IG 201003C2) and the program of Beijing Talent Training Plan (No. 2012D 0020 2200 0002). Thanks a lot for them.

In addition, I want to thank all our colleagues, both past and present, for their assistance during the progression of this research.

References

1. Apache Software Foundation: The Apache HttpComponents project. Apache Foundation, March 30, 2013. http://jakarta.apache.org/commons/httpclient/. Accessed 10 Feb 2013
2. Baidu Encyclopedia: The introduction of Htmlparser. Savagert. http://baike.baidu.com/view/1174491.htm. Accessed 10 Feb 2013
3. CodePlexProject Hosting for Open Source Software: Htmlparser. http://htmlparser.codeplex.com/. Accessed 20 Mar 2013
4. Lin, L.: The Chinese word V2012 user manual for IKAnalyzer. 30 Mar 2012

Development of Control System of Wheel Type Backhoe Loader

Luwei Yang

Abstract The control system of wheel type backhoe loader plays an important role in its operations. In order to enhance the reliability and efficiency of corresponding construction machine, a variety of field control bus, for example, CAN bus, LIN bus, and KWP2000 bus, are widely applied in this field. In this paper, the CAN bus technology is applied in the control system of wheel type backhoe loader. Its features are explored. The designing scheme of typical control system is introduced. Therefore, the architecture of the control system of wheel type backhoe loader using CAN bus technology is constructed. The control principles and functions of loading and excavating operation as well as failure alarm scenario of wheel type backhoe loader are analyzed. Then, the characteristics of the control system of wheel type backhoe loader are illustrated. Finally, the practical implementation of CAN bus control system in wheel type backhoe loader is carried out.

Keywords CAN bus technology · Wheel type backhoe loader · Control system

1 Introduction

Wheel type backhoe loader, shown in Fig. 1, is mainly applied in the earth moving of civil engineering construction. This kind of whole products is one of the chains of construction machinery and is also a kind of typical completed machine. The excavating and loading functions of wheel type backhoe loader typically feature one of the construction machinery operation patterns. With the rapid development of computer technology, network communication technology, and mechatronics technology, more and more control elements and control components are installed

L. Yang (✉)
School of Computer Science and Technology,
Beijing University of Technology, Beijing 100124, China
e-mail: ycx10@163.com

S. Patnaik and X. Li (eds.), *Proceedings of International Conference on Soft Computing Techniques and Engineering Application*, Advances in Intelligent Systems and Computing 250, DOI: 10.1007/978-81-322-1695-7_11, © Springer India 2014

Fig. 1 Wheel type backhoe loader

in wheel type backhoe loader. Hence, it gives rise to the fact that longer wire and cable need to be fixed on the chassis of wheel type backhoe loader. As a result, the reliability of completed machine will be decreased and failure maintainability will be difficult. By introducing advanced CAN bus technology and using controllers, sensors, and implementers in CAN bus control system, control accuracy and reliability of wheel type backhoe loader will be improved. Moreover, the productivity of wheel type backhoe loader will also be increased.

2 The Features of CAN Bus Control System

The controllers of CAN bus control system are used in the operation control of wheel type backhoe loader. The CAN bus control system is comprised of a kind of distributed control architecture, and its topological structure can, respectively, opt for bus model, star model, and loop model, etc. The control system can also use twisted pair, cable, wireless, and infrared ray for transmitting data. These features are as follows [1]:

- Open and interoperation

  Because of the open and interoperation of control system, plug-and-play function is realized to make control system suitable for different types of elements;

- Rebuilding and expandability

  Because of the disperse characteristic of control system, it is very easy for the control system of wheel type backhoe loader to be rebuilt and expanded. So CAN bus control system adapts to the control of operation systems of different construction machinery;

- Low costs

  Using the opening architecture and OEM technology, new product developmental period will be shortened and development costs will go down. The opening and distributed structure can convert one-to-one analog signals into one-to-*N* digital signals. Therefore, control system will not require for the *A/D* and *D/A*

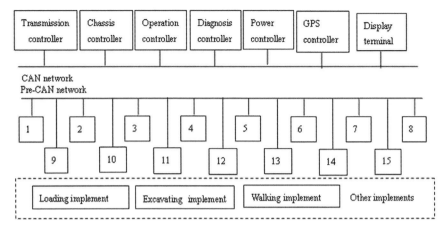

Fig. 2 Typical architecture of the CAN control system of construction machinery. *1* Differential control unit, *2* gun pedal control unit, *3* steering control unit, *4* ignition control unit, *5* electrical control unit, *6* operation control unit, *7* braking control unit, *8* cooling control unit, *9* walking sensor, *10* braking sensor, *11* speed sensor, *12* chassis sensor, *13* steering sensor, *14* operation sensor, and *15* engine sensor

instruments, and the costs of cable installation and system maintenance will also be remarkably decreased in the process of signal transmission. Consequently, the control system using CAN bus technology is very economic for wheel type backhoe loader.

3 Design of the Architecture of CAN Bus Control System

3.1 Typical Architecture of CAN Bus Control System

Typical architecture [2] of the control system of construction machinery based on CAN bus technology, shown in Fig. 2, consists of five intelligent controllers with CAN interfaces, a GPS controller for communication, an intelligent display terminal, eight different function pre-control units with CAN interfaces, and seven embedded intelligent sensors. Intelligent controller system is linked to pre-control unit system through two-level CAN network. These intelligent sensors with CAN interfaces can directly be linked to CAN bus; however, the intelligent sensors that have not CAN interfaces can be linked to related intelligent controllers or pre-CAN network to realize the exchange of data and control orders. The control system is powered by special auto-power supply to output 12 or 5V voltages. This kind of power is not interfered by other electrical systems. The power of control system using twisted pair wire similar to CAN bus cable will supply electrical energy for control units and test instruments.

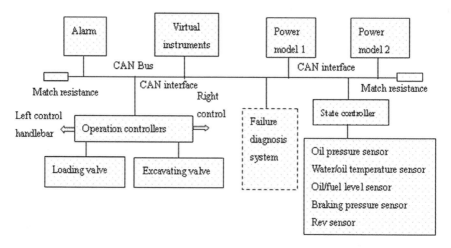

Fig. 3 Architecture of CAN bus control system of wheel type backhoe loader

Multi-functional intelligent controllers are selected as the control system hardware of construction machinery for applications. The intelligent control system is a real-time and opening system. Developmental platform equipped with intelligent controllers is used for application software. The MC9S12DG128BCPV chip made by Motorola company is employed as a microcontroller in the control system. The MSP430 series mixed signal controllers made by TI Company in the recent years are applied as a microcontroller of intelligent sensors. Special programmable tool IAR C430 in MSP430 series mixed signal controllers is directly used for programming, and JTAG is used for loading and debugging of program.

3.2 The Architecture of the CAN Bus Control System of Wheel Type Backhoe Loader

The architecture [3] of the CAN bus control system of wheel type backhoe loader is shown in Fig. 3, and the hardware of the CAN bus control system consists of electrical control unit (ECU) and its interfaces, implements, such as solenoid valves and electrical motor, and sensors. However, the software of the CAN bus control system is encapsulated in ECU and controls ECU for real-time testing. Even the software has also the functions of data acquisition, processing computation, controlling output, and monitoring and diagnosing system.

The control system can realize the loading operation control, excavating operation control, mechanical steering control, state monitoring for complete machine, and failure diagnosis. Controlling functions in detail are shown in Table 1.

Table 1 The functions of the CAN bus control system of wheel type backhoe loader

Hydraulic system control for excavating operation and leg support	Transmission control	Start control
Hydraulic system control for loading operation	Lighting system control	Diagnosis indicator
Advanced hydraulic system control for automatically excavating and loading control	Locking functions of transmission, lighting, and hydraulic system	Driven by two wheels or driven by four wheels
Hydraulic system control for accessories and drag implement	Heating, ventilating, and air condition control	Suspension control
Emergency model operation	Alarm and display of gauges and lights	Failure diagnosis
Other kinds of controls and monitors, such as window wiper, horn, and buzz		

4 Principle and Functions of the CAN Bus Control System of Wheel Type Backhoe Loader

4.1 The Control of Loading Operation

• Principle of control

Principle of loading operation control of wheel type backhoe loader is shown in Fig. 4.

• Description of functions

The closed control circuit is comprised of VCM3520 controller, loading solenoid valve, and sensors [4]. The VCM3520 controller receives control signals and feedback signals from sensors, and then, after processing input signals, the controller sends control signal to loading solenoid valve to make wheel type backhoe loader perform specified actions.

During lifting arm, the position of arm lift is automatically tested by position sensor. The position of arm lift is automatically determined through the controller judging, disposing, and controlling loading valve. Meanwhile, loading operation model can automatically be set and perform position memory operation. And it is convenient to exchange loading operation and excavating operation.

4.2 The Control of Excavating Operation

• Principle of control

Principle of excavating operation control of wheel type backhoe loader is shown in Fig. 5.

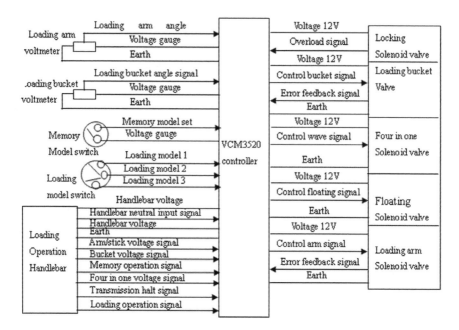

Fig. 4 Control principle of loading operation

- Description of functions

Similar to loading operation control, the closed control circuit [5] is comprised of VCM3520 controller, excavating solenoid valve, and sensors. Excavating operation is set at low speed, mid speed, and high speed. Excavating operation is also divided into three kinds of models as follows:

- Heavy load model (*H*-model): Output electric current from VCM3520 controller ranges from 0 to Imax;
- Standard load model(*S*-model): Output electric current from VCM3520 controller is constant to drive electrohydraulic proportional valve;
- Light load model (*L*-model): Output constant and a little electric current from VCM3520 controller drives electrohydraulic proportional valve.

While handlebar is, after 1 min, at neutral position, engine will automatically go to low speed and idle state in excavating operation period.

4.3 System Indication and Failure Alarm Control

- Principle of control

Principle of system indication and failure alarm control of wheel type backhoe loader is shown in Fig. 6.

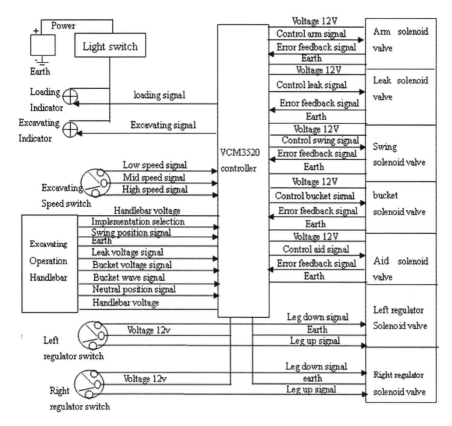

Fig. 5 Control principle of excavating operation

- Description of functions

The VCM3520 controller receives the control signals from control buttons and the state feedback signals from sensors. After processing input signals, the VCM3520 controller will transmit indications and alarm signals into LIM24 alarm through CAN bus to realize the alarm function of excavation, load, bucket indication, cooling, oil pressure, oil filter jam, oil level, fuel level, and braking pressure, etc. Data displayed in panel and alarm information are transmitted into virtual instruments for display and alarm. In addition, the control system can also perform the display and test of parameters, such as water temperature, oil temperature, oil pressure, braking pressure, oil level and fuel level, and real-time display failures, engine temperature, and overload alarm.

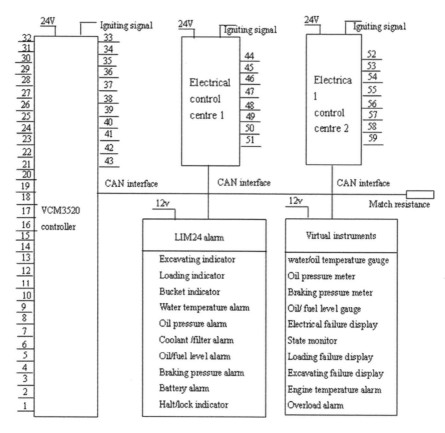

Fig. 6 Control principle of system indication and failure alarm. *1* Excavating model, *2* loading model, *3* fork model, *4* engine rev, *5* start input, *6* horn, *7* nozzle motor, *8* wiper low/mid/high speed, *9* aid braking pressure, *10* rear braking pressure, *11* front braking pressure, *12* water level, *13* fuel level, *14* air filter, *15* oil pressure, *16* coolant temperature, *17* oil level, *18* oil filter jam, *19* adjusting light input 1, *20* adjusting light input 2, *21* rear light, *22* headlight, *23* ceiling light switch, *24* halt light switch, *25* main switch, *26* indicating light switch, *27* walking light switch, *28* emergency alarm, *29* right turn, *30* left turn, *31* high-intensity light, *32* low-intensity light, *33* halt light, *34* escort light, *35* walking light, *36* braking light, *37* readjusting light, *38* hour meter, *39* fuel igniting valve, *40* horn, *41* alarm, *42* ceiling light, *43* sensor power, *44* left front operation light, *45* left rear operation light, *46* reverse light, *47* left front indicator, *48* right front indicator, *49* low-intensity light, *50* high-intensity light, *51* drag halt light, *52* right drag light, *53* front nozzle motor, *54* wiper at low speed, *55* wiper at mid speed, *56* wiper at high speed, *57* fuel igniting valve, *58* start motor relay, *59* start solenoid valve

5 The Features of the Control System of Wheel Type Backhoe Loader

- All modules in the control system are linked to J1939 CAN network, exchanging information among modules; data from any one module are accessed by other modules, which complete their functions according to accessed data. The data from all modules can be centralized to display.
- Microcontroller achieves all logic controls; for instance, adding new functions need only to modify software, and changing functions are also completed by only amending software.
- The controller can monitor the operation state of wheel type backhoe loader. Control interfaces will be reserved for adding new functions without changing any hardware configurations.
- The controller can monitor the transmission state of wheel type backhoe loader. Lights and hydraulic system are controlled by the transmission state signals, such as temperature, pressure, and oil level signal acquired by the control system.
- Alarm levels can be readjusted, and alarm information is shown in the virtual display terminal.
- Power module can monitor the state of fuse and relay, and the state is shown in the virtual display terminal. When fuse and relay are abnormal, the controller can control wheel type backhoe loader in normal operation.

6 Conclusions

Design for the control system of wheel type backhoe loader is based on CAN bus technology, and it will not only enormously increase operational accuracy and efficiency, but also remarkably decrease failure rate of the machine and greatly improve reliability of the machine. It has been proved by engineering applications that using CAN bus technology to design the control system of wheel type backhoe loader is provided with economy and is worthy of wide applications.

References

1. Zhao, F.: Developmental trend and situation of CAN bus technology. J. Electr. Appliance Ind., 8, 22–25 (2007) (in Chinese)
2. Zhou, J., Chen, H.Y.: Application of CAN bus technology in military vehicle. J. Veh. Power Technol., 23, 47–51 (2001) (in Chinese)
3. Zhang, J., Yin, P.L., Liao, X.M., et al.: Design on electrical system of mid-size excavator based on CAN bus technology. Chin. J. Constr. Mach., 8, 303–308 (2010) (in Chinese)
4. Yu, Z.G., Li, N.M., Nie, S.L., et al.: A general control solution of engineering machines based on CAN bus. J. Southwest Univ. Sci. Technol., 23, 19–21 (2004) (in Chinese)
5. Liao, W.L.: Design of automobile electric control system based on CAN bus. J. Foreign Electron. Meas. Technol., 27, 36–38 (2008) (in Chinese)

Some Results on Fuzzy Weak Boolean Filters of Non-commutative Residuated Lattice

Wei Wang, Yang Xu, Dan Tong, Xiao-yan Cheng and Yong-fei Li

Abstract After we present some basic definitions and results on non-commutative residuated lattice and several kinds of filters of it, we extend the concept of fuzzy filter to non-commutative residuated lattice. We introduce and investigate the properties of fuzzy weak Boolean filters of residuated lattice and further characterize the fuzzy weak Boolean filters.

Keywords Non-commutative logical algebras · Non-commutative residuated lattice · Fuzzy filter · Fuzzy weak Boolean filters

1 Introduction

The rapid development of computing science, technology, and mathematical logic put forward many new requirements, thus contributing to the non-classical logic and the rapid development of modern logic [1]. The non-commutative logical algebras are the algebraic counterpart of the non-classical logic. Non-commutative residuated lattice are algebraic counterparts of non-commutative monoidal logic [2]. Pseudo-*BL*-algebras and pseudo-*MTL*-algebras are non-commutative residuated lattices [3].

The theory of filters functions well not only in non-classical logic, but also in Computer Science. From logical point of view, various filters correspond to

W. Wang (✉) · Y. Xu
College of Electrical Engineering, Southwest Jiaotong University, Chengdu, China
e-mail: wwmath@xsyu.edu.cn

W. Wang · D. Tong · X. Cheng · Y. Li
Department of Applied Mathematics, Xi'an Shiyou University, Xi'an, China

S. Patnaik and X. Li (eds.), *Proceedings of International Conference on Soft Computing Techniques and Engineering Application*, Advances in Intelligent Systems and Computing 250, DOI: 10.1007/978-81-322-1695-7_12, © Springer India 2014

various sets of provable formulae [4]. Hájek introduced the notions of filters and prime filters in *BL*-algebras and proved the completeness of Basic Logic *BL* [5]. In [6], Turunen proposed the notions of implicative filters and Boolean filters of *BL*-algebras and proved the equivalence of them in *BL*-algebras. In [7–13], filters of pseudo-*MV* algebras, commutative residuated lattice, triangle algebras, pseudo-effect algebras, and pseudo-hoops were studied.

Fuzzy sets were introduced in 1965 by Zadeh [14]. At present, fuzzy filters ideas have been a useful tool to obtain results on classical filters and been applied to other algebraic structures. In recent years, fuzzy logic-based reasoning seems to be active and in-depth study of these domestic and international research works carried out to explore some new, mainly related to the formalization of fuzzy logic and related algebraic structure.

We find that the structures of the non-commutative logical algebras can be described by the tools of fuzzy filters. Therefore, in this paper, the theory of fuzzy weak Boolean filters in non-commutative residuated lattices is studied, which lays a good foundation for the further research in non-commutative logical algebras.

2 Review of Preliminaries

First, we recall some basic definitions and results which will be needed later (see details in [2, 3, 10, 13, 15–17]).

Definition 1 A lattice-ordered residuated monoid is an algebra $(K, \vee, \wedge, \star, \rightarrow, \hookrightarrow, e)$ satisfying the following conditions:

(1) $(K, \vee, \wedge)$ is a lattice,
(2) $(K, \star, e)$ is a monoid,
(3) $x \star y \leq z$ iff $x \leq y \rightarrow z$ iff $y \leq x \hookrightarrow z$ for all $x, y, z \in K$.

A lattice-ordered residuated monoid K is called integral if $x \leq e$ for all $x \in K$. In an integral lattice-ordered residuated monoid, we use "1" instead of e.

Definition 2 A residuated lattice is a bounded and integral residuated lattice-ordered monoid, i.e., a residuated lattice is an algebra $(K, \vee, \wedge, \star, \rightarrow, \hookrightarrow, 0, 1)$ satisfying the following conditions:

(1) $(K, \vee, \wedge, \star, \rightarrow, \hookrightarrow, 0, 1)$ is a bounded lattice,
(2) $(K, \star, 1)$ is a monoid,
(3) $x \star y \leq z$ iff $x \leq y \rightarrow z$ iff $y \leq x \hookrightarrow z$ for all $x, y, z \in K$.

Lemma 1 (Jipsen, Tsinakis and Blount, Zhang [3, 15, 16]) *In a non-commutative residuated lattice K, the following properties hold for all $x, y, z \in K$*

(1) $$(x \star y) \to z = x \to (y \to z), \ (y \star x) \hookrightarrow z = x \hookrightarrow (y \hookrightarrow z),$$

(2) $$x \leq (x \to y) \hookrightarrow y, x \leq (x \hookrightarrow y) \to y,$$

(3) $$x \leq y \text{ iff } x \to y = 1 \text{ iff } x \hookrightarrow y = 1,$$

(4) $$x \star y \leq x \wedge y \leq x, y, \ 1 \to x = x = 1 \hookrightarrow x,$$

(5) $$x \leq y \to y \to z \leq x \to z \text{ and } y \hookrightarrow z \leq x \hookrightarrow z,$$

(6) $$x \leq y \to z \to x \leq z \to y \text{ and } z \hookrightarrow x \leq z \hookrightarrow y$$

(7) $$x \to y \leq (z \to x) \to (z \to y), \ x \hookrightarrow y \leq (z \hookrightarrow x) \hookrightarrow (z \hookrightarrow y),$$

(8) $$x \to y \leq (y \to z) \hookrightarrow (x \to z), \ x \hookrightarrow y \leq (y \hookrightarrow z) \to (x \hookrightarrow z),$$

(9) $$x \vee y \leq ((x \to y) \hookrightarrow y) \wedge ((y \to x) \hookrightarrow x)$$

(10) $$x \vee y \leq ((x \hookrightarrow y) \to y) \wedge ((y \hookrightarrow x) \to x).$$

In the sequel, we shall use K to denote a non-commutative residuated lattice and define $x^- = x \to 0$, $x^\sim = x \hookrightarrow 0$ for any $x \in K$.

Definition 3 A filter of K is a non-empty subset N of K such that for all $x, y \in K$, one of the following holds

(1) *if $x, y \in N$, then $x \bigstar y \in N$ and*
    if $x \in N$ and $x \leq y$, then $y \in N$,
(2) $1 \in N$ *and $x, x \to y \in N$ imply $y \in N$,*
(3) $1 \in N$ *and $x, x \hookrightarrow y \in N$ imply $y \in N$.*

Definition 4 For any $x, y \in K$, a filter N of K is called Boolean if $x \vee x^- \in N$ and $x \vee x^\sim \in N$.

Definition 5 Let N be a subset of K. Then N is called a weak Boolean filter of K if for all $x, y, z \in K$, the following conditions hold

(1) $1 \in N$,
(2) $(x \rightarrow y) \star z \hookrightarrow ((y \hookrightarrow x) \rightarrow x)$ and $z \in N$ implies $(x \rightarrow y) \hookrightarrow y \in N$,
(3) $z \star (x \hookrightarrow y) \rightarrow ((y \rightarrow x) \hookrightarrow x)$ and $z \in N$ implies $(x \hookrightarrow y) \rightarrow y \in N$.

3 Fuzzy Weak Boolean Filters of Non-commutative Residuated Lattice

In this section, we introduce and investigate the properties of fuzzy weak Boolean filters of non-commutative residuated lattice and further characterize the fuzzy weak Boolean filters as an extension work of [11].

Definition 6 Let T be a fuzzy set of K. T is called a fuzzy filter of K if for all $t \in [0, 1]$, T_t is either empty or a filter of K.

Definition 7 Let T be a fuzzy filter of K. Then T is a fuzzy Boolean filter of K, if for all $x \in K$, $T(x \vee x^-) = T(1)$ and $T(x \vee x^\sim) = T(1)$.

Definition 8 Let T be a fuzzy subset of K. Then T is called a fuzzy weak Boolean filter of K if for all $x, y, z \in K$, the following conditions hold

(1)
$$T(1) \geq T(x),$$

(2)
$$T((x \rightarrow y) \hookrightarrow y) \geq T((x \rightarrow y) \star z \hookrightarrow ((y \hookrightarrow x) \rightarrow x)) \wedge T(z),$$

(3)
$$T((x \hookrightarrow y) \rightarrow y) \geq T(z \star (x \hookrightarrow y) \rightarrow ((y \rightarrow x) \hookrightarrow x)) \wedge T(z).$$

Inspired by [11], we can get the following results.

Theorem 1 Let T be a fuzzy filter of K. T is a fuzzy weak Boolean filter of K if and only if for each $t \in [0, 1]$, T_t is either empty or a weak Boolean filter of K.

Theorem 2 Let T be a fuzzy filter of K. T is a fuzzy weak Boolean filter of K if and only if $T_{T(1)}$ is a weak Boolean filters of K.

Corollary 1 Let N be a non-empty subset of K. N is a weak Boolean filter of K if and only if χ_N is a fuzzy weak Boolean filter of K.

Next, we characterize the fuzzy weak Boolean filters.

Theorem 3 *Let T be a fuzzy filter of K. Then the followings are equivalent:*

(1) T is a fuzzy weak Boolean filter,
(2) $T((x \to y) \hookrightarrow y) = T((x \to y) \hookrightarrow ((y \hookrightarrow x) \to x)))$ for any $x, y \in K$,
(3) $T((x \hookrightarrow y) \to y) = T((x \hookrightarrow y) \to ((y \to x) \hookrightarrow x)))$ for any $x, y \in K$.

Proof (1) $\Rightarrow$ (2). Let $T((x \to y) \hookrightarrow ((y \hookrightarrow x) \to x))) = t$, then $((x \to y) \hookrightarrow ((y \hookrightarrow x) \to x)) \in T_t$. $1 \in T_t$, $(((x \to y) \star 1) \hookrightarrow ((y \hookrightarrow x) \to x) \in T_t$, hence $(x \to y) \hookrightarrow y \in T_t$, i.e., $T((x \to y) \hookrightarrow y) \geq t = T(((x \to y) \hookrightarrow ((y \hookrightarrow x) \to x))$. The inverse inequation is obvious since T is isotone and $(x \to y) \hookrightarrow y \leq (x \to y) \hookrightarrow ((y \hookrightarrow x) \to x)$.

(1) $\Rightarrow$ (3). Similar to (1) $\Rightarrow$ (2).
(2) $\Rightarrow$ (1). Let $T(((x \to y) \star z) \hookrightarrow ((y \hookrightarrow x) \to x) \wedge T(z) = t$, then $((x \to y) \star z) \hookrightarrow ((y \hookrightarrow x) \to x, z \in T_t$. Since T_t is a filter, $(x \to y) \hookrightarrow ((y \hookrightarrow x) \to x \in T_t$, and $(x \to y) \hookrightarrow y \in T_t$, i.e., $T((x \to y) \hookrightarrow y) \geq T(((x \to y) \star z) \hookrightarrow ((y \hookrightarrow x) \to x) \wedge T(z)$.
(3) $\Rightarrow$ (1). Similar to (2) $\Rightarrow$ (1).

Theorem 4 *Every fuzzy weak Boolean filter of K is a fuzzy filter.*

Proof Let $T(x \hookrightarrow y) \wedge T(x) = t$, then $x \hookrightarrow y, x \in T_t$. Since we have $((y \to y) \star x) \hookrightarrow ((y \hookrightarrow y) \to y) = x \hookrightarrow y \in T_t$, we have $y = (y \to y) \hookrightarrow y) \in T_t$. Hence, $T(y) \geq T(x \hookrightarrow y) \wedge T(x)$ and T is a fuzzy filter.

Theorem 5 *Let T be a fuzzy filter of K. Then the followings are equivalent*

(1) T is a fuzzy weak Boolean filter,
(2) $T(y) \geq T((y \to z) \hookrightarrow (x \to y)) \wedge T(x)$ for any $x, y, z \in K$,
(3) $T(y) \geq T((y \hookrightarrow z) \to (x \hookrightarrow y)) \wedge T(x)$ for any $x, y, z \in K$.

Proof (1) $\Rightarrow$ (2). Let $T((y \to z) \hookrightarrow (x \to y)) \wedge T(x) = t$, then $(y \to z) \hookrightarrow (x \to y) = x \to ((y \to z) \hookrightarrow y), x \in T_t$. We get $(x \to y) \hookrightarrow y \in T_t$. And $z \leq y \to z$, $(x \to z) \hookrightarrow y \leq z \hookrightarrow y \in T_t$. hence $z \hookrightarrow y \in T_t$. And we have $(y \to z) \hookrightarrow y \leq (y \to z) \hookrightarrow ((z \hookrightarrow y) \to y) \in T_t$. Then we get $(y \to z) \hookrightarrow z \in T_t$ since T is a fuzzy weak Boolean filter of K. And $(y \to z) \hookrightarrow z \leq (z \hookrightarrow y) \to ((y \to z) \hookrightarrow y)$; then $(z \hookrightarrow y) \to ((y \to z) \hookrightarrow y) \in T_t$. So $(z \hookrightarrow y) \to y \in T_t$, since $z \hookrightarrow y \in T_t$, $y \in T_t$, i.e., $T(y) \geq T((y \to z) \hookrightarrow (x \to y)) \wedge T(x)$.

(1) $\Rightarrow$ (3). Similar to (1) $\Rightarrow$ (2).
(2) $\Rightarrow$ (1). Let $T((x \to y) \hookrightarrow ((y \hookrightarrow x) \to x)) = t$ for all $x, y \in K$, then $(x \to y) \hookrightarrow ((y \hookrightarrow x) \to x) \in T_t$. Since $(((x \hookrightarrow y) \hookrightarrow y) \to 1) \hookrightarrow (((x \to y) \hookrightarrow ((y \hookrightarrow x) \to x))) \to ((x \to y) \hookrightarrow y)) = ((x \to y) \hookrightarrow ((y \hookrightarrow x) \to x)) \to (((x \to y) \hookrightarrow y) \hookrightarrow ((x \to y) \hookrightarrow y)) = (x \to y) \hookrightarrow ((y \hookrightarrow x) \to x) \in T_t$, $(x \hookrightarrow y) \hookrightarrow y \in T_t$.
(3) $\Rightarrow$ (1). Similar to (2) $\Rightarrow$ (1).

Remark 1 Let T be a fuzzy filter of K. Then the followings are equivalent:

(1) T is a fuzzy weak Boolean filter,

(2) $T((y \hookrightarrow x) \to x) \geq T((x \to y) \hookrightarrow ((y \to x) \hookrightarrow x))$ for any $x, y \in K$,

(3) $T((y \to x) \hookrightarrow x) \geq T((x \hookrightarrow y) \to ((y \hookrightarrow x) \to x))$ for any $x, y \in K$.

Theorem 6 *Let T be a fuzzy filter of K. Then the followings are equivalent*

(1) *T is a fuzzy weak Boolean filter,*

(2) *$T((x \to y) \hookrightarrow x) = T(x)$ for any $x, y \in K$,*

(3) *$T((x \hookrightarrow y) \to x) = T(x)$ for any $x, y \in K$.*

Proof (1) $\Rightarrow$ (2). Let T be a fuzzy weak Boolean filters of K and $T((x \to y) \hookrightarrow x) = t$. Then $(x \to y) \hookrightarrow x \in T_t$. And $1 \in T_t$, $(x \to y) \hookrightarrow (1 \to x) = (x \to y) \hookrightarrow x \in T_t$, we get $x \in T_t$, i.e., $T(x) \geq t = T((x \to y) \hookrightarrow x)$. The inverse inequation is obvious since T is isotone and $x \leq (x \to y) \hookrightarrow x$.

(1) $\Rightarrow$ (3). Similar to (1) $\Rightarrow$ (2).

(2) $\Rightarrow$ (1). Let $T((y \to z) \hookrightarrow (x \to y)) \wedge T(x) = t$ for all $x, y, z \in K$, then $(y \to z) \hookrightarrow (x \to y), x \in T_t$. Since $(y \to z) \hookrightarrow (x \to y) = x \to ((y \to z) \hookrightarrow y), x \in T_t$ and T_t is a filter, $(y \to z) \hookrightarrow y \in T_t$, and $T((y \to z) \hookrightarrow y) = T(y) \geq t = T((y \to z) \hookrightarrow (x \to y)) \wedge T(x)$, and we have T is a fuzzy weak Boolean filter.

(3) $\Rightarrow$ (1). Similar to (2) $\Rightarrow$ (1).

Remark 2 Let T be a fuzzy filter of K. Then the followings are equivalent

(1) T is a fuzzy weak Boolean filter,

(2) $T(x^{-} \hookrightarrow x) = T(x)$ for any $x, y \in K$,

(3) $T(x^{\sim} \to x) = T(x)$ for any $x, y \in K$.

Theorem 7 *Let T, g be two fuzzy filters of K which satisfy $T \leq g, T(1) = g(1)$. If T is a fuzzy weak Boolean filters of K, so is g.*

Proof Let $T((x \to y) \hookrightarrow x) = t$ for all $x, y \in K$, then we let $z = (x \to y) \hookrightarrow x, z \to z = 1 \in T_t$. Since $z \to ((x \to y) \hookrightarrow x) = (x \to y) \hookrightarrow (z \to x) = 1 \in T_t$, we have $(x \to y) \hookrightarrow (z \to x) \leq ((z \to x) \to y) \hookrightarrow (z \to x) \in T_t$.

Since T_t is a weak Boolean filter of K, we get $g(1) = T(1) \leq T(z \to x) \leq g(z \to x)$. Thus $z \to x \in g_t$ and $x \in g_t$ since $z \in g_t$. Then we have g as a fuzzy weak Boolean filter.

Remark 3 Every fuzzy filter T of a residuated lattices K is a fuzzy weak Boolean filter if and only if χ_K is a fuzzy weak Boolean filter.

Theorem 8 *Every fuzzy weak Boolean filter T of K is equivalent to a fuzzy Boolean filter.*

Proof Let T be a fuzzy weak Boolean filter of K. Since $T((x \vee x^{-})^{-} \hookrightarrow (x \vee x^{-})) = T((x^{-} \wedge x^{--}) \hookrightarrow (x \vee x^{-})) = T(1)$, we have $T(x \vee x^{-}) = T(1)$. Similarly, we can get $T(x \vee x^{\sim}) = T(1)$, so we have T as a fuzzy Boolean filter.

Conversely, suppose T is a fuzzy Boolean filter of K and let $T(x^- \hookrightarrow x) = t$. Then $x^- \hookrightarrow x \in T_t$ and T_t is a Boolean filter since $(x \lor x^- \hookrightarrow x = (x \hookrightarrow x) \land (x^- \hookrightarrow x) = x^- \hookrightarrow x \in T_t$. And T_t is a Boolean filter, so $x \lor x^- \in T_t$; then we have $x \in T_t$ and $T(x) \geq T(x^- \hookrightarrow x) = t$. The inverse inequation is obvious, so we get $T(x) = T(x^- \hookrightarrow x)$. Dually, we can get $T(x) = T(x^\sim \to x)$. Thus T is a fuzzy weak Boolean filters.

Acknowledgments This work is partially supported by China Postdoctoral Science Foundation funded project (Grant No.2013M540716); the National Natural Science Foundation of China (Grant No. 60875034, 61175055); the project TIN-2009-0828; Sichuan Key Technology Research and Development Program of China (Grant No. 2011FZ0051); Wireless Administration of Ministry of Industry and Information Technology of China ([2011]146); the Natural Science foundation of Shaanxi Province (Grant No. 2012JQ1023); and doctor initial fund of Xi'an Shiyou University of China (Grant No. 2011BS017)

References

1. Hájek, P.: Observations on non-commutative fuzzy logic. Soft. Comput. **8**, 38–43 (2003)
2. Nola, A.D., Georgescu, G., Iorgulescu, A.: Pseudo-*BL* algebras I, II. Multiple-Valued Logic. **8**(673–714), 717–750 (2002)
3. Zhang, X.H.: Fuzzy Logic and its Algebric Analysis. Science Press, Beijing (2008)
4. Zhang, X.H., Li, W.H.: On pseudo-*BL* algebras and *BCC*-algebras. Soft. Comput. **10**, 941–952 (2006)
5. Hájek, P.: Metamathematics of fuzzy logic. Kluwer, Dordrecht (1998)
6. Turunen, E.: Boolean deductive systems of *BL*-algebras. Arch. Math. Logic **40**, 467–473 (2001)
7. Mertanen, J., Turunen, E.: States on semi-divisible generalized non-commutative residuated lattices reduce to states on *MV*-algebras. Fuzzy Sets Syst. **159**, 3051–3064 (2008)
8. Ciungu, L.C.: Algebras on subintervals of pseudo-hoops. Fuzzy Sets Syst. **160**, 1099–1113 (2009)
9. Zhang, X.H., Fan, X.S.: Pseudo-*BL* algebras and pseudo-effect algebras. Fuzzy Sets Syst. **159**, 95–106 (2008)
10. Gasse, B.V., Deschrijver, G., Cornelis, C., Kerre, E.E.: Filters of non-commutative residuated lattices and triangle algebras. Inf. Sci. **180**, 3006–3020 (2010)
11. Wang, W., Xin, X.L.: On fuzzy filters of pseudo-BL algebras. Fuzzy Sets Syst. **162**, 27–38 (2011)
12. Liu, L.Z., Li, K.T.: Boolean filters and positive implicative filters of non-commutative residuated lattices. Inf. Sci. **177**, 5725–5738 (2007)
13. Ghorbani, S.: Weak Boolean filters of non-commutative residuated lattice. World Appl. Sci. J. **12**, 586–590 (2011)
14. Zadeh, L.A.: Fuzzy sets. Inf. Control **8**, 338–353 (1965)
15. Jipsen, P., Tsinakis, C.: A survey of non-commutative residuated lattices. Ordered Algebraic Structures. Kluwer Academic Publishers, Dordrecht (2002)
16. Blount, K., Tsinakis, C.: The structure of non-commutative residuated lattices. Int. J. Algebra Comput. **13**, 437–461 (2003)
17. Georgescu, G., Leustean, L.: Some classes of pseudo-*BL* algebras. J. Aust. Math. Soc. **73**, 127–153 (2002)

CGPS: A Collaborative Game in Parking-Lot Search

Peng Li, Demin Li and Xiaolu Zhang

Abstract Searching for parking lot can be very complicated due to the rapid movement of vehicles and the change of network topology, thus how to increase the parking efficiency becomes very important to this day. In this paper, we explore a parking-lot discovery algorithm to find the best and most reachable parking lot by evaluating the neighbors' strategy based on game theory. Without the assistance of central server, vehicles exchange information by inner opportunistic communication based on distributed system of vehicular ad hoc networks (VANET). As the vehicles drive around the parking area, they continuously evaluate the competing neighbors' cost, with different tendencies of cooperation, they can cooperate with the competitors in some degree, which is also good to raise the parking efficiency of the whole parking area. We formulate the parking-lot discovery problem as an instance of resource selection game, in order to achieve the cooperation between the vehicles, we change the strategy cost using a new pricing policy that each driver's choice over the competing parking lot should reflect the extra cost of opponent's lost, based on this principle to derive the Nash equilibrium, which reflects the total efficiency of all competitors. Finally, the simulations of the algorithms are presented to provide hint for parking-lot management that aims to raise the efficiency of the parking area.

Keywords Parking-lot discovery · Collaborativegame · Opponent selection · Vehicular ad hoc networks

P. Li · D. Li (✉) · X. Zhang
College of Information Science and Technology, Donghua University,
Shanghai 201620, China
e-mail: deminli@dhu.edu.cn

P. Li
e-mail: LiPengMaster@126.com

X. Zhang
e-mail: xiaoludhu@126.com

S. Patnaik and X. Li (eds.), *Proceedings of International Conference on Soft Computing Techniques and Engineering Application*, Advances in Intelligent Systems and Computing 250, DOI: 10.1007/978-81-322-1695-7_13, © Springer India 2014

1 Introduction

During the parking-lot search process, vehicles often face with the situation of competing for a certain parking lot with other neighbors without knowing the chance of winning, tough competition for a limited number of parking spaces compels drivers to arrive well in advance of their preferred time [1]. Thus, invalid driving may happen when the vehicle is in a disadvantage. Considering other vehicles'-specific parking-lot choices, the disadvantaged vehicles also have the chance of winning the competing parking lot, thus we have the collaborative game in parking-lot search.

To achieve better flexibility in vehicles' communication, vehicular ad hoc network (VANET) presents a promising way to build up a decentralized parking guidance system [2]. Without the aid of central server, many pieces of information can be exchanged in the context of inter-vehicle communications [3]. Vehicles exchange data with the neighbors within their network range to gain the knowledge of the environment such as real-time traveler information and traffic control data [4]. VANET relies on the use of short-range networks (about a hundred meters), such as IEEE802.11 or ultra-wide band (UWB) standards for vehicles to communicate [5].

A queue theory model was introduced in [1], the authors found out a way to calculate the possibility of finding an available parking lot given a certain situation. Likewise, the authors in [6] used continuous-time Markov chain to evaluate the possibility of parking-lot model, further they gave the formula of transition possibility matrix. The opportunistically assisted parking search (OAPS) algorithm was introduced in [7], the authors gave comparison simulations between OAPS and non-assisted parking search (NAPS), which drew to the conclusion that OAPS performs with good flexibility and especially well with high density and randomness of nodes. Subsequent research in [8] focused on the parking lot selection strategy, with the infra facility to obtain parking information, cars are treated as the players with different costs for different parking lots (resources). Authors in [9] evaluated the impact of different pricing policies drawn to drivers' strategies by exploiting the position information of others, which is irrespective of the drivers' actual destinations.

This paper seeks to explore a way to benefit the drivers' parking-lot selection and the efficiency of the whole parking area. To this end, we assume that drivers become aware of the neighbors within their VANET area and the detected parking lots; through information spread, the vehicles may also know some parking lots beyond their detection range. Based on these information, we formulate the parking-lot selection as a resource game, when a vehicle run into a round of parking-lot competition, firstly, he finds out the best winnable parking lot, then choose the most possible opponents, based on the analysis of each other's strategies and costs, finally calculate the best parking lot to run for. Pricing policy is the key to the parking-lots selection, in order to achieve the cooperation between the vehicles, we view the cost as the weighted sum of each competitor's gain and the others' lost to evaluate the group decision that represents the partial total parking efficiency. Though vehicles make their decisions selfishly, but due to the collaborative cost that they evaluate, collaborative game can be finally achieved. For a better representation, we name our algorithm as collaborative game in parking-lot search (CGPS).

We give the description of parking-lot search game and some definitions of the math model in Sect. 2, basically drivers choose the most advantageous parking lot with minimum cost selfishly. Most importantly in Sect. 3, we derive the equilibrium behaviors of the drivers under the cooperation situation. With the consideration of OAPS, we derive the comparative conclusion between OAPS and CGPS in total efficiency. In Sect. 4, we give the simulations of different parking-lot search approaches. Finally, we conclude the math model and algorithm in Sect. 5.

2 The Parking-Lot Searching Scenario

The typical scenario of parking-lot search is introduced in Kokolaki et al. [9], usually drivers are facing with two choices, on the one hand, they can compete for the nearest parking lot but with the risk of failing to occupy the spot and suffering the additional cruising time; on the other hand, they can run for the less expensive parking lot but with much higher chance of winning the spot.

We defined the parking-lot discovery game as follows:

Definition 1
- $N = \{N_1, \ldots N_i, \ldots N_n\}, 1 < i \leq n$ is the set of drivers who seek for the usable parking lot,
- $M_i = \{M_i^1, \ldots M_i^j, \ldots M_i^m\}$, $1 \leq j \leq m < n$, $M_i^j \in N$ is the neighbors of N_i,
- $P = \{P_1, \ldots P_i, \ldots P_p\}, 1 < i \leq p$ is the usable parking lots in the parking area,
- N_i' where $i \in \{1, \ldots n\}$ and $N_i' \neq N_i$ is the opponent of N_i,
- P_{N_i} where $i \in \{1, \ldots n\}, P_{N_i} \in P$ is the nearest parking lot for N_i,
- P_{N_i}' where $i \in \{1, \ldots n\}, P_{N_i}' \in P$ is the less nearest parking lot for N_i,
- $S_{N_i} = \{P_{N_i}, P_{N_i}'\}$ is the strategy set of N_i,
- $C(N_i, P_j)$ where $N_i \in N, P_j \in P$ is payoff of N_i for parking lot P_j,
- $E(N_i, N_i')$ where $N_i, N_i' \in N$ is the total payoff of the strategy for N_i and N_i', which is used to evaluate the public efficiency.

The payoff of a vehicle's strategy when choosing a specific parking lot is formally evaluated by the driving time. With the geography knowledge, we consider the driver can choose the shortest path from one position to another. We abstract the geography map as a graph, so the cost can be calculated by *Dijkstra* algorithm as follows, where $\bar{V}$ is the average driving speed in the parking area and $C(N_i, P_j)$ is the cost of time for vehicle N_i to drive to parking lot P_j.

$$C(N_i, P_j) = Dijkstra(N_i, P_j)/\bar{V} \tag{1}$$

3 Algorithm Analysis

Vehicles make their parking-lot decisions under the knowledge of VANET information that comes from the detected vacant parking lots and the extended parking lots known from neighbors. To achieve cooperation, collaborative pricing policy and strategy analysis are very important to the final parking-lot selection.

3.1 Opponent Selection

As for competitors of a specific driver, not all neighbors are qualified to participate in the competition. When considering the different parking lots as a target, its neighbors have different advantages that will cause different degrees of impact on driver's decision-making. The real competitor can be derived in the theorem below.

Theorem 1 There's only one opponent for N_i when considering P_j, which is the N_i'.

Proof Since N_i' is the most advantaged player in N_i's neighbors, then we have $N_i' = \min(M_i)$;

(a) If N_i is to win P_j, then $N_i = \min(M \cup N_i)$, which is determined by $N_i < N_i'$;
(b) If N_i is to lose P_j, then as long as $N_i \neq \min(M \cup N_i)$, which is determined by $N_i > N_i'$;

Thus, to confirm whether a driver has the chance of winning a specific parking lot, the result relies on the competition with N_i'. $\square$

A driver is willing to pay a premium to park in a garage that is closer to his destination since doing so reduces his walking time costs [10]. Thus, the nearest parking lot is the top choice for each player to evaluate, thus the component is also chosen according to this, define $P_{N_i}^{\text{nearest}}$ as the nearest parking lot for N_i, the possible opponent set O_{N_i} is given by:

$$O_{N_i} = \bigcup_{j=1}^{m} M_j \cup N_i \quad \text{where } P_{M_j}^{\text{nearest}} = P_{N_i}^{\text{nearest}}. \tag{2}$$

Only the top two players have the right to participate in the game, thus the best player O_0 and the second player O_1 is given by:

$$O_0 = \min\left(O_{N_i}^j\right) \quad \text{where } j \in \{1, \ldots |O_{N_i}|\}$$

$$O_1 = \min\left(O_{N_i}^j\right) \quad \text{where } j \in \{1, \ldots |O_{N_i}|\} \quad \text{and } O_{N_i}^j \neq O_0$$

By comparing N_i, O_0 and O_1, N_i' is calculated as follows:

$$N_i' = \begin{cases} O_0 & N_i = O_1 \\ O_1 & N_i = O_0 \\ \text{null} & N_i \neq O_0, N_i \neq O_1 \\ \text{null} & O_1 = \text{null} \end{cases} \tag{3}$$

3.2 Game Analysis

In each round of gaming, player N_i has to choose the parking lot from the nearest parking lot P_{N_i} and the less nearest parking lot P_{N_i}' by evaluating the costs compared with N_i', thus each player has its strategy set $S_{N_i} = \left\{ P_{N_i}, P_{N_i}' \right\}$, its opponent N_i' also has its similar strategy set $S_{N_i'} = \left\{ P_{N_i'}, P_{N_i'}' \right\}$.

Due to the difference between $C(N_i, P_{N_i})$ and $C(N_i', P_{N_i})$, there are two possibilities as $C(N_i, P_{N_i}) \leq C(N_i', P_{N_i})$ and $C(N_i, P_{N_i}) > C(N_i', P_{N_i})$, which differs the payoffs. Below is the payoff when $C(N_i, P_{N_i}) \leq C(N_i', P_{N_i})$:

		N_i'	
		P_{N_i}	$P_{N_i'}'$
N_i	P_{N_i}	$C(N_i, P_{N_i}), C(N_i', P_{N_i}) + C\left(N_i', P_{N_i'}'\right)$	$C(N_i, P_{N_i}), C\left(N_i', P_{N_i'}'\right)$
	P_{N_i}'	$C\left(N_i, P_{N_i}'\right), C(N_i', P_{N_i})$	$C\left(N_i, P_{N_i}'\right), C\left(N_i', P_{N_i'}'\right)$

In regular game, $\{N_i, N_i'\}$ has the Nash equilibrium as

$$S_{(N_i, N_i')}^{\text{game}} = \begin{cases} S_{N_i} & C(N_i, P_{N_i}) \leq C\left(N_i, P_{N_i}'\right) \\ S_{N_i'} & C(N_i, P_{N_i}) > C\left(N_i, P_{N_i}'\right) \end{cases} \tag{4}$$

By game analysis, player can prevent the invalid driving by evaluating the total payoff, but sometimes bad competition happens that can cause serious damage to the less advantaged player, the feature of the bad competition is to cause large total payoff, thus one player's decision cause another player's huge loss. Also, if player can compromise to its alternative parking lot without much loss but bring the huge benefit to the opponent, we call it cooperated competition, which can significantly raise the whole efficiency of the parking area.

When two players are competing for a common parking lot, one player's gain become other's loss, thus we define the extra payoff as the opponent's loss due to the common parking lot. Considering the different drivers' cooperate inclination, we use cooperate parameter α to adjust the payoff matrix. Here, the new payoff matrix is used to re-evaluate the strategies, as to the real cost it is still according to the cost formula (1). The new payoff matrix is given below:

		N_i'	
		P_{N_i}	$P_{N_i'}'$
N_i	P_{N_i}	$C(N_i, P_{N_i}) + \alpha * \left\{ C\left(N_i', P_{N_i'}'\right) - C(N_i', P_{N_i}) \right\},$ $C(N_i', P_{N_i}) + C\left(N_i', P_{N_i'}'\right) + \alpha * \left\{ C\left(N_i, P_{N_i'}'\right) - C(N_i, P_{N_i}) \right\}$	$C(N_i, P_{N_i}) + \alpha * \left\{ C\left(N_i', P_{N_i'}'\right) - C(N_i', P_{N_i}) \right\},$ $C\left(N_i', P_{N_i'}'\right)$
	P_{N_i}'	$C\left(N_i, P_{N_i'}'\right), C(N_i', P_{N_i}) + \alpha * \left\{ C\left(N_i, P_{N_i'}'\right) - C(N_i, P_{N_i}) \right\}$	$C\left(N_i, P_{N_i'}'\right), C\left(N_i', P_{N_i'}'\right)$

According to the different strategies, we have the total payoff set E as:

$$E = \left\{ C(N_i, P_{N_i}) + C\left(N_i', P_{N_i'}'\right), C\left(N_i, P_{N_i'}'\right) + C(N_i', P_{N_i}) \right\}, E_{(N_i, N_i')}^{CGPS}, E_{(N_i, N_i')}^{OAPS}$$
$$\in E.$$

(5)

The total payoff of OAPS $E_{(N_i, N_i')}^{OAPS}$ can be calculated as follows:

$$E_{(N_i, N_i')}^{OAPS} = \begin{cases} C(N_i, P_{N_i}) + C\left(N_i', P_{N_i'}'\right), C\left(N_i, P_{N_i'}'\right) \le C(N_i, P_{N_i}) \\ C\left(N_i, P_{N_i'}'\right) + C(N_i', P_{N_i}), C\left(N_i, P_{N_i'}'\right) > C(N_i, P_{N_i}) \end{cases} E_{(N_i, N_i')}^{OAPS} \in E.$$

(6)

The total payoff of CGPS $E_{(N_i, N_i')}^{CGPS}$ can be calculated as follows:

$$E_{(N_i, N_i')}^{CGPS} = \begin{cases} C(N_i, P_{N_i}) + C\left(N_i', P_{N_i'}'\right), C\left(N_i, P_{N_i'}'\right) - C(N_i, P_{N_i}) \le \alpha * \left(C\left(N_i', P_{N_i'}'\right) - C\left(N_i', P_{N_i'}'\right) \right) \\ C\left(N_i, P_{N_i'}'\right) + C(N_i', P_{N_i}), C\left(N_i, P_{N_i'}'\right) - C(N_i, P_{N_i}) > \alpha * \left(C\left(N_i', P_{N_i'}'\right) - C\left(N_i', P_{N_i'}'\right) \right) \end{cases}$$

(7)

When $\alpha = 0$, it degenerates to $\min\left(C(N_i, P_{N_i}), C\left(N_i, P_{N_i'}'\right) \right)$, which is the regular game situation. When $\alpha = 1$, it is the maximum cooperation situation.

Theorem 2 When considering the total payoff of OAPS and CGPS, we have $E_{(N_i, N_i')}^{CGPS} \le E_{(N_i, N_i')}^{OAPS}$.

Proof When $\alpha = 1$, $E_{(N_i, N_i')}^{CGPS} = \min\left(C(N_i, P_{N_i}) + C\left(N_i', P_{N_i'}'\right), C\left(N_i, P_{N_i'}'\right) + C(N_i', P_{N_i}) \right)$, which is $\min(E)$. As $E_{(N_i, N_i')}^{CGPS}, E_{(N_i, N_i')}^{OAPS} \in E$, thus $E_{(N_i, N_i')}^{CGPS} \le E_{(N_i, N_i')}^{OAPS}$.

When $\alpha = 0$, we have

$$E^{CGPS}_{(N_i,N'_i)} = \begin{cases} C(N_i, P_{N_i}) + C\left(N'_i, P'_{N'_i}\right), C\left(N_i, P'_{N_i}\right) \le C(N_i, P_{N_i}) \\ C\left(N_i, P'_{N_i}\right) + C\left(N'_i, P_{N_i}\right), C\left(N_i, P'_{N_i}\right) > C(N_i, P_{N_i}) \end{cases}, \quad \text{which is}$$

equals to $E^{OAPS}_{(N_i,N'_i)}$. Thus, we have $E^{CGPS}_{(N_i,N'_i)} = E^{OAPS}_{(N_i,N'_i)}$.

When $0 < \alpha < 1 E^{CGPS}_{(N_i,N'_i)}$ lies in between as to the linearity of the formula.

In conclusion, we have $E^{CGPS}_{(N_i,N'_i)} \le E^{OAPS}_{(N_i,N'_i)}$. $\square$

As to the range of α, α cannot be negative or greater than 1, when $\alpha < 0$ each vehicle is forced to choose the expensive parking lot that is against the rule of fairness, when $\alpha > 1$ the opponent's advantage is overestimated, thus it is too hard for the vehicle to win the parking lot that is better to itself.

4 Simulation Analysis

We have developed a simulation environment in Java language that implements the three parking-lot search approaches as OAPS, CGPS, and the selfish competition algorithm with no system assistance. The simulation environment is in the parking area of 2×2 km with various vehicle amount 25–85 and parking lot amount 20–80, the average speed is 40 km/h and the cooperation parameter $\alpha = 1.0$. By comparison between OAPS and CGPS, further conclusions are drawn.

The analysis in Sect. 3 suggests that through cooperated gaming, vehicles can achieve win–win situation that is relative to the cooperation parameter α. Below is the simulation of the three parking-lots selection algorithms such as selfish competition, OAPS, and CGPS ($\alpha = 1$). The results are shown in the figure below:

Figures 1 and 2 are from two similar simulation environments with different placing of vehicles and parking lots, each parking-lot selection value for a given amount of cars and parking lots (for better illustration, we use the equal amount for cars and parking lots) is the average result of ten simulations, thus to eliminate the influence of different geographies. Y axis is the average time of parking-lot search for all vehicles, and the X axis is the amount of vehicles and parking lots from 5 to 100.

Comparing the two results, we can see that the selfish competition has the very unstable result, which is highly related to the positions of cars and parking lots, thus the parking-lot search without any system assistance can be very inefficient. While COAPS and CGPS have the pretty stable results due to the cars' cost analysis that can avoid the invalid driving, while CGPS have the lower cost, which is due to the mechanism of collaborative parking-lot selection shown in Sect. 3.2.

Fig. 1 Comparative
simulation 1

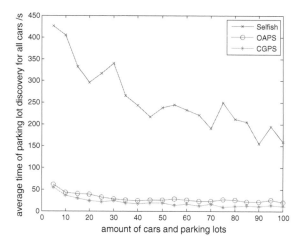

Fig. 2 Comparative
simulation 2

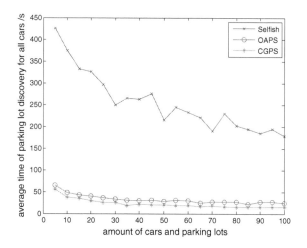

5 Conclusions

In this paper, we focus on the efficiency of decentralized parking-lot search system
based on VANET, especially the efficiency of the whole parking area under the
cooperated gaming within the vehicles. Besides the selfish parking-lot competition, the corresponding damage brought to the neighbors is also considered, thus
with the changed payoff matrix and cooperation parameter α, cooperated gaming
has the chance to be achieved, which is useful to decrease the total cost of the
players participated in the parking-lot competition, further, the parking efficiency
of the overall parking area can also be increased.

References

1. Arbatskaya, M., Mukhopadhaya, K., Rasmusen, E.: The parking lot problem. Available at SSRN 571101, (2006)
2. Caliskan, M., Barthels, A., Scheuermann, B., et al.: Predicting parking lot occupancy in vehicular ad hoc networks. In: IEEE 65th Vehicular Technology Conference 2007. VTC2007-Spring, pp. 277–281. IEEE (2007)
3. Delot, T., Cenerario, N., Ilarri, S., et al.: A cooperative reservation protocol for parking spaces in vehicular ad hoc networks. Proceedings of the 6th International Conference on Mobile Technology, Application and Systems. ACM 30 (2009)
4. Mousannif, H., Khalil, I., Al Moatassime, H.: Cooperation as a service in VANETs. J. Univ. Comput. Sci. **17**(8), 1202–1218 (2011)
5. Luo, J., Hubaux, J.P.: A survey of research in inter-vehicle communications. Embedded Security in Cars, pp. 111–122. Springer, Berlin Heidelberg (2006)
6. Klappenecker, A., et al.: Finding available parking spaces made easy, Ad Hoc Netw. http://dx.doi.org/10.1016/j.adhoc (2012). Accessed 02 Mar 2012
7. Kokolaki, E., Karaliopoulos, M., Stavrakakis, I.: Opportunistically assisted parking service discovery: Now it helps, now it does not. Pervasive Mob. Comput. **8**(2), 210–227 (2012)
8. Ayala, D., Wolfson, O., Xu, B., et al.: Parking slot assignment games. Proceedings of the 19th ACM SIGSPATIAL International Conference on Advances in Geographic Information Systems. ACM, New York 299–308 (2011)
9. Kokolaki, E., Karaliopoulos, M., Stavrakakis, I.: On the efficiency of information-assisted search for parking space: A game-theoretic approach. Proceedings of 7th International Workshop on Self-Organizing Systems, IFIP IWSOS 13, Palma de Mallorca (2013)
10. Arnott, R.: Spatial competition between parking garages and downtown parking policy. Transp. Policy **13**(6), 458–469 (2006)

A Fuzzy-Based Context-Aware Privacy Preserving Scheme for Mobile Computing Services

Eric Ke Wang and Yunming Ye

Abstract Currently privacy issues are challenging mobile computing services. We proposed a new privacy protection scheme for mobile computing services that is able to adapt to context. The accurate context is inferred by fuzzy reasoning. The experiment has been executed, and preliminary results have been encouraging.

Keywords Context aware · Privacy preserving · Fuzzy logic

1 Introduction

With the increasing availability of mobile devices, users enjoy more and more mobile services due to the users have much more mobilities than before. For example, when a user comes to a place, he can discover the nearest restaurant or ATM around his location, which is called location-based services. When users move, the environment changes. The original rules applied in one environment may not be applied in the other. For example, when users sit inside home, the environment is simple and gentle; the privacy protection can be home level. However, when users are in shopping mall, the environment becomes more complicated, the privacy protection should be high level.

Therefore, it is necessary to enforce different privacy policies according to context. That is called context-aware privacy preserving mobile computing that is different from the traditional privacy preserving computing. Traditionally, privacy

E. K. Wang (✉) · Y. Ye
Shenzhen Graduate School, Harbin Institute of Technology, Shenzhen Key Laboratory of Internet Information Collaboration, HIT Campus, Shenzhen University Town, Shenzhen, China
e-mail: wk_hit@hitsz.edu.cn

Y. Ye
e-mail: yym@hitsz.edu.cn

S. Patnaik and X. Li (eds.), *Proceedings of International Conference on Soft Computing Techniques and Engineering Application*, Advances in Intelligent Systems and Computing 250, DOI: 10.1007/978-81-322-1695-7_14, © Springer India 2014

Fig. 1 Context-aware
privacy protection model

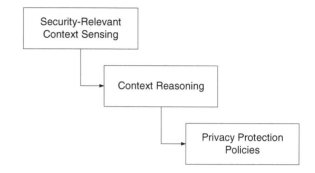

protection can only solve problems in common scenario, or users should manually
update protection policies. However, in context-aware privacy preserving mobile
computing, the system enforces reasoning process based on the context data to get
accurate inference result to adjust the privacy protection policies.

Context is the information that can determine an application's behavior or in
which an application event occurs [1]. The context can be from various context
information providers and can be in varying forms from temperature to user
behavior. Besides, context-aware computing includes active context awareness
that can automatically adapt the application to current context, and passive context
awareness that can only give the users the context data to let users to decide the
next step.

In our framework, we mainly tackle active context awareness and privacy-
relevant context, which consists of the set of contextual attributes that can be used
to characterize the situation of an entity, whose value affects the choice of the most
appropriate controls (measures). Figure 1 shows the basic context-aware privacy
protection model.

2 Privacy Challenge of Mobile Computing

Actually, many people besides friends and acquaintances are interested in the
information that people post or exist on mobile services. Privacy issues for mobile
computing are increasingly challenging users. Identity thieves, scam artists, debt
collectors, stalkers, and corporations looking for a market advantage are using
location-based applications to gather information about consumers [2]. Companies
that provide mobile services are themselves collecting a variety of data about their
users, both to personalize the services for the users and to sell to advertisers.

For example, more and more people are willing to use location-based services to
make queries related to locations. As it is well known, the values at the core of
services are precise and connecting [3]. However, the more precisely and convenient
we become with these services, the more apt we are to share personal details about

Fig. 2 System architecture

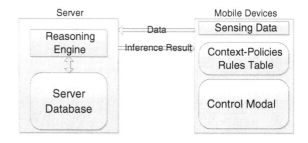

ourselves and let our guard down as we interact with others. Actually, the facts tell that the majority of mobile users post risky information online, without giving due diligence to privacy and security concerns. Unfortunately, the native core value of openness, connecting and sharing are the very aspects that allow cyber-criminals to use these services as a vector for various kinds of bad online behavior. At the same time, cyber-criminals are targeting location-based services with increasing amounts of malware and online scams, honing to this growing user base.

Although people's attitudes of privacy protection are various, most people hope that privacy protection measures would be smarter to let human be free while, currently most privacy protections for mobile services have to be handled and updated manually.

Therefore, how to protect privacy in an intelligent way is the main problem need to be solved for privacy issues of mobile computing. In order to tackle the above privacy problems, we proposed a privacy protection framework that can be able to adapt privacy protection polices by current environment. In this framework, the most important part is context awareness. However, because most data captured are ambiguous and scattered, a precise context is very hard to be achieved. In order to solve the problem, we adopt a reasoning engine based on fuzzy-logic theory [4] to solve the uncertain and vague data collected by mobile devices.

3 System Overview

3.1 Architecture

Considering the constraint resources of mobile devices, we design Server-Mobile Client model. The complicated computing tasks are done by server while, Mobile Client only solves the data collections and devices controls.

Figure 2 shows the architecture of the fuzzy-based context-aware privacy preserving scheme. It mainly includes two parties, one is the server, which solves the reasoning procedures, the other is the client side, which collects context data and execute controls. After users' mobile devices collect the sensing data, data will be sent to the server. Inside the server, there is a reasoning engine that receives the

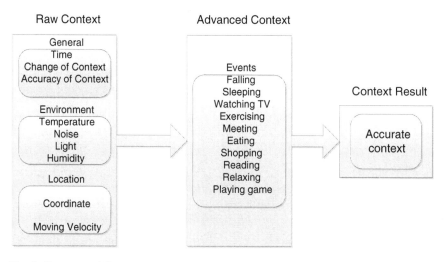

Fig. 3 Context modeling

data and executes reasoning process to send back the inference result to the users' mobile devices. After that, the mobile devices can read the inference result as context to adjust rules.

3.2 Context Modeling

Context is information that is used to describe the situation of an entity. However, how to model context is a crucial part in the system. Commonly, the context can be categorized into four types: system context (e.g., wireless network status, etc.), user context (e.g., location, emotion, medical history, etc.), physical environment context (e.g., lighting, temperature, weather, etc.), and time context (e.g., time) [5].

In our scheme, we model the context as raw model and advanced model. Raw context model includes some low-level data such as environment context data and user context data. Advanced context model includes high-level context such as behavior context. Figure 3 shows the context modeling.

All data collected by sensors in the mobile devices can be as the features of the context entities. In our scheme, we adopt <key, value> to map these features into individual fuzzy set in the fuzzy-logic framework. These fuzzy sets can be used for high-level context interpretation and further decision inference.

Some of the attributes associated with entities in our context model and their fuzzy sets are detailed as follows:

```
<Humidity, wet>
<Light, dark>
 <Location, Home>
```

Fig. 4 Common fuzzy-logic system

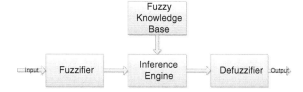

```
<Temperature, cold>
```

3.3 Fuzzy-Reasoning Process

A fuzzy-logic system commonly includes three parts: fuzzy sets, rules and inference engine [6].

The procedure of fuzzy logic is as follows: (1) fuzzification: converts input data to a fuzzy set by fuzzy variables, membership functions, and fuzzy terms. (2) Inference: makes an inference based on a set of rules. (3) defuzzification: maps the fuzzy output to a crisp output. We adopt Takagi–Sugeno Fuzzy modeling [7] to realize the processes. The reason why we select T–S fuzzy modeling is because that it provides a simple method to achieve a definite conclusion based on imprecise, ambiguous data since it employs a linear combination of input variables as a rule-consequent variable.

The fuzzification process is as follows:

If x is the member of the physical context sets, $\tilde{A}$ and y are the members of the user behavior set $\tilde{B}$;

Then $z = f(x, y)$;

The physical context sets $\tilde{A}$ and the user behavior set $\tilde{B}$ are the fuzzy sets of modus ponens [8], and $z = f(x, y)$ is the precise function of modus tollens [9] and it is the fusion relation of the two contexts. $z = f(x, y)$ is the polynomial of the input variable x and y. Figure 5 is the fuzzification process.

Figure 6 shows the flow in the reasoning engine. The raw data collected by the mobile devices with context information is processed in build stage, which can produce context fuzzy sets. Then, fuzzy rules loaded from the inference rules database are utilized to generate advanced level context such as user activity. At last, the rule engine identifies the current state of the user based on the combination of advanced level context and gets the advanced result.

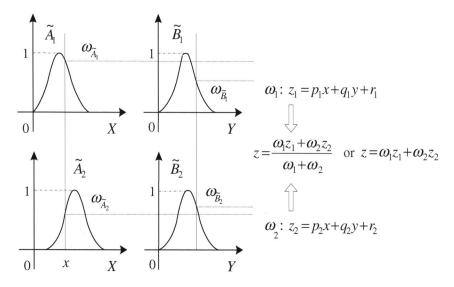

Fig. 5 Fuzzification process

Fig. 6 Reasoning engine

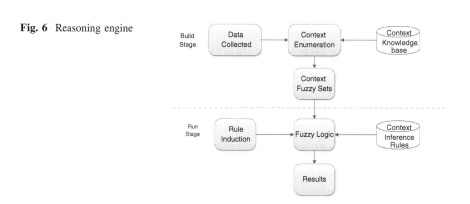

4 Evaluation

In order to evaluate the scheme, we implemented a prototype in a smart phone (Samsung Galaxy Note I with android 4.0) and made experiment. Figure 7 shows the implementation structure. The server runs a context acquire service that receives the GPS-location information, Wi-Fi connection information, and users'

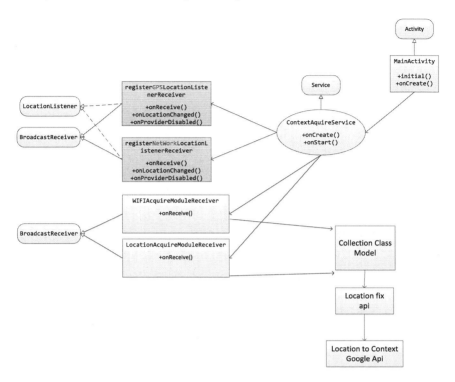

Fig. 7 Implementation modules

activities. The class (register GPS-location listener receiver) is used to get the GPS information, the class (Wi-Fi acquire module receiver) functions to get the Wi-Fi hotspot location information.

5 Conclusions

We proposed a fuzzy-based context-aware privacy preserving scheme that can dynamically adapt privacy protection policies to current scenario. It enforces fuzzy-reasoning based on context data to achieve accurate scenario. We built an experiment prototype to evaluate the scheme. The results show to be encouraging.

Acknowledgments This research was supported by National Natural Science Foundation of China (No. 61100192), Research Fund for the Doctoral Program of Higher Education of China (No. 20112302120074), and was partially supported by Shenzhen Strategic Emerging Industry

Development Foundation (No. JCYJ20120613151032592 and ZDSY20120613125016389), National Key Technology R&D Program of MOST China under grant no. 2012BAK17B08, and National Commonweal Technology R&D Program of AQSIQ China under grant no. 201310087. The authors thank the reviewers for their comments.

References

1. Dey, A.K.: Understanding and using context. Pers. Ubiquit. Comput **5**(1), 4–7 (2001)
2. Gong, Z., Sun, G.-Z., Xie, X.: Protecting privacy in location-based services using K-anonymity without cloaked region. In: 11th International Conference on Mobile Data Management, pp. 366–371, New York (2010)
3. Zhang, W., Cui, X., Li, D., Yuan, D., Wang, M.: 2010 18th International Conference on Geoinformatics. Beijing, China (2010)
4. Zadeh, L.A.: Fuzzy logic. Computer **21**(4), 83–93 (1988)
5. Hong, J., Suh, E., Kim, S.J.: Context-aware systems: a literature review and classification. Expert Syst. Appl. **36**(4), 8509–8522 (2009)
6. Liang, Q., Mendel, M.: Interval type-2 fuzzy logic systems: theory and design. IEEE Trans. Fuzzy Syst. **8**(5), 535–550 (2000)
7. Su, X., Wu, L., Shi, P., Song, Y.D.: H∞ model reduction of Takagi–Sugeno fuzzy stochastic systems. IEEE Trans. Syst. Man Cybern. B Cybern. **42**, 1574–1585 (2012)
8. Zardini, E.: Naive modus ponens. J. Philos. Logic **42**(4), 575–593 (2013)
9. Yalcin, S.: A counterexample to modus tollens. J. Philos. Logic **41**(6), 1001–1024 (2012)

Research and Application of Trust Management System

Fengyin Li and Peiyu Liu

Abstract Trust status of a service provider is critically important to potential buyers in electronic commerce (e-commerce), particularly when they are unknown to each other, while trust establishment and trust accumulation are still key problems need to be solved. By defining new trust architecture and introducing new concepts of trades' turnover and trust decline, a trust management model was proposed in this paper. In the new architecture, events were divided into two categories and rules were predefined to determine which formula to use according to the feature of the events occurred during the electronic transaction. In the new trust calculation method, the concepts of trades' turnover and trust decline were first introduced to further depict the dynamic feature of trust status. Further, to adapt to different application domains, formulae with variable parameters were adopted. Simulation results show that the proposed model precisely implements the long and incremental characteristics of trust establishment process.

Keywords Trust management · Trust evaluation · Electronic commerce

1 Introduction

The concept of trust management was proposed by Blaze in 1996 [1]. It aimed to solve the security problems in Internet applications and provided a security policy framework adapted to open and dynamic network application systems. In

F. Li (✉) · P. Liu
School of Information Science and Engineering, Shandong Normal University, Jinan, China
e-mail: lfyin318@126.com

P. Liu
e-mail: liupy@sdnu.edu.cn

F. Li
School of Computer Science, Qufu Normal University, Rizhao, China

S. Patnaik and X. Li (eds.), *Proceedings of International Conference on Soft Computing Techniques and Engineering Application*, Advances in Intelligent Systems and Computing 250, DOI: 10.1007/978-81-322-1695-7_15, © Springer India 2014

electronic commerce (e-commerce), a variety of electronic transactions (e-trans-actions) are implemented in a loosely coupled network environment. In most cases, both sides of the e-transaction do not know each other, neither of them is willing to be cheated, and this makes trust management a very challenging and critical issue.

In e-transaction, the dynamic trust status of a seller always plays a decisive role in real-time e-transaction systems, and more factors (such as trades' turnover, trust decline, and goods types) should be considered to precisely reflect the latest trust status of each seller. Furthermore, a better incentive mechanism should be studied to better reflect the evolution of trust values. In trust evolution, positive events and negative events play completely different roles. So, not only incentive rules but also punishment rules should be included in the incentive mechanism. Further, the incentive scale and the penalty scale should vary from event to event and from domain to domain. The most important is that trust establishment is a long and incremental process. Even if a "good" service provider behaves very well, it takes an enough long time for him to obtain a high reputation value.

To solve the problems discussed above, a trust management model for e-com-merce applications was proposed in this paper. The new model was composed of a centralized trust management architecture and a trust calculation method. In the new architecture, events were divided into two categories and rules were predefined to determine which formula to use and what arguments to use according to the events occurred during the e-transaction. By first introducing the concepts of trades' turn-over and trust decline, based on formulae, a new trust calculation method was pro-posed. The used formulae were designed to adapt to different application domains by revising corresponding arguments. Simulation results show that the proposed model addresses the long and incremental characteristics of trust establishment process.

The rest of this paper was organized as follows. In Sect. 2, we presented the rule-based and event-driven trust management architecture. In Sect. 3, the for-mula-based trust computation method was proposed. Some empirical study results were illustrated in Sect. 4. In Sect. 5, we concluded our work.

2 Trust Management Architecture

Centralized trust management architecture (like eBay) for e-commerce applica-tions was proposed in this section (Fig. 1). The trust management server was responsible for the storage and management of all trust data and other related information. System clients (buyers or sellers) reported the rating of the other side of the transaction to the trust management server after every transaction.

We assumed the centralized trust management server was independent of any system client. This makes it feasible to apply the unified evaluation approach to the same domain of applications. In addition, it ensures that the trust computation could be completed on relatively complete trust data, which is difficult in a decentralized architecture. Certainly, the centralized architecture is vulnerable to a

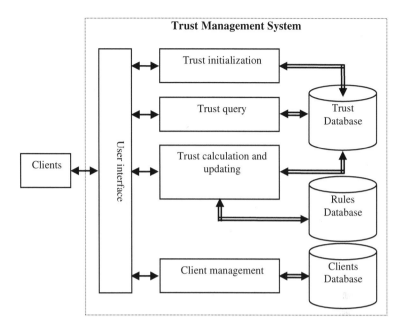

Fig. 1 Trust management architecture

single point failure due to the scale expansion problem, which is less risky in the decentralized environment.

In Fig. 1, all clients' requests were sent to the user interface module, which would distribute the requests to the trust initialization module, the trust query module, the trust calculation, and updating module or the client management module according to the feature of each request. The trust initialization module was in charge of the allocation of an ignorance trust value (e.g., 0.1) to each new comer. The final trust value of each new comer depended on its subsequent behavior performance. The trust query module mainly responded the trust requests by querying the trust database. The trust calculation and updating module afresh calculated and updates the trust value of a client as long as a new valid rating of that client was produced. When positive cases or negative cases happened, the incentive or punishment rules could be used to reward or punish a client during the trust calculation process. The incentive scale and penalizing scale were also pre-defined as variable parameters.

3 Trust Evaluation Method

In this section, a formula-based trust calculation method was proposed. In the new trust calculation method, the concepts of trades' turnover and trust decline were first introduced into trust management to depict the dynamic feature of trust status.

In different domains, there may be different policies for trust evaluation. The used formulae could be applied into different domains by setting different arguments. In addition, in the case of penalty, the decrement was determined by arguments, which were selected according to the nature of negative events.

Definition 1 Let T_i^x denote the trust value of target service provider x at current time period t_i, R_{i+1}^x be the rating of x at time period t_{i+1}, and $\Delta = R_{i+1}^x - T_i^x$. The trust value of x at time period t_{i+1} was defined in Eq. (1).

$$T_{i+1}^x = \begin{cases} \min\left(1, T_i^x + \lambda_+ \cdot \theta \cdot \Delta\right) & \text{if } \Delta \geq 0 \\ \max\left(0, T_i^x + \lambda_- \cdot \theta \cdot \Delta\right) & \text{if } \Delta < 0 \end{cases} \tag{1}$$

where $0 \leq \theta < 1$ was the impact factor function determining the impact of recent change (i.e., Δ) on trust calculation, and $\lambda_+ \leq 1$ (or $\lambda_- \geq 1$) factor determining the incentive (penalty) scale of trust change. From the arguments, we can see that the penalty scale is no less than the incentive scale. That is, the improvement of trust value is more difficult than the drop of it.

Formula (1) yielded a value in the range of [0, 1]. To obtain the trust value in period t_{i+1}, the trust value in period t_i and the latest rating value were used.

We assumed that in time period t_i the service provider x received such valid rating values as $R_{i_1}^x, R_{i_2}^x, \ldots, R_{i_m}^x$ and the corresponding transaction prices as $P_{i_1}^x, P_{i_2}^x, \ldots, P_{i_m}^x$, and then the ultimate rating R_i^x of x in period t_i could be obtained as Eq. (2).

$$R_i^x = \sum_{j=1}^m \left(R_{i_j}^x \cdot P_{i_j}^x / \sum_{k=1}^m P_{i_k}^x \right) \tag{2}$$

That is, we used the rating value and the corresponding trades' turnover of each transaction to calculate the ultimate trades rating in every time period. In this way, we first considered price feature in trust calculation to promote the accuracy of trust evaluation system.

According to trust principles, the impact factor function should be a decreasing function whose decrement is decreasing. There may be more than one function to depict the feature, and we just selected a simple one of them.

The impact factor function was defined as follows.

Definition 2 If the current trust value was T_i^x, the *impact factor function* was defined in Eq. (3).

$$\theta\left(T_i^x\right) = C/(T_i^x + C)^2 \tag{3}$$

where $C > 0$ is a constant parameter affecting the curve shape of function θ.

Based on the same principles, we selected a decreasing function whose decrement was increasing as our trust decline function, which was defined as Eq. (4).

Definition 3 If a service provider got no valid rating values after time period t_i, the trust value decline function was defined as Eq. (4).

$$T(t) = T(t-1) - (t - t_i)/D \tag{4}$$

where t stands for time, and D is a constant parameter.

From the above discussion, we can see that, when $\Delta \geq 0$, according to formulae (1)–(3), there would be an increment in the trust calculation, namely $T_{i+1}^x \geq T_i^x$. In this case, we set $\lambda_+ = 0.9$. Thus, the increment would be $\lambda_+ \cdot \theta(T_i^x) \cdot \Delta$. When $\Delta < 0$, there would be a decrement in the trust calculation. By definition, we set $\lambda_- = 1$. As $\lambda_+ \leq \lambda_-$, assuming the same $|\Delta|$ and T_i^x, the decrement $\lambda_- \cdot \theta(T_i^x) \cdot |\Delta|$ was no less than the increment $\lambda_+ \cdot \theta(T_i^x) \cdot \Delta$. This indicates that it is not easier to improve the trust value than to worsen it. Therefore, generally it takes longer time to reach a high-level trust value (e.g., 0.98) than to drop from a high level to a low level. The value of λ_+ (or λ_-) was determined by the predefined rules in different applications. In addition, if a severely negative event happened, λ_- would be applied for decrement cases. This also results a harder trust improvement process.

4 Simulation and Application

To further study the properties of our proposed approaches, a set of empirical studies were conducted in this section, and the results were illustrated and analyzed. For trust computation, the formulae discussed in Sect. 3 were adopted.

4.1 A Simulation

In this section, we compared the trust establishment process of several service providers with different system entry periods and different trust levels, and we aimed to compare their reputation levels delivered by the proposed model. We studied the trust establish process of new service providers and the trust accumulating process of old service providers. To avoid the whitewashing behaviors, the initial trust value was set to $T_0 = 0.1$, an ignorance initial value. We assumed that there were 5 service providers and they were marked as S_1, S_2, S_3, S_4, and S_5. In each time round, S_1 got a static rating of 0.95. S_2 got a static rating of 0.9 in the former 149 rounds, and at round 150 it got a rating of 0.2 due to an extremely negative case, and then, its ratings stayed at 0.6. S_3 got a static rating of 0.8 in the former 100 rounds, but at the following rounds, its trust value decayed based on the formula (4) due to lacking of valid ratings, while S_4 and S_5 attended at round 150 and got their static ratings of 0.95 and 0.7.

The simulation curve with above 5 service providers is shown in Fig. 2, and Fig. 3 shows the same simulation curve with a rating deviation of $|\varepsilon| \leq 0.1$. In the above figures, we obtained the curves by setting the arguments $\lambda_+ = 0.9$, $\lambda_- = 2$, and $C = 20$.

Fig. 2 Simulation curves

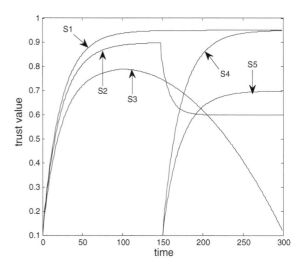

Fig. 3 Simulation curves
with R deviation $|\varepsilon| \le 0.1$

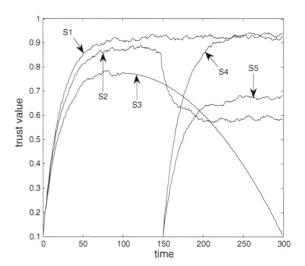

From the simulation curves, we can see that the trust curves could depict the dynamic trust status of every service provider, that is, the service providers with the same trust values have the same trust levels. Obviously, our trust model could accurately depict the trust changes in different situations.

4.2 Performance Discussion

Compared with trust evaluation method in the literature [2–4], the proposed trust evaluation method could differentiate positive events and negative events by set scale control factors λ_+ and λ_-, and further the scale control factors could vary from domain to domain by setting different parameters. Compared with the trust management model in the literature [5–7], the new trust model introduced such factors as trades' turnover, incentive scale, penalty scale, and trust decline into trust evaluation process and could more precisely depict the trust establishment process. Compared with the scheme in the literature [8–10], the new trust model selected a simple function with variables to avoid the hyperbolic tangent transformation and so greatly improved the efficiency of trust evolution. By setting different variable parameters, the proposed trust evaluation method could be applied into different fields. Furthermore, the new trust model implemented the long-term and incremental trust establishment process and could solve the open trust problems discussed above.

In the new trust model, due to the implement of the long-term and incremental process of trust establishment and accumulation process, the trust value of each client precisely depicted its trust status. That is, the service providers with the same trust value must have the same trust reputation level, and vice versa. And so in the new trust model, we do not need to differentiate the clients with the same trust values and the above trust issues has been completely solved.

5 Conclusions and Future Work

In this paper, a trust management architecture and a trust evaluation method were proposed. The new trust model has some advances in the following aspects. The features of the new model are critically important to a trust management service authority with a large pool of system clients.

1. It is incentive to good service providers with good service quality for a long service period.
2. It is incentive to new service providers.
3. It penalizes service providers with poor service quality.
4. For service providers who are not active, their trust values will decay in a special way, until eventually into zero or a valid rating is obtained.

All these properties aim to provide clearer information to service clients and prevent some service providers from cheating clients after obtaining a good reputation rank.

For future work, more impact factors (e.g., the varieties and properties of different goods) will be studied in trust management model. The distributed trust management architecture for e-commerce applications will also be studied. In addition, in order to better ensure the fairness of the trust model, the trust status of service clients (buyers) will be further studied in e-commerce applications.

Acknowledgments This work is partially supported by National Nature Science Foundation of China (61373148), National Social Science Foundation of China (12BXW040), Nature Science Foundation of Shandong Province of China (ZR2012FM038), and Science & Technology

Foundation planning Project of Colleges and Universities of Shandong Province of China (J12LN07). The authors also gratefully acknowledge the helpful comments and suggestions of the reviewers, which have improved the presentation.

References

1. Blaze M., Feigenbaum Jand Lacy, J.: Decentralized trust management. In: Proceedings of the 17th Symposium on Security and Privacy, pp. 164–173. IEEE Computer Society Press, Oakland (1996)
2. Qureshi, B., Min, G., Kouvatsos, D.: A distributed reputation and trust management scheme for mobile peer-to-peer networks. Comput. Commun. 35(5), 608–618 (2012)
3. Denko, M.K., Sun, T., Woungang, I.: Trust management in ubiquitous computing: a Bayesian approach. Comput. Commun. 34(3), 398–406 (2011)
4. Omar, M., Challal, Y., Bouabdallah, A.: Certification-based trust models in mobile ad hoc networks: a survey and taxonomy. J. Netw. Comput. Appl. 35(1), 268–286 (2012)
5. Cho, J.H., Swami, A., Chen, I.R.: Modeling and analysis of trust management with trust chain optimization in mobile ad hoc networks. J. Netw. Comput. Appl. 35(3), 1001–1012 (2012)
6. Manzanares-Lopez, P., Malgosa-Sanahuja, J., Muñoz-Gea, J.P.: The importance of considering unauthentic transactions in trust management systems. J. Parallel Distrib. Comput. 72(6), 809–818 (2012)
7. Sun, J., Sun, Z., Li, Y., et al.: A strategic model of trust management in web services. Phys. Procedia 24(B), 1560–1566 (2012)
8. Wang, Y., Varadharajan, V.: Interaction trust evaluation in decentralized environments. In: Proceedings of the 5th International Conference on Electronic Commerce and Web Technologies (EC-Web'04). Lecture notes in Computer Science, vol. 3182, pp. 144–153. Zaragoza, Spain (2004)
9. Zacharia, G., Maes, P.: Trust management through reputation mechanisms. Appl. Artif. Intell. 14(9), 881–908 (2000)
10. Wang, Y., Lin, K.J., Wong, D., et al.: Trust management towards service-oriented applications. SOCA 3(2), 129–146 (2009)

Ranaad-Xek: A Prototype Design of Traditional Thai Musical Instrument Application for Android Tablet PC

Kasikrit Damkliang, Chawee Kaeoaiad and Sulkiplee Chehmasong

Abstract This paper proposes a prototype of an architectural, algorithms, and graphical user interface (GUI) design of "Ranaad-Xek," a traditional Thai musical instrument application for Android tablet PCs. The application provides percussion methods for a player as real as a physical instrument with both of traditional Thai and universal organology. The application supports both of soft and hard sound tone, which generated by percussion mallet types. The player can freely multi-touches on wooden bar to produce instrument sounds and record the user own songs.

Keywords Architectural software design · Algorithms design · GUI design · Traditional instrument · Thai musical instrument · Percussion instrument · Android · Tablet PC · Table computer

1 Introduction

Thai classical music is an integral part of the lives of Thais. It is the focal point of their way of life, tradition, culture, arts, education, religion, and philosophy. Throughout Thai history, a valuable and impressive musical foundation was set up with a variety of instruments, compositions, and playing techniques that could

K. Damkliang (✉) · C. Kaeoaiad · S. Chehmasong
Information and Communication Technology Programme, Prince of Songkla University, Hat Yai, Thailand
e-mail: kasikrit.d@psu.ac.th

C. Kaeoaiad
e-mail: 5110210716@email.psu.ac.th

S. Chehmasong
e-mail: 5110210608@email.psu.ac.th

S. Patnaik and X. Li (eds.), *Proceedings of International Conference on Soft Computing Techniques and Engineering Application*, Advances in Intelligent Systems and Computing 250, DOI: 10.1007/978-81-322-1695-7_16, © Springer India 2014

Fig. 1 Wooden bars or
Rang-Ranaad of Ranaad-Xek
instrument

clearly express a range of emotions and feelings [1, 2]. Thai Government gives an important on development of pre-primary school children to improve their knowledge and motivate them to learn by themselves. Urgently, public policy of the Thai Government exposed on August 2011, to support and distribute tablet PCs to schools throughout the country. The project initialized with pilot primary schools on grade 1 or *Prathomseuksa* 1 students in academic year 2012 [3]. Nowadays, the project is an ongoing implementation, but there is a lack of performance handwriting application for the students. It is a good chance for us to propose better application to the students.

In our previous work [4], we have proposed a prototype of "Ranaad-Xek." The Ranaad-Xek application supports iOS and runs on Apple trademark tablet computer: the iPad [5]. In addition, we distribute it for iPhone and it can compatible with other Apple's iDevices. This paper proposes a prototype of an architectural and GUI design and implementation of Ranaad-Xek for Android table PCs. We have improved the application based on the Ranaad-Xek's iOS. The player can freely multitouches on wooden bar to produce instrument sounds and record the players' own songs.

2 Ranaad-Xek

Ranaad is a traditional Thai musical trough-resonated keyboard percussion instrument generally played with two mallets in Thai classical music and performance [6]. The traditional Thai system of organology classifies "Ranaad-Xek" as a higher-tone xylophone with bars usually made of hardwood. In addition, the Ranaad-Xek is also a symbolic representative of Thai classical music [2].

2.1 Ranaad-Xek's Composition

Ranaad-Xek has evidently occurred since The Kingdom of Ayutthaya, 1350–1767. The Ranaad-Xek composes of three parts [1, 2, 7]. The first part is boat-shaped resonated trough for amplifying reverberation and bright sounds. Ranaad-Xek consists of 21 or 22 wooden bars shown in Fig. 1.

Fig. 2 The soft mallets and the hard mallets

Fig. 3 Basic or 8-double percussion

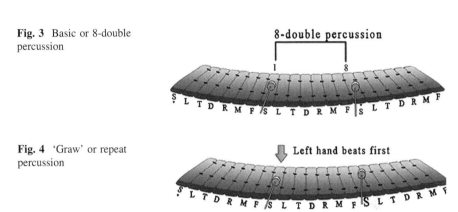

Fig. 4 'Graw' or repeat percussion

Tone adjustment is achievable using a combination between lead (malleable metal) and paraffin to attach under each wooden bar in both leading ends. Moreover, the tones of some wooden bars are used as reference tones for other instrument in the ensemble. We play Ranaad-Xek with two mallets. There are two types of mallets. The soft mallets provide relaxed, silky, and softer tones for playing slow songs. Meanwhile, the hard mallets provide sharp bright sounds when wooden bars are being percussed for faster playing. Both of the mallets types are shown in Fig. 2.

2.2 Ranaad-Xek Percussions

Basic Ranaad-Xek percussion is beating methods. When the player holds mallets and strikes double wooden bars concurrently, both sound tones are generated concurrently. There are many ways to play Ranaad-Xek. Main and basic percussions are classified into four methods: gep (Thai: ตีเก็บ), graw (Thai: ตีกรอ), seaw (Thai: ตีเสี่ยวมือ), and gwaad (Thai: ตีกวาด) [1, 7]. Gep in Thai means 'to keep or clean up' and it is the most important basic percussion when a player holds mallets and hits double wooden bars concurrently as shown in Fig. 3.

The second percussion, 'graw' or repeat, is for playing long notes when the player frequently swaps hitting double wooden bars with equal weight of left and right hands, as shown in Fig. 4. The third percussion, 'seaw' or portion, is a way to

Fig. 5 'Seaw' or portion percussion

Fig. 6 'Gwaad' or sweep percussion

percuss wooden bars when a hand is repeatedly hitting the same bar while another hand synchronizing hits through wooden bars producing resonated sound tones and harmony melodies, as shown in Fig. 5. The last one is 'gwaad' or sweep percussion, as shown in Fig. 6. In music, it refers to the player running the mallets along the entire wooden bars keyboard in one long draw, slow or fast, same or different directions, generating charming sounds. These basic percussions are fundamental practice for the beginning player. These also support the capable player for later applying in adaptive and advanced percussion styles or more interesting rhythms. Therefore, our application will support and focus on these main and basic percussions.

3 Application Specification and Architecture

3.1 Framework

The Cocos2d-x is an Object-oriented APIs framework. Cocos2d-x is a cross-platform game engine branched from Cocos2d-iPhone, which consists of expanding supported platforms, with multiple choices of programming languages that shares the same API structure [8]. Our application supports 2D graphical interfaces and implement following the OpenGL ES standard with Cocos2d-x.We choose the Android NDK environment development with C++ programming language that helps us to embed native machine code compiled from our C++ source files into application packages.

3.2 Application Specification

The application provides both of traditional Thai and universal tuning and notes. The application provides recording system that consists of record, pause, play, save, and list recorded song functions by user's interaction. The application provides sample songs. The player can play these bundled songs on the playing mode. It consists of pitches and delay times. The sample songs provide note sheets and video clips for percussions suggested by The Arts Cultural Center of Prince of Songkla University. The note sheets are in Thai musical systematic style. However, the user can play the songs with different tunings of percussions. The sample songs named one-class rhythm *Khae-Bor-Ra-Ted* (*Thai:* แขกบรเทศชั้นเดียว), two-class rhythm *Lao-Siang-Tian* (*Thai:* ลาวเสี่ยงเทียน 2 ชั้น), *Lao-Kruan* (*Thai:* ลาวครวญ 2 ชั้น), and famous *Moonlight Serenade* or *Lao-Duang-Duen* (*Thai:* ลาวดวงเดือน 2 ชั้น).

3.3 Application Architecture

We have revised our previous architecture because the practicing mode is not well as we design and we have hardcoded that initialized with only one song, which is the *Khae-Bor-Ra-Ted*. We have decided to remove practice mode and will investigate in the future. For this work, we propose recoding system. We have to improve all related modules for recording system consists of State Saving and Resuming Engine (SSRE), Rhythms Synchronizing Engine (RSE), Sound Engine (SE), Multitouch Detecting Engine (MDE), Presentation Engine (PE), and Playing Manager Engine (PME).We design the application architecture and divide it into 6 modules, as shown in Fig. 7.

The first module, MDE, is a module for detecting all concurrent touches from the user. There are wooden bars in our application starting from the far left to the far right. Each produces an identical sound tone of organology. For traditional Thai organology, it is from the lowest Sal to the highest Far tone. For universal organology, the wooden bar is the virtual instrument tone bridge of Bb (or *A#*) cord tuning. Therefore, in this work, there are 21 monotones (or 14 8-double percussion tones) in traditional Thai organology and there are 21 pitches of Bb's cord in universal organology. Both of the organologies also support the note sounds of soft and hard mallets. Therefore, there are 48 monotones in the application.

The second module, SE, is a module for playing each sound of each wooden bar. The third module, RSE, is for managing the song rhythm that is classified according to the speed of playing into one-class, two-class, and three-class in the traditional Thai system of organology. In previous work [4], our application supports the sounds of *ching*, a traditional Thai musical instrument for controlling rhythm. The forth module, SSRE, is a module for collecting pitches and delay times of a recording song in case that the player records, pauses, stop, and save it into a file. The fifth module, Song note Inputting Engine (SIE), is a module for inputting song pitches and delay times. Therefore, it transforms of these data into the PME. The application supports all percussion methods as a minimum requirement.

Fig. 7 The architecture of
the application

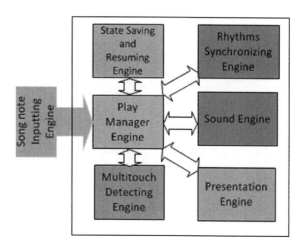

The last module is Presentation Engine (PE). Our application will run in user-event-driven orientation such that the interaction between the application and the player will achieve by PE module.

Due to the public policy of the Thai Government, the application distribution platform is Android tablet PCs. The device should run on a fitting 7.0 inches and $600 \times 1,024$ pixels at least 170 ppi pixel density resolutions. User inputs are established via multitouch in order to do Ranaad-Xek percussions that the user concurrently touches double wooden bars with two fingers.

3.4 The Playing Mode

The player can choose soft or hard mallets and rhythm classes, respectively. The player can freely multitouches. Even though they play wrong percussions, the application still generates note sounds located on the touched wooden bars. Furthermore, we enhance the recording module into the playing mode. The state diagram is shown in Fig. 8. In this work, we improve and propose the playing mode. On the main screen, a player can select to enable playing mode or to view tutorial information. The On Recording state is responsible to record the player's percussions consisting of pitches and delay times. When the player has pressed a stop recording button, the application has finalized a record timer. Then, the application is into the On Waiting to the Save a Song state. If the player has pressed a save button, the application will save percussion information into a file.

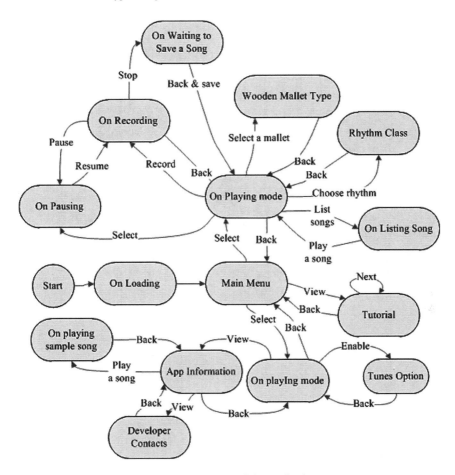

Fig. 8 The overview of improved-state diagram of the application

3.5 Recording System

We propose preliminary of algorithms for recording system in user-driven event oriented using timers to count notes' delay times and dynamic array to store all notes' information such as pitches and delay time as shown below. Due to the application supports 84 monotones, we configure button's tag for each wooden bar. The tag identifies a note's sound for sending it to play and record which depends on the user's percussion. The algorithm is shown in the *hit Sound Button () function.

Fig. 9 The Rang-Ranaad screen in the playing mode of the application

3.6 GUI Design

We are still keeping the golden motif theme and supporting landscape orientation 21 wooden bars from the previous work. The design in this work is for 7 inches screen due to the OTPC Project of Thai Government [3]. Figure 9 shows a wooden bars or *Rang-Ranaad* in the Thai organology sound system of the application when the user has pressed the playing mode. It consists of two button groups. The first set consists of five buttons such as home, mallet type, class rhythm, Thai, and universal organography, from the left to the right, respectively. The second set is control buttons for the recording system. There are six buttons consisting of record, stop, play, save, list, and information, respectively. The information screen provides application information and demonstrates song video footages and developer team contacts.

4 Conclusions

This paper proposes a prototype of an architectural, algorithms and GUI design of Ranaad-Xek for Android tablet PCs. The player can freely multitouches on wooden bar to produce instrument sounds and record the players' own songs. The full implementation and testing are on progress.

```
Function: record()
Start
IfplayTimerisValidThen
  SetplayTimerinvalidated
End If
Begin:recordTimer
  Setincrementation += 1
  timeHitremoveAllObject
  noteHitremoveAllObject
End:recordTimer
Setincrementation = 0
Stop
```

```
Function:playSounds(
)
Begin:playTimer
  Setincrementation
+= 1
  IftimeHit != NULL
Then
    Set index =
timeHit
    at
incrementation
    Set note =
noteHit
      at index
    Play note sound
  End If
  Setincrementation
+= 1
EndplayTimer
Setincrementation =
0;
Stop
```

```
Function:hitSoundButton(button
)
Start
Set note = button tag
Set time = incrementation
IfrecordTimerisValidThen
  Add time to timeHit
  Add note to noteHit
End If
Play note sound
Stop
```

```
Function:saveRecord(
)
Required:noteHit
Start
Setpath, fileName
IfnoteHitisValidThen
  SavenoteHit to
fileName
  If save
isCompleteThen
    SetfileName
Attribute
  Else  Return error
  End If
End If
Stop
```

Acknowledgments This research work is a part of senior students' project at Information and Communication Technology Programme. Faculty of Science, Prince of Songkla University supports this work. The authors are also thankful for The Arts Cultural Center of Prince of Songkla University.

References

1. "Divine Music", First English Edition, 1st Printed. Bangkok Printing Co., Ltd, Bangkok (1984, 2003). ISBN 974-91812-6-3
2. The Royal Institute, "Thai Classical Music Encyclopedia: Kred Song and Sing Plays History Section", 1st Printed. The Royal Institute, Bangkok (2007). ISBN 9789749588819
3. OTPC Project (One Tablet per Child). Available online: August 2013, http://www.otpc.in.th/
4. Damkliang, K., Chanlert, S., Thongnuan, A.: Traditional Thai musical instrument for tablet computer—Ranaad EK. In: Proceedings of the International Conference on Computer, Communication and Information Sciences, and Engineering (ICCCISE 2012), Paris, 2012
5. Ranaad-Xek App for iPad retina display. Available online: August 2013, https://itunes.apple.com/th/app/ranaad-xek/id569446973?l=en&mt=8
6. Suraporn Suwan, "Thai Classical Music in Thai Culture," 1st Printed. Chulalongkorn University Press, Bangkok (2006). ISBN 974-9941-57-8
7. Physical and development of Ranaad-Xek. Office of the National Culture Commission. Available online: August 2011, http://www.culture.go.th/knowledge/story/ranad/direct5.htm
8. Developers Manual, Cocos2d-x. Available online: August 2013, http://www.cocos2d-x.org/projects/cocos2d-x/wiki

Identifying Accurate Refactoring Opportunities Using Metrics

Yixin Bian, Xiaohong Su and Peijun Ma

Abstract Cloned code, also known as duplicated code, is among the bad "code smells." Refactoring can be used to remove clones and makes a software system more maintainable. However, there is a problem that causes the output results of the clone code detection tool cannot be directly refactored. The problem is not all the clone groups are suitable for refactoring. To address it, we propose a metric method to identify clone groups that are suitable for refactoring. The results of several large-scale software system studies indicate that our method can significantly increase the accuracy of identifying clone groups that are suitable for refactoring. It is not only beneficial to the following study of refactoring, but also it connects the entire process from clone detection to clone refactoring.

Keywords Cloned code · Metric · Refactoring

1 Introduction

Code clones are code fragments similar to one another in syntax and semantics [1]. One location is copied and paste it to another location with or without modifications during software development. This kind of activity causes multiple copies of exact or closely similar code fragments to coexist in software systems. These code fragments are known as clones [2]. In most cases, cloned codes are harmful in software

Y. Bian (✉) · X. Su · P. Ma
Department of Computer Science, Harbin Institute of Technology, Harbin, China
e-mail: bianyu79@163.com

X. Su
e-mail: sxh@hit.edu.cn

P. Ma
e-mail: ma@hit.edu.cn

S. Patnaik and X. Li (eds.), *Proceedings of International Conference on Soft Computing Techniques and Engineering Application*, Advances in Intelligent Systems and Computing 250, DOI: 10.1007/978-81-322-1695-7_17, © Springer India 2014

maintenance and evolution [3–8]. Although code cloning can help developers to quickly reuse existing design and implementation, it also incurs a significant increase in development and maintenance cost because programmers need to apply repetitive edits when the common logic among clones changes. Furthermore, failing to apply those changes can result in defects and field failures. On the other hand, there has been a good number of empirical evidence in favor of clones concluding that clones are not harmful [9–12].

Refactoring improves code structure without changing program behavior [13]. Fowler introduced many techniques for refactoring in his book, which is widely read by practitioners [14]. One of the most frequently performed refactoring techniques is "Extract Method," which means extracting one part of an existing method as a new method and replacing the extracted part with a procedure call [15]. This technique, a common way of reducing repetitions in writing code, is also known as "extract function" or "extract procedure." The commonly used refactoring tools on various IDEs, such as Eclipse, support procedure extraction to a certain degree in order to help programmers in dealing with this common and recurring situation.

Refactoring is widely used to delay the degradation effects of software aging and facilitate software maintenance [16]. However, there is a problem that causes the output results of the clone code detection tool which is not to be directly refactored. The problem is all code clones detected by a code clone detection tool are not appropriate for refactoring [17]. So far, no study has mentioned the method of eliminating false positives of cloned code-related bugs. The chief contribution to this paper is as follows: A metric method is developed to identify clone groups that are suitable for refactoring.

The rest of the paper is organized as follows: Sects. 2 and 3 provide the background and the clone analysis algorithm developed in our research. Section 4 outlines the directions for future work.

2 Relate Work

2.1 Cloned Code

Cloned code also known as duplicated code is similar code fragments to one another in syntax and semantic. Programmers' copy–paste-modification practice is regarded as one of the main reasons for majority of clones. There are four types of cloned codes up to now:

- Type-1 clones: Identical code fragments except for variations in white-space and comments.
- Type-2 clones: Similar code snippets, where identifiers/variables can be renamed.

- Type-3 clones: Code fragments may be one or more statements added/modified/ deleted beyond the syntactic similarity.
- Type-4 clones: Code fragments that perform the same calculation with different syntax.

Previous studies reported that software systems may have 5–15 % duplicated code [18], up to 50 % [19]. Based on the level of analysis applied to the source code, the techniques can roughly be classified into four main categories: textual, lexical, syntactic, and semantic [5].

2.2 The Difficulties of Identifying Refactoring Opportunities

Code clone detection can be perceived as the identification of code fragments to be refactored [3]. However, not all clone groups are suitable for refactoring. Usually, large-scale software systems have complicated intertwining logics, which makes it difficult to identify which code clones can be merged and how best to merge them [3].

3 An Approach to Identifying Refactoring Opportunities with Metrics

Other than computing resources, refactoring via function extraction incurs some software maintenance costs by resulting in dependencies. Each dependency means a contract that needs to be maintained by the development team. On the other hand, refactoring via procedure extraction also provides a benefit by resulting in a size reduction, i.e., a smaller number of code lines to maintain for the team. In this section, we derive a method to identify clone groups which are suitable for refactoring by analyzing costs and benefits of refavoring via procedure extraction. This cost–benefit analysis method makes an assumption by assigning the same weight to a dependency and a line of code. These weights can be adjusted by software developers or managers depending on their particular context and needs.

3.1 Benefits

The benefits of Extract Method refactoring are the reduction in the length of cloned code. Herein, we assume that clone group F includes code fragments $f_1, f_2,$ ..., f_m. As a result, the benefit of extracting clone group F can be represented as

$$\text{Benefit}(F) = m \times (|cf| - 1) \tag{1}$$

where |cf| is the number of statements which can be extracted in each fragment of group F. In some cases, there are some non-cloned code which cannot be moved outside the cloned code statements for the dependencies. Therefore, the statements which can be extracted may include both cloned code and non-cloned code. However, procedure extraction produces a procedure call in the original method. Therefore, actually, the length of reduction is equal to |cf| − 1.

3.2 Costs

Coupling is used to indicate the cost of procedure extraction. The principle of strategy for merging code clones is migration of duplicated code to another place. To migrate implemented code, it is desirable that the code has low coupling with its surrounding code [3]. In this paper, we mainly focus on data coupling. Consequently, we calculate the coupling between the original method and the new method (result of Extract Method refactoring) by counting how many parameters are needed by the new method. The detailed formula is shown as follows:

$$\text{Coupling}(F) = \sum_{i=1}^{m} \left(|P(i)_{\text{in}}| \right) + \left(|P(i)_{\text{out}}| \right) \tag{2}$$

where $|P(i)_{\text{in}}|$ and $|P(i)_{\text{out}}|$ are the amounts of the input parameters and output parameters of the new method if clone fragments are extracted from their inclosing method.

For each fragment, we denote the externally defined variables and modified by it as V_w, and externally defined variables accessed but not modified by it as V_r. The variables that appear before the fragments are denoted as V_b, and the variables that appear after the fragments are denoted as V_a. If the fragment is extracted as a new method and called in the original place, variables which appear before the fragment and accessed by the fragment (no matter read or write) should be passed in as input parameters. Those modified by the fragment and accessed by following fragments should be returned as output parameters.

The formulas are shown as follows:

$$P(i)_{\text{in}} = V_b \cap (V_w \cup V_r) \tag{3}$$

$$P(i)_{\text{out}} = V_a \cap V_w \tag{4}$$

In this paper, the $|P(i)_{\text{out}}|$ is 1 or less for we acquire the return value of the new method is no more than 1 in C programming language. If the value is more than 1, then the fragment is not suitable for extracting.

Table 1 The results of identifying clone groups that are feasible for refactoring

Products	The total clone groups (n1)	The clone groups that are feasible for refactoring (n2)	n2/n1 (%)
Linux 2.6.6/arch	5,534	4,573	82.6
Linux 2.6.6/net	2,543	1,939	76.2
Linux 2.6.6/sound/drivers	75	61	81.3
Unix/make 3.82	68	57	83.8
http2.2.2/server	121	81	66.9

3.3 Evaluation of the Benefit and Cost

The ratio of benefit/cost can be represented as

$$R(F) = \begin{cases} \dfrac{\text{Benefit}_{(F)}}{\text{Coupling}_{(F)}} = \dfrac{m \times (|cf| - 1)}{\sum_{i=1}^{m} (|P(i)_{\text{in}}| + |P(i)_{\text{out}}|)}, (\text{Coupling}(F) > 0) \\ \text{Benefit}(F) = m \times (|cf| - 1) \quad (\text{Coupling}(F) = 0) \end{cases} \quad (5)$$

If $R(F) > 1$. then this clone group can be suitable for refactoring or it is not.

In addition, some cloned statements are only composed of declaration statements. These cloned codes are not feasible for refactoring because of the high coupling between the original method and the new method extracted from the original one. We have evaluated all cloned code in the selected open-source programs. The results are shown in Table 1.

4 Future Work

Our results indicate that our approach accurately identify clone groups that are feasible for refactoring. In future work, we hope our study motivates IDEs such as Eclipse CDT and Microsoft Visual Studio to provide functionality to automatically analyze cloned code. We will replicate this study using more systems. In particular, we will extend our study on cloned code analysis to prune more kinds of false positives.

Acknowledgments The research is supported by the National Natural Science Foundation of China under Grant No. 61173021.

References

1. Cai, D., Kim, M.: An empirical study of long-lived code clones. In Proceedings of the 14th International Conference on Fundamental Approaches to Software Engineering: Part of the Joint European Conferences on Theory and Practice of Software, FASE'11/ETAPS'11, pp. 432–446 (2011)
2. Rahman, M.S., Saha, R.K., Krinke, J. Schneider, K.A., Mondal, M., Roy, C.K.: Comparative stability of cloned and non-cloned code: an empirical study. SAC12 March 25–29 (2012)
3. Inoue, K., Higo, Y., Kusumoto, S.: Identifying refactoring opportunities for removing code clones with a metrics-based approach. Create Space (Nov 30, 2011)
4. Juergens, E., Deissenboeck, F., Hummel, B., Wagner, S.: Do code clones matter? In: Proceedings of the 31st International Conference on Software Engineering, ICSE'09, pp. 485–495 (2009)
5. Roy, C.K., Cordy, J.R., Koschke, R.: Comparison and evaluation of code clone detection techniques and tools: a qualitative approach. Sci. Comput. Program. **74**(7), 470–495 (2009)
6. Li, Z., Lu, S., Myagmar, S., Zhou, Y.: CP-miner: finding copy-paste and related bugs in large-scale software code. IEEE Trans. Software Eng. **32**(3), 176–192 (2006)
7. Lozano, A., Wermelinger, M.: Tracking clones' imprint. In: Proceedings of the 4th International Workshop on Software Clones, IWSC'10, pp. 65–72 (2010)
8. Lozano, A., Wermelinger, M.: Assessing the effect of clones on changeability. In: IEEE International Conference on Software Maintenance ICSM, pp. 227–236 (2008)
9. Aversano, L., Cerulo, L., Di Penta, M.: How clones are maintained: an empirical study. In: 11th European Conference on Software Maintenance and Reengineering, CSMR'07, pp. 81–90, March (2007)
10. Hotta, K., Sano, Y., Higo, Y., Kusumoto, S.: Is duplicate code more frequently modified than non-duplicate code in software evolution? An empirical study on open source software. In: Proceedings of the Joint ERCIM Workshop on Software Evolution (EVOL) and International Workshop on Principles of Software Evolution (IWPSE), IWPSE-EVOL'10, pp. 73–82 (2010)
11. Kim, M., Sazawal, V., Notkin, D., Murphy, G.: An empirical study of code clone genealogies. SIGSOFT Softw. Eng. **30**, 187–196 (2005)
12. Saha, R.K., Asaduzzaman, M., Zibran, M.F., Roy, C.K., Schneider, K.A.: Evaluating code clone genealogies at release level: an empirical study. In: 10th IEEE Working Conference on Source Code Analysis and Manipulation (SCAM), pp. 87–96 (2010)
13. Maruyama, K., Omori, T.: A security-aware refactoring tool for Java programs. In Proceedings of the 4th Workshop on Refactoring Tools, WRT'11, pp. 22–28 (2011)
14. Fowler, M.: Refactoring: Improving the Design of Existing Code. Addison Wesley, USA (1999)
15. Murphy-Hill, E., Parnin, C., Black, A.P.: How we refactor, and how we know It. In: Proceedings of the 31st International Conference on Software Engineering, ICSE'09, pp. 287–297 (2009)
16. Liu, H., Ma, Z., Shao, W., Niu, Z.: Schedule of bad smell detection and resolution: a new way to save effort. IEEE Trans. Softw. Eng. **38**(1), 220–235 (2012)
17. Ishio, T., Inoue, K., Sano, T., Choi, E., Yoshida, N.: Finding code clones for refactoring with clone metrics: a case study of open source
18. Roy, C.K., Cordy, J.R.: A survey on software clone detection research. School of Computing TR 2007-541, Queens University, pp. 115 (2007)
19. Rieger, M., Ducasse, S., Lanza, M.: Insights into system-wide code duplication. In: Proceedings of the 11th Working Conference on Reverse Engineering, pp. 100–109 (2004)

Research on Neural Network Predictive Control of Induction Motor Servo System for Robot

Chaofa Yu, Zelong Zhou, Zhiyong Chen and Xiangyong Su

Abstract Neural network control has some applications in many areas; the neural network has strong capability of self-learning, adaptive and fault tolerance, and predictive control for complex systems that has strong adaptability. Through combining the approximation ability of neural network for nonlinear objects and optimization strategy of predictive control, predictive control scheme based on BP neural network has been proposed. In this paper, predictive control algorithm design idea based on BP neural network is proposed: Firstly, by the use of BP neural network model predictive control, the controlled object prediction model can be established, then by taking advantage of the prediction model, based on the input and output information of the current system and the future output values of predict objects, by the use of feedback correction, so as to overcome the model prediction error due to other uncertain disturbance in the system, more accurate predictive value of the object can be obtained. On this basis, based on the future corrected predicted value of the object, with given system output values, the control variable can be scrolling optimized to obtain future system control sequence according to the defined quadratic performance standard. The predictive control has achieved good control effect based on BP neural network; it has proved the feasibility and superiority of this control scheme.

Keywords Induction motor · Vector control · Predictive control · Neural network

C. Yu · Z. Zhou (✉) · Z. Chen · X. Su
Wuhan Mechanical Technology College, No. 1038, Luoyu Road, Hongshan,
Wuhan, Hubei, China
e-mail: zhouzelong2005@163.com

C. Yu
e-mail: cfyu@163.com

Z. Chen
e-mail: zychen@163.com

X. Su
e-mail: xysu@163.com

S. Patnaik and X. Li (eds.), *Proceedings of International Conference on Soft Computing Techniques and Engineering Application*, Advances in Intelligent Systems and Computing 250, DOI: 10.1007/978-81-322-1695-7_18, © Springer India 2014

1 Introduction

Currently, permanent magnet synchronous motor is widely used in the world, but compared with asynchronous motors, high cost, difficult to start, manufacture, use, maintenance requiring a relatively high level of technology are shortcomings, what is more, the induction motor has high operational reliability. Generally, a robot consists of four parts: mechanism, sensors, motor drives, and computer; wherein the motor drive in the robot is the basis for moving; the neural network has the capability of strong self-learning, adaptive and fault tolerance, and predictive control for complex systems that has strong adaptability. Through combining the neural network approximation ability for nonlinear objects and predictive control optimization strategy, the predictive control scheme based on BP neural network has been proposed and good control effect has been obtained, proving the feasibility and superiority of this control scheme.

2 Servo System Modeling

According to the robot dynamics equation:

$$\tau_i = \sum_{j=1}^{n} H_{ij}\ddot{q} + \sum_{j=1}^{n}\sum_{k=1}^{n} h_{ijk}\dot{q}_j\dot{q}_k + G_i \quad (i = 1, 2, \ldots, n) \tag{1}$$

Here,

i stands for robot joint coordinate number;
H stands for robot link inertia matrix;
h stands for coupling matrix between the robot rods;
q stands for robot generalized coordinates;
G stands for robot link gravity matrix.

$$T_{mi} = (J_{ai} + J_{mi} + \eta_i^2 J_{li})\frac{d\omega_i}{dt} + (B_{mi} + \eta_i^2 B_{li})\omega_i \quad (i = 1, 2, \ldots, n) \tag{2}$$

Here

ω_i stands for the ith joint angular velocity of the motor;
$J_{ai} + J_{mi} + \eta_i^2 J_{li}$ stands for the equivalent moment of inertia on the motor shaft of the ith joint;
T_{mi} stands for the motor electromagnetic torque of the ith joint;
η stands for the gearbox reduction ratio on the ith joint ($\eta < 1$);
$B_{mi} + \eta^2 B_l$ stands for the equivalent damping coefficient of the motor shaft on the ith joint.

Assuming the rotating coordinate system is aligned with the stator flux vector ψ_s, then $\psi_{sd} = \psi_s$, $\psi_{sq} = 0$.

System state equations can be expressed as follows:

$$\frac{d\omega}{dt} = \frac{n_p L_m i_{sq} \psi_s}{(J_{ai+}J_{mi} + \eta_i^2 J_{li})L_r} - \frac{(B_{mi} + \eta_i^2 B_{li})}{(J_{ai+}J_{mi} + \eta_i^2 J_{li})}\omega \tag{3}$$

$$\frac{d\psi_s}{dt} = -R_s i_{sd} + u_{sd} \tag{4}$$

$$0 = -R_s i_{sq} - \omega_1 \psi_s + u_{sq} \tag{5}$$

$$\frac{d i_{sd}}{dt} = \frac{1}{\sigma L_s T_r}\psi_s - \frac{R_s L_r + R_r L_s}{\sigma L_s L_r}i_{sd} + (\omega_1 - \omega)i_{sq} + \frac{u_{sd}}{\sigma L_s} \tag{6}$$

$$\frac{d i_{sq}}{dt} = -\frac{1}{\sigma L_s}\omega\psi_s - \frac{R_s L_r + R_r L_m}{\sigma L_s L_r}i_{sq} - (\omega_1 - \omega)i_{sd} + \frac{u_{sq}}{\sigma L_s} \tag{7}$$

Here,

σ stands for the motor leakage coefficient, $\sigma = 1 - \frac{L_m^2}{L_s L_r}$;

T_r stands for the electromagnetic time constant of the rotor, $T_r = \frac{L_r}{R_r}$.

The torque equation is Eq. (2).

3 The Predictive Control Servo System Based on BP Neural Network

3.1 Control Structures

From Fig. 1, this predictive control consists of four parts, of which the model prediction and optimization calculation section have made use of two feed-forward neural networks.

3.2 Neural Network Predictive Model

In predictive control algorithm, it requires an object model to describe the dynamic characteristics of the controlled objection which can predict system output state in the future according to the past and future input and output information of the object. A three-layer BP network can approximate any continuous function and non-continuous function in arbitrary precision, suitable to be used for the controlled object modeling in predictive control.

$$y_m(k) = f[y(k-1), y(k-2), \ldots, y(k-n), u(k), u(k-1), \ldots, u(k-m)] \tag{8}$$

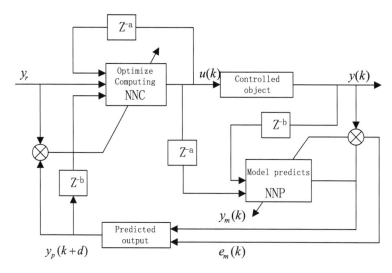

Fig. 1 The neural network predictive control system structure

Here, a predictive model has been taken a three-forward network shown in Fig. 2. There are about $n_s = n + m + 1$ figure nodes in the input layer S, the hidden layer A with p nodes, R as the output layer is a single node, and n and m are the orders of the output value and the input value, respectively. At the time of k, $y(k)$ and $u(k)$ stand for the system output and input, respectively.

Here, the neural network hidden layer output can be expressed as

$$A(k) = f_1[V \cdot S(k)] = [\alpha_1(k), \alpha_2(2), \ldots, \alpha_p(k)]^t \qquad (9)$$

$$\alpha_i(k) = f_1\left(\sum_{l=1}^{n_s} v_{il} s_l(k) + v_{i,n_s+1}\right) \qquad (10)$$

where
A	stands for the hidden layer output vector;
$V \in R^{p \times n_s}$	stands for the connection weight matrix between the input layer and the hidden layer;
$S = [s_1, s_2, \ldots, s_{n_s}]^T$	stands for the input vector;
$f_1(\cdot)$	stands for the transfer function of the hidden layer.

The output of the output layer can be expressed as:

$$y_m(k) = f_2(W \cdot A(k)) = f_2\left(\sum_{i=1}^{p} w_i \alpha_i(k) + w_{p+1}\right) \qquad (11)$$

where
$y_m(k)$ stands for the hidden layer output vector;

Fig. 2 Three-forward neural
network model

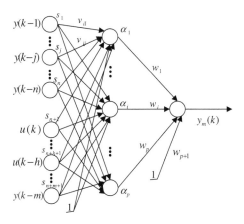

w_i stands for the connecting weight vector between the hidden layer and the
output nodes;

$f_2(\cdot)$ stands for the transfer function of the output layer.

For a complex system with an unknown nonlinear model, after been selected
the neural network model NNP, the input and output sample data can go through
the system training. The step prediction model to describe the system can be
represented by formulas (2) and (4). As for system, multi-step prediction can be
obtained through iteration:

$$y_m(k+d) = f[y(k+d-1), \, y(k+d-2), \ldots, y(k+d-n), \, u(k+d), \\ u(k+d-1), \ldots, u(k+d-m)] \tag{12}$$

In the process of calculating the expression of $y_m(k+d)$, let the weights of the
neural network model be unchanged at the time of k, only consider changing the
size of the control u, in the subsequent step d, the control remains unchanged, that
is $u(k+d) = \cdots = u(k+d-m) = u(k)$, by the way, the expression of some
important parameters cannot be obtained. For example, $y(k+d-1), \ldots, y(k+
d-n)$ cannot be calculated if k is determined. Therefore, the neural network
model predictions $y_m(k+d-1), \ldots, y_m(k+d-n)$ can be used to approximate
the output value of the corresponding object.

3.3 Feedback Correction

In the actual object, time-varying, nonlinear, and various random interferences are
inevitable; therefore, it is impossible to forecast model output fully consistent with
the actual object, while BP neural network is a global network. It is required to use
global information during training; however, the training samples of the real object

are difficult to cover all the circumstances; therefore, between the predicted results and the actual output, there is a certain error between the bound. This requires the predictive models that can be corrected through feedback. In the neural network predictive control algorithm, the actual output in each step must be detected, and the output is compared with the model to form the error message, and then, this error message can be sent to the neural network model for learning and correction. This process can be carried out online or offline. To meet the requirements of real-time control systems, online learning and correction are usually used. Correction algorithm can be expressed as follows:

$$y_p(k+d) = y_m(k+d) + h[y(k) - y_m(k)] \tag{13}$$

where
$y_p(k+d)$ stands for the model predicted output after feedback correction;
h stands for the error correction coefficient; generally, h can be obtained through the trial and error experience.

3.4 Scroll Optimization

Predictive control algorithm optimization calculation is based on the prediction model above; the controlled every step is required to propose optimal requirements to the future finite number of steps in order to achieve optimal control. The standard of the optimum performance is the rolling optimization with the change in time. The most common method is taking the reference point trajectory prediction and outputs, where the squared error is minimized in the optimization process. Its standard form is minimized:

$$\min J_p = \sum_{i=1}^{d} \frac{1}{2} [y_r(k+1) - y_p(k+i)]^2 \tag{14}$$

where
$y_r(k+i)$ stands for the reference trajectory output of the timing of $k+i$;
$y_p(k+i)$ stands for the process output prediction of the timing trajectory of $k+i$.

By formula (14) to calculate the optimal control law using the plurality of future predicted value, the so-called multi-value prediction control algorithm, the calculation of which is a complicated process, it cannot meet the requirements of real-time system. Therefore, a single-value predictive control algorithm has been used, which selects only the d step future predicted values to calculate the optimal control law. Performance index can be expressed by the following formula (15):

$$J_p = \frac{1}{2} \left[y_r(k+d) - y_p(k+d) \right]^2 \tag{15}$$

A forward indicator network (NNC) can be used to optimize formula (15), that is, through online correction of NNC weights to make formula (15) of the optimization index reach its minimum. At this point, the controlled object $u(k)$ can be expressed as formula (16).

$$u(k) = NNC \left[y_r(k+d-1), y_p(k+d-1), \ldots, y_p(k+d-n), u(k-1), \ldots u(k-m) \right] \tag{16}$$

3.5 Neural Network Predictive Control Algorithm Steps

1. Initialize NNC and NNP network parameters and variables.
2. At the time of k, by formula (16) to determine $u(k)$.
3. Take $u(k)$ as the input object to obtain the output $y(k)$.
4. Taking advantage of formula (8) to obtain $y_m(k)$, under the premise of not changing the NNP weights and $u(k)$, through formula (12) to obtain $y_m(k+d)$, and the NNP weights can be obtained in accordance with the gradient correction method.
5. According to the performance indicators of Eq. (15) to change the NNC weights through the gradient correction method, which uses the gradient information $\partial y_p(k+d)/\partial u(k)$ provided by the NNP.
6. Set $k = k + 1$, return to step 2 to continue.

4 MATLAB Simulation and Analysis

The simulation experiment is carried out in MATLAB 7.1. During the experiment, in order to test the performance of the designed neural network predictive control system, a lot of simulation experiments have been completed. The following experiment takes three-phase squirrel-cage induction motor as the controlled object to carry out the simulation experiments. Induction motor nameplate parameters: rated power $P_N = 50 \times 746$ VA; rated voltage $U_N = 460$ V; rated frequency $I_N = 60$ Hz; stator resistance $R_S = 0.087\,\Omega$; stator inductance $L_S = 0.0008$ H; the initial value of the rotor resistance $R_r = 0.228\,\Omega$; rotor inductance $L_r = 0.0008$ H; mutual inductance $L_m = 0.0347$ H; and number of pole pairs $n_p = 2$.

In the experiment, in order to test the performance of the neural network predictive control system, the motor speed output test records have been obtained.

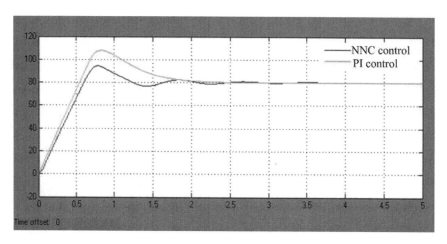

Fig. 3 The normal PID and NN controller predicted response curves at low speed

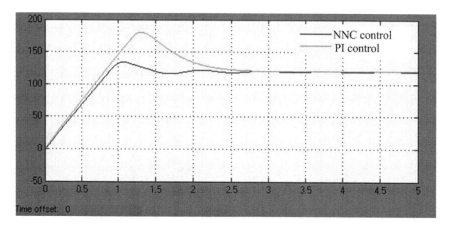

Fig. 4 The normal PID and NN controller predicted response curves at middle speed

Under different conditions, the simulation of the motor has been recorded, the changes in the motor speed, load torque, and the output have been observed.

At a given speed of low, medium, and high speed, the speed curve comparison of the traditional PI control method and the neural network predictive controller have been recorded in Figs. 3, 4, and 5, respectively. The system speed response curve of the traditional PI control has been marked green; the system controller speed response curve of the neural network to predict has been marked blue.

1. Low speed

At the specified speed of 80 rad/s, two induction motor speed controller output curves are shown in Fig. 3.

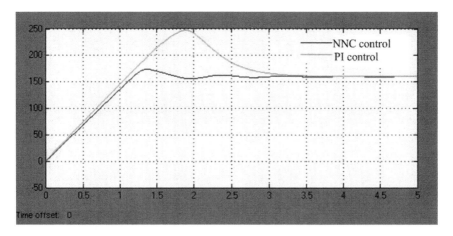

Fig. 5 The normal PID and NN controller predicted response curves at high speed

2. Middle speed

At the specified speed of 120 rad/s, two induction motor speed controller output curves are shown in Fig. 4.

3. High speed

At the specified speed 160 rad/s, two induction motor speed controller output curves are shown in Fig. 5.

From the above comparison response curves and performance indicators, for the regulation time aspect, the neural network predictive control should be faster than the PID control during the motor operation with low, middle, and high speed, what is more, the overshoot is also far less than PID control in the low-, medium-, and high-speed operations. Comparing the neural network prediction system with the conventional PID control system, it is easy to find that the former control quality is better than the latter.

Acknowledgments The work and results discussed in this paper were supported by Youth Science and Technology Foundation of Wuhan Mechanical Technology College.

References

1. Boshi, C.: Electric Drive Control System, pp. 63–64. China Central Radio and TV University Press, Beijing (1998)
2. Licheng, J.: Neural Network System Theory, pp. 21–22. Xi'an University of Electronic Science and Technology Press, Xi'an (1990)
3. Diqian, S.: Predictive Control System and Its Application, pp. 11–12. Mechanical Industry Press, Beijing (1996)

4. Bo, C., Feng, Q., Man Dan, L.: BP network based predictive control algorithm and its application. East China Univ. Technol. 400–404 (2003)
5. Yong, G., Hongye, S., Jian, C.: Recurrent neural network modeling and its application in nonlinear predictive control. Control Decis. 254–256 (2000)
6. Xu, L.: Neural Network Control, 1st edn, pp. 36–37. Electronic Industry Press, Beijing (2003)
7. Seyab, R.K.A., Cao, Y.: Differential recurrent neural network based predictive control. Comput. Chem. Eng. 1533–1545 (2008)
8. Kodogiannis, V.S., Lisboa, P.J., Lucas, J.: Neural network modeling and control for underwater vehicles. Artif. Intell. Eng. 203–212 (1996)
9. Zhu, J.: Intelligent Predictive Control and Its Application, pp. 54–55. Zhejiang University Press, Zhejiang (2002)
10. Yan, D., Ping, L., Xia, L.: Based on BP network model nonlinear predictive control strategy. Comput. Simul. 152–154 (2004)

Research on Scale-Out Workloads and Optimal Design of Multicore Processors

Qiong Wang, Li Shen and Zhiying Wang

Abstract In recent years, cloud computing has been emerging as an infrastructure of online services. Most of the applications deployed on the data center have the typical scale-out features, such as Google search engine, MapReduce, and media streaming. However, while social demand for cloud computing continues to grow, the infrastructure in the data center cannot meet the needs. The inherent characteristics of scale-out workloads place them into a distinct workload from desktop, parallel, and traditional server workloads. Therefore, data center efficiency should be improved by matching the processor design to the needs of the scale-out workloads. In this paper, we test some representative benchmarks of scale-out workloads and find out their performance under different core counts and CPU frequencies. Our work verifies its needs both in low latency and in high throughput. Moreover, we analyze the results and propose several ways to improve the performance of multicore processors.

Keywords Scale-out workloads · Throughput · Latency · Multicore processors

1 Introduction

Driven by application, semiconductor technology, and architecture improvements, multicore processors have been widely used in various fields from high-performance computing to servers, desktop computing, and embedded systems.

Q. Wang (✉) · L. Shen · Z. Wang
State Key Laboratory of High Performance Computing, School of Computer, National University of Defense Technology, 410073 Changsha, China
e-mail: wangqiong@nudt.edu.cn

L. Shen
e-mail: lishen@nudt.edu.cn

S. Patnaik and X. Li (eds.), *Proceedings of International Conference on Soft Computing Techniques and Engineering Application*, Advances in Intelligent Systems and Computing 250, DOI: 10.1007/978-81-322-1695-7_19, © Springer India 2014

Meanwhile, in some fields which have higher standard for throughput, power, and performance, multicore processors have been playing a more important role. For instance, several multicore accelerators have been used in high-performance computing such as NVidia GPU and Intel Xeon Phi. In the latest TOP 500 ranking list [1], most of the supercomputers ranking the top 10 have adopted the "multicore CPU + many core accelerator" heterogeneous systems such as Tianhe 2, Titan, and Stampede. Moreover, the servers based on Tilera Gx [2] series multicore processors can fill the needs of digital media, Internet communication, and other same fields. Undoubtedly, multicore processors will be the first choice of high-performance computing in the long run. However, the diversification in the type of applications brings new challenge to the design of multicore processors. One of them is scale-out workloads. Most of such applications deployed on the data center, such as data serving, MapReduce, media streaming, SAT solver, Web front end, and Web search, all have the typical scale-out features. Along with the data center becoming the economic infrastructure as well as transportation and energy, those applications are playing a vital role in our life.

According to the traditional design methods, modern processors can be classified into two types. One is processor which has less number of cores but complex in order to decrease the latency such as Intel/AMD multicore processors. Another type has more cores but simpler to improve the throughput such as Sun Niagara processors. However, both general-purpose and traditional server processors are all targeting for the characteristics of scale-up workloads. These workloads pursue either a high performance of single threads or a high throughput. Moreover, most of the multicore processor designs follow a trajectory that benefits scale-up workloads, which means, to meet the increasing high-performance demand by adding more computational resources in a single node. However, the scale-out workloads deployed in the data center have distinct characteristics, which bring challenges to the design and optimization of multicore processors. Distinct from existing desktop, parallel, and traditional server workloads, scale-out workloads have some brand new features such as high I-cache miss ratio, low instruction level parallel (ILP), large work sets, and low demand for on-chip and off-chip bandwidth.

In this work, our tests show that scale-out workloads have both needs in single thread efficiency and the number of threads, which on the other side verifies that exiting modern processors cannot support the scale-out workloads efficiently. We test and analyze several representative scale-out workloads' benchmark and find out the influences of core numbers and frequencies on the latency and throughput, proving that scale-out workloads have both needs in single thread efficiency and number of thread numbers. One step further, we propose some methods to optimize the existing multicore processors based on the results to match the needs of scale-out workloads.

The rest of the paper is organized as follows. Section 2 introduces the representative scale-out workloads we test to. Section 3 performs the methodology and test results. Section 4 proposes several methods to the optimal design of multicore processors for scale-out workloads. Section 5 summarizes the related works, and Sect. 6 concludes the work carried out in the paper.

2 Scale-out Workloads

As cloud computing becomes ubiquitous, the number of scale-out workloads based on cloud platform increases at the same time. They all share some similar characteristics as follows: (1) based on the large amount of data sets in the clusters; (2) handle independent user requests having no interact data; and (3) designed specifically for use in cloud infrastructure.

CloudSuite [3] is a benchmark suite of scale-out workloads. It is chosen based on the popularity of online services. It consists of eight most popular application benchmarks in the data center, including data serving, data analytics, Web serving, Web search, media streaming, data caching, graphic analytics, and software test. Those benchmarks operate on real software stack and represent the real system configuration. Our work has chosen two of them to introduce and test.

Data Serving Most of the online services use NoSQL as its huge data storage such as Cassandra, HBase, and PNUTS. They split data into fragments and scale out to clusters. Cassandra is a mixed no-relation database. One of its features is that it is not a database in fact, but a distributed network service that consists of a lot of data nods. For a write operation to the database, the data will be replicated to other nodes. And also a read operation will be routed to some node in the cluster. For a Cassandra cluster, it is easy to improve the performance through adding more nodes. Yahoo! Cloud Serving Benchmark [4] (YCSB) is a framework to test data storage system, which provides test interface for various popular data service system. Its main purpose is to test the cloud service infrastructure to promote comparing emerging cloud data service system. Data serving is a benchmark that consists of YCSB and Cassandra. YCSB sends read or write operation to Cassandra and tests its performance.

Data Analytics Mahout is a framework of machine learning and data mining. Building on Hadoop distinguishes itself from other open source data mining software. It provides some extensible classical algorithm realization in data learning field to support researchers to create intelligent applications. Hadoop is an open source software framework that supports data-intensive distributed applications. It supports the running of applications on large clusters of commodity hardware. It has the reliability, scalability, efficiency, and high fault tolerance features. Since it assumes that the computation elements and storage can fail, it maintains several working data copy to ensure the reconstruction of failure nodes. Moreover, it does its work in parallel to speed up the process. Its bottom layer consists of Google File System (GFS) and Google's MapReduce. Data analytics is a Mahout implementation of machine learning and data mining constructed on Hadoop cluster.

3 Experiment and Analysis

Scale-out workloads deployed on the cloud platform process hundreds and thousands of independent requests from user terminals, which have no sharing data. Therefore, the data center should create enough threads to response and handle the messages. So we conjecture that the more the cores and parallel threads, the higher the throughput we can get. We can verify it by testing the relationship between the number of cores and the application performance. Meanwhile, as users, they would not want to wait too long to get response. Therefore, the performance of a single thread cannot be ignored.

According to the analysis above, we test the needs of two scale-out workloads benchmarks in the throughput and latency respectively. We conduct our study in an Intel Xeon E5350 machine with eight cores and an Intel SandyBridge with four cores. Both of them use CentOs 6.4 with 2.6.2 kernel.

3.1 Throughput

The definition of the throughput is the number of user requests completed within a unit time. Under the same number of user requests, the less time it takes, the higher the throughput. Therefore, we can get the relationship between throughput and cores through testing the time under different number of cores. In the experiment of testing the needs of scale-out workloads in throughput, we increase the number of cores gradually and obtain the time of dealing with all the requests. Figure 1 shows the execution time of data serving and data analytics, respectively, under different number of cores. From the chart, we can see that the time decreases with core count. To make it clearer, we calculate the speedup normalized to one core as shown in Fig. 2. In Fig. 2a, when core number is above eight, the curve has the trend to continue increasing. But for data analytics, the speedup tends to be flat when it reaches the point of eight cores. The reason is that in addition to a lot of threads created to map and reduce, it has hundreds of terabytes of data to deal with.

As for these two applications, under same user requests, the speedup increases as the core number increases, which shows that scale-out workloads have a high demand for core numbers. One of the reasons is that it is determined by the huge amount of users it target to when it is designed, and the other is by its inherent characteristics. As to the scale-out workloads deployed on the cloud platform, it has to create large numbers of threads to response to the requests sent from the users around the world. By increasing the core numbers, the number of hardware threads increases and the throughput increases as well, which reflects its needs in core amount. Modern existing processors are designed specifically for scale-up workloads, which cannot support scale-out workloads efficiently. First of all, almost all the processors develop ILP through out-of-order (OoO) implementation. Researchers increase the depth of assembly line and enlarge the instruction

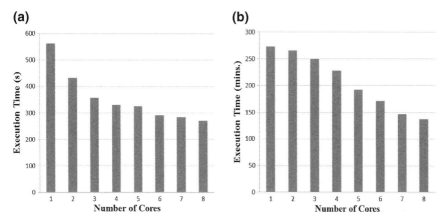

Fig. 1 Execution time varying the number of cores. **a** Data serving. **b** Data analytics

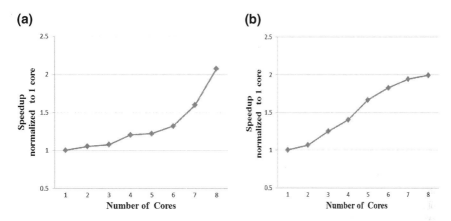

Fig. 2 Speedup varying the number of cores. **a** Data serving. **b** Data analytics

window to develop more ILP, but get trivial benefits since scale-out workloads have limited ILP. Deep assembly line and OoO implementation request large number of hardware to support such as multiple branch prediction, instruction schedulers, forwarding paths, many-port register banks, load–store queues (LSQ), ALUs, reorder buffers (ROB), and other on-chip structures. However, within the limit chip area, the more complex the core is, the less the number of cores and the less the hardware threads. Since scale-out workloads need the multicore processors which are highly computation intensive and power efficient, designers have to distribute the resources to the processors, cache on-chip, and core interconnect rationally.

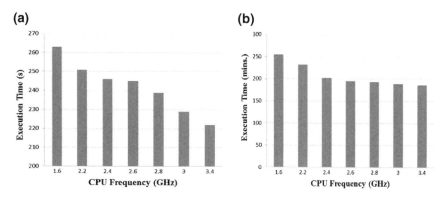

Fig. 3 Execution time varying frequency. **a** Data serving. **b** Data analytics

3.2 Latency

In the experiment of examining the scale-out workloads' needs in single thread performance, we test the time under different frequencies since CPU frequency has influence on its computation speed, and therefore, we can obtain the relationship between core computing performance and the applications. Figure 3 demonstrates the execution time of data serving and data analytics under different frequencies. Figure 4 plots data serving and data analytics speedup under different frequencies normalized to the lowest 1.6 GHz. In Fig. 4a, it shows that for data serving, the speedup increases as the frequency increases, but the highest point does not exceed 1.2 yet. For data serving benchmark, we can find that the highest point in Fig. 4a is not as much as in Fig. 1a, which to some extent shows that the number of cores has more influence on this application than core computation performance. The reason is that Cassandra needs huge amount of threads to respond but relatively simple process to handle. In Fig. 4b, even under the highest frequency 3.4 GHz, the highest point is merely 1.4, which shows limited benefit from high frequency. Above all, the speedup increases when the frequency becomes higher, but not sharp, which means that the better core performance brings trivial benefits to these applications. Therefore, existing deep assembly line and complex core occupies the die area but cannot benefit scale-out workloads. When devise a multicore processor for scale-out workloads, we cannot choose either from the complex core designed for computer intensive applications or simple core designed for traditional server applications.

Scale-out workloads have high demand both in response latency and in throughput, but these two, to some extent, contradict each other. In order to shorten the response latency, computation resources in the core should be increased (such as the number of computation units), which would take more die area at the same time. And in the interest of high throughput, more cores (or hardware threads supported by per-core) should be added in the processors, which would decrease

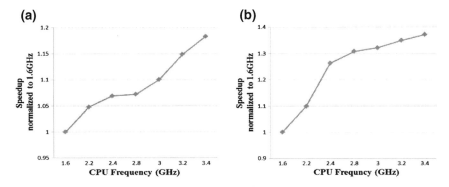

Fig. 4 Speedup varying frequency. **a** Data serving. **b** Data analytics

the per-core areas on chip. Therefore, researchers need to design a multicore architecture, specifically for the scale-out workloads on the limited die area. The architecture researchers have summarized some design experience in pursing the best performance under given area or (and) power constraints.

By far, the automatic design process based on fine-grained accelerator has been built, which can adjust existing instruction set and design a core that meets the demand of single thread application and other corresponding tools. For instance, application-specific instruction set architecture [5] (ASIP) can automatically do the confirmation of extensive instruction, the design of the function unit, and the modification of software tools. Scale-out workloads contain a lot of data level parallel so that performance can be improved by SIMD. Fine-grained accelerator can benefit scale-out workloads in area of function unit, power, and latency and meet its needs in latency, throughput, and power.

4 Optimal Design of Multicore Processors for Scale-out Workloads

Scale-out workloads have huge data sets and complex instruction stream and need high throughput and low latency. However, most existing multicore processors cannot meet its needs. Under such condition, we need to design and optimize the multicore architecture, specifically for scale-out workloads. In future work, we should start from following aspects.

Microarchitecture Most of the scale-out workloads are online services demanding for short response latency and high throughput. However, they cannot fully take advantage of deep assembly line or OoO core due to their limited ILP [6], which conversely wasted many transistor resources on chip. Moreover, in order to improve the throughput, more cores (or threads) should be added, which put constraints on the die area. Therefore, constraints and demands should be considered at the same time to provide enough cores and improve the thread performance. Scale-out workloads contain lots of data level parallel so that performance can be improved by SIMD. Fine-grained accelerator can benefit scale-out workloads in area of function unit, power, and latency and meet its needs in latency, throughput, and power.

Cache hierarchy Scale-out workloads do not have obvious temporality of instruction, leading to a high I-cache miss ratio. Moreover, the instruction working set considerably exceeds the last-level cache, which causes high access latency. We have to find out an optimal cache hierarchy for scale-out workload and consider its impact on power efficiency.

Network-on-chip (NoC). Threads created by scale-out workloads are mostly independent and have few data coherence operations, which makes it need no high bandwidth interconnect on-chip. Simple crossbar construction is enough for the communication between cores when the number of cores is small. However, with the improvement in cores and caches on-chip, the number of cores on the chip can be increased to tens or hundreds. Therefore, a well-suited on-chip interconnect for scale-out workloads should be considered. Dynamic reconfigurable topology NoC should be a good option since it supports reconfiguration of network paths, which can gain a high throughput and decrease network latency at the same time.

Memory controller Modern processors generally have integrated high-performance memory controller as a medium between processors and memory to overcome the timing and resource constraints brought by the storage devices and to realize the access efficiency to the memory. Current DDRx memory controllers basically adopt fixed hardware logic units including complex address mapping logic, request scheduling logic, power management, and updated algorithms. However, fixed hardware logic implementations cannot fill the needs of scale-out workloads. We propose to use a programmable memory controller, which can enhance the memory system ability of adjusting to the various applications.

5 Related Work

Existing multicore processors are designed specifically for scale-up workloads satisfying the emerging needs by adding more computation resources. However, scale-out workloads deployed on the data center have shown some distinct characteristics that bring new challenge to the design and optimization of multicore processors. Kgil et al. [7] show that modern processors is power inefficiently for the Web applications which emphasize more on high throughput. As scale-out workloads become ubiquitous, researchers start to analyze its features. Some of them are from system levels [4, 8–10], and the others are from the microarchitecture [11, 12] levels. There are some methods of design oriented to applications in multicore processors and use it in the design of commercial processors, but no matter the product construction itself or usage efficiency all have space to improve. More importantly, those methods are all aimed at improving the scale-up workloads, which cannot be used to design scale-out workloads' processors directly.

Until now, researchers have found some preliminary characteristics of scale-out workload and proposed some custom-made strategy of multicore processors gaining some certain benefits, but there is still a long way to summarize the relative design and optimization theory. Therefore, we need to analyze the scale-out workloads and explore the design and optimization method of multicore processors. Ferdman et al. [6] test and analyze the I-cache miss ratio, ILP, and bandwidth usage of scale-out workloads. They point out that the huge amount of data set exceeds the size of cache on-chip, but existing hierarchical cache takes a lot of die area but cannot support its efficient implementation. Oh et al. [13] show that the time spent on the last level occupies half of the data stall, which means that existing cache hierarchy is not rational. Our results corroborate these findings, showing that we should increase more cores to improve the throughout on the limited die area.

6 Conclusions

Scale-out workloads have both demands in low latency and in high throughput, and its inherent characteristics distinct from traditional workloads bring new challenges and opportunities to the design and optimization of multicore processors. As scale-out workloads become ubiquitous, its impact on social life is growing as well. However, existing multicore processors are designed specifically for scale-up workloads, which cannot meet its needs in computation density and power efficiency. To design multicore processors for scale-out workloads becomes a challenge researchers confronted to. In this work, we test several representative benchmarks of scale-out workloads and prove that scale-out workloads have the needs both in low latency and in high throughput, which provides experience in the multicore architecture design for scale-out workloads in the further study. And we propose several aspects of optimization and design of multicore architecture, specifically for scale-out workloads.

Acknowledgments This work is supported by the National Basic Research Program of China (863 Program) under Grant No. 2012AA0-10905, National Natural Science Foundation of China under Grant No. 61272143.

References

1. TOP500 supercomputer sites. http://www.top500.org (2013)
2. Tilera Inc. http://www.tilera.com (2013)
3. Cloudsuite. http://parsa.epfl.ch/cloudsuite (2013)
4. Cooper, B.F., Silberstein, A., Tam, E., Ramakrishnan, R., Sears, R.: Benchmarking cloud serving systems with YCSB. In: Proceedings of the 1st ACM Symposium on Cloud Computing, June 2010

5. Keutzer, K., Malik, S., Newton, A.R.: From ASIC to ASIP: the next design discontinuity. In: ICCD'02, pp. 84–90 (2002)
6. Ferdman, M., Adileh, A., Kocberber, O., et al.: Clearing the clouds, a study of emerging scale-out workloads on modern hardware. In: ASPLOS (2012)
7. Kgil, T., D'Souza, S., Saidi, A., Binkert, N., Dreslinski, R., Mudge, T., Reinhardt, S., Flautner, K.: PicoServer: using 3D stacking technology to enable a compact energy efficient chip multiprocessor. In: Proceedings of the International Conference on Architectural Support for Programming Languages and Operating Systems, Oct 2006
8. NVIDIA Tesla Computing Processor. http://www.nvidia.com/docs/IO/43395/NV_DS_Tesla_C1060_US_Jan10_lores_r1.pdf
9. Li, A., Yang, X., Kandula, S., Zhang, M.: CloudCmp: comparing public cloud providers. In: Proceedings of the 10th Annual Conference on Internet Measurement, Nov 2010
10. Kozyrakis, C., Kansal, A., Sankar, S., Vaid, K.: Server engineering insights for large-scale online services. IEEE Micro **30**(4), 8–19 (2010). (July–Aug)
11. Janapa Reddi, V., Lee, B.C., Chilimbi, T., Vaid, K.: Web search using mobile cores: quantifying and mitigating the price of efficiency. In: Proceedings of the 37th Annual International Symposium on Computer Architecture, June 2010
12. Tang, L., Mars, J., Vachharajani, V., Hundt, R., Soffa, M.L.: The impact of memory subsystem resource sharing on datacenter applications. In: Proceeding of the 38th Annual International Symposium on Computer Architecture, June 2011
13. Oh, T., Lee, H., Lee, K., Cho, S.: An analytical model to study optimal area breakdown between cores and caches in a chip multiprocessor. In: Proceedings of the IEEE Computer Society Annual Symposium on VLSI, May 2009

Study of Modified Montgomery's Algorithm and Its Application to 1,024-bit RSA

Yulin Zhang and Xinggang Wang

Abstract The modified Montgomery algorithm made the modular multiplication simple using addition and shifting. We introduce four-to-two CSA architecture to design the modified Montgomery's algorithm. It cannot convert the carry-save form of an operand into its binary representation at each end of modular multiplication. At the end of RSA, we use the basic 4-CPA to accomplish the data form conversion. As a result, our design can reduce the operating time. This architecture adapts to the single chip for key lengths in excess of 1,024 bits of RSA.

Keywords RSA · Modified montgomery · Four-to-two CSA

1 Introduction

Safe and quick encryption algorithms have got more and more favor of people. RSA [1] algorithm is becoming gradually popular for its safety and ease-of-use functions. There is wide application prospect especially for realizing RSA arithmetic with hardware. However, operation at a slower speed has become its biggest defect. Although, the RSA algorithm's demand of processor speed is not high, in order to achieve rapid decrypted operation, the design must be inevitable complex and costs higher. Now the FPGA technology is developing rapidly, and slice-loaded resources grow into geometric progression; these conditions have laid solid foundation for realizing RSA arithmetic with hardware.

Y. Zhang (✉) · X. Wang
Shandong Provincial Key Laboratory of Network Based Intelligent Computing,
250022 Jinan, People's Republic of China
e-mail: ise_zhangyl@ujn.edu.cn

X. Wang
e-mail: wxg23@163.com

S. Patnaik and X. Li (eds.), *Proceedings of International Conference on Soft Computing Techniques and Engineering Application*, Advances in Intelligent Systems and Computing 250, DOI: 10.1007/978-81-322-1695-7_20, © Springer India 2014

Modular multiplication is the core of RSA algorithm. In order to guarantee the password intensity, usually needs large number of calculation, such as 512- or 1,024-bit integer. It is hard to improve performance using hardware to realize RSA algorithm because the modulus by large number costs a lot of hardware resources. In numerous modulus exponentiation algorithms, Montgomery algorithm [2] is the most effective one. It uses a lot of additions and divisions by a power of 2 instead of the trial division. This advantage makes Montgomery algorithm in hardware implementation more effective.

A number of hardware implementations of Montgomery algorithm have been reported using CSA [3–6] (carry-save-additions/adders). These can be divided into two categories. In the first method, the intermediate results are kept in carry-save form to avoid carry propagation. But at the end of each modular multiplication, it is needed to convert the data of carry-save form to binary representation in [3, 4]. On the contrary, some implementations used five-to-two or four-to-two carry-save-additions/adders [5, 6]. In our design, we use the modified Montgomery algorithm [3–6], which avoids the final comparison a subtraction and the four-to-two CSA.

The rest of the paper is organized as follows. Section 2 deals with the RSA cryptosystem. Section 3 presents our design and analyzes our performance. Finally, Sect. 4 concludes the paper.

2 RSA Cryptosystem

In 1978, R. L. Rivest, A. Shamir, and L. Adleman developed the RSA algorithm [1]. It is an encryption system based on modular exponentiation and widely used public key cryptosystem. The modulus N is the product of two large primes p and q. The relationship between public key e and the private key d is $ed \bmod (p - 1)(q - 1)$. The encryption operation using the public key e is as follows:

$$C \equiv M^e \bmod N \ (0 < M < N)$$

The M is the plaintext, and C is the ciphertext. So the C is decrypted using the private key d, as follows: $M \equiv C^d \bmod N$.

2.1 Montgomery Algorithm and Modified Montgomery Algorithm

The procedure of RSA using Montgomery algorithm is shown in Fig. 1. It has three operations [6]: mapping, modular exponentiation, and remapping.

Each of the operations should be processed by Montgomery modular multiplication. The factors of Montgomery modular multiplication are N-residues A,

Fig. 1 RSA uses Montgomery algorithm

N-residues B and r such that $r = 2^n$. The result $Mont(A, B, N)$ [3] is $ABr^{-1} \bmod N$. The modular multiplication is shown as Algorithm I:

where the A is $ar \bmod N$, and the B is $br \bmod N$. In this algorithm, the operation of "$if (R > N) \ R = R - N$" can be removed without changing any other steps. The method is that the size of r must be two bigger than N.

Algorithm I $Mont(A, B, N)$

$$R = 0$$
$$for \quad i = 0 \ to \ n - 1 \ do \backslash \{$$
$$q_i = (R_0 + A_i B_0) \bmod 2;$$
$$R = (R + A\_\{i\}B + q\_\{i\}n);$$
$$R = R/2; \backslash \}$$
$$if (R > N) \ R = R - N;$$
$$return(R);$$

So it calls the modified Montgomery modular multiplication. The result $MMont(A', B', N)$ is $A'B'2^{-(n+2)} \bmod N$ (A': $a2^{(n+2)} \bmod N$; B': $b2^{(n+2)} \bmod N$). The details of this algorithm are as follows:

Algorithm II $MMont(A', B', N)$

$$R = 0$$
$$for \quad i = 0 \ to \ n + 1 \ do \{$$
$$q_i = (R_0 + A_i B_0) \bmod 2;$$
$$R = (R + A_i B + q_i n);$$
$$R = R/2; \backslash \}$$
$$return(R);$$

2.2 Four-to-two CSA

The critical delay of Algorithm I and Algorithm II is the addition of tree input data.

$$R = R + A_i B + q_i n \tag{1}$$

The carry propagation will cost lot of time using serial adder. We can calculate (1) using the CSA. At the end of each modular multiplication, we do not change the data from carry-save form to its binary form, because this will cost a lot of system clocks to do. In our design, we choose the carry-save form data as the input of modular multiplication. Therefore, the number of input data of modular multiplication is changed from three to five. We can use the four-to-two CSA or five-to-two CSA [3].

In our design, we use the four-to-two CSA to rewrite the Algorithm II as the Algorithm III. The input operands A' and B' are represented in carry-save format as $A1$ and $A2$, $B1$ and $B2$. The output operand R is represented as S and C. Carry-save representation is short for CSR. The Algorithm III is given below.

2.3 Modified R-L Binary Algorithm

There are two methods to implement the RSA in [2]. One is calling the L-R binary method and another is R-L method. The L-R binary method uses one *MMont* modular multiplication. This reduces the hardware resources, but the speed improvement is slow. Another method implements two *MMont* modular multiplications. It can get higher speed but engrosses much more hardware resources.

The Algorithm III: Four-to-two $(A1, A2, B1, B2, N)$

$$(M1, M2) = CSR(B1 + B2 + N + 0);$$
$$S = 0;$$
$$C = 0;$$
$$AC = 0;$$
$$for\ i = 0\ to\ n + 1\ do\{$$
$$\quad A_i = A1_i \otimes A2_i;$$
$$\quad AC = A1_i A2_i + A1_i AC + ACA2_i;$$
$$\quad q_i = (S[0] + C[0]) + (A_i(B1[0] + B2[0]))\ \text{mod}\ 2;$$
$$\quad if\ A_i = 0\ and\ q_i = 0$$
$$\quad\quad (S, C) = CSR(S + C + 0 + 0)div2;$$
$$\quad elseif\ A_i = 1\ and\ q_i = 0$$
$$\quad\quad (S, C) = CSR(S + C + B1 + B2)div2;$$
$$\quad elseif\ A_i = 0\ and\ q_i = 1$$
$$\quad\quad (S, C) = CSR(S + C + N + 0)div2;$$
$$\quad else$$
$$\quad\quad (S, C) = CSR(S + C + M1 + M2)div2;$$
$$\}$$
$$return(S, C);$$

In our design, we use the R-L method to implement the RSA algorithm. So the RSA algorithm can be rewritten as the Algorithm IV using the four-to-two algorithm. CPA stands for carry propagation adder. It is shown in below.

Algorithm IV: *Modified R − L Multiplier Modular*

$$Exponentition(M, e, N)$$

$$K = 2^{2(n+1)} \bmod N; (computed \ externally)$$
$$(P_s, P_c) = Four - to - two(K, 0, M, 0, n);$$
$$(R_s, R_c) = Four - to - two(K, 0, 1, 0, n);$$
$$for \ i = 0 \ to \ e_k \ do\{$$
$$\quad (P_s, P_c) = Four - to - two(P_s, P_c, P_s, P_c, n);$$
$$\quad if \ e[i] = 1$$
$$\quad \quad (R_s, R_c) = Four - to - two(R_s, R_c, P_s, P_c, n);$$
$$\}$$
$$(C_s, C_c) = Four - to - two(1, 0, R_s, R_c, n);$$
$$C = CPA(C_s + C_c);$$

3 Algorithm Implementation

3.1 Four-to-two Architecture

In our design, we use the four-to-two CSA to calculate (1). The important parts are the architecture of calculating A_i and the CSR architecture.

The determination of the correct A_i values means that we can implement a full adder. The input operands are $A1_i$, $A2_i$, and AC. The register AC stores the carry. We can use a full adder and shift register. It is shown in Fig. 2. The operand of A_i must be ready when it is needed in the loop. If the current A_i is used, the shift register should be shifted one place to the right, which prepares for the next operation of calculating A_i [3]. This operation does not have any extra clock cycles. In a word, it takes only 1,027 clock cycles in the Algorithm III.

The architecture of CSR is shown in Fig. 3, and the part of CSA is shown in Fig. 4. In our design, the CSA has n full adders so it can calculate the result in one clock cycle. The logic of the FA is as follows:

$$s_i = x_i \otimes y_i \otimes z_i \tag{2}$$

$$c_{i+1} = x_i y_i + x_i z_i + y_i z_i \tag{3}$$

Fig. 2 Calculate the operand
of A_i

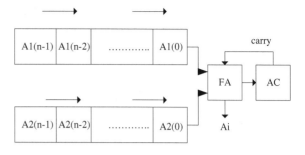

Fig. 3 Architecture of CSR

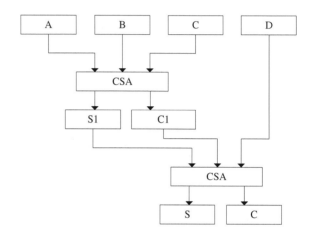

Fig. 4 Architecture of CSA

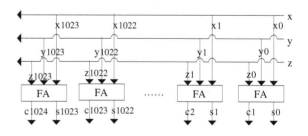

In order to make the implementation consistently, the value of c_0 is 0. Because the value of c_{n-1} is 0, it can be lost.

In Algorithm III, we take more than two register to store the values of $CSR(B1 + B2 + N + 0)$ at the beginning. The four input operand can be got by judging the value of A_i and q_i. So Algorithm III should take 1,027 clock cycles.

Fig. 5 Architecture of RSA

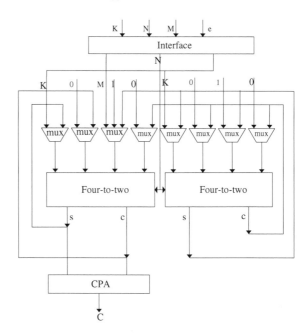

3.2 Modified R-L Binary Architecture

The RSA using the Montgomery algorithm has three processes. In R-L binary method, the architecture has two four-to-two architecture. These are performed in parallel [3]. Figure 5 illustrates that the RSA uses the modified four-to-two Montgomery multiplier in R-L method. The step of mapping can be performed after only one multiplication because the two multipliers are in parallel. At the modular exponentiation stage, we can control the input operands to accomplish the modular exponentiation. At the same time, the multiplication and squaring is performed independently. We compute the temporary result after the remapping step. Finally, we use the CPA to carry out the final result.

The final step in the RSA is converting the carry-save form of an operand into its binary representation. We can use the common 1,024-adder, but it exist the delay of carry propagation during the addition stages. So we implement the basic 4-CPA to accomplish the data form conversion. The detail of CPA is shown in Fig. 6.

In our design, each multiplication takes only $(n + 3)$ cycles. In mapping and remapping processes, the total number of modular multiplication is 2. Modular exponentiation stage is n. Converting the carry-save form of an operand into its binary representation takes $(n/4)$ clock cycles. So accomplishing the RSA, we should take $(n + 3)(n + 2) + n/4$ clock cycles. It takes about 1.05 M clock cycles where n is 1,024. As a result, the total operating time is 21 ms at 50 MHz. Table 1 shows the comparison between our design and [3].

Fig. 6 Detail of CPA

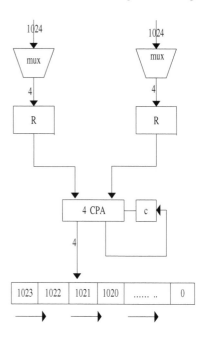

Table 1 Comparative results of RSA

Process	Items	
	[3]	Ours
Mapping	$(n + 2 + 32)$	$(n + 3)$
Modular exponentiation	$n(n + 2 + 32)$	$n(n + 3)$
Remapping	$(n + 2 + 32)$	$(n + 3)$
Form conversion	0	$(n/4)$

4 Conclusions

In this paper, we studied modified Montgomery's algorithm and its application to 1,024-bit RSA. The modified Montgomery algorithm made the modular multiplication simple using addition and shifting. We introduce four-to-two CSA architecture to design the modified Montgomery's algorithm. It cannot convert the carry-save form of an operand into its binary representation at each end of modular multiplication. At the end of RSA, we use the basic 4-CPA to accomplish the data form conversion. As a result, our design can reduce the operating time comparison [3]. This architecture adapts to the single chip for key lengths in excess of 1,024-bits of RSA.

References

1. Rivest, R.L., Shamir, A., Adleman, L.: A method for obtaining digital signature and public-key cryptosystems. Commun. ACM **21**, 120–126 (1978)
2. Koc, C.K., Acar, T., Kaliski Jr, B.S.: Analyzing and comparing Montgomery multiplication algorithms, vol. 16. IEEE Micro, pp. 26–33 (1996)
3. Kwon, T.W., You, C.S., Heo, W.S., Kang, Y.K., Choi, J.R.: Two implementation methods of a 1024-bit RSA cryptoprocessor based on modified Montgomery algorithm. Proceedings of IEEE International Symposium on Circuits System, vol. 4. pp. 650–653 (2001)
4. Cilardo, A., Mazzeo, A., Romano L., Saggese, G.P.: Carry-save Montgomery modular exponentiation on reconfigurable hardware. IEEE Proceedings of the conference on Design, automation and test in Europe-vol. 3. pp. 206–211 (2004)
5. McIvor, C., McLoone, M., McCanny, J.V.: Fast Montgomery modular multiplication and RSA cryptographic processor architectures. 37th Asilomar Conference on Signals, Systems and Computers, vol. 1. pp. 379–384 (2003)
6. Shieh, M.D., Chen, J.H., Wu, H.H., Lin, W.C.: A new modular exponentiation architecture for efficient design of rsa cryptosystem. IEEE Transactions on Very Large Scale Integration (VLSI) Systems, vol. 16. pp. 1151–1161 (2008)

A MVS-Based Object Relational Model of the Internet of Things

Huijuan Zhang and Ran Xu

Abstract Currently, there are lacks of relational networks consistent with the characteristics of data of the Internet of things. According to this situation, this paper studies on current complex network model and proposes a network model, which is based on data relationship of the Internet of things. The construction of this data relational model focuses on improving the MVS model. It implements the construction and expansion of basic relational network. Meanwhile, the properties of the Internet of things' calculation is considered in algorithmic model make the model well reflect the intimacy of the relationship between objects in network. It is a relatively simple calculation. Experiments show that the relational network model can achieve the purpose of the network growth within a certain range, replacement of network nodes and better reflecting the intimacy of the relationship between objects.

Keywords Internet of things · Relational network · MVS model · Social network

1 Introduction

The perspective of the Internet of things [1] development is organic integrate the physical space, information space and social space. Current information perception, communication, computers, and other technological developments make it possible for physical space information sensing, communications, storage, and analysis [2–5]. The key problem of implementation these three spaces information

H. Zhang (✉) · R. Xu
The School of Software Engineering, Tongji University, Shanghai 201804, China
e-mail: mszhj@tongji.edu.cn

R. Xu
e-mail: xuran2334@gmail.com

S. Patnaik and X. Li (eds.), *Proceedings of International Conference on Soft Computing Techniques and Engineering Application*, Advances in Intelligent Systems and Computing 250, DOI: 10.1007/978-81-322-1695-7_21, © Springer India 2014

integrate is mining characteristics and analyzing the regular pattern of relationship between objects in physical space based on data of the Internet of things. Constructing the object relational network model [6] is the core solution of the problem.

Currently, relational network model study focuses on social network and computer network. Its universal idea is to use the complex network model to deal with the corresponding problems in these fields, especially in social network [7]. Compared with traditional regular network, complex network model has characteristics of randomness and growth [8]. Randomness reflects the uncertainty of network objects' relationship, and growth reflects the dynamic change in network objects' relationship. Watts and Strogatz described that the complex network has small-world property [9]. The network has short average path length. Barabasi add BA model proposed by Albert [10–12]. This model describes that the complex network's degree distribution is power law distribution and the mechanism of node contact is degree priority. Toivonen et al. [13] used the theory that people prefer to communicate with people who have same properties with them to improve the BA model and built the THOSK model. This model aims at a broad degree distribution and a high clustering coefficient. It successfully describes the community structure of social network [14–16].

Above methods of data relational models construction which based on complex network mainly used in social networks, communication networks and transportation networks and other fields. These fields tend to rely on single data source. Interaction between data emphasizes the communication between various people and objects in the physical world. Currently, there are very little researches on construction of complex network model in this field. In this paper, according to the above characteristic of the Internet of things, study the current complex network model and construct the data relational model. The model that is proposed in this paper has the following features: (1) Implement the relational network expansion and nodes' link replacement; (2) well reflect the relational intimacy of objects in relational network; and (3) the calculation of model is relatively simple.

2 MVS-Based Object Relational Model of the Internet of Things

This paper proposes an object rational model, which is based on the characteristics of data of the Internet of things. The thinking and the process of implementation the model: Phase (1) Construction of basic data model: According to the construction algorithm of MVS model, first contact nodes with random probability. After that, use the triadic closure mechanism that nodes make new contacts through searching its neighbor's neighbor nodes to expand the existing network. With this mechanism, the rational network is constructed. Phase (2) Calculate the characteristics of the Internet of things: calculate the properties of objects in the

Fig. 1 Three nodes rational closure

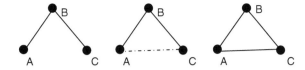

Internet of things. Calculate the properties of objects in the Internet of things. For example, make data of time, space, and physical space situational information as parameters. Use these parameters in calculated process of construction of basic data rational model in phase (1).

In the process of constructing rational network, we need to quantify the changes of objects' properties in the Internet of things. Using these quantifications in calculation can make the rational model suitable for the Internet of things.

2.1 Construction of Basic Data Model

The theories of construction of basic data model are based on triadic closure mechanism: Node A is friend of node B, and node B is friend of node C, making it possible that A and C are friends. According to this theory, the three nodes rational closure can be built as following figure.

In this paper, the algorithm of construction of basic data model is improved from MVS model algorithm. The theories of algorithm: Use the MVS model algorithm to construct basic rational network. In the process of model's evolution, the new nodes are constantly added in the network. Meanwhile, the new rational contacts are inserted between original nodes and new nodes to form new triadic closure. It implements the expansion of relationship between nodes and network. The algorithm uses random selection and prioritized selection in the construction process. Typically, the new nodes and the nodes that have less degree use the random selection. It let the calculation become simple. On the other hand, the node that has relatively large degree uses prioritized selection. It ensures that the node can make contact with the node, which has intimacy relationship with itself (Fig. 1).

In the above constructed algorithm, the nodes in network represent the object in the Internet of things. Each node has a contact list to store its neighbor nodes' information. Rational contact represents the relationship between two nodes. When a new contact is created, the connected two nodes add the other node in its contact list; when a contact is deleted, the two nodes remove the other node from its contact list.

Specifically, the implementation of MVS-based object relational model of the Internet of things constructed algorithm is below:

(1) Initially, there are N individual nodes in the network and every nodes' contact list is null.

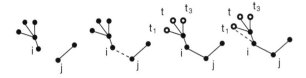

Fig. 2 (1) Randomly contact node i and j; (2) Search node i's neighbor's neighbors t_1, t_2, and t_3. For each node, make prioritized selection; (3) Contact node t_1 with biggest probability ξ (i, t_1)

(2) Randomly select node i and node j, make a contact between them, and add each nodes' information into their contact list.
(3) Search every node i's neighbors' neighbors t_i. For every node t_i which does not have contact with i, calculate their contact probability $\xi(i, t_i)$. Select the node t_i which has the biggest $\xi(i, t_i)$ and make contact with node i. Update their contact lists.
(4) Repeat step (3) and step (4) to construct a rational network.

(Figure 2 show the process)
Nodes' contacts removed and replaced mechanism must be included in the algorithm, to avoid ending up with a fully connected network. Relationship intimacy of nodes should be considered during the process. This problem will be described in the next section.

2.2 Calculation of Properties' Characteristics of the Internet of Things

To better reflect the characteristics of data exchange and relationship intimacy of nodes, the properties of the Internet of things are added in model calculation. Data of time, space, and physical space situational information as parameters and use them in construction of data rational model in Sect. 2.1.

The study of this paper focuses on the impacts of time information and distance information on nodes' relationship. The basic theories: Calculate these two properties separately to get these two nodes' contact relationship probability function. Use this function to calculate the probability of contacting.

Basic properties calculation

Definition 1 *Distance information $d(i, j)$* It means the distance between node i and node j (A node represents an object in network). The $d(i, j)$ is smaller, show that the nodes are closer. It shows that they have higher relationship intimacy.

Definition 2 *Time information 1 $\Delta T(i, j)$* It describes the time information of node i and node j. Situation 1: The differences in the moments of these two nodes appear at the same location. If $\Delta T(i, j)$ is smaller, then their relationship intimacy is higher.

Definition 3 *Time information 2* $\Phi T(i, j)$ It describes the time information of node i and node j. Situation 2: The time length of these two nodes' interaction. If ΦT (i, j) is higher, then their relationship intimacy is higher.

Function 1: The Relationship intimacy function based on definitions 1, 2, and 3:

$$F(i,j) = 1 - e^{\frac{n(i,j) \times \Phi T(i,j)}{d(i,j)}}$$

Typically, the original relationship intimacy function is $F(i, j) = 1 - e^{-n(i, j)}$. It reflects the relationship intimacy of node i and node j. $n(i, j)$ means the number of mutual neighbors of node i and node j. The higher the $n(i, j)$ and the higher the $F(i, j)$, the higher the relationship intimacy. According to the characteristics of the Internet of things, this paper improves the original relationship intimacy function to propose the new relationship intimacy function (Function 1) as definitions 1, 2, 3.

Function 2: The contact probability function based on nodes' basic properties:

$$\eta(i, t) = \frac{F(i,t)}{\sum_k F_i(i,k)}$$

This function describes the relationship intimacy of node i and its specific neighbor's neighbor node t divides the sum of node i and all its neighbors' neighbors' relationship intimacies. If $\eta(i, t)$ is higher, the node t has higher relationship intimacy with i in all of i's neighbor's neighbors' nodes.

As above definitions and functions, $\eta(i, t)$ can be used as the contact probability function for random selection in 3 step of the model constructed algorithm (Sect. 2.1). Use the triadic closure mechanism and relationship intimacy function to insert new contacts to expand the network.

The description at the end of 2.1 section shows contacts in relational network unlimitedly increase will end up with a fully connected network (all nodes contact each other). The original MVS algorithm randomly deletes some contacts in network to avoid this. This paper chooses the replaced mechanism, which complies with the characteristics of the Internet of things to prevent the fully connected network. Contacts' replacement relies on the relationship intimacy function (Function 1). When the size of nodes in contact list reaches a fixed count, compare the relationship intimacy of new contact node with originals' to decide the contact replacement.

Specifically, the steps of replacing the contacts are below:

1. Node i and node j will contact, and the size of nodes in contact list has received max count M.
2. Traverse node i's contact list. For each r_t ($0 < t \leq M$) node in contact list, compare the $F(i, j)$ and $F(i, r_t)$
3. For all of the nodes r_t for which $F(i, r_t)$ is smaller than $F(i, j)$, get the smallest $F(i, r_t)$ and replace node r_t by j
4. No $F(i, r_t)$ smaller than $F(i, j)$. The node j is rejected.

Each node's contact list maximum count M is decided by some properties of object which represented by node.

Situational properties calculation During analysis of the physical world by the theories of the Internet of things, and important element should be considered: situation. For example: Umbrellas used to shelter the rain in rainy but housing the sunlight in sunny day.

The situational parameter set is θ. It can be used as one of the properties to determine two nodes' relationship intimacy.

Function 3: Situation 'm' ratios' function:

$$\Theta(i)m = \frac{\theta m}{\sum_k \theta_k}$$

Ratio of situation m: $\Theta(i)_m$ describes the proportion of situation 'm' weight in all of node i's situations. The bigger the $\Theta(i)_m$, the object occurs in this situation more frequently.

The study of situation factors impact on objects' relationship intimacy focus on two points: The number of these two objects' mutual situations and the approaching degrees of their mutual situations' $\Theta(i)_m$. The mutual situations are more and the approaching degrees are higher, their relationship intimacy is higher. So this paper uses $\lambda(i, j)$ to represent two nodes' situational relationship function:

Function 4: Two nodes' situational relationship function:

$$\lambda(i,j) = \frac{\sum_{N(\Theta(i),\Theta(j))} |\Theta(j)_a - \Theta(j)_a|}{N(\Theta(i),\Theta(j))}$$

In this function, the denominator $N(\Theta(i), \Theta(j))$ means the count of node i's and node j's mutual situations and the numerator $\sum_{N(\Theta(i),\Theta(j))} |\Theta(i)_a - \Theta(j)_a|$ describes the sum of absolute value of situational ratio differences in their mutual situations. The numerator is smaller and the denominator is larger, which indicates that the $\lambda(i, j)$ is smaller and they have higher relationship intimacy.

In summary, the situational relationship function $\lambda(i, j)$ and nodes' relationship intimacy have monotone decreasing relationship.

According to the above-mentioned research, the node's situation influence function can be defined as below:

Function 5: Node's situation influence function:

$$g(i,j) = \frac{1}{1 - e^{-1}}\left(e^{-\lambda(i,j)} - e^{-1}\right)$$

Calculating basic properties (time data, distance data and number of mutual neighbors) and situational properties can get a new contact probability function: $\xi(i,j) = \eta(i,t)g(i,j)$. It combines the contact probability function based on nodes' basic properties (Function 2) and the node's situation influence function (Function 5). We can apply it at step 3 of the constructed algorithm as the $\xi(i, j)$ to make prioritized selection.

The constructed algorithm of object relational model of the Internet of things not only implement the network's expansion and contact replacement but also add objects' basic properties, data's characteristics and situational properties in network calculation. It let the constructed rational network consistent with characteristics of objects in the Internet of things.

3 Experimental Verification and Conclusion

This paper uses 200 nodes which have above parameters to design experiment for theories verification.

For these 200 data nodes, this paper adjusts their range of parameters by different locations, time, and situations. The experimental result: Network constructed by algorithm of MVS-based object relational model of the Internet of things is below:

The experimental result show: The 200 data nodes which divided into 6 categories can form a completed rational network after calculation. The nodes in same category relatively aggregate in network. It indicates the nodes in same category have higher relationship intimacy. Figure 3 describes the relationship between different nodes.

This paper proposes the MVS-based object relational model of the Internet of things. After improving the MVS model algorithm, the network model that is constructed by the algorithm not only corresponds with characteristics of the Internet of things but also can fulfill the network expansion and contact replacement and well reflects the nodes' relationship intimacy. The calculation is relatively simple. Through the experimental verification, these purposes are proved to be achieved. This paper just considers the common properties of the objects represented by nodes. In the future work, the specific properties of different nodes should be studied to make the constructed network to better reflect the objects' characteristics. It is helpful to more precisely define the nodes' relationship.

Fig. 3 Experimental result:
relational network
constructed by 200 data
nodes

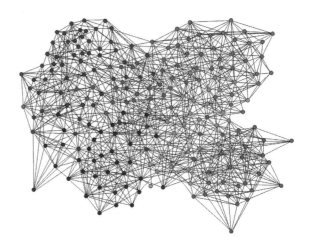

References

1. ITU Internet Reports: The Internet of Things. ITU Publications (2005)
2. Olguin, D.O., et al.: Sensible organizations: Technology and methodology for automatically measuring organizational behavior. IEEE Transactions on Systems, Man, and Cybernetics Part B: Cybernetics (2009)
3. Eagle, N., Pentland, A.: Social serendipity-mobilizing social software. IEEE Pervasive Computing (2005)
4. Nijholt, A., Rienks, R.J., Zwiers, J., Reidsman, D.: Online and off line visualization of meeting information and meeting support. The Visual Computer (2006)
5. Calabrese, F., Ratti, C.: Real time Rome. Networks and Communication Studies (2006)
6. Chen, P.: The entity-relationship model toward a unified view of data. ACM Transactions on Database Systems (1976)
7. Tang, J., Wang, T.: Research on the approximation algorithms for the betweenness property computation on complex social networks. Computer engineering and Science (2008)
8. Wang, X., Li, X., Chen, G.: Complex network theories and its application. Tsinghua University Press, Beijing (2006)
9. Watts, D.J., Strogatz, S.H.: Collective dynamics of 'Small World' networks. Nature (1988)
10. Barabási, A.L., Albert, R.: Emergence of scaling in random networks. Science (1999)
11. Jian, A., Xiaolin, G., Wendong, Z., Jinhua, J.: Nodes social relations cognition for mobility-aware in the internet of things. IEEE International Conferences on Internet of Things, and Cyder, Physical and Social Computing (2011)
12. Fortunato, S., Castellano, C.: Community structure in graphs. In: Meyers, R.A. (ed.) Encyclopedia of Complexity and System Science. Springer, Berlin (2009)
13. Toivonen, R., Onnela, J.P., Saramaki. J., Hyvonen, J., Kaski, K.: A model for social networks. Physica A 371(2), (2006)
14. Toivonen, R., Kovanen, L., Kivela, M., Onnela, J.P., Saramaki, J., Kaski, K.: A comparative study of social network models: network evolution models and nodal attribute models (2009)
15. Granovetter, M.: The strength of weak ties. Am. J. Sociol. 78 (1973)
16. Marsili, M., Vega-Redondo, F., Slanina, F.: The rise and fall of a networked society: A formal model. Proceedings of the National Academy of Sciences (PNAS) (USA) 101 (2004)

Rateless Code-Based Unequal Loss Protection for Layer-Coded Media Delivery

Xuan Dong, Shaohe Lv, Hu Shen, Junquan Deng, Xiaodong Wang and Xingming Zhou

Abstract Layered media delivery, such as multiview video coding (MVC) or scalable video coding (SVC), provides new opportunities for mobile multimedia services. In this paper, we propose a novel approach called unequal loss protection (ULP)-Strider based on Strider codes [1], which provides extra protection for layered media delivery. ULP-Strider introduces minimum distance transmission (MDT) in symbol level and soft hint for transmission to improve the probability of decoding process. We implement the algorithm in MATLAB with a four-node topology. The result shows that ULP-Strider can provide extra protection and remove 14 % time cost for recovering first-layer content in the scenes of low SNR, while our approach also has better first-layer protection than Strider's codes in high SNR.

Keywords ULP-Strider · Soft hint · MDT · ULP

X. Dong (✉) · S. Lv · H. Shen · J. Deng · X. Wang · X. Zhou
National Laboratory of Parallel and Distributed Processing, College of Computer,
National University of Defense Technology, Changsha 410073,
People's Republic of China
e-mail: dongxuan@nudt.edu.cn

S. Lv
e-mail: shaohelv@nudt.edu.cn

H. Shen
e-mail: hushen@nudt.edu.cn

J. Deng
e-mail: junquandeng@nudt.edu.cn

X. Wang
e-mail: xiaodongwang@nudt.edu.cn

X. Zhou
e-mail: xingmingzhou@nudt.edu.cn

S. Patnaik and X. Li (eds.), *Proceedings of International Conference on Soft Computing Techniques and Engineering Application*, Advances in Intelligent Systems and Computing 250, DOI: 10.1007/978-81-322-1695-7_22, © Springer India 2014

1 Introduction

Layered media delivery, such as multiview video coding (MVC) or scalable video coding (SVC), provides new opportunities for mobile multimedia services. In some situations, all the received packets are regarded equal at the application layer. However, this is not the case in layered media content, where lower layers are always more important than higher ones, i.e., a certain layer i is useless unless all the lower layers from layer 1 to layer $i - 1$ are correctly received [2]. Therefore, an efficient transmission in layered media requires a different protection depending on the priority of each layer. Many researches have focused on the area of unequal loss protection (ULP) in layered media delivery, which provides a high probability in receiving the basic layer. In [3], Bouabdallah proposes dependency-aware (DA) UEP, where the existing frames including I, P, and B video frames are protected depending on its importance, and different redundant protection codes are arranged based on their impact on the reconstruction quality.

Rateless code is a novel coding technology, which ignores the channel conditions when deciding to transmit packets in wireless network. A rateless encoder generates an (near) unlimited stream of coded packets from a given size of message bits. The transmitter progressively sends these packets to the receiver, until the receiver successfully decodes the source message and sends an ACK to the transmitter [4]. At present, Raptor codes have been exploited in ITPV applications of DVB-H standard and standardized by Third-Generation Partnership Project (3GPP) in the context of multimedia broadcast multicast services (MBMS). Although many measures have been discussed in layered media delivery, such as FEC codes, bit rate adaption, LT codes, and Raptor codes, the main challenge for layered media transmission still remains.

- Forward error correction (FEC) codes [2, 3, 5] exploit different redundancy data to improve the different reliable level of the layered content. Although feedback is not needed in FEC, it performs poorly at misunderstanding channel conditions, i.e., wasting bandwidth at better channel conditions while failing to recover the original data at worse channel conditions, where retransmission in still needed.
- Bit rate adaption [6] matches the sending bit rate to the channel condition to provide reliable transmission. However, bit rate adaption requires seamless information about channel condition. To pick the proper bit rate in the time-varying wireless channels, nodes should successively inspect channel quality via probing or by receiving the channel state feedback from receiver. Probing is a coarse measure in estimating the channel condition, since the sender can only detect the channel state around itself, which is inaccurate for evaluating the channel condition at receiver side.
- LT codes [2, 7, 8] or Raptor codes [9] are kinds of rateless codes and automatic repeat request (ARQ) schemes for reliable communication over lossy network [7]. Although LT codes and Raptor codes can achieve the capacity of an erasure channel without knowing the channel state, however,

these techniques require the correct decoding with the received packets, which may lead the waist of channel resources [1], even when there is only a small part of mistake in the received packets.

In this paper, we build on [1] and provide a twofold contribution. Firstly, we propose ULP-Strider codes for providing extra protection, with few overhead in low SNR, for layered media content based on a novel rateless code called Strider codes [1]. ULP-Strider codes make full use of the channel resources and do not need to balance the trade-off and overhead as the conventional strategy in bit rate adaption. Secondly, ULP-Strider codes improve the minimum distance transmission (MDT) between symbols by using soft hint and enhance the protection in poor channel condition. Besides, ULP-Strider codes can also decode the information of lower layer in advance and generate intermediate results for layered media delivery.

The remainder of the paper is organized as follows. ULP-Strider coding system is introduced in Sect. 2. Section 3 presents the implementation for ULP-Strider and provides experimental results on the performance and scalability of the system. Finally, in Sect. 4, we highlight our conclusions.

2 ULP-Strider Coding System

In this section, we describe the coding system of ULP-Strider to enhance former rateless codes with ULP capability, including encoding process and decoding algorithm, and then discuss the design of the redundancy symbol blocks.

2.1 Decoding Algorithm

The key insight of ULP-Strider decoding is that the algorithm not only improves ULP capability, but also introduces little affection for decoding of other sub-packets. We enhance the decoder of Strider to provide the extra protection for the whole sub-packets in different level based on priorities of sub-packets. Hence, ULP-Strider decodes the data from first code block and correlative redundancy layer. If decoder can decode successfully, then ULP-Strider resends the first sub-packet and redundancy sub-packets to the encoding process; after subtracting the re-encoding sub-packets from the original received packet, decoder tries to decode following code blocks using the same steps.

Obviously, the above intuition implies that ULP-Strider attempts to decode the first code block that relied on both original first code block and redundancy ones, while treats other sub-packets' as interference. Using the techniques SIC, we can recover the original code block and continue to decode the following blocks, as shown in Fig. 1.

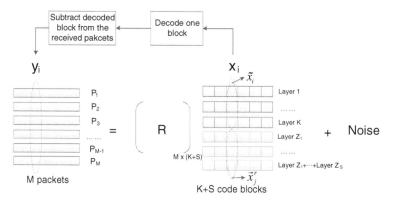

Fig. 1 ULP-Strider's decoding algorithm

The decoding algorithm of ULP-Strider can be described as follows.

$$\vec{y}_i = R\vec{x}_i + \vec{n}_i = [\tilde{R}, R']\,[\tilde{x}_i, \vec{x}_j^{r}]^T + \vec{n}_i = R\tilde{x}_i + R'\vec{x}_j^{r} + \vec{n}_i \qquad (1)$$

where $\tilde{R}$ is the $M \times k$ coefficients matrix of each original transmission, which is a complex with different phase but same magnitude, and R' is the $M \times S$ coefficients matrix of redundancy code blocks. We carry out the decoding process as the following algorithm.

$$r_1^*\vec{y}_i = r_1^*r_1\tilde{x}_{i1} + r_1^*r_2\tilde{x}_{i2} + \cdots + r_1^*r_k\tilde{x}_{ik} + r_1^*r_1'\tilde{x}_{i1} + \cdots + r_1^*r_S'\tilde{x}_{iS} + r_1^*\vec{n}_i = \|r_1\|^2\tilde{x}_{i1} + I$$

$$r_1'^*\vec{y}_i = r_1'^*r_1\tilde{x}_{i1} + r_1'^*r_2\tilde{x}_{i2} + \cdots + r_1'^*r_k\tilde{x}_{ik} + r_1'^*r_1'\tilde{x}_{i1} + \cdots + r_1'^*r_S'\tilde{x}_{iS} + r_1'^*\vec{n}_i = \|r_1'\|^2\tilde{x}_{i1} + I$$

$$(2)$$

where $I = \sum_{j=2}^{K} r_1^*r_j\tilde{x}_{ij} + \sum_{j=1}^{S} r_1^*r_j'\vec{x}_{ij} + r_1^*\vec{n}$, and R is the random coefficients matrix; any two columns in the matrix will be uncorrelated. The magnitude of the dot product of two uncorrelated complex vectors of equal magnitude will be less than the squared magnitude of either victor, so $I \ll \|r_1\|^2\tilde{x}_{i1}$ and $I \ll \|r_1'\|^2\tilde{x}_{i1}$; we can easily get $\tilde{x}_{i1} = r_1^*\vec{y}_i/\|r_1\|^2 - I/\|r_1\|^2 \approx r_1^*\vec{y}_i/\|r_1\|^2$ and $\tilde{x}_{i1} = r_1' \times \vec{y}_i/\|r_1'\|^2$. While $\tilde{x}_{i1}$ and $\vec{x}_{i1}$ carry the same information but in different orders which will be discussed carefully in Sect. 3.3, the decoder tries to use the soft hint of these redundant code blocks to recover the original code one.

2.2 Encoding Process

ULP-Strider's encoding and decoding algorithm is based on [1]; however, we provide extra protection for layered media content. Currently, we just give reader an overview of the end-to-end protocol. When a node wants to transmit, it follows the next four steps (Table 1).

Table 1 ULP-Strider encoding algorithm

Step 1: According to the priority, dividing packet into K code blocks $[l_1, l_2, \ldots, l_K]$ and exploiting the redundancy code production algorithm to produces redundancy code blocks

Step 2: Each of $K + S$ data blocks is passed through a fixed channel code and constellation (currently we use a 1/5 rate channel code and a BPSK constellation), and we map all the $K + S$ code blocks into complex symbols each

Step 3: The transmitted sub-packet p_j is a combination of N symbols $[p_{j1}, p_{j2}, \ldots, p_{jN}]$ and each $P_{jt} = \sum_{i=1}^{K} r_{it} x_{it} + \sum_{i=1}^{S} r'_{it} x_{(K+i)t}$, $1 \leq t \leq N$ can be produced through ULP-Strider as needed

Step 4: For j from 1 to infinite, if sender does not receive ACK or $j <$ threshold, sender transmits sub-packet p_j, while if sender receives ACK, then sender transmits the further one, and move to step 1. At last, if $s >$ threshold, sender gives up the packet, and submits an error to the upper layer of protocol

Redundancy code production algorithm in step 1 is the fundamental encoding step for ULP-Stride, which provides a different protection depending on the importance of each layer and guarantees that code block i can decode successfully only if the whole layers from 1 to $i - 1$ have been decoded. The redundancy code block Z_i, $1 \leq i \leq S$ design is as bellow:

$$Z_i = \sum_{j=1}^{i} l'_j \tag{3}$$

where l'_j is the jth code block but in a redundant sequence, and how to choose the proper sequence will be discussed carefully in Sect. 3.3. And then, according to the current redundancy code block Z_i, the right half of the equation is the summation of all the lower code blocks.

2.3 Analyzing the Construction of Redundancy

The design of redundancy code block is the key point for layered media content protection, which improves the probability of packet recovery. For simplification, we just discuss the media content which only has two different layers. Thus, the redundancy only contains two code blocks: layer 1 and layer 1 + layer 2. In this section, we discuss the arrangement of redundancy code in different order carefully and find the optical sequence of redundancy code block to enhance the robustness of important layer content.

Random sequence algorithm Assuming that the number of symbols in each code block is N and the original sequence is $[1, 2, \ldots, N]$, to enhance the robustness of first code block, the redundancy code blocks are added into the encoding process. In intuition, in order to improve the average distance of different symbol level, we just copy the original sequence and re-arrange it randomly to acquire the gains. In this paper, we propose a random symbol arrangement algorithm (as shown in Fig. 2). And the algorithm is as follows.

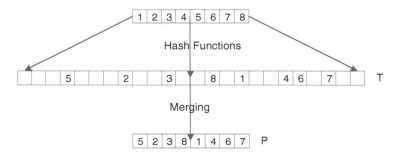

Fig. 2 Multihash random symbol arrangement

Fig. 3 Fine-designed symbol arrangement algorithm

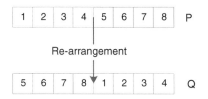

Step 1: Based on hash functions, the original symbols on layer 1 is split to a much larger sequence array T with no collisions (if the space which one symbol want to take is occupied by other symbol, the symbol can use another hash function to choose a new place), and the size of the new sequence is S ($S \gg N$).

Step 2: Collect the symbols from the new sequence T and merge them into a new array P with the same size of layer 1.

In this way, we can rearrange the symbols randomly and make the redundant sequence P of the important data in layer 1. And the average distance between symbols in the redundant sequence P and the first-layer code block is

$$\bar{d} = \frac{1}{N}\sum_{i=1}^{N}\left(\frac{1}{N}\left(\sum_{k=1}^{i-1}k\right) + \sum_{k=1}^{N-i}k\right) \qquad (4)$$

Fine-designed symbol arrangement algorithm To guarantee that the higher-priority layer can be decoded correctly, we have to increase the average symbol distance between the original code block and the redundancy one. The biggest average distance between these two code blocks is $N/2$. So we propose a simple approach to meet the threshold, which just moves the sequence $N/2$ symbols to the right side circularly (Fig. 3).

Under the method of fine-designed symbol arrangement algorithm, the average distance between the symbols in two layers is $\bar{d}$:

$$\bar{d} = \frac{N}{2} \qquad (5)$$

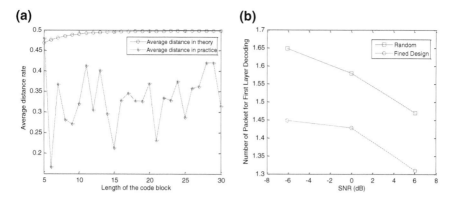

Fig. 4 a Average distance comparing in both theory and practice. **b** Number of received packets for decoding the first code block

3 Simulation

In this section, we evaluate the ULP-Strider through MATLAB by comparing the first-layer protection and throughput with Strider codes. We discuss the distinction between random sequence algorithm and fine-designed algorithm for providing extra protection. Moreover, we analyze the soft hint of symbols, which increases the probability for decoding with accumulating received packet. At last, we compare our approach with the basic Strider code and highlight our advantages.

3.1 Random Sequence Versus Fine-Designed Sequence

The redundancy code block is the key point for layered media protection, which will affect the robustness and efficiency of coding system. In this paper, we propose two kinds of algorithm: random sequence algorithm and fine-designed sequence algorithm. Both of them can construct the redundancy code block; however, their protection for the first-layer content is different. Average distance is a core factor for robustness; Fig. 4a shows the average distance for random sequence algorithm in both theory and practice, where random algorithm could not acquire the steady average distance. While fine-designed sequence algorithm is the better choice, as shown in Fig. 4b, the number of received packets for first-layer content is lower than that of random sequence algorithm, receiving approximately 10 % lesser packets to recover the first code block, which is more important for layered media delivery.

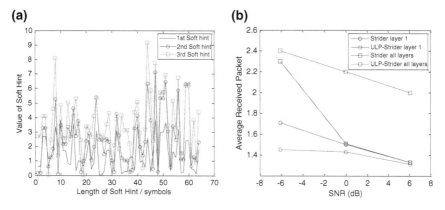

Fig. 5 **a** Soft hint for three successive transmissions. **b** The comparisons of first-layer protection and throughput between Strider codes and ULP-Strider codes

3.2 Extra Information from Soft Hint

Soft hint is the confidence information calculated before demodulation, which contains more abundant information for successful decoding. Not like the error bit rate (EBR), soft hint represents the distance between the received point and the nearest standardization point in constellation. Especially in rateless codes, ULP-Strider can enhance the soft hint with accumulating packets from the send node, which will improve the decoding probability. Supposing ULP-Strider has received three sub-packets before successfully decoding, as shown in Fig. 5a, the soft hint in the third sub-packet is better than that of the second and first one, which represents that we have a higher probability in decoding the third sub-packet than the former ones.

3.3 Strider Versus ULP-Strider

ULP-Strider is an enhanced rateless code based on Strider, which provides extra protection for higher-priority code block. However, the additive redundancy code block will decrease the throughput of decoding process. In Fig. 5b, ULP-Strider provides 14 % extra protection for the first code block with only 4.3 % loss in total throughput when SNR is equal to −6.06 dB. However, when the SNR increases linearly, the cost of Strider decreases rapidly, while the average received packets of ULP-Strider remains plateau with the protection for first code block slightly better than that of Strider. However, in the layered media delivery, the lower layers are much more important than the higher ones, so extra protection for those layers is necessary, regardless of a little throughput loss for the higher layers.

4 Conclusions

In this paper, we propose ULP-Strider codes for layered media delivery, which requires extra protection for the important content. ULP-Strider codes are based on novel rateless codes called Strider codes and enhance the ability of extra protection with low overhead in the scenes of low SNR. We provide two different kinds of algorithms to construct redundancy code blocks and prove that fine-designed sequence algorithm is one of the best combination redundancy sequences in coding system, whose erroneous symbols are continuous. By using the soft hint, ULP-Strider can enhance the decoding probability with accumulating packets from the senders.

Acknowledgments This work is supported by the National Natural Science Foundation of China under Grant No. 61070201, No. 61070203, No. 61202484 and Excellent Graduate Innovation Foundation of National University of Defense Technology (NUDT), China, under Grant No. B120608.

References

1. Gudipati, A., Katti, S.: Strider: Automatic rate adaptation and collision handling. SIGCOMM Comput. Commun. Rev. **41**(4), 158 (2011)
2. Lu, H., Cai, J., Foh, C.: Joint unequal loss protection and LT coding for layer-coded media delivery. IEEE Global Telecommunications Conference (GLOBECOM), pp. 1–5 (2010)
3. Bouabdallah, A., Lacan, J.: Dependency-aware unequal erasure protection code. In: Proceedings of PV, 15th International Packet Video Workshop (Apr 2006)
4. Sun, Y., et al.: Network control without CSI using rateless codes for downlink cellular systems (2012)
5. Hellge, C., Schierl, T., Wiegand, T.: Receiver driven layered multicast with layer-aware forward error correction. In: Proceedings of IEEE International Conference on Image Processing (ICIP), pp. 2304–2307 (Oct 2008)
6. Vutukuru, M., Balakrishnan, H., Jamieson, K.: Cross-layer wireless bit rate adaptation. In ACM SIGCOMM (2009)
7. Sejdinovic, D., Vukobratovic, D., Doufexi, A., Senk, V., Piechocki, R.: Expanding window fountain codes for unequal error protection. IEEE Trans. Commun. **57**(9), 2510–2516 (2009)
8. Chen, Z., Yin, L., Xu, M., Lu, J.: Rateless codes with progressive recovery for layered multimedia delivery (2012)
9. Cataldi, P., Grangetto, M., Tillo, T., Magli, E., Olmo, G.: Sliding-window raptor codes for efficient scalable wireless video broadcasting with unequal loss protection. IEEE Trans. Image Process. **19**(6), 1491–1503 (2010)

Construction of the Grade-3 System for GJB5000A-2008

Yonggang Li, Jinbiao Zhou, Jianwei He, Xiangming Li and Libing Guo

Abstract Software capability maturity model and software engineering are two inseparable aspects of the software process, this paper analyzes the need for the both integration, proposed a kind of construction method of grade-3 system for GJB5000A-2008 based on the software engineering process, and instance the compliance and effectiveness of this method with the software engineering process and software capability maturity model.

Keywords Software engineering · Software capability maturity model · GJB5000A-2008

1 Current Software Engineering Process Overview

The department implemented the software engineering process began from 1996, the current software engineering process executed in accordance with the software engineering standards of the general armament department, after practice and

Y. Li (✉) · J. Zhou · J. He · X. Li · L. Guo
China Satellite Marine Tracking and Controlling Department, 103#503,
Jiangyin 214431 Jiangsu, China
e-mail: ice_leo@sina.com.cn

J. Zhou
e-mail: zjb1101@163.com

J. He
e-mail: hjw1301@163.com

X. Li
e-mail: lxm1302@163.com

L. Guo
e-mail: guoguo_leo@163.com

S. Patnaik and X. Li (eds.), *Proceedings of International Conference on Soft Computing Techniques and Engineering Application*, Advances in Intelligent Systems and Computing 250, DOI: 10.1007/978-81-322-1695-7_23, © Springer India 2014

improvement many years later, it has formed a software engineering system effectively, successfully adapt to the demand for space launch monitoring and control tasks, completed the research and development of the monitoring and control software for the Shenzhou spacecraft, Chang'e and big dipper tasks satisfactorily [1, 2]. Current software engineering process is divided into two kinds: one is the software development process and the other is the software maintenance process.

In the process of implementation of software development and maintenance, set up the quality assurance members and the configuration manager, to supervise and inspect the software process and product quality strictly, ensure that the software process accord with the engineering requirements, software product meets the design requirements of the task; in order to manage the configuration items and control the software change implementation, established the development libraries, controlled libraries and product libraries in the development process, to ensure the consistency and integrity of the software configuration items.

2 Analysis of the Correlation and the Necessity of Fusion Between GJB5000A-2008 and Software Engineering Process

Understanding the overall relationship between the software capability maturity model and the life cycle model of software engineering process will determine the direction of the software development activities. We focus on the process of software development projects in the field of software engineering [3], and the software capability maturity model is more concerned about the overall software development activities. How to process all aspects of the software development activities, including the process of engineering activities, the organic integration of technology and management ensure activities constitute two important support of the software process [4].

In the GJB5000A-2008 military standard of military software development capability maturity model released in 2008 [5, 6], each maturity level process domain is shown in Table 1, and there are 7 areas correlated with grade-2 process areas, which is closely related to the engineering process except supplier agreement management and project monitoring process; there are 11 areas correlated with grade-3, including five engineering process areas, and there are 3 process areas of the project management and support classes directly related to the engineering process except the organized process management class [7]. It shows the important position of the software engineering in the software development capability maturity model, at the same time, for the effective implementation of the GJB5000A, must improve the implementation and execution efficiency, combine and merge with the GJB5000A process system and software engineering [8, 9].

Table 1 The maturity level process area of GJB5000A-2008 distribution

N	Class	Process areas	G	N	Class	Process areas	G
1	Project management	Project planning	2	12	Engineering	Requirements management	2
2		Project monitoring and control		13		Requirements development	3
3		Supplier agreement management		14		Technical solution	
4		Integrated project management	3	15		Product integration	
5		Risk management		16		Verification	
6		Quantitative project management	4	17		Validation	
7	Process management	Organizational process definition	3	18	Support	Process and product quality assurance	2
8		Organizational process focus		19		Configuration management	
9		Organizational training		20		Measurement and analysis	
10		Organizational process performance	4	21		Decision analysis and resolution	3
11		Organizational innovation and deployment	5	22		Causal analysis and resolution	5

N is number, G is grade level

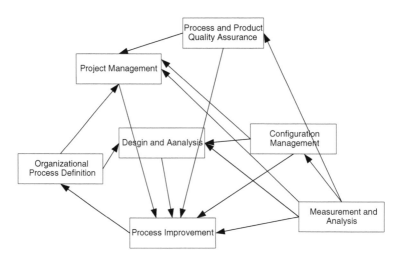

Fig. 1 Grade-3 process framework of GJB5000A-2008

3 Built Process System Based on the Software Engineering

According to the requirements of the GJB5000A-2008 standard model, at the time of the implementation of the grade-3 system of maturity, requirements covering the grade-2 maturity process areas, therefore, requirements covering 18 process areas in the process of implementing grade-3 system, including 3 process areas of process management class, 5 process areas of project management class, 6 process areas of engineering class, 4 process areas of support classes. We can see from the distribution of these process areas, there are 6 process areas of engineering class, and some process areas of project management classes and support classes are closely related to the activities require of the software engineering process, so these process areas can be integrated into the software engineering process according to its own attributes and the correlation between these processes and the software engineering process, formed the development and design process; The integration project management, risk management, supplier agreement management, project monitoring and control process areas of project management process area focus management, could be formed with the project management process; configuration management and process and product quality assurance process areas could be formed of two separate processes in order to ensure product quality and strengthen the integrity of the product, measurement analysis also form a separate process in order to ensure the objectivity of the data of process improvement required; the organizational process definition areas and the organization training areas of the process management class merged into the organizational process definition process for software organizations follow and use, organizational process focus as organizational identified strengths, weaknesses and continuous improvement process through the established goals, formed a separate process improvement process. Based on the above principles, the 18 process areas can be divided into seven processes (Fig. 1).

Organizational Process Definition: including organizational process definition and organizational training process areas, one is establishing and maintaining the organizational process assets set and work environment, the organizational process assets set include the organizational policy, standard processes, life cycle instructions, tailoring guidelines, measurement library, process assets library and work environment standards; two is determining the requirements, procedures methods of the organizational training.

Project management: including integrated project management, risk management, supplier agreement management and project monitoring and control, the main purpose is to establish the project defined process and working environment, and control requirements to ensure consistency, eliminate and mitigate project risk, obtained the desired product from a qualified supplier, monitored and controlled the process of project implementation, corrected the deviation.

Development and design: including project planning, requirements development, requirements management, technical solutions, product integration, decision analysis and resolution, verification, validation. According to the software engineering process, used of the organizational process asset library, planned project, analysis and developed requirements, developed and designed work products in accordance with the selected technical solutions, completed product integration according to a certain order, formed of the larger product, donned decision analysis and resolution on the problems encountered in the process, verified the compliance of the product and the requirements, validated the compliance of the product and user requirements.

Measurement and Analysis: measured objectively of the project operation process in accordance with the measurement items determined according to the needs and goals of management information, analysis of the measured data, recognized the problem of the process, to achieve the purpose of the process improvement.

Configuration management: It implements configuration management according to the control level determined, establishes and maintains the integrity of the work product, and controls the changes.

Process and product quality assurance: In the whole process of the project, the process activities and work products are objectively evaluated based on the process descriptions, standards, and procedures, the non-conformity is resolved, and the process and product visibility of all level staff are ensured.

Process Improvement: including organizational process focus process area evaluated and identified the current strengths and weaknesses of the process according to the needs and goals of the organization, make process improvement plan, implemented and deployed of organizational improvement, and brought the process of product and the lessons into the organizational process asset library.

Then, in the analysis of the composition of the design and development process system, the software engineering process can be described by the following eight components at any stage:

Design and development planning: It includes planning, decision analysis and resolution on the integrated management, planning, risk and supplier decision in the software stage;

Product realization: the product realization process, developed the input as this stage requirements, exported customer demand, product requirements and product component requirements, proposed multiple alternatives for product realization, determined technical solutions, detailed design to meet development requirements of current stages, integrated product according to the product integration rules and a predetermined sequence of design and development, decision analysis and resolution on the issues encountered in the development process in accordance with the guidelines.

Review: assessed the process and product, monitored the process of the design and development, throughout the whole process of development and design, especially in the critical control points of the phases and milestones, to ensure its quality satisfy the process standards and needs of the target, and resolve non-conformity discovered.

Verification: The compliance of products and demands is confirmed, and the requirements are traced.

Validation: The conformity of the product and user requirements is confirmed.

Change control: It includes implementation of configuration management and control of the change implementation phase of product and process activity.

4 Process Instance Analysis

According to the requirement analysis process as example, illustrated the integration of the software engineering process and the activities of process areas:

- Tailored standard process of the organization, established the defined process of the project, used process asset library, estimated the project scale, workload, resource, schedule, cost, formed the development plan (configuration management plan, process and product quality assurance plan, project training plan, project measurement plan, project risk management strategies etc.).
- Acquainted the planning resources and established the working environment.
- Participated in organization-level training and project-level special training, to ensure the project team roles understood properly, have the skills to perform their duties.
- Participated in the assessment of system design specifications and software developed mission statement and receiver them, and communicate with the user, assessment and determine the software functionality, performance and other requirements.
- Developed user demand into product requirements; detailed each product needs; determined the functions, interfaces, performance, and design constraints; analyzed the necessity and adequacy of demand; and formed a number of alternative

implementation; according to the criteria select the best scheme to form a product component requirements; analysis of the interface demand of the product and the external interface and the in the interface of product components, integrated product components requirements accordance with the decomposition components and the functional interface relationship, formed a software requirements specification (software interface requirements specification).

- Analysis the obtain way of product or product parts, managed the product or product components obtained out of the organization developed in accordance with the supplier agreement management.
- Established and maintained requirement traceability matrix and implemented the requirements two-way track.
- Peer-reviewed the requirement specification by the trade experts and ensured that the software requirement specifications meet the requirements of the software system design specification requirements and the software-developed mission statements.
- Reviewed the requirement specification together with users and stakeholders to determine the consistency of the product and user needs and confirmed the correctness of the product runs in the environment expected by the method of analysis, simulation, and presentation.
- Planned configuration tests and drew up configuration test plan.
- Collected and analyzed the measurement data in accordance with the project measurement plan and put forward the improvement proposal.
- Implemented the quality assurance activities on the products and process in accordance with the process and product quality assurance plan, found and solved the non-conformity of the product and process activities.
- Implemented the configuration management of product and process assets in accordance with the configuration management plan and controlled the item changes;
- Monitored the process in accordance with the plan, analysis the deviations and the factors of the schedule, cost, effort, resources, scale, risk with the plan, and implemented milestone review after the end of the requirements analysis phase, analysis phase deviations, identified the problem needed to correct and improve, make the adjustment programs and corrective measures to improve productivity, and improved the process, solved the deviation problems, alleviated or eliminated the risk.
- Assessed and identified the strengths, weaknesses and items to be improved of the process implementation in this stage according to the process improvement plan, analysis process improvement needs, made process action plan, deployed and implemented process improvement, fitted experience process-related into the organization process asset library.

Through the above fusion requirements analysis phase of the software process analysis, it showed that it combined better with the original software engineering process, at the same time, but also comprehensive coveraged the require of the process of GJB5000A grade-3 at this stage, as shown in Table 2.

Table 2 Process areas covered in the requirement analysis phase

No.	Process areas covered	Process areas related
1	Integrated project management, project planning, risk management, measurement and analysis, configuration management, process and product quality assurance	Organizational process definition
2	Integrated project management	Organizational process definition
3	Organizational training	
4	Requirements management, requirements development	
5	Requirements development, product integration, decision analysis and resolution, technical solution	
6	Technical solution	
7	Requirements management	
8	Verification	
9	Validation	
10	Project planning	
11	Measurement and analysis	
12	Process and product quality assurance	
13	Configuration management	
14	Project monitoring and control, risk management, project planning	
15	Organizational process focus	

5 Conclusions

Based on existing software engineering process, this paper puts forward the integration method of software process and software capability maturity model according to the characteristics of GJB5000A-2008 software capability maturity model grade-3 process area, constructed a rough framework, take requirements analysis phase for example, proved that the system constructed by this method not only meet GJB5000A-2008 requirements, but also meet the software engineering requirements. It hope that play a role on improve software efficiency and improve the quality of software work and promote GJB5000A-2008 promotion through this article.

References

1. Yan, Y.: Military Software Development Capability Maturity Model of GJB 5000A-2008. General Armament Department Military Standard Publishing Department, Beijing (2009)
2. Shi, Z.: Military Software Development Capability Maturity Model and Application. Standards Press of China, Beijing (2009)
3. Li, X.: Research on project requirements management process based on CMM. J. Comput. Knowl. Technol. **5**(30), 8434–8437 (2009)
4. Zhang, N., Kun, S., Zhang, P., Jiang, Y.: Research on process improvement for small and medium software enterprises based on CMMI. J. Softw. Tribune **1**, 30–33 (2011)

5. Zhou, J., Zhang, P.: Research on software process improvement based on CMMI. J. Comput. Eng. Des. **11**, 45–48 (2003)
6. Wang, W.: How to implement the software quality assurance process and product audits. J. Softw. Tribune **3**, 38–40 (2011)
7. Fan, X.: Software Development Process of Small and Medium Software Enterprises Based on CMMI3. Kunming University (2012)
8. Liu, X., Shao, W., Huang, W.: Research and design on enterprise project management system based on CMMI3. J. Comput. Eng. Des. **13**, 50–52 (2009)
9. Xie, D.: Dual iterative model of agile software development. J. Comput. Appl. Softw. **6**, 20–24 (2012)

Virtual Training System of Assembly and Disassembly Based on Petri Net

Xiaoqiang Yang, Jinhua Han and Yi Pan

Abstract Petri net and human–machine interactive function have been integrated into the virtual disassembly/assembly system in EON studio. The system consists of four function modules: virtual environment, interactive control, dedicated simulation, and data management as well as image generation modules of assembly/disassembly. The static part model is constructed by Pro/Engineer platform. It is subsequently imported to EON virtual environment from the Pro/Engineer software output file. The human–machine interface is implemented by event-driven and routing mechanism together with sensor node. The assembly/disassembly sequence planning in Petri net is constructed such that the real-time collision detection can be performed. The critical points on assembly/disassembly path are recorded and transferred to data chained list of assembly/disassembly sequence. In addition, the motion control of virtual machine tools is conducted by place node and script node of EON. Consequently, the development of virtual assembly/disassembly training system is carried out by EON studio, Visual C++6.0, and script language.

Keywords Virtual training · Assembly and disassembly · EON studio · Petri Net

X. Yang (✉) · J. Han
PLA University of Science and Technology, Houbiaoying 88, Nanjing, China
e-mail: yanglab@126.com

Y. Pan
PLA No. 93968 Troop, Nanjing, China
e-mail: youngpaper@126.com

S. Patnaik and X. Li (eds.), *Proceedings of International Conference on Soft Computing Techniques and Engineering Application*, Advances in Intelligent Systems and Computing 250, DOI: 10.1007/978-81-322-1695-7_24, © Springer India 2014

1 Introduction

Virtual maintenance technology has become one of the key technologies of national defense science and technology development. With the development of virtual reality technology, maintenance has already developed from the manual maintenance in the traditional scene to virtual maintenance environment of "field," "real time" based on virtual reality technology. This method is safe, economic, controllable, repeatable, risk-free, but not restricted by climate conditions and venue space. The training of high efficiency, high benefit, and other unique advantages makes it valued by the military all over the world.

The current domestic and international mainstream virtual reality platform mainly include some kinds of development software, such as Vir Tools, Quest 3D, VRP, Java 3D, and EON. All of these can realize the basic virtual simulation effect. Among them, the EON software compatibility is good, it can provide real physics simulation of 3D model and support multiple CAD format with a strong extensibility and secondary development ability. Therefore, it is a good virtual reality development platform of the best integration and the ductility [1].

In this work, a virtual training system of assembly and disassembly based on EON virtual platform is developed. The key technologies for the system design and development are also been studied.

2 Framework of Virtual Assembly/Disassembly System

2.1 System Structure

The design objective of the virtual assembly/disassembly system is described in the following:

- Firstly, a favorable virtual repair assembly and disassembly environment is built in EON studio with assembly/disassembly planning and analysis of repair object. This environment should have interactive interface and facilitate human to exercise his initiative and creativity.
- Secondly, the sequence planning of repair assembly and disassembly for parts with reciprocal constraints is conducted based on Petri net. The path of assembly/disassembly in EON environment is adjusted and planned through human–machine interaction. In addition, the assembly and disassembly sequence is verified by real-time interference checking so that it ensures the validity of assembly and disassembly path.
- Thirdly, the assembly/disassembly data chained list is established by its sequence. The key points of path are transferred to sequence chained list. The assembly/disassembly sequence is transmitted to the components of diesel engine so as to perform simulation of operation process.

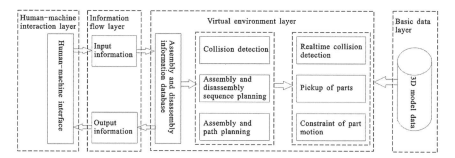

Fig. 1 Block diagram of virtual training system structure

The assembly and disassembly system can be divided into four layers in accordance with the system requirement and data flow. The layers are human–machine interface, information flow, virtual environment, and basic data layers, as shown in Fig. 1.

2.2 Function Module

Consequently, the function structure of assembly and disassembly consists of five modules, namely virtual assembly/disassembly environment, interactive control, simulation, data management, and image generation module, as shown in Fig. 2.

Virtual maintenance environment module mainly accomplishes the virtual maintenance scene, the modeling of repair objects, maintenance tools, and maintenance personnel in the CAD system. CAD model is to adopt the accurate mathematical formula to describe the geometry information of 3D model. This model which contained large amounts of data information in a VR system takes a lot of time to fulfil graph display and makes virtual scene count for large amount of calculation. So we need preprocess the CAD model in a distortionless condition to a greatest extent to reduce the amount of data model so that maintenance objects can be easily handled in the virtual maintenance system.

Interactive control module includes viewpoint control and disassembly operation control to the repair objects. Virtual maintenance training system should make operating personnel having an immersive feeling and make the virtual maintenance environment more authentic. Operators should be easily able to browse the virtual maintenance scenarios through the interactive manipulator. Maintenance disassembly objects can be observed through various angles. Viewpoint control is to control the objects in virtual maintenance environment by the position and direction of user's viewpoint. Virtual disassembly objects control mainly represents by the motion control of six degrees of freedom (DOF). By manipulating the motion along the translation and rotation axes, the pickup, drag operation, and alteration of object space pose are carried out.

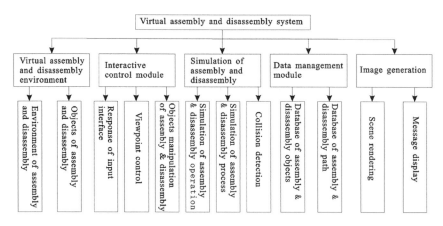

Fig. 2 Function block diagram

The module of maintenance, dismantling, and simulation is mainly to complete adjustment and optimization of disassembly path. The real-time records of disassembling objects' moving position are key points for disassembling paths. The simulation of disassembling process is realized according to the reading of the disassembling path information; the collision experiment is carried out in the process of disassembling in real time, if a maintenance disassembling objects is pick up, it will be highlighted and moved with the movement of the mouse. In the process of virtual maintenance disassembling operation, we use hybrid collision detection algorithm for inspection, which ensures the feasibility of dismantling process. The organization of disassembling objects adopts a hierarchical structure, which reduces the number of collision detection and improves the efficiency of process of disassembling and simulation.

The module of data management is to manage spare parts in the virtual scene, such as increase, delete, modify the maintenance objects, and manage the maintenance object motion path.

The formation of visual is to render virtual dismounting environment, finding out the fault information, and timely couple back to the operator in the process of virtual dismounting.

3 Key Technologies

3.1 Assembly/Disassembly Model in Petri Net

As long as the resources in Petri net are used by a transition, they will not be used by other transitions until they are released by the owner. Only the part to be disassembled removing its constraints to other ones could it be handled by

Fig. 3 Parts constraint structure

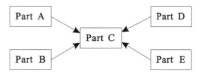

Fig. 4 Disassembly model in Petri net

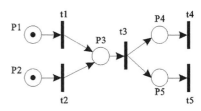

disassembly operation. Disassembly operation at a time and the changes corresponded can only be triggered once. The subsequent constraint condition shall be changed accordingly, if one constraint is removed. Constraints condition change has the characteristics of multi-path transmission. Because of the complexity of the connection of the maintenance equipment parts, the disassembling problem of diesel engine is not only related to the constraint condition between the parts, but also related to the change process of constraint condition. Constraint relation determines its dismantling process.

The constraint relation of every part is analyzed before assemble and disassemble operation to engine. The constraint relation can represent by Petri net model. The constraint relationship between parts is shown in Fig. 3 in which arrow represents the constraint relationship between parts. If you want to tear down/fit on part E, you must tear down/fit on A, B, C, that means solving the parts A, B, C to its constraints. The parts in Fig. 3 correspond to those of parts library in Fig. 4 where t is used to express parts disassembling operation. The constraint structure parts can be converted to disassembling process model of Petri net. Library with black spots stands for its parts in unconstrained condition, which can tear open outfit.

Disassembly Petri net is defined: $PN = (P, T, F, S, W, R, M0)$, among them, the $(P, T, F) = N$ is oriented base network of DPN.

Definition:

1. $P = \{p1, p2, \ldots, PM\}$, disassembling Petri net library collections, m (natural number) is the number of parts, and PI refers to the ith kind of parts or components;
2. $T = \{t1, t2, \ldots, tn\}$ is change of finite set, n (natural number) is the number of changes, ti refers to tear open outfit without the constraint state of the ith type of parts, $1 \leq i \leq m$. Library and changes are two different types of elements in Petri net, so $P \cap T = \varphi, P \cup T \neq \varphi$ indicates that the net should have at least one element;

3. $F \subseteq (P \times T) \cup (T \times P)$ represents the flow relation in disassembling Petri net, and it is actually the oriented arc between the library and change. Positive arc F indicates the direction of constraints between components, and inverse arc F indicates the direction of constraints between removed components;

4. $S = \{S(p1), S(p2), S(p3), \ldots\}$ is targeted at the constraint condition of a single part in different time, $S(\mathrm{PI}) = 1$ indicates that parts are in unconstrained condition, and $S(\mathrm{PI}) = 0$ indicates parts are in the bound state;

5. W is an integer greater than 1, also the weight function from changes to the library, which indicates the number of the same parts or components in a dismantling of the disassembling operation in this article;

6. $R = \{R(t1), R(t2), R(t3), \ldots\}$ indicates changes have occurred, which refers to the parts have been dismantling in this article;

7. $M0 : N0 = \{0, 1\}$ is defined as initial logo of Petri net, which indicates the bound state of all parts before equipment have been disassembled in this article, and $M0$ is in vector form, namely $M0 = \{a1, a2, \ldots, ai, \ldots, am\}T$.

3.2 Path Planning of Assembly/Disassembly

In the last section, we analyzes the relationship between the assembly and dismantling, which is based on the idea of "detachable can be assembled" for path planning, and we think that disassembling is the inverse process of assembly. Because of the domestic existing mature path automatic search algorithm can only be used in the exceedingly simple process of dismantling, with considerable gap from the actual equipment dismantling, therefore, this path planning is undoubtedly a simple and applicable choice to solve practical application of disassembly path planning.

The dismantling of path planning in EON environment is on the basis of disassembly sequence planning which is based on the Petri net, and it also establishes the mounting parts data chain table, planning path based on the concept of "detachable can be assembled." And with the use of powerful capabilities provided by EON studio, we employ the way with a combination of man–machine interaction and collision detection to get disassembly path, which is based on the assumption to pack without the detachable. And the disassembly path planning in the EON basically has the following characteristics:

1. With the human–computer interaction, maintenance personnel conduct the path optimization for disassembly path point by the means of linear interpolation and curve interpolation to make the path smooth and intuitive.

2. With the human–computer interaction, give full play to the initiative of people and the ability of computer, record the disassembling process and historical information, using the interference detection technology to guarantee the effectiveness of the loading path.

The disassembly path planning process in EON environment is shown in Fig. 5.

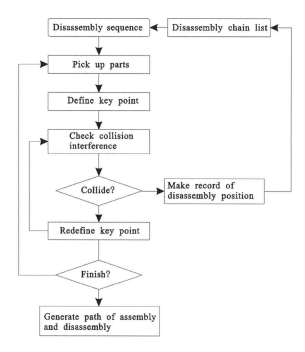

Fig. 5 Path planning diagram of assembly and disassembly

3.3 Collision Detection

EON studio supplies three nodes for collision detection: Collision Geometry Type, Collision Object and Collision Manager. The Collision Geometry Type Node is used to specify the type of bounding box for collision objects. Currently, there are six types of bounding volumes for user selection. They include TBounding Speres, Bounding Boxes, Bounding Boxes, Bounding Box Convex, Hull Convex and Geometry Axis. The Collision Object Node holds properties and report collisions for individual collision objects. These nodes are linked to a Frame node by adding a frame reference in the Object Frame field or by placing it below a Frame node and thereby identify the Frame as collision object in the simulation. The Collision Manager node holds global properties for a set of collision objects between which collisions should be tested.

The interactive collision detection and response detection between equipment parts, tools, work stand, and workplace can be realized by adding Collision Object node and Rigid Body Node below sub-tree of parts model frame node, work stand frame node, and wall frame node, along with the insertion of Collision Manage node and Dynamics Manager Node at any position below Scene sub-tree as well as relevant configuration. The immersion and reality sensor may be further enhanced.

4 Conclusions

In this paper, the key technologies of constructing virtual maintenance assembly/ disassembly training system have been studied for the application on the process engineering equipment maintenance training. The main contributions of this work could be briefly summarized as follows:

- The interactive control techniques between engineering equipment parts, work-place, work stand, tools, and others are studied. It resolved the problem of parts assembly/disassembly and demonstration of work principle and maintenance guidance.
- The collision detection and response detection of object in virtual scene are described. The penetration of machine parts is implemented effectively.
- The work in this paper can meet a wide variety of virtual training system requirement of military engineering support, and it also provides wide range of benefits and functionality to the other development of virtual training system.

Reference

1. Liu, B., Li, Z., Li, Z.: An image hiding algorithm based on bit plane. Comput. Intell. Secur. **3802**, 611–616 (2005). doi:10.1007/11596981_90

The Design of Visual RBAC Model Based on UML and XACML Integrating

Baode Fan and Mengmeng Li

Abstract Security is an indispensable part of modern software system. Access control is an important measure to guarantee the security of a system; however, the modeling of modern software system does not deal with security, which brings great hidden danger in later security maintenance and the system integration. In this paper, we use a score management system as an example, describe the visualization modeling of RBAC model using the visualized UML, and provide the details about the modeling method and concrete realization of the RBAC model into this system. At the same time, we describe the permissions between users and objects in the RBAC model, combined with eXtensible Access Control Markup Language (XACML), realized the formal description of the access control policy, simplified authorization, and increased the security of the model.

Keywords XACML · RBAC · Model integration · Role-permission · Score management

1 Introduction

The modeling of modern software systems mainly refers to requirements, structure, components, and deployment, while ignoring the safety aspects. At present, to increase interactivity and security of resource sharing, it still needs a method of integrating RBAC features into the software design model and a standard language to describe the access control policy, although there are a lot of research on RBAC

B. Fan (✉) · M. Li
School of Computer Science, Yantai University, Shandong, China
e-mail: fanbaode@aliyun.com

M. Li
e-mail: 15064559661@sina.cn

S. Patnaik and X. Li (eds.), *Proceedings of International Conference on Soft Computing Techniques and Engineering Application*, Advances in Intelligent Systems and Computing 250, DOI: 10.1007/978-81-322-1695-7_25, © Springer India 2014

model. In this chapter, we use a score management system as an example, describe the method and process of integrating the RBAC features into the specific system, and describe the authorization process between users and objects with eXtensible Access Control Markup Language (XACML), which not only simplifies authorization management, but also increases the security and interoperation of the model.

2 RBAC Model

In August 2001, NIST published a RBAC reference model; the model combines the basic features, defines the main components, and unifies the related terms of the RBAC model; it fully embodies the safe principles of the least privilege, separation of duties, and data abstraction. It has an important role in promoting the development of RBAC technology. As shown in Fig. 1, the RBAC reference model contains the following parts: Core RBAC, Hierarchical RBAC, and Constrained RBAC [including static separation of duties (SSD) and dynamic separation of duties (DSD)].

Core RBAC is the essence of RBAC model, which defines the minimum set of elements that can constitute a RBAC system. The model, shown in Fig. 1, mainly consists of six kinds of basic elements: Users, Roles, Objects, Operations, Permissions, and Sessions. Users and their access permissions are decoupled by roles; the grant and cancelation of user's permission are completed through distribution and cancelation of role in RBAC model; it also provides role assignment rules. Security managers define various kinds of roles as needed and set appropriate access permissions for these roles. Users are assigned to different roles in accordance with their responsibilities or positions. So, the whole access control process is divided into two parts, that is, the assignment of access permissions to roles and roles to users. Thus, it implements the logical separation between users and their access permissions and greatly facilitates the management of permissions. A user establishes a session when the user stimulates some subsets of his roles. The available permissions of the user are permissions sets of all roles which are activated by the current session. A session can be established by a single user, while a user can open multiple sessions at the same time.

The definition of the Core RBAC is as follows:

1. U (Users), R (Roles), P (Permissions), and S (Sessions)
2. $PA \subseteq P \times R$: The assignment of permissions to roles is many-to-many authorization association.
3. $UA \subseteq U \times R$: The assignment of roles to users is many-to-many authorization association.
4. User: $S \rightarrow U$, Each session is mapped to a single user.
5. Roles: $S \rightarrow 2^R$, A session is mapped to a role set roles(si) $\subseteq \{r|(user(si), r) \in UA\}$, and the permissions of this session are $\cup_{r \in role(si)}\{p|(p, r) \in PA\}$.

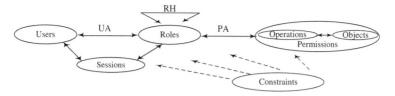

Fig. 1 RBAC reference model

In order to improve efficiency and avoid repeat setting of the same permissions, the role inheritance strategy is introduced in the RBAC model. If a role r1 inherits a role r2, then the role r1 will obtain all the permissions of the role r2. In addition, the role r1 also can define its own permissions.

SSD specifies the mutually exclusive relationship among roles during the assignment process of roles to users; it avoids the same user getting the mutually exclusive roles for a user.

DSD specifies the mutually exclusive relationship among roles during the role activation phase; it allows the same user to get some mutually exclusive roles, but these roles are not activated by the user at the same time.

3 The Basic Knowledge of XACML

XACML is an open standard language based on XML. It is designed to describe the security policy and access permissions of network service, digital rights management, and enterprise security application information. It also defines a common policy language and access decision language used to protect the resources. The policy language allows administrators to define access control requirements in order to obtain the needed application resources; and access decision language is used to describe the request of resources at runtime. After defining the policy of protected resources, the function will compare the attribute in the request to which in the policy rule and ultimately generate a permit or deny. Launched in February 2007, XACML 3.0 is an access control policy language, which is scalable, signature, encrypted, and independent of application. It can provide authorization and access control platform for network application systems which are multi-level, large, and complex. The provided platform is unified and effective, is unrelated to development and management of the specific application systems, has high security strength, and is flexible and easily scalable.

The framework of XACML consists of multiple node components, shown in Fig. 2.

These node components communicate with each other in the system and collect attribute information of environment and target to participate in policy decision, to complete the control of the requested access to resources.

When the client puts forward a resource request to the Server, PEP will receive and transmit the access request to Context Handler which proposes attribute

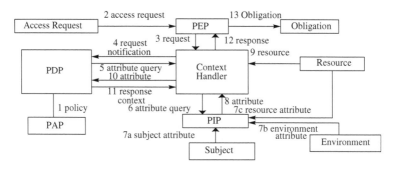

Fig. 2 The component diagram of XACML

request to PIP; PIP obtains relevant attributes from subjects, resources, and environment and returns them to Context Handler; Context Handler does some processing on these attributes, resource, and request information and proposes decision request containing necessary information to PDP; PDP makes decision for decision request based on the policy provided by PAP according to established logic and returns the decision result to Context Handler which then returns to PEP; PEP executes the decision result and submits obligation requirement to the obligation service or denies user's request.

4 The RBAC Model Based on UML and XACML

We use the concept of class in the UML to describe each component of the core RBAC model. The pattern class diagram of RBAC is described by UML class diagram, shown in Fig. 3, including class, class association, and associated multiplicity. The assignment of roles to users and permissions to roles is many-to-many authorization association. The model has a special class, session hour, which is used when a session is created by a user to activate his roles.

In order to achieve the reuse of the RBAC model, we regard this model as a pattern which is described by UML class diagram. This class diagram can be instantiated to implement the reuse.

We can obtain an instantiated class diagram of specific system by instantiating the classes and relations in the pattern, namely instance model. Combined with a specific system, the score management system, we will show how to instantiate the RBAC model and realize this model using XACML.

Fig. 3 RBAC pattern class
diagram

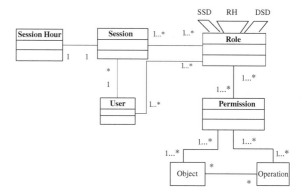

Fig. 4 The initial model of
the score system

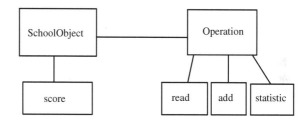

4.1 RBAC Instance Model

The initial model of the score management system is shown in Fig. 4; it shows the main function of this system, student can read his score, teacher can add student's score, and administrator can make statistics of student's score.

The features of RBAC can be integrated into this initial model by instantiating RBAC pattern shown in Fig. 3. The integrated process mainly consists of the following two steps:

Firstly, to instantiate RBAC pattern, the elements in RBAC pattern are bound to the corresponding elements of the actual system. Figure 5 shows the RBAC instance model of this system. SchoolUser, SchoolRole, and SchoolObject in this model, respectively, correspond to User, Role, and Object in the RBAC pattern class diagram.

Secondly, the method of integrating the initial model into instance model is as follows: First, find the elements having the same name in these two models, and if these two elements have the same semantics, such as SchoolObject in the initial model and also in the RBAC instance model, they can be merged into one element in the integration model. Then, their respective elements are added to the integration model.

The integrated model has the characteristics of RBAC; Figure 6 shows the integration model of the score management system; it can be seen that School-Object and Operation in the RBAC instance model and also in the initial model are

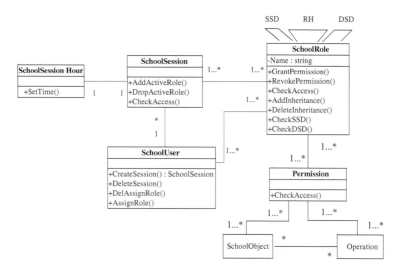

Fig. 5 RBAC instance model score system

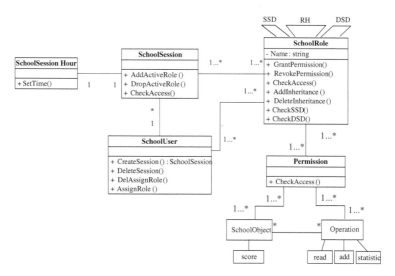

Fig. 6 The integration model of the score system

merged in this integration model, while SchoolUser, SchoolRole, SchoolSession, and Permission are unique classes of RBAC instance model, so they are directly added to the integration model. The unique elements in the initial model are also directly added to the combination model.

The integration model cannot clearly show the assignment of roles to users and permissions to roles; these problems can be solved by UML object diagram and collaboration diagram. Object diagram can clearly describe the assignment

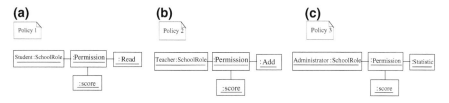

Fig. 7 Object diagram: role-permission assignment

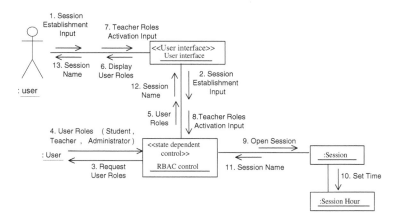

Fig. 8 Collaboration diagram: Session Establishment

relations between roles and permissions, while collaboration diagram clearly shows the assignment relations between users and roles.

Figure 7 shows the assignment method between roles and permissions in the score management system.

The roles of this system are Student, Teacher, and Administrator. The assignment policies of permissions are as follows:

(a) Student can read his score object, shown in Fig. 7a;
(b) Teacher can add student's score object, shown in Fig. 7b;
(c) Administrator can make statistics of student's score object, shown in Fig. 7c.

We use collaboration diagram to describe the assignment relations between users and roles in this system. The collaboration diagram of Session Establishment, shown in Fig. 8, is used to activate a Teacher role:

Firstly, a user submits the information (to activate the Teacher role) in order to establish a session; secondly, the system displays the roles (Student, Teacher, and Administrator) that can be activated by the user; thirdly, the user selects a role (Teacher) to be activated; and finally, the system establishes a session depending on the activated role.

Table 1 The relationship between RBAC and XACML

RBAC element	XACML element
Users	Subjects
Roles	Subject attributes
Objects	Resources
Operations	Actions
Permissions	Role<PolicySet>
	Permission<PolicySet>

4.2 The Realization of RBAC Using XACML

In the Core RBAC, a user can become a member of some roles; a role can obtain some users. Similarly, a role can get multiple permissions; permission can be assigned to multiple roles. The above-mentioned relationships and the elements of RBAC can be described by XACML, shown in Table 1.

Figure 9 shows the RBAC model based on XACML. The access permission between user and object is described by XACML; the whole process can be divided into two aspects of authentication and authorization, to get authorization must be certified, the whole process is as follows:

The user firstly establishes a session with the Role Server to determine the user's identity and obtain the user's role; if the user is legal, then the authentication information and role information will be stored on the client machine. Next, the user establishes a session with the Authorization Server and sends corresponding information to the Authorization Server for verification. If verification is successful, then Authorization Server will execute role-based access control according to user's request and then send the resource information of the request's result to the policy enforcement point of the Policy Server; then, the policy enforcement point executes policy evaluation and returns the evaluation result to the Authorization Server. Finally, to permit or deny the user's request is returned to the user.

The access control policy that user requests to add student's score can be described by XACML as follows:

```
<Policy>
<Rule Effect=''Permit''>
<Description><teacher can add student's score>
</Description>
<Target>
<Subject><Attribute>
<Attribute Value>teacher</Attribute Value>
</Attribute></Subject>
<Action><Attribute>
<Attribute Value>add</Attribute Value>
</Attribute></Action>
<Resource><Attribute>
```

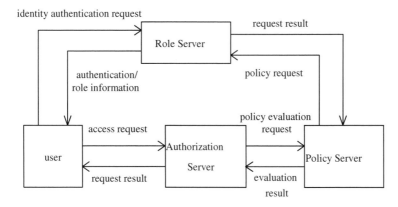

identity authentication request

Fig. 9 RBAC model based on XACML

```
<Attribute Value>score</Attribute Value>
</Attribute></Resource>
</Target>
</Rule>
</Policy>
```

The access request that user requests to add student's score using Teacher role formed a decision request by PEP; then, PEP sends it to Context Handler to form a request context:

```
<Request>
<Target>
<Subject><Attribute>
<Attribute Value>teacher</Attribute Value >
</Attribute></Subject>
<Resource><Attribute>
<Attribute Value>score</Attribute Value>
</Attribute></Resource>
<Action><Attribute>
<Attribute Value>add</Attribute Value>
</Attribute></Action>
</Target>
</Request>
```

5 Conclusions

In this chapter, we briefly describe the RBAC model and combine a specific score management system to instantiate the model. We use visualized UML to describe the model and, at the same time, use XACML to describe the access permission between user and object in this model. Our work can help system developers understand the RBAC model and integrate its features into a specific system more easily. We also recognize that the described RBAC model is static, and some useful function and constraints are excluded in the early stage of the model. Therefore, in the future, we will continue to study the dynamic RBAC model and how to express the role inheritance and constraints using UML extension and XACML.

References

1. Li, M., Fan, B.: The modeling of RBAC model based on UML and XACML. In: Proceedings of the 2012 International Conference on Systems and Informatics, p. 5. (2012)
2. Extensible access control markup language version 3.0. OASIS standard (2007)
3. Yin, Z.: The Application of UML and its Modeling Tools, pp. 10–95. Tsinghua University Press, Bei Jing (2004)
4. XACML Profile for Role Based Access Control (RBAC) Committee Draft 01. OASIS. 13 Feb 2004
5. T, Guo, Z, Yin, B, Fan: Integration of systematic RBAC security model and UML model. Comput. Eng. Des. **28**(4), 789–791 (2007)
6. Y, Gao, J, Zhang, M, Wu: Access control system based on XACML and RBAC. Comput. Appl. Softw. **23**(8), 65–67 (2006)

A New Approach to Reproduce Traffic Accident Based on the Data of Vehicle Video Recorders

Hong Li, Qing Kang and Jing He

Abstract The identification of the responsibility of traffic accident has become a major problem because of the rising accident rates. Limited surveillance video data often restrict its accuracy. More believable data source could be helpful to improve the efficiency and accuracy of the traffic accident identification. In this paper, we propose a new method for traffic accident identification which collects large amount of data through the vehicle video recorder, classifies the data, and then reorganizes the accident video. We design the system architecture and main framework, analyze the concrete realization steps, and explain its feasibility.

Keywords Video reorganization · Vehicle video recorder · Cloud server · Video frame · Image similarity

1 Introduction

The rates of traffic accident are rising with the increase in private cars. How to identify the responsibility of the accident has become a major problem for The Department of Transportation and Information Technology. Generally, we can

H. Li
School of Information Science and Engineering, Yunnan University,
Kunming 650091, China
e-mail: lh_1985@163.com

Q. Kang
No.59 BeiMen Street, Kunming 650091, Yunnan, China
e-mail: kangqing595@163.com

J. He (✉)
School of Computer Science and Engineering, University of Electronic
Science and Technology of China, Chengdu 611731, China
e-mail: ynuhj@163.com

S. Patnaik and X. Li (eds.), *Proceedings of International Conference on Soft
Computing Techniques and Engineering Application*, Advances in Intelligent Systems
and Computing 250, DOI: 10.1007/978-81-322-1695-7_26, © Springer India 2014

collect data from vehicle owners around or street surveillance video. But it is very difficult to ensure the accuracy of the accident or violation identification due to the small amount of data as well as some wrong data. There may exist fake if we use just one or two videos to help with the accident identification. In many cases, traffic police even need to distinguish which evidence is believable and which is not. More believable spot data could be needed to improve the efficiency and accuracy of the traffic accident identification.

Recent development of high-speed wireless networks and embedded systems has enabled the recording and delivery of high-performance multimedia to heterogeneous mobile users. Mobile devices can connect to cloud server through networks easily. Vehicle video recorder has been used by most of the vehicle owners. With the development of next-generation automotive IT, the vehicle video recorder can be used as connected mobile devices.

In this paper, we propose a new method for traffic accident identification. The method is to collect large amount of data through the vehicle video recorder in every vehicle. Combining the analysis of these data with GPS technology, it can compute multiple values which represent the similarity of different frames. Based on these values, we can classify the video frames into different parts. Each part may represent the recorded data from certain angle. Video reorganization technology can help us to organize these different parts into a full video. It can ensure the accuracy of accident identification.

The rest of the paper is structured as follows: Section 2 describes the main idea of the approach, Sect. 3 presents the main architecture and framework of the approach, Sect. 4 explains the concrete realization steps, Sect. 5 describes the related work and background, and Sect. 6 concludes our work.

2 The Main Idea of Our Approach

When a traffic accident happens, traffic police need to collect data from the scene. These data can be of different types such as street surveillance video, testimony from people around, scattered pictures, and so on. But in many cases, we do not have the surveillance video or the video is blur. So it will be very difficult for traffic police to identify the truth and responsibility of the accident. There is nearly no feasibility for traffic police to do the identification just according to the testimony from people around or some scattered pictures. We need some ways to collect enough data from the accident scene and reproduce the accident.

In order to help solving this problem, we propose a novel solution which can collect large amount of real-time video frames with geographical location. These data were collected from vehicle video recorders. Then, we transfer the video frame data to cloud server. After the accident happened, we can analyze the video data recorded around the accident geographical area, compute similarity of different video frames, classify these frames into different sets according to the calculated similarity, and then use the classified frames to reorganize a full video of the accident. The full video is used to help with the accident identification work. That may greatly improve the accuracy and reliability of accident identification.

The main functions of our approach is to collect real-time video frame data through vehicle video recorders, search video frames according to parameters, compute similarity, classify them, and reorganize the frames.

Each vehicle video recorder can send its video frames to cloud server in real time, and to the video frames can be added a geographical location mark and time can be recorded. When received some accident parameters such as the accident geographical location and time, the cloud video server searches out the video frames according to the accident parameters and then computes the similarity of different video frames, and we design an algorithm to classify these frames using the similarity value. This produced a list of frame sets, each set contains some frames with high similarity. Finally, these frame sets are used to reorganize the full video which could help traffic police to reproduce the accident.

The solution contains the following five steps:

1. Collect real-time video frame data marked with geographical location from vehicle video recorder
2. Receive accident parameters such as geographical location, happen time, and so on
3. Search out related video frames according to the accident parameters
4. Classify frames based on calculating frame similarity value
5. Use the final set of frames to reorganize the full video and output it to traffic police.

3 System Architecture

3.1 Main Architecture

The main architecture of the approach is described in Fig. 1.

Every vehicle video recorder connects to the cloud server via wireless network. We deem that the video can be marked with geographical location information for every frame. The recorded video frames are automatically sent to the cloud server.

On the cloud server, we can store all the video frames through geographical location. So if we input a certain geographical location and time as keyword, the server will find out the video frames with same or similar geographical location as the keyword.

We use an algorithm to classify the frames searched out by inputted accident geographical location and accident time. This algorithm is based on image similarity calculation method. This means we put the high-similarity frames together. The classify algorithm will generate a list of frame sets. Each set contains some frames with high similarity.

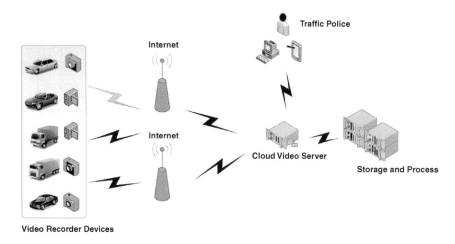

Fig. 1 Main architecture

Finally, we use video reorganization technology to reproduce the accident scene video according to the list of frame sets generated before. And the reorganized video can be used for accident identification.

3.2 General Framework

The general framework of our approach is illustrated in Fig. 2.

In Step 1, each vehicle video recorder transfers the real-time video frames to the cloud server through Internet. Of course, each frame is marked with geographical location and recorded time.

In Step 2, when some accident happens, the traffic police should input the geographical location and time of the accident.

Step 3 will search out related video frames according to the data received in Step 2. The related video frames have the similar geographical location data and time. This means we get the video frames which are recorded at the nearby place of accident during the accident time.

Step 4 uses an algorithm to classify the frames searched out in Step 3. This algorithm is based on image similarity calculation method. We put the high-similarity frames together. The classify algorithm will generate a list of frame sets. Each set contains some frames with high similarity.

The final step of our solution is to reorganize the classified frames sets into a full video. Then, the video is outputted to traffic police to help them for the accident identification.

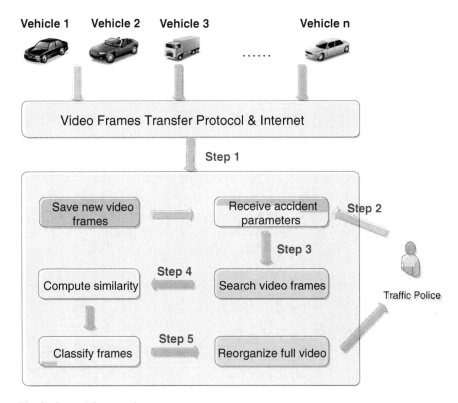

Fig. 2 General framework

4 Realization Steps

4.1 Some Basic Definitions

We use symbol D to denote the set of devices (vehicle video recorder) which transferred real-time video frame data to cloud server, where $D = \{d_1, d_2, \ldots, d_n\}$. Use symbol T to denote the set of time, where $T = \{t_1, t_2, \ldots, t_n\}$, use symbol G to denote the set of GPS location, where $G = \{g_1, g_2, \ldots, g_n\}$, and use symbol F to denote the set of video frames transferred to the server, where $F = \{f_1, f_2, \ldots, f_n\}$, and each frame is related to device, record time, and GPS location; this is denoted as the following formula:

$$f_i = \{d_j, t_k, g_m\}, f_i \in F, d_j \in D, t_k \in T, g_m \in G. \tag{1}$$

We define $s(f_i, f_j)$ as the similarity of two frames f_i and f_j.

4.2 Main Steps

Step 1: Receive real-time video frames from vehicle video recorder devices

The cloud server receives a new frame f_i from the devices and adds f_i into our video frame set F.

$$F = F \cup f_i. \tag{2}$$

Step 2: Receive accident parameters (accident time, geographical location of accident) from traffic police

When an accident happens, traffic police will input the traffic parameters which will be sent to the server through interface. The parameters include the following: the time of accident t_a, the geographical location of accident g_a, the time area t_r, and the location area g_r

Step 3: Search out related video frames according to the traffic parameters received from step 2.

We define two ranges:

- Time range: $t_{condition} = [t_a - t_r, t_a + t_r]$
- Geographical location range: $g_{condition} = [g_a - g_r, g_a + g_r]$.

The search result is a set of video frames F_{search} which accord with the search condition.

$$F_{search} = \{f_1, f_2, \ldots, f_n\}. \tag{3}$$

Each video frame in the search result can be denoted as follows:

$$f_i = \{d_j, t_k, g_m\}, f_i \in F_{search}, d_j \in D, t_k \in t_{condition}, g_m \in g_{condition}. \tag{4}$$

Step 4: Calculate the similarity and classify the frames according to the similarity value

We define $s(f_i, f_j)$ as the similarity of two frames f_i and f_j. $s(f_i, f_j) < 1$.

Choose the first frame f_1 from F_{search}, and calculate the similarity of f_1 and other frames in F_{search}. This will generate a list of similarity values. According to the similarity, we can classify F_{search} into two parts: one part F_{high}^1 has the high similarity with f_1 and another set F_{low}^1 with low similarity with f_1 (we can set a standard value for similarity, such as 0.85).

Then, choose the first frame f_1' from F_{low}^1, and calculate the similarity of f_1' and other frames in F_{low}^1. Repeat the above steps, until F_{low}^i has only one element.

This algorithm is a recursion which can be described as follows:

```
Sub Classification (F_search)
{
while (count (F_search) > 1)
{
m++;
choose f₁ as frame from F_search
dim S = {s(f₁,fᵢ), s(f₁,fᵢ₊₁), ..., s(f₁,fₙ)}
dim F¹_high as the frames set with high similarity with f₁
dim F¹_low as the frames set with low similarity with f₁
foreach (fᵢ(i ≠ 1) in F_search) {
calculate s(f₁,fᵢ), add it to S
}
Foreach (s(f₁,fᵢ) in S) {
if s(f₁,fᵢ) ≥ 0.85 {
add fᵢ to F¹_high
} else {
add fᵢ to F¹_low
}
}
Save F¹_high to server
Classification (F¹_low)
}
}
```

This step will generate an array containing multiple sets. Each set contains a number of video frames with high similarity and can be considered as one part of the scene. We denote it as the following:

$$F_{\text{final}} = \{F^1_{\text{similar}}, F^2_{\text{similar}}, F^3_{\text{similar}}, \ldots, F^n_{\text{similar}}\} \tag{5}$$

$$F^i_{\text{similar}} = \{f_1, f_2, \ldots, f_m\}, s(f_k, f_j) > 0.95 \text{ and } f_k, f_j \in F^i_{\text{similar}}. \tag{6}$$

Each frame of video can be seen as an image. In this paper, we consider a video frame as a picture with recorded time and geographical location. So calculating video frame similarity is same as the calculation of image similarity. There exist a number of image similarity calculation algorithms that can do retrieval on the whole image content [1–4]. In this paper, we choose the image similarity calculation method from Huang and Chin-Jung, which is amply described in the Ref. [1].

Step 5: Reorganize full video with the set of frames classified in Step 4.

Use F_{final} to reorganize the full video. This video is the result of our solution which can be used to help traffic police do the accident identification works.

The reorganization method uses the image similarity value as the base data. It uses a self-adapting reorganization algorithm which can choose frames according to the degree of image variety [5]. In this algorithm, we use image similarity value to represent the degree of image variety.

The main idea of the algorithm is as follows:

Choose f_K as current frame, f_L as the next frame of f_K, and T_h as the standard value. Dim a linked list H to store the reorganized video frames list.

(1) In each classified frame set F_{similar}, set $f_K = f_i, f_i \in F_{\text{similar}}$, so $f_L = f_{i+1}, f_{i+1} \in F_{\text{similar}}$. In this step, linked list H has only one element f_i.
(2) Get the similarity value of f_K and f_L, that is, $s(f_K, f_L)$.
(3) If $s(f_K, f_L) \leq T_h$, then $f_K = f_K, f_L = f_L + 1$;

If $s(f_K, f_L) > T_h$, then $f_K = f_L, f_L = f_K + 1$, add f_K into linked list H.

(4) If f_K is not the last frame of F_{similar}, jump to repeat (2); otherwise, end the algorithm.

The generated linked list H (equal to F_{final}) is the reorganized video frame.

5 Related Work

In this paper, we propose a new solution to collect video data, analyze, and reorganize videos in traffic accident identification, data collect tool is vehicle video recorder installed in most cars. This solution is supported by automotive IT technology and video reorganization [6–8].

Multi-source traffic data analysis technology also plays an important role in the areas of traffic analysis and management. Paper [9] uses the multi-source traffic data of the intelligent transportation system (ITS) applications to analyze and evaluate the road network performance. The multi-source traffic data-based computation algorithms and models on travel speed calculation, short-term forecasting, and automatic incident detection were introduced in this paper. Paper [10] discusses data fusion algorithms for computing the "time to lane crossing" (TLC) of a vehicle traveling along a lane on the basis of road images, collected by an on-board video camera and kinematic data coming from car sensors.

The video comparison and video reorganization technology can make a video by numbers of video frames from different videos [11–13]. This has been widely used in many areas such as criminal detection, video copy detection, and so on.

Paper [14] presented the Argos system which is an improved in-vehicle data recorder (IVDR) that allows recording many kinds of alphanumerical data such as the speed (vehicle data), the point of gaze (driver data), or the current distance to lateral road marks (environmental data). Argos can also record up to nine simultaneous video images which are synchronized with the alphanumerical data. Based on these data, a better understanding of driver actions will help in determining the most common reasons for car accidents and help researchers in the study of car driver behaviors.

6 Conclusions

Traffic accident identification is an important work for traffic police when accident happens. Generally, most of the traffic identification is based on the street surveillance video or testimony from people around. These data are insufficient and fragment; sometimes, it is difficult to organize them. Toward this problem, we propose a new approach to reproduce the traffic accident based on the vehicle video recorders' data. Most cars have vehicle video recorder which could be used as data collect tools. We provided a general architecture of a system to do data collection, data classification, and use the video frames with high similarity to reorganize a full video. The reorganized video can be used as the reproduction of the accident. We also plan to work on the use of big data analysis and cloud computing technology to improve the speed of data processing.

Acknowledgments This work is supported by "CDIO-based Data Engineering Research and Implementation" (Grant No. Rj010) and National Natural Science Foundation of China (Grant No. 61263043, 61063044).

References

1. Huang, C.J.: A new approach of image similarity calculation. International Conference on Management and Service Science (MASS) (2010)
2. Chechik, G., Sharma, V., Shalit, U., Bengio, S.: Large scale online learning of image similarity through ranking. J. Mach. Learn. Res. 11, 1109–1135
3. Stejic, Z.: Image similarity computation using local similarity patterns generated by genetic algorithm. Proceedings of the 2002 Congress on Evolutionary Computation, CEC '02. (vol. 1). pp. 771–776 (2002)
4. Chalom, E., Asa, E., Biton, E.: Measuring image similarity: an overview of some useful applications. IEEE Instrum. Meas. Mag. 16(1), 24–28
5. Cheng, Y., Shen, Y., Wang, Y., Luo, R.: A video reorganization method based on contents. J. Image Graphics. 6(4) (2001)
6. Nolte, T., Hansson, H., Bello, L.L.: Implementing Next Generation Automotive Communications (2004)
7. Lim, S.H., Choi, M., Jeong, Y.S.: Data reorganization for scalable video service with embedded mobile devices, in: ACM Transactions on Embedded Computing Systems (TECS)—Special issue on embedded systems for interactive multimedia services, vol. 12, Issue 2, Feb 2013, Article No. 27 (2013)
8. Shenoy, P., Vin, H.M.: Efficient support for interactive operations in multi-resolution video server. ACM Multimedia Syst. 7(3), 241–253 (1999)
9. Weng, J., Rong, J., Liu, L., Zhai, Y.: Applications of multi-source traffic data on mobility analysis for urban road network. Comput. Intell. Traffic Mobility Atlantis Comput. Intell. Syst. 8, 267–296 (2013)
10. Cario, G., Casavola, A., Franze, G., Lupia, M.: Data fusion algorithms for lane departure warning systems. American Control Conference (ACC), pp. 5344–5349 (2010)
11. Boreczky, J.S., Rowe, L.A.: Comparison of video shot boundary detection techniques. J. Electron. Imaging. 5(2), 122–128 (1996)

12. Hampapur, A., Hyun, K., Bolle, R.M.: Comparison of sequence matching techniques for video copy detection (2001)
13. Michalopoulos, P.G.: Vehicle detection video through image processing: the Autoscope system. Vehicular Technology, IEEE Transactions (1991)
14. Perez, A., Garcia, M.I., Nieto, M., Pedraza, J.L., Rodriguez, S., Zamorano, J.: Argos: An advanced in-vehicle data recorder on a massively sensorized vehicle for car driver behavior experimentation. IEEE Trans. Intell. Transp. Syst. 11(2), 463–473

Improved RNS Montgomery Modular Multiplication with Residue Recovery

Tao Wu, Shuguo Li and Litian Liu

Abstract Finite field arithmetic in residue number system (RNS) necessitates modular reductions, which can be carried out with RNS Montgomery algorithm. By transforming long-precision modular multiplications into modular multiplications with small moduli, the computational complexity has decreased much. In this work, two implementation methods of RNS Montgomery algorithm, *residue recovery* as well as *parallel base conversion*, are reviewed and compared. Then, we propose a new residue recovery method that directly employs binary system rather than mixed radix system to perform RNS modular multiplications. This improvement is appropriate for a series of long-precision modular multiplications with variant operands, in which it is more efficient than parallel base conversion method.

Keywords Montgomery algorithm · Residue number system · Binary system

T. Wu (✉)
Department of Microelectronics and Nanoelectronics, Tsinghua University,
Beijing 100084, People's Republic of China
e-mail: twu03ster@gmail.com

S. Li · L. Liu
Institute of Microelectronics, Tsinghua University,
Beijing 100084, People's Republic of China
e-mail: lisg@tsinghua.edu.cn

L. Liu
e-mail: liulitian@tsinghua.edu.cn

S. Patnaik and X. Li (eds.), *Proceedings of International Conference on Soft Computing Techniques and Engineering Application*, Advances in Intelligent Systems and Computing 250, DOI: 10.1007/978-81-322-1695-7_27, © Springer India 2014

1 Introduction

Modular multiplication is essential for popular Public Key cryptography that is defined in a finite field: RSA, Diffie-Hellman, and elliptic curve cryptography. In fact, a cryptosystem with enough security level often necessitates frequent large-operand modular multiplications.

As is known, additions, subtractions, and multiplications are parallel and carry-free in residue number system (RNS) [1–4]. Therefore, RNS arithmetic accelerates long-precision operations. However, modular reduction in RNS is harder than that in binary system.

An effective way to tackle RNS modular multiplication is to extend Montgomery algorithm from binary system to RNS [5–12]. Regular Montgomery algorithm breaks modular reductions in binary system into sequential additions and right shifts [13, 14], while the RNS Montgomery modular multiplication utilizes parallelism of RNS. As far as cryptography is concerned, the parallelism also provides natural immunity to side attacks, when the integral information is distributed into several channels. Besides, Phillips et al. [15] have proposed an algorithm directly based on the chinese remainder theorem, in which the constants are also represented in RNS.

In this paper, we will discuss the less-efficient RNS Montgomery algorithm with residue recovery [7, 10, 16], after an overview of the parallel base conversion method. It has been found that the usual residue recovery method can be built on binary system and RNS, without transformation into mixed radix system.

The remaining part of this paper is organized as follows: Sect. 2 reviews the RNS Montgomery algorithm with parallel base conversion; Sect. 3 discusses the RNS Montgomery algorithm with residue recovery; in Sect. 4, we propose our residue recovery method with binary system; and the last section concludes this paper.

2 RNS Montgomery Algorithm with Parallel Base Conversion

There are several approaches to implement RNS Montgomery modular multiplications, all of which obtain the quotients in parallel and then perform modular reduction of $M^{-1} \bmod P$ in an auxiliary RNS [5–9, 11, 17–19]. Among these methods, the parallel base conversion algorithm is the most efficient in computation.

2.1 Montgomery Algorithm

Usually, Montgomery algorithm is implemented in an interleaved form, and the multiplier is counted bit by bit while the multiplicand enters as a whole [14].

Algorithm 1 Montgomery algorithm [13]

Input: Integers A, B, and P satisfy $0 \leq A < 2^n$, $0 \leq B < P$, $2^{n-1} < P < 2^n$, and $\text{GCD}(P, 2) = 1$.
Let $R = 2^n$ and R^{-1} be the modular inverse of R modulo P. Then, $RR^{-1} - P\rho = 1$, or
$\rho = (-P^{-1}) \bmod R$.

Output: $S \equiv A \cdot B \cdot R^{-1} \pmod{P}$, where $0 \leq S < 2P$.

    1: $T = A \cdot B$;
    2: $Q = T\rho \bmod R$;
    3: $S = (T + PQ)/R$;
    4: **return** S.

2.2 RNS Montgomery Algorithm with Parallel Base Conversion

At Line 1, $P^{-1} = \text{RNS}\left((p_1)_{m_1}^{-1}, (p_2)_{m_2}^{-1}, \ldots, (p_n)_{m_n}^{-1} \right)$.

At Line 3, $M^{-1} = \left(|M^{-1}|_{m_{n+1}}, |M^{-1}|_{m_{n+2}}, \ldots, |M^{-1}|_{m_{2n}} \right)$. There are two base conversions in such an RNS Montgomery modular multiplication: (1) Conversion of $Q = (q_1, q_2, \ldots, q_n)$ from Ω to $Q' = (q_{n+1}, q_{n+2}, \ldots, q_{2n})$ in Γ; (2) Conversion of $R' = (r_{n+1}, r_{n+2}, \ldots, r_{2n})$ from Γ to $R = (r_1, r_2, \ldots, r_n)$ in Ω.

Algorithm 2 RNS Montgomery algorithm with parallel base conversion [5, 7]

Input: The moduli for residue number system Ω are $\{m_1, m_2, \ldots, m_n\}$, with $M = \prod_{i=1}^{n} m_i$.
Meanwhile, an auxiliary residue number system Γ are defined by the other moduli
$\{m_{n+1}, m_{n+2}, \ldots, m_{2n}\}$, with $N = \prod_{i=n+1}^{2n} m_i$. Integers A, B, and P are represented in $\Omega \bigcup \Gamma$ as
$A = (a_1, a_2, \ldots, a_{2n})$, $B = (b_1, b_2, \ldots, b_{2n})$, $P = (p_1, p_2, \ldots, p_{2n})$. Additionally, $A \cdot B < M \cdot P$,
$2P < M < N$.

Output: $R \equiv A \cdot B \cdot M^{-1} \pmod{P}$ with $R < 2P$.

    1: $Q = (A \times B) \times (-p^{-1}) : \Omega$;
    2: $Q' \leftarrow Q : \Gamma \leftarrow \Omega$;
    3: $R' = (A \times B + P \times Q') \times M^{-1} : \Gamma$;
    4: $R \leftarrow R' : \Omega \leftarrow \Gamma$.
    5: **return** $R \bigcup R' : \Omega \bigcup \Gamma$.

Set $M_j = M/m_j$, $\sigma_i = \left| q_i |M_i^{-1}|_{m_i} \right|_{m_i}$, and $G_{i,j} = |M_j|_{m_{n+i}}$, then

$$
\begin{pmatrix} q_{n+1} \\ q_{n+2} \\ \vdots \\ q_{2n} \end{pmatrix} = \begin{pmatrix} G_{11} & G_{12} & \cdots & G_{1n} \\ G_{21} & G_{22} & \cdots & \vdots \\ \vdots & \vdots & \ddots & G_{n-1,n} \\ G_{n1} & G_{n2} & \cdots & G_{n,n} \end{pmatrix} \begin{pmatrix} \sigma_1 \\ \sigma_2 \\ \vdots \\ \sigma_n \end{pmatrix} - \alpha \cdot \begin{pmatrix} |M|_{m_{n+1}} \\ |M|_{m_{n+2}} \\ \vdots \\ |M|_{m_{2n}} \end{pmatrix}, \quad (1)
$$

where q_j is obtained as modulo m_j, and the integer factor α comes from the improved chinese remainder theorem [20]. Suppose that the moduli m_i have k binary bits, i.e., $2^{k-1} < m_i < 2^k$, for $i = 1, 2, \ldots, n$. With $M_i = M/m_i$, the chinese remainder theorem can be written as

$$
\begin{aligned}
x &= \left| \sum_{i=1}^{n} M_i \left| x_i M_i^{-1} \right|_{m_i} \right|_M = \sum_{i=1}^{n} M_i \left| x_i M_i^{-1} \right|_{m_i} - \alpha M \\
&= \sum_{i=1}^{n} M_i \sigma_i - \alpha M,
\end{aligned}
\tag{2}
$$

where $\sigma_i = \left| x_i M_i^{-1} \right|_{m_i}$, and the indefinite factor α can be fixed by an approximation method [5, 8] or by an extra modulus [7, 20].

3 RNS Montgomery Algorithm with Residue Recovery

Although a direct map of Montgomery algorithm to RNS suffers from a problem in losing residues, it is a good guide or bridge for subsequent algorithms.

Algorithm 3 Direct map of Montgomery algorithm in residue number system [16]

Input: The integers A, B, and P are represented with residue number system Ω: $\{m_1, m_2, \ldots, m_n\}$, with $M = \prod_{i=1}^{n} m_i$ and $M_i = M/m_i$. $A = (a_1, a_2, \ldots, a_n)$, $B = (b_1, b_2, \ldots, b_n)$, and $P = (p_1, p_2, \ldots, p_n)$. Meanwhile, A is also expressed in the mixed radix system as $A = a_1 + a_2 \cdot m_1 + \cdots + a_n \cdot \prod_{i=1}^{n-1} m_i$. In addition, $0 \leq A < \frac{m_n - 1}{2} \cdot \frac{M}{m_n}$, $0 \leq B < 2P$, $0 < P < \frac{M}{\max_{1 \leq i \leq n}\{m_i\}}$.

Output: $R = A \cdot B \cdot M^{-1} = \text{RNS}(r_1, r_2, \ldots, r_n)$.

  1: $R := (0, 0, \ldots, 0)$;
  2: **for** $i = 0$ to $n - 1$ **do**
  3:   $q_i' := (r_i + a_i' \cdot b_i)(m_i - p_i)_i^{-1} \bmod m_i$;
  4:   $R := R + a_i' \cdot B + q_i' \cdot P$;
  5:   $R := R \div m_i$;
  6: **end for**
  7: **return** R.

3.1 Direct Map of Montgomery Algorithm into RNS

At Line 3, we make sure that

$$
(R + a_i' \cdot B + q_i' \cdot P) \bmod m_i = 0.
\tag{3}
$$

Given that $R \bmod m_i = r_i$, $B \bmod m_i = b_i$, and $P \bmod m_i = p_i$, we have

$$R + a_i' \cdot B + q_i' \cdot P \equiv r_i + a_i' \cdot b_i + q_i' \cdot p_i$$
$$= (r_i + a_i' b_i) + p_i \cdot (r_i + a_i' b_i)(m_i - p_i)^{-1} \bmod m_i$$
$$\equiv (r_i + a_i' b_i)\left(1 - (m_i - p_i)(m_i - p_i)^{-1}\right)$$
$$= 0 (\bmod\, m_i).$$

Therefore, at Line 4 of the ith iteration, R is a multiple of m_i. At the end of all iterations, as is shown in [7, 16], we have

$$R = \frac{1}{m_1 \cdot m_2 \cdots m_n}\left(B \cdot \sum_{i=1}^{n} a_i' \prod_{j=1}^{i-1} m_i + P \cdot \sum_{i=1}^{n} q_i' \prod_{j=1}^{i-1} m_i\right). \qquad (4)$$

Notice that $M = m_1 \cdot m_2 \cdots m_n$ and $A = \sum_{i=1}^{n} a_i' \prod_{j=1}^{i-1} m_i$, the above equation yields $R \equiv A \cdot B \cdot M^{-1}(\bmod\, P)$. At the last step of each loop, R will keep below $3P$, which can be examined by induction [16]. Given that $R_{n-1} < 3P$ and $A < \frac{m_n-1}{2} \cdot \frac{M}{m_n}$, it should yield [16] $R_n < 2P$.

However, there is a problem hidden at line L5: The residue r_i gets lost when $r_i \div m_i$ appears [7]. In RNS, r_i/m_j is equivalent to $r_i \times (m_j)_{m_i}^{-1} \bmod m_i$. However, $(m_i)_{m_i}^{-1}$ does not exist because of $m_i \times (m_i)_{m_i}^{-1} \equiv 0 (\bmod\, m_i)$ (The product of an integer and its inverse modular m_i should be equal 1 modulo m_i). Therefore, one more residue becomes meaningless after each cycle of the loop.

3.2 RNS Montgomery Algorithm with Residue Recovery

With the improved chinese remainder theorem [20], the lost residue can be recovered by the help of the redundant residue r_{n+1}. At Line 4,

---
Algorithm 4 RNS Montgomery algorithm with residue recovery [16]

Input: The group of moduli: $\{m_1, m_2, \ldots, m_n\} \bigcup \{m_{n+1}\}$ defines the residue number system Ω, with $M = \prod_{i=1}^{n} m_i$ and $M_i = M/m_i$. A is represented in mixed radix system: $A = a_1' + \sum_{i=2}^{n} a_i' \prod_{j=1}^{i-1} m_j$. Meanwhile, $B = (b_1, b_2, \ldots, b_n, b_{n+1})$ and $P = (p_1, p_2, \ldots, p_n, p_{n+1})$. Also, $A < \frac{m_n-1}{2} \cdot \frac{M}{m_n}$, $B < 2P$.
Output: $R \equiv A \cdot B \cdot M^{-1}(\bmod\, P)$.
    1: $R := (0, 0, \ldots, 0)$;
    2: **for** $i = 1$ to n **do**
    3: $q_i' := (r_i + a_i' \cdot b_i)|(m_i - p_i)^{-1}|_{m_i} \bmod m_i$;
    4: $R' := R + a_i' \cdot B + q_i' \cdot P$;
    5: $R := R' \div m_i$;
    6: $r_i := \mathrm{restoreRNS}(r_1, \ldots, r_{i-1}, r_{i+1}, \ldots, r_{n+1})$;
    7: **end for**
    8: **return** R.

$$R + a'_i \cdot B + q'_i \cdot P = (r_1, \ldots, r_{n+1}) + a'_i \cdot (b_1, \ldots, b_{n+1}) + q'_i \cdot (p_1, \ldots, p_{n+1})$$
$$= \left(\left| r_1 + a'_i \cdot b_1 + q'_i \cdot p_1 \right|_{m_1}, \ldots, \left| r_{n+1} + a'_i \cdot b_{n+1} + q'_i \cdot p_{n+1} \right|_{m_{n+1}} \right).$$

At Line 5, $R \div m_i$ is conducted for $j \neq i, j \in \{1, 2, \ldots, n+1\}$ by multiplying the modular inverse, i.e., $r_j := r_j \cdot \left| m_i^{-1} \right|_{m_j} \bmod m_j$ for $j \neq i$.

At Line 6, the function 'restore RNS' just restores the lost residue r_i at Line 5.

Set $\sigma_{i,j} = \left| r_j \cdot \left| \left(\frac{M_{t,i}}{m_j} \right)^{-1} \right|_{m_j} \right|_{m_j}$, then r_i is obtained by the improved chinese remainder theorem as follows:

$$r_i = \left| \sum_{j=1, j \neq i}^{n} \left| \sigma_{i,j} \cdot \left| \frac{M_i}{m_j} \right|_{m_i} \right|_{m_i} - \left| \gamma_i \cdot \left| M_i \right|_{m_i} \right|_{m_i} \right|_{m_i}, \tag{5}$$

where

$$\gamma_i = \left| \left| M_i^{-1} \right|_{m_{n+1}} \cdot \left(\left| \sum_{j=1, j \neq i}^{n} \left| \sigma_{i,j} \cdot \left| \frac{M_i}{m_j} \right|_{m_{n+1}} \right|_{m_{n+1}} \right| - r_{n+1} \right) \right|_{m_{n+1}}. \tag{6}$$

Taking the computation of a basic modular multiplication with m_i as the unit, i.e., $a_i \cdot b_i \bmod m_i$, then there are $(6n + 5)$ modular multiplications at each iteration of the above algorithm. Then, a total Montgomery modular multiplication with residue recovery costs $6n^2 + 5n$ basic modular multiplications.

4 Proposed Algorithm with Residue Recovery

The idea of RNS Montgomery algorithm is to keep the final result below some constants such as $2P$ with parallel or sequential modular reductions. The afore-mentioned residue recovery method performs sequential k-bit modular multiplications, in which one operand should be represented in mixed radix system. However, it is not convenient for conversions among binary system, RNS, and mixed radix system. Therefore, we propose a new residue recovery algorithm that avoids the use of mixed radix system.

4.1 Initial Algorithm

Algorithm 5 Proposed RNS Montgomery algorithm with residue recovery I

Input: The residue number system Ω has a group of moduli: $\{m_1, m_2, \ldots, m_n\} \bigcup \{m_{n+1}\}$, where $m_1 = 2^t$, $2^{k-1} < m_i < m_j < 2^k$ for $2 \leq i < j \leq n + 1$, and the integer $t \leq k - 1$. All the moduli are relatively prime, with $M = \prod_{i=1}^{n} m_i$ and $M_i = M/m_i$. Also, $A = \sum_{i=1}^{n} A_i \cdot 2^{(i-1)t}$, where $0 \leq A_i \leq 2^t - 1$ and $A < 2^{nt-1}$. B and P are represented in residue number system: $B = (b_1, b_2, \ldots, b_n, b_{n+1})$, and $P = (p_1, p_2, \ldots, p_n, p_{n+1})$, with $T = 2^{nt}$, $A < T$, $P < T$, and $B < 2P$.

Output: $R \equiv A \cdot B \cdot T^{-1} \pmod{P}$, where $0 \leq R < 2P$.

    1: $R := (0, 0, \ldots, 0)$;
    2: **for** $i = 1$ to n **do**
    3: $\quad q_i' := (r_1 + A_i \cdot b_i) \big| (m_1 - p_1)^{-1} \big|_{m_1} \bmod m_1$;
    4: $\quad R' := R + A_i \cdot B + q_i' \cdot P$;
    5: $\quad R := R' \div m_1$;
    6: $\quad r_1 := \text{restoreRNS}(r_2, r_3, \ldots, r_{n+1})$;
    7: **end for**
    8: **return** R.

With $i = 1$,

$$R = (A_1 \cdot B + q_1' \cdot P)/m_1$$
$$< (2^t - 1)B + (2^t - 1)P/2^t$$
$$< 3P.$$

Assuming $R < 3P$ at the end of the ith iteration, then the $(i + 1)$-th iteration yields

$$R < (3P + (2^t - 1)2P + (2^t - 1)P)/2^t = 3P.$$

Then, the induction shows that $R < 3P$ at last. However, we expect to get the final result within a smaller range $[0, 2P)$. As

$$R < (3P + A_n \cdot B + q_n' \cdot P)/2^t$$
$$\leq (3P + A_n \cdot 2P + (2^t - 1)P)/2^t,$$

we can set

$$(3P + A_n \cdot 2P + (2^t - 1)P)/2^t < 2P.$$

The above equation yields $A_n < 2^{t-1} - 1$. Notice that $A - A_n \cdot 2^{(n-1)t} < 1 \cdot 2^{(n-1)t}$, then as long as $A < (A_n + 1) \cdot 2^{(n-1)t} < 2^{nt-1}$, there will be $R < 2P$.

At Line 6, we have

$$r_1 = \left| \left| \sum_{j=2}^{n} \sigma_{1,j} \cdot \left| \frac{|M_1|}{m_j} \right|_{m_1} \right|_{m_1} - \left| \gamma_1 \cdot |M_1|_{m_1} \right|_{m_1} \right|_{m_1}, \tag{7}$$

where $\sigma_{1,j} = \left| r_j \cdot \left| \left(\frac{M_1}{m_j} \right)^{-1} \right|_{m_j} \right|_{m_j}$, and

$$\gamma_1 = \left| |M_1^{-1}|_{m_{n+1}} \cdot \left(\left| \sum_{j=2}^{n} \sigma_{1,j} \cdot \left| \frac{|M_1|}{m_j} \right|_{m_{n+1}} \right|_{m_{n+1}} - r_{n+1} \right) \right|_{m_{n+1}}. \tag{8}$$

4.2 Improved Algorithm

While Algorithm 5 is more efficient than Algorithm 4, it still requires more sequential steps to compute one RNS Montgomery modular multiplication than parallel base conversion method with Algorithm 2. However, we found out that a little revision will double the efficiency of our proposal, which is shown in Algorithm 6.

At first glance, Algorithm 6 differs from Algorithm 5 in the choice of t and m_{n+1}. By setting $t = 2k$ rather than $t = k$, the number of loops is reduced by a half, while the computational complexity only increases a little. The range of m_1 then gets much larger than 2^k and other RNS moduli, but it does not impact the validity of the algorithm.

By setting $m_{n+1} = 2^s - 1$, then the modular reduction in $x \cdot y \bmod m_{n+1}$ can be simplified. Take Eq. 6 as example, set $x = \sigma_{i,j} < 2^k$, $y = \left| \frac{M_i}{m_j} \right|_{m_{n+1}} < 2^s$. Since $k \gg s$, one can set $l = \lceil k/s \rceil$, and there will be $x = x_{l \cdot s - 1..0}$, with $x_i = 0$ for $i \geq k$. As $2^{j \cdot s} \bmod (2^s - 1) = 1$, then

$$x \bmod m_{n+1} = \sum_{j=1}^{l} x_{j \cdot s - 1...j \cdot s - s} \cdot 2^s \bmod (2^s - 1) = \sum_{j=1}^{l} x_{j \cdot s - 1...j \cdot s - s}.$$

Furthermore, the additions modulo $2^s - 1$ can be simplified by setting the highest carry out as the lowest carry in. Assuming that $x' = x \bmod m_{n+1}$, then $x \cdot y \bmod m_{n+1} = x' \cdot y \bmod (2^s - 1)$ can be computed by one $s \times s$-bit multiplication and one s-bit addition.

The sequential steps and computational complexity of Algorithm 6 can be measured as follows:

Algorithm 6 Proposed RNS Montgomery algorithm with residue recovery II

Input: RNS $\Omega : \{m_1, m_2, \ldots, m_n\} \bigcup \{m_{n+1}\}$, where $t = 2k$, $m_1 = 2^t$, $2^{k-1} < m_i < m_j < 2^k$ **for**
$2 \leq i < j \leq n$, $m_{n+1} = 2^s - 1$, $s \ll k$. $GCD(m_i, m_j) = 1$ for $i \neq j$. $M = \prod_{i=1}^{n} m_i$, $M_i = M/m_i$.
$n' = \lceil n/2 \rceil$, $A = \sum_{i=1}^{n'} A_i \cdot 2^{(i-1)t}$, where $0 \leq A_i \leq 2^t - 1$ and $A < 2^{n't-1}$. B and P are
represented in residue number system: $B = (b_1, b_2, \ldots, b_n, b_{n+1})$, and
$P = (p_1, p_2, \ldots, p_n, p_{n+1})$, with $T = 2^{2n't}$, $A < T$, $P < T$, and $B < 2P$.
Output: $R \equiv A \cdot B \cdot T^{-1} (\text{mod } P)$, where $0 \leq R < 2P$.

    1: $R := (0, 0, \ldots, 0)$;
    2: **for** $i = 1$ to n' **do**
    3:     $R' := R + A_i \cdot B$;
    4:     $q_i' := q_1' \cdot \left| (m_1 - p_1)^{-1} \right|_{m_1}$ $\text{mod } m_1$;
    5:     $R'' := R' + q_i' \cdot P$;
    6:     $R := R'' \div m_1$;
    7:     $r_1 := \text{restoreRNS}(r_2, r_3, \ldots, r_{n+1})$;
    8: **end for**
    9: **return** R.

---

1. Assuming that the complexity of a k-bit modular multiplication is 1, and the complexity of a $k \times k$ multiplication is $1/3$. As is well known, modular multiplication by Montgomery algorithm and Barrett modular reduction requires three multiplications. In addition, the computational complexity of a $k \times 2k$ multiplication is $2/3$, and the complexity of an s-bit modular multiplication is ε, $\varepsilon \ll 1$.

2. Neglecting the complexity of modular additions and modular subtractions, since it is small compared with modular multiplications.

3. The computation of $A_i \cdot B$ requires one sequential modular multiplication, and the total complexity is $(n + \varepsilon)$ k-bit modular multiplications.

4. The determination of q_i' requires $2/3$ sequential modular multiplication, and the total complexity is just $2/3$. At Line 4, the value $(m_1 - p_1)^{-1}$ can be precomputed.

5. The computation of $q_i' \cdot P$ requires $5/3$ sequential modular multiplications, and the total complexity is $(5n/3 + 2\varepsilon)$ k-bit modular multiplications. The increased computational complexity of $2/3$ and ε corresponds to the modular reduction in q_i' to k-bit and s-bit moduli m_i, $i = 2, 3, \ldots, n + 1$.

6. The computation of $R_i'' \div m_1$ can be performed by modular multiplications of $m_1^{-1} = 2^{-t}$ modulo m_i. It requires one sequential modular multiplication, and the total complexity is $(n + \varepsilon)$ k-bit modular multiplications.

7. As are shown in Eqs. (5) and (6), the recovery of residue r_1 after division of m_1 requires $(2/3 + \varepsilon)$ sequential modular multiplication, and the total complexity is $(2n/3 + 2\varepsilon)$ k-bit modular multiplications. With $i = 1$ in Eq. 5, modular reduction over m_1 can be performed by truncation instead of multiplications, which leads to the complexity of $2/3$ rather than 1.

In total, Algorithm 6 merely needs

$$
\begin{aligned}
K &= (1 + 2/3 + 5/3 + 1 + (2/3 + \varepsilon)) \cdot n' \\
&= (15/3 + \varepsilon) \cdot n' \\
&\approx (2.5 + \varepsilon)n
\end{aligned}
\tag{9}
$$

sequential modular multiplications, and the total computational complexity is about

$$
\begin{aligned}
N &= ((n + \varepsilon) + 2/3 + (5n/3 + 2\varepsilon) + (n + \varepsilon) + (2n/3 + 2\varepsilon)) \cdot n' \\
&= (13n/3 + (2/3 + 6\varepsilon))n' \\
&\approx (13n/6 + 1/3 + 3\varepsilon)n
\end{aligned}
\tag{10}
$$

k-bit modular multiplications.

4.3 Comparison with Parallel Conversion Method

By contrast, the parallel base conversion method in Algorithm 2 will need about $2n + 8$ sequential k-bit modular multiplications [7, 15], and the total computational complexity is about $(2n^2 + 10n + 4)$ k-bit modular multiplications. While this result seems better than our proposals, it should be noticed that all the inputs of Algorithm 2 should be in residue number system. If only one input lies in RNS, the other should be transformed from binary number system to RNS [8], then the parallel base conversion method will require $3n + 8$ sequential k-bit modular multiplications and $(3n^2 + 9n + 4)$ k-bit modular multiplications. Therefore, if one input of RNS modular multiplication keeps in binary number system, our proposed Algorithm 6 will be better than parallel base conversion method. The above comparison is shown in Table 1, where 'Seq. Num. Mod-mult' denotes sequential number of k-bit modular multiplications and 'Tot. Num. Mod-mult' denotes the total number of k-bit modular multiplications.

If there are a series of modular multiplications by variant long-precision integers, i.e., $\tau_1, \tau_2, \ldots, \tau_h$, with h be some regular index. The modulus is denoted as P. All τ_i and P are long-precision binary numbers, e.g., between $(2^{L-1}, 2^L]$, with $L \gg 1$. Then, we intend to compute $R = \prod_{i=1}^{h} \tau_i \bmod P$.

We just need τ_1 and P in RNS $\Omega : \{m_1, m_2, \ldots, m_n\}$, where $\tau_1 = U_{\text{rns}}^{(0)} = (u_1, u_2, \ldots, u_n)$, $P = (p_1, p_2, \ldots, p_n)$. By the new residue recovery method 'MR'(Algorithm 6), one gets

$$
U_{\text{rns}}^{(1)} = \tau_1 \cdot \tau_2 \cdot T^{-1} \bmod P \equiv \text{MR}(U_{\text{rns}}^{(0)}, \tau_2)(\bmod P),
\tag{11}
$$

Table 1 Comparison of proposed residue recovery method to parallel base conversion method, with one input in RNS and the other in binary number system

Method	Residue recovery (this work)	Parallel conversion [7]
Seq. num. mod-mult	$(2.5 + \varepsilon)n$	$3n + 8$
Tot. num. mod-mult	$(13n/6 + 1/3 + 3\varepsilon)n$	$3n^2 + 9n + 4$

$$U_{\text{rns}}^{(2)} = \prod_{i=1}^{3} \tau_i \cdot T^{-2} \bmod P \equiv \text{MR}(U_{\text{rns}}^{(1)}, \tau_3)(\bmod P),$$ (12)

$$\cdots, \quad \cdots,$$

$$U_{\text{rns}}^{(h-1)} = \prod_{i=1}^{h} \tau_i \cdot T^{-(h-1)} \bmod P \equiv \text{MR}(U_{\text{rns}}^{(h-2)}, \tau_h)(\bmod P).$$ (13)

At the last step, there is

$$U_{\text{rns}}^{(h-1)} = \left(\prod_{i=1}^{h} \tau_i \cdot T^{-(h-1)} \right) \bmod P.$$

Suppose that $c = T^h \bmod P$ has been precomputed, then

$$R = \prod_{i=1}^{h} \tau_i = \text{MR}(U_{\text{rns}}^{(h-1)}, c).$$

If it is necessary, R can also be converted from RNS to binary system by the chinese remainder theorem:

$$R = \sum_{i=1}^{n} M_i \left| r_i M_i^{-1} \right|_{m_i} - \alpha M,$$ (14)

where r_i is the ith component of R, and α is obtained with the aforementioned approximation method [8].

Nevertheless, the proposed method is not less efficient than parallel conversion method [7] for modular exponentiation, in which the computational complexity can be greatly decreased in the latter since only one RNS-to-binary conversion with the base number is enough.

5 Conclusion

Long-precision modular multiplications can be performed in RNS with RNS Montgomery algorithm, for which there are mainly two implementation methods: parallel base conversion as well as residue recovery. The first method performs modular reductions with respect to the dynamic range in parallel, and base conversions between two RNSs are required. The second method, however, performs modular reduction in a number of sequential steps and requires more sequential steps.

Then, we propose a new residue recovery method that is based on binary system rather than mixed radix system, which is supposed to suit modular multiplications of many various large integers. In this case, it is more efficient than parallel base conversion method since little binary-to-RNS conversion is required.

Acknowledgments The authors would like to thank the editor and the reviewers for their comments. This work was partly supported by the National High Technology Research and Development Program of China (No.2012AA012402), the National Natural Science foundation of China (No.61073173), and the Independent Research and Development Program of Tsinghua University (No. 2011Z05116).

References

1. Soderstrand, M., Jenkins, W., Jullien, G., Taylor, F. (eds.): Residue Number System Arithmetic: Modern Applications in Signal Processing, pp. 1–185, IEEE Press, New York (1986)
2. Mohan, P.A.: Residue Number Systems: Algorithms and Architectures. Kluwer Academic Publishers, Boston (2002)
3. Wu, A.: Overview of Residue Number Systems. National Taiwan University, Taipei (2002)
4. Taylor, F.: Residue arithmetic: a tutorial with examples. Computer **17**(5), 50–62 (1984)
5. Posch, K., Posch, R.: Modulo reduction in residue number system. IEEE Trans. Parallel Distrib. Syst. **6**, 449–454 (1995)
6. Ciet, M., Neve, M., Peeters, E., Quisquater, J.: Parallel FPGA implementation of RSA with residue number systems-can side-channel threats be avoided? In: 46th IEEE International Midwest Symposium on Circuits and Systems, vol. 2. pp. 806–810 (2003)
7. Bajard, J., Didier, L., Kornerup, P.: An RNS Montgomery modular multiplication algorithm. IEEE Trans. Comput. **47**, 766–776 (1998)
8. Kawamura, S., Koike, M., Sano, F., Shimbo, A.: Cox-rower architecture for fast parallel Montgomery multiplication. In Preneel, B. (ed.) Advances in Cryptology-EuroCrypt'00. Volume 1807 of Lecture Notes in Computer Science, pp. 523–538, Springer-Verlag, Berlin (2000)
9. Nozaki, H., Motoyama, M., Shimbo, A., Kawamura, S.: Implementation of RSA algorithm based on RNS Montgomery modular multiplication. In: Third International Workshop on Cryptographic Hardware and embedded systems. Volume 2162 of Lecture Notes in Computer Science, pp. 364–376. Springer, Berlin (2001)
10. Bajard, J., Didier, L., Kornerup, P.: Modular multiplication and base extensions in residue number systems. In: 15th IEEE Symposium on Computer Arithmetic, pp. 59–65 (2001)
11. Bajard, J., Imbert, L.: A full RNS implementation of RSA. IEEE Trans. Comput. **53**, 769–774 (2004)
12. Bajard, J., Imbert, L., Liardet, P., Yannick, T.: Leak resistant arithmetic. In: Cryptographic Hardware and Embedded Systems (CHES 2004). Volume 3156 of Lecture Notes in Computer Science, pp. 62–75 Springer, Berlin (2004)
13. Montgomery, P.: Modular multiplication without trial division. Math. Comput. **44**, 519–521 (1985)
14. Orup, H.: Simplifying quotient determination in high-radix modular multiplication. In: 12th IEEE Symposium on Computer Arithmetic, pp. 193–199 (1995)

15. Phillips, B., Kong, Y., Lim, Z.: Highly parallel modular multiplication in the residue number system using sum of residues reduction. Appl. Algebra Eng. Commun. Comput. **21**, 249–255 (2010)
16. Bajard, J., Didier, L., Kornerup, P.: An RNS Montgomery modular multiplication algorithm. In: 13th IEEE Sympsoium on Computer Arithmetic, pp. 234–239 (1997)
17. Yang, T., Dai, Z., Yang, X., Zhao, Q.: An improved RNS Montgomery modular multiplier. In: 2010 International Conference on Computer Application and System Modeling (ICCASM 2010), Vol. 10, pp. 144–147
18. Guillermin, N.: A high speed coprocessor for elliptic curve scalar multiplications over $\mathbb{F}_p$. In: Cryptographic Hardware and Embedded Systems, CHES 2010. Vol. 6225 Lecture Notes in Computer Science, pp. 48–64 (2010)
19. Wu, T., Liu, L.: Elliptic curve point multiplication by generalized Mersenne numbers. J. Electro. Sci. Technol **10**(3), 199–208 (2012)
20. Shenoy, A., Kumaresan, R.: Fast base extension using a redundant modulus in RNS. IEEE Trans. Comput. **38**, 292–297 (1989)

Functionally Equivalent C Code Clone Refactoring by Combining Static Analysis with Dynamic Testing

Xiaohong Su, Fanlong Zhang, Xia Li, Peijun Ma and Tiantian Wang

Abstract Software with code clones is difficult for maintenance. It increases the cost of software maintenance. To solve the key problems of function optimization and parameter matching during the process of functionally equivalent code clone refactoring, this paper puts forward an approach for restructuring the fourth type (functionally equivalent) code clone by combining static analysis and dynamic testing. First, two kinds of function optimization strategy are proposed, i.e., running time and static characteristics. Then, determine the optimization function in each functionally equivalent code clone group according to the proposed optimization strategy. Finally, use the method of static analysis and dynamic testing to match the parameter matching for the replacement of procedure. On the basis of parameter matching, replace other clones with the optimization function and then complete the C code clone refactoring. Functionally equivalent C code clone refactoring system prototype is developed. Experimental results on the open source program show that the method can be accurately and effectively refactor the functionally equivalent C clone code.

Keywords Code clone · Code refactoring · Static analysis · Dynamic testing

1 Introduction

Code clone refers to the same or similar code fragments in source code file [1]. Code clone is widespread in the practice of software development, such as the "Copy–Paste–Modify" operation of the programmers, as well as the thinking

X. Su (✉) · F. Zhang · X. Li · P. Ma · T. Wang
School of Computer Science and Technology, Harbin Institute of Technology, Harbin, China
e-mail: sxh@hit.edu.cn

F. Zhang
e-mail: zhangfanlong@hit.edu.cn

S. Patnaik and X. Li (eds.), *Proceedings of International Conference on Soft Computing Techniques and Engineering Application*, Advances in Intelligent Systems and Computing 250, DOI: 10.1007/978-81-322-1695-7_28, © Springer India 2014

mode of developers and the impact of development capabilities [2]. Studies have shown that code clones account for about 7–23 % [3]. In most cases, code clone is harmful to the system because it is easy to introduce defects [4, 5]. Refactoring is an internal adjustment to the structure of the software, to improve the intelligibility and reduce the cost of modification without changing the observable behavior [6].

Roy [3] divides code clone into four types as follows, based on the similarity of text or function. In code clone detection, a number of methods have been proposed, and many detection tools have been developed [7–12]. The tools can detect the code clone of four types, while the refactoring is mainly focused on Type 1, 2 and 3 [13, 14], and it is lack of the refactoring method of functionally equivalent code clone. To solve this problem, this paper proposes a method of functionally equivalent C code clone refactoring by combining static analysis with dynamic testing. By analyzing the actual C open sources, refactoring experimental results show that the proposed method can refactor the functionally equivalent C code clones effectively and safely.

2 Related Work

2.1 Detection of Functionally Equivalent Code Clones

Definition 1 (*Functionally equivalent code clone* [15]):

If for the randomly generated set of inputs Π, there are two test input data arrangements $p_1, p_2: \Pi \rightarrow \Pi$ satisfies $\forall I \in \Pi$, $OC1(p_1(I)) = OC2(p_2(I))$, the two code fragments $C1$ and $C2$ are functionally equivalent.

In Definition 1, Π means all input sets, $OC1$ and $OC2$ denote the output sets of $C1$ and $C2$. Figure 1 shows the example of functionally equivalent code clone.

2.2 Program Static Metrics

Code metrics aim to measure the quality of the code by quantifying the various properties of various entities in code. The content of software measures is called metrics. Huang [16] uses three metrics including line metrics, McCabe metrics, and Halstead metrics. Line metrics are simple but rough, and McCabe metrics need to draw a flowchart; Halstead metrics have high reliability. This paper uses Halstead metrics to calculate the static characteristics of the program.

Code Block a	Code Block b

void Insert_Sort(int n, int a[])
Bubble_Sort(int num, int p[])
{
int i, j, temp;
for (i=1; i<n; i++)
1; i++)
{
temp = a[i];
j>=i+1; j--)
for (j=i; j>0 && a[j-1]>temp; j--)
{
a[j] = a[j-1];
}
a[j] = temp;
}
}
temp;
 	
}
}
}
	void
	
	{
	int i, j;
	for (i=0; i<num-
	
	for (j=num-1;
	
	{
	if (p[j] < p[j-1])
	{
	int temp;
	temp = p[j];
	p[j] = p[j - 1];
	p[j] =
	
	}

Fig. 1 Example of functionally equivalent code clone

3 The Approach of Functionally Equivalent Code Clone Refactoring

3.1 The Refactoring Model by Combining Static Analysis with Dynamic Testing

The framework of functionally equivalent C code clone refactoring method proposed in this paper is shown in Fig. 2. The functionally equivalent C code clone refactoring involves five subprocesses: clone detection, choosing the optimal function, parameter matching, procedure replacement, and correctness verification.

1. Clone detection

Using the functionally equivalent code clone detection method, Kong [15] proposed to obtain the multiple sets of functionally equivalent code clones.

2. Choose the optimal function

Each set of functionally equivalent code clones contains multiple code fragments, and each set is not functionally equivalent to the other sets. In order to select one fragment as the optimal fragment to replace the other functionally equivalent code fragments, this paper presents two optimal criterions, running-time-based optimal strategy and static-characteristics-based optimal strategy.

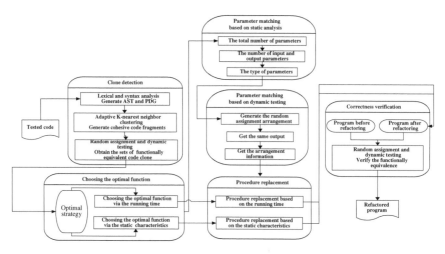

Fig. 2 The model to refactor functionally equivalent clone by combining static analysis with dynamic testing

3. Parameter matching

For each set of functionally equivalent code clones, make the parameter matching before replacement to ensure the safety of refactoring.

4. Procedure replacement

Since the optimal function has been chosen and the correct parameter arrangement information has been obtained, we can replace the non-optimal functions not with the optimal function.

5. Correctness verification

Make the program before and after refactoring as the input of functionally equivalent code clone detection. The random assignment and dynamic testing are combined to determine whether the program before and after refactoring is functionally equivalent.

3.2 Choose the Optimal Function

Determine the optimal function according to the principle of running time precedence. The main idea of running-time-based optimal algorithm is as follows: First, record the running time of each code clone in every group. The function with the shortest running time is selected as the optimal function. Then, use the optimal function to replace rest of the functions.

Determine the optimal function according to the principle of static characteristics precedence. The main idea of static-characteristics-based optimal

Table 1 Determine parameter matching according to the number of input and output parameters

Function name	Number of input parameters	Number	Parameters' name	Number of input parameters	Number	Parameters' name
name_1	1	1	input	1	2	output
name_2	1	1	in	1	2	out

Table 2 Determine parameter matching according to the type of input parameter and output parameter

Function name	Input parameters	Output parameters							
		Counting results				Matching results			
		Type	Number	Type	Number	Position	Name	Position	Name
name_3()	0	int	1	int	1	1	n	2	a
name_4()	0	int	1	int	1	2	n	1	a

algorithm is as follows: Use the Halstead metric to calculate the static characteristics. The Halstead metric provides the total number of operators and operands in one code fragment. For each code clone in functionally equivalent code sets, calculate the value of Halstead metric. The function with the smallest Halstead metric value is regarded as the optimal function. Then, use the optimal function to replace rest of the functions.

3.3 Parameter Matching

In this section, we propose an IOT-behavior-based parameter matching algorithm. The first step is parameter matching via static analysis, and the other one is the parameter matching via dynamic testing, which is shown in Table 1.

Parameter Matching via Static Analysis. Static analysis is carried out according to the function's IOT behavior.

Step 1: Determine matching parameters according to the parameter number. If the optimal function has only one parameter, match the parameter directly.

Step 2: For the functions with parameters not matched in Step 1, count the number of input and output parameters. The parameter matching information of name_1() and name_2().

Step 3: For the functions with parameters not matched in Step 2, make determinations according to parameters' type. Table 2 shows the matching in name_3() and name_4().

Parameter Matching via Dynamic Testing. Dynamic testing is for those functions that failed in parameter matching via static analysis. For example, there are functions named name_7 and name_8 (shown in Table 3).

Table 3 Determine parameter matching according to dynamic testing

Function name	Parameter-matched (Step 1)			Parameter to be matched	Output	Parameter arrangement	Parameter matching results					
	Position	Name	Position	Name	Name			Position	Name	Position	Name	Position	Name
name_7()	4	n	5	a[]	i, j, temp	12,34,25	1,2,3	1	i	2	j	3	temp
name_8()	4	n	5	a[]	i, temp, j	12,34,25	1,3,2	1	i	3	j	2	temp

```
Algorithm Parameter-Matching (P, R)
Input: Optimal function P, Un-optimal function R
Output:  Parameter matching of P and R
Begin
1: if the numbers of parameter in P is 1
2:    return 1
3: else
4:    if the numbers of input parameter is 1
5:          input parameter of P matches input parameter of
R
6:    if the numbers of output parameter is 1
7:          output parameter of P matches output parameter of
R
8:    else if the numbers of input parameter or number of
output parameter! = 1
9:                 if the numbers of input parameter is
not 1
10:               if the numbers of type of parameter i is 1
11:                    P.i matches R.i
12:               else if the numbers of output parameter
is not 1
13:               if the numbers of type of parameter o is 1
14:                    P.o matches R.o
15:                 if Judge (P, R) = 1//Judge means to judge
the parameter matching, if success, return 1
16:           dynamic testing
17:           parameters of P match parameters of R
18:    output the parameter matching information
End
```

Fig. 3 The algorithm description of IOTPM

The process of parameter matching algorithm is shown in Fig. 3. Assume that the total number of parameters in a function is n, the time complexity of direct dynamic testing is $O(n!)$, while static analysis is $O(1)$. The dynamic testing with static analysis is $\alpha O(1) + \beta O(n!)$, where $0 \leq \alpha \leq 1, 0 \leq \beta \leq 1, \alpha + \beta = 1$. It is clear that $\alpha O(1) + \beta O(n!) \leq O(n!)$. This indicates that algorithm in this paper is more efficient.

4 Experimental Results and Analysis

Take the code in Fig. 1 as a sample code to verify the parameter matching algorithm. Table 4 shows the experimental results. It will achieve the same effect for the open source code. As Kong et al. [15] described, we select the open source code Simens as an example. The experiment to verify the correctness of parameter

Table 4 Experimental result of sample code in Fig. 1

Function name	Function parameter	Running Time(ms)	Static characteristics			Parameter matching									
			Number of operator	Number of operand	Halstead metrics	N	P	N	P	N	P	N	P	N	P
Insert_Sort	int&i,int&temp,int& j,int&a[],int n	4415	24	27	51	i	1	temp	2	j	3	a[]	4	n	5
Bubble_Sort	int&i,int&j,int&temp, int&p[],int num	4435	28	29	57	i	1	temp	3	j	2	p[]	4	num	5

Notes N represents the parameter name, and P represents its position in the original parameter list of the function

Table 5 The experimental results of open source codes

Open source code	LOC	Number of functionally equivalent code clone groups	The success rate of parameter matching(%)	Manual confirmation on functionally equivalence(Y/N)
Simens	2,890	20	100	Y
UCC	3,140	50	100	Y
nginx-0.8.15/src/core	17,126	70	100	Y
devecot 2.0.8/src/auth	18,243	70	100	Y

matching algorithm which have the steps include artificially inject 20 groups functionally equivalent code blocks to each file of Simens to detect, do experiment 100 times for each groups, determine the correct rate of final parameter matching.

In addition, this paper uses the open source code such as Simens[1], UCC[2], nginx[3], and dovecot[4] to perform the experiments. The experimental results of open source codes are shown in Table 5.

5 Conclusions

This paper presents an approach against the fourth type (functionally equivalent) code clone. The following solutions are proposed to solve the problems about function optimization and parameter matching.

1. Running time priority and static characteristics priority are proposed as two strategies to determine which function is relatively optimal to replace the other functionally equivalent code clones in each group.
2. An IOT-behavior-based parameter matching algorithm is proposed to achieve parameter matching in procedure replacement, in order to ensure the correctness and safety of code refactoring.

The experimental results show that the approach refactors functionally equivalent C code clones correctly and effectively. Future works will apply the proposed approach to other language codes.

Acknowledgments This research is supported by the National Natural Science Foundation of China (Grant No. 61173021) and the Research Fund for the Doctoral Program of Higher Education of China (Grant Nos. 20112302120052 and 20092302110040).

---

[1] git clone http://git.code.sf.net/p/simens/codesimens-code

[2] http://www.unset.usc.edu/research/CODECOUNT/

[3] http://www.nginx.org/

[4] http://www.dovecot.org/

References

1. Cai, D., Kim, M.: An empirical study of long-lived code clones. In 14th International Conference on Fundamental Approaches to Software Engineering: Part of the Joint European Conferences on Theory and Practice of Software, FASE'11/ETAPS'11, pp. 432–446 (2011)
2. Roy, C.K., Cordy, J.R.: A survey on software clone detection research. Technical Report, Kingston, Ontario: Queen's University (2007)
3. Roy, C.K., Cordy, J.R., Koschke, R.: Comparison and evaluation of code clone detection techniques and tools; A qualitative approach. Science of Computer Programming, 74(7):470–495 (2009). Special Issue on Program Comprehension (ICPC 2008)
4. Wang, Q.: C Code Clone and Related Software Defect Detection Based on Sequence Mining, pp. 1–47. Harbin Institute of Technology, Harbin (2009)
5. Chou, A., Yang, J., Chelf, B., Hallem, S., Dawson, E: An empirical study of operating systems errors. In: Proceedings of the Eighteenth ACM Symposium on Operating Systems Principles, SOSP'01, pp. 73–88 (2001)
6. Fowler, M.: Refactoring: Improving the Design of Existing Code. Addison Wesley, Boston (1999)
7. Baker, B.S.: On finding duplication and near-duplication in large software systems. In: Proceedings of the Second Working Conference on Reverse Engineering, WCRE'95, p. 86 (1995)
8. Baxter, I.D., Yahin, A., Moura, L., Sant'Anna, M., Bier, L.: Clone detection using abstract syntax trees. In: International Conference on Software Maintenance, pp. 368–377 (1998)
9. Jiang, L., Misherghi, G., Su, Z., Glondu, S.: Deckard: Scalable and accurate tree-based detection of code clones. In: 29th International Conference on Software Engineering, 2007. ICSE 2007, pp. 96–105 (2007)
10. Kamiya, T., Kusumoto, S., Inoue, K.: CCFinder: A multilinguistic token-based code clone detection system for large scale source code. IEEE Trans. Software Eng 28(7), 654–670 (2002)
11. Kim, H., Jung, Y,. Kim, S., Yi, K.: MeCC: Memory comparison-based clone detector. In: 33rd International Conference on Software Engineering (ICSE), pp. 301–310 (2011)
12. Li, Z., Lu, S., Myagmar, S., Zhou, Y.: CP-Miner: Finding copy-paste and related bugs in large-scale software code. IEEE Trans. Software Eng. 32(3):176 –192, Mar (2006)
13. Feng, J.: Code Clone Restructuring of C Programs Via K-nearest Neighbor Algorithm, pp. 1–57. Harbin Institute of Technology, Harbin (2011)
14. Yu, D., Peng, X., Zhao, W.: Automatic refactoring method of cloned code using abstract syntax tree and static analysis. J Chinese Comput. Syst. 30(9), 1752–1760 (2009)
15. Kong, D., Su, X., Wu, S.: Detect functionally equivalent code fragments via k-nearest neighbor algorithm
16. Lina, Q.: Research on Static Prediction of Software Defects, pp. 1–56. Central China Normal University, Wuhan (2011)

Architecture Designing of Astronaut Onboard Training System Based on AR Technology

He Ning, Hou Quanchao and Hu Fuchao

Abstract In order to reduce the burden of training and dependence on ground support and improve the astronaut's operation proficiency, considering space station requirements in the future, the astronaut onboard training technology based on augmented reality (AR) technology was proposed. The AR technology application and current research status in astronaut training was introduced; then, the architecture of onboard astronaut training system based on AR technology was developed; lastly, the collaborative AR technology of astronaut robotic arm training was discussed. The astronaut onboard training technology based on AR technology could reduce the burden of training and dependence on ground support and improve the astronaut's operation proficiency.

Keywords Augment reality (AR) · Onboard training · Architecture

1 Introduction

Augmented reality (AR) technology constructed a space with the superposition of virtual object and real scene, to help people interact in three dimensions with real and virtual objects around them timely and naturally by the joint action of hardware and software. AR system has many new features, such as combine the virtual and real objects, interactive real time, and the 3D registration [1]. AR technology

H. Ning (✉) · H. Quanchao · H. Fuchao
National Key Laboratory of Human Factors Engineering, China Astronaut
Research and Training Center, Beijing, China
e-mail: hening1980@126.com

H. Quanchao
e-mail: hou_isme@126.com

H. Fuchao
e-mail: hfc_acc1205@163.com

S. Patnaik and X. Li (eds.), *Proceedings of International Conference on Soft Computing Techniques and Engineering Application*, Advances in Intelligent Systems and Computing 250, DOI: 10.1007/978-81-322-1695-7_29, © Springer India 2014

is a research hot spot in recent years and will be widely used in future, which will help us percept and interact with the present world. There are many key technical problems in building an AR system, for instance, display, tracing registration, virtual-real synthesis, and man–machine interactive technique [2]. A desired AR system should combine virtual 3D objects with surrounding real world perfectly and computer-generated virtual vision changing synchronize with real vision following user's viewpoint.

The astronauts confronted with complicated flight programs and space environment in onboard space mission; furthermore, they took many complex operations and onboard training on the zero-gravity condition. These activities cost a lot of time and effort and required astronauts taking many special training before mission execution.

So it is necessary to seek a new kind of efficient onboard astronaut training technology, which can reduce the burden of training and dependence on ground support and improve the astronaut's operation proficiency. Considering space station requirements in the future, the astronaut onboard training technology based on AR technology was proposed, and then the AR technology application and current research status in astronaut training was introduced; the feasibility of combined AR technology with astronaut training was analyzed, and then, the architecture of onboard astronaut training system based on AR technology was developed; lastly, the collaborative AR technology of astronaut robotic arm training was discussed.

2 Applications and Current Research

National Aeronautics and Space Administration (NASA) and European Space Agency (ESA) are developing the AR technology, which can reduce the amount of exercise time and expenses and improve the training effect of astronauts.

In 2004, NASA Ames Research Center and University of California developed a prototype airflow hazard visual display for use in helicopter cockpits to alleviate this problem [3, 4]. In 2012, NASA Johnson Space Center researched and developed AR electronic procedure system (AReProc). The hardware, software, and procedure content were combined to create a new user experience to perform maintenance repair tasks that could dramatically improve human performance and situational awareness [5]. In 2010, ESA developed the Wearable AR Station Development Technology Objectives (SDTO). The Wearable AR SDTO is a system that aims to assist the astronauts when performing onboard procedures by allowing them to operate the system and consult procedures and manuals hands-free [6]. In 2012, the ESA developed the Computer Assisted Medical Diagnosis and Surgery System (CAMDASS) [7]. The new AR system could help astronauts take care of each other, overlaying computer graphics over a real patient to guide diagnoses or even surgery. It could even improve telemedicine in developing countries or remote spots. For now, CAMDASS, only works with ultrasound, has already been available on the International Space Station.

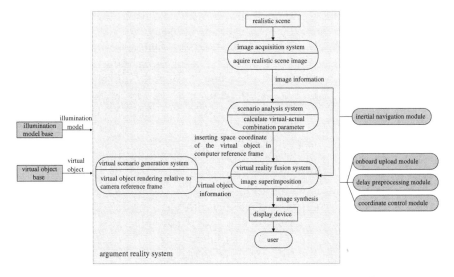

Fig. 1 Architecture of the astronaut onboard training system based on AR technology

3 Architecture Designing of Astronaut Onboard Training System

In the future space station task, astronauts confront with complicated flight programs and space environment in onboard space missions, and they take many complex operations and onboard training on the zero-gravity condition. The AR system needs to provide a great deal of graphics and multimedia information for the astronauts. The AR system gets the AR information from remote server instead of display real time rendering through wireless LAN. The astronaut can share the multiuser resource cross different space with no restrict.

Since the position particularity of the space station, the onboard astronaut training will face with space-ground network transmission delay, visual tracking failures, and illumination inconsistencies, etc. Therefore, the architecture of the astronaut onboard training system based on argument reality technology was composed of the argument reality system master module, delay preprocessing module, onboard upload module, inertial navigation module, coordinate control module, illumination model base and virtual object base, etc. The architecture of onboard astronaut training system based on AR technology is shown in Fig. 1.

3.1 Transmission Delay

The registration, tracking, and output displaying would make the argument reality system larger time-delay and virtual-actual combination error. In order to avoid

time delay, the possible position prediction was taken. Beginning with the registration algorithm, the prediction algorithm was introduced to predict the next time
viewpoint information. If the prospective location of the object was determined,
the virtual-actual scenario can be drawn and synthesized according to the prospective location information, rather than drawing after measuring the position. In
this way, the same time integration virtual-actual scenario was displayed to the
user.

3.2 Onboard Upload

To avoid space-ground network interconnected transmission delay, the orbit
synchronous prediction and the motion and posture prediction in cabin were
adopted. Currently, space-ground network interconnected transmission delay were
caused by two factors, the relay processing delays and path propagation delay.
The relay processing delays could be regarded as fixed value, while the path
propagation delay was changing with the transmission path distance. Countering
on the transmission delay problem, the orbital dynamics was adopted to predict the
orbit and the orbital delay random parameter was set.

3.3 Visual Tracking Failures

When the camera moved too fast or the marker was blocked, the tracking technology basing on vision would be failure, so the proposed vision-based and inertial
sensor data fusion method was presented to resolve the vision sensor could not get
the attitude angle and position information of the reality scene markers.

3.4 Illumination Inconsistencies

To make the computer-generated virtual environment realistic and allow the users
to convince that the virtual object was a part of its surroundings, it should keep
consistence of illumination between the virtual objects and the real environment.
In order to keep consistency of illumination, the real scene illumination model
needed to be restored. The real light source was used to correct on the virtual
object illumination, and then the real scene illumination impact on the virtual
objects was calculated, which was including shading and reflection effects. The
argument reality system needs a real-time illumination synthesis technology to
meet the needs of photorealistic rendering; therefore, the real illumination conditions were registered, and the real environment illumination model was
established.

3.5 Coordinate Control

The application of the coordinate control methods in the collaborative AR system faced with three key issues, which were including the physical clock synchronization, hysteresis value determination, and inconsistent state reconditioning. In the coordinate control method, the time stamp was set physical mark, the hysteresis value was set larger than the network delay, the same operation in each node was ensured to be executed at the same, so the system could remain consistent.

4 Collaborative AR Technology of Astronaut Robotic Arm Training

In space station mission, robotic arm was a very important tool of collaboration, and astronauts regularly need cooperation of robotic arm to complete tasks, such as the extravehicular activity (EVA) during assemble the International Space Station. In EVA progress, one astronaut stood on the fixing device in front of robotic arm end, and another astronaut operated robotic arm to transform partner to destine and carried on follow-up repair job.

Currently, the astronaut onboard training methods of robotic arm were limited. The new collaborative AR astronaut onboard training techniques can be combined with robotic arm simulator to conduct joint training. Remote control of robotic arm was hard to conduct, especially the much longer moving route of robotic arm and larger time delay. Astronauts designed and assigned the robotic arm operations by the virtual models of robotic arm, while the result of control displays in the real scene. After verifying and confirming the design, astronauts performed on the real robotic arm. In this way, it will avoid the instability of robotic arm's motion path due to delay.

5 Conclusions

Considering space station requirements in the future, the astronaut onboard training technology based on AR technology was proposed. The AR technology application and current research status in astronaut training was introduced; the feasibility of combined AR technology with astronaut training was analyzed, and then, the architecture of onboard astronaut training system based on AR technology was developed; lastly, the collaborative AR technology of astronaut robotic arm training was discussed.

AR technology can be applied in astronauts' activities and in-orbit complex operations training during the space station mission. The applications of AR in astronaut training could assist the astronauts when performing onboard procedures by allowing them to operate the system and consult procedures and manuals hands-free, expand the scope of astronauts activities, significantly improve astronauts' independent operation ability, reduce the astronauts' dependence on the ground support and electronic and paper form procedures manual, improve astronauts to complete the task efficiency and performance, reduce incorrect operation, improve astronaut location awareness, and reduce cognitive load.

References

1. Shen, K.: Research of the human-computer interaction system for augmented reality based on ARToolKit. Hefei University of Technology, Hefei (2007)
2. Yuan,Y.: Studies on implement technology of augmented reality, Zhejiang University,(2006)
3. Aragon, C.R.A.: Prototype flight-deck airflow hazard visualization system. Engineering **113**, 1607–1613 (2004)
4. Aragon, C.R., Long, K.R.: Airflow hazard visualization for helicopter pilots: Flight simulation study results. http://www.ti.arc.nasa.gov/pub-archive/965h/0965%20(Arag-on).pdf
5. Meeting NASA's needs with creativity and innovation "TECH & TELL"-A FY'12. Report from the JSC office of the chief technologist. http://www.nasa.gov/centers/johnson/pdf/703052main_2012_JSC_CIF_Annual_Report.pdf
6. Operation of European SDTO at Col-CC. http://www.nasa.gov/pdf/191855main_exp16_press_kit.pdf
7. http://www.impactlab.net/2012/02/13/augmented-reality-will-help-astronauts-in-the-future-perform-surgery-on-each-other/

Design and Implementation of Bibliometrics System Based on RIA

Geyang Han and Bing Sun

Abstract With the growth of attention on e-Science and collaborative environment for scientific research, bibliometrics software needs to adopt a new form of architecture instead of traditional desktop software architecture. In this paper, the design of a bibliometrics system based on Rich Internet Application (RIA) and its core module's implementation are presented. The system's browser-side application is built upon Adobe Flex, and its server-side application is built upon J2EE. PureMVC framework is introduced to create the Flex application based on MVC concept. The system's main function is to analyze the literature information based on mathematical and statistical laws of bibliometrics, providing good visualization as well.

Keywords Bibliometrics system · RIA · Flex · PureMVC

1 Introduction

Bibliometrics is a set of methods to analyze scientific and technological literature through mathematical and statistical methods. Originating from library and information science, it has been widely used by researchers from different fields to quantitatively study the progress and development trends of their research areas and assess the impact of particular fields, scholars, or literature. Bibliometrics is also an important approach to evaluate academic output.

Existing bibliometrics analysis systems are either desktop applications or simply additional tools of particular online databases [1]. With the development of network

G. Han (✉) · B. Sun
School of Electronic and Information Engineering, Beihang University, Beijing, China
e-mail: sweeroty@gmail.com

B. Sun
e-mail: bingsun@buaa.edu.cn

S. Patnaik and X. Li (eds.), *Proceedings of International Conference on Soft Computing Techniques and Engineering Application*, Advances in Intelligent Systems and Computing 250, DOI: 10.1007/978-81-322-1695-7_30, © Springer India 2014

technology, e-Science and collaborative environment for scientific research have been paid attention increasingly. Under this circumstance, bibliometrics software needs to adopt a new form of architecture. Browser/Server (B/S) architecture is becoming the mainstream solution due to its fast deployment, good maintainability, and high expansibility.

Among many options of browser-side technologies, Rich Internet Application (RIA) is an appealing one, which incorporates many advantages of desktop applications such as high interactivity and short reaction. It works within the browser, providing users richer user experience than ordinary Web applications.

Besides, most current bibliometrics software is incapable of analyzing the literature using mathematical and statistical laws of bibliometrics and only counts its basic information. In view of all these above, we propose a bibliometrics system based on B/S architecture. In particular, we apply RIA technology to the browser-side application in order to provide good interactivity and visual effects.

2 Related Technology Solution

Common RIA technology solutions include Adobe Flex, Java FX, Microsoft Silverlight, and so on. Adobe Flex, with the purpose of developing cross-platform RIAs based on the Adobe Flash platform, is composed of three components: Adobe Flex Software Development Kit (SDK), MXML, and ActionScript. Flex program source code, consisting of ActionScript files and MXML files, is complied by Flex complier of Flex SDK into SWF format file which can be run in Flash platform. Then, the SWF file is deployed within the HTML file to the Web server. Users need to install Adobe Flash Player before using Flex applications.

Compared to the other technologies, Adobe Flex has three major advantages: (1) It is well integrated with Adobe Flash which offers powerful visualization and interactivity; (2) it has the largest market share owing to a wide user base Flash Player owns [2]; (3) it provides an abundant component class library which makes development of Flex applications easier and faster. Therefore, we select Flex as our RIA solution.

3 System Design

3.1 System Function Analysis

Bibliometrics system described in this paper is designed for researchers to analyze the literature information from five aspects: publication time, source, authors, organs, and keywords on the basis of mathematical and statistical laws of bibliometrics. Its main function includes management and analysis of the literature

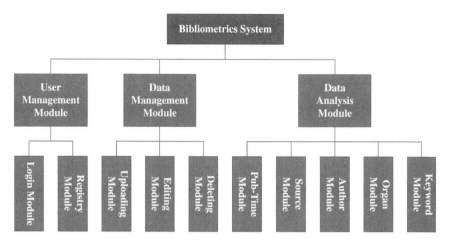

Fig. 1 System function structure

data. And the latter part is the core of the system. According to its function, the system is divided into three modules, as shown in Fig. 1.

Registry and login of users are implemented in User Management Module. In Data Management Module, users can upload new metadata files, edit and modify the literature information, or delete the unwanted information. The analysis results of the literature information are presented in Data Analysis Module.

3.2 System Architecture

The system's overall architecture is shown in Fig. 2. The server-side application is implemented based on J2EE, the most popular enterprise software solution. In addition, Hibernate, an open source framework which provides powerful and high-performance object data persistence in a relational database table service [3], is used in the persistence layer of the server-side application to facilitate the storage and retrieval of Java domain object via object/relational mapping [4] with MySQL which is selected as the database of the system.

Adobe Flex, as discussed in Sect. 2, is used to construct the browser-side application. PureMVC framework is introduced to create applications based on classical model-view-controller architecture pattern. The communication between the Flex application and the server-side application is built upon BlazeDS, a server-based Java remoting and Web messaging technology which enables Flex to invoke java remote methods in the business layer.

Fig. 2 System architecture

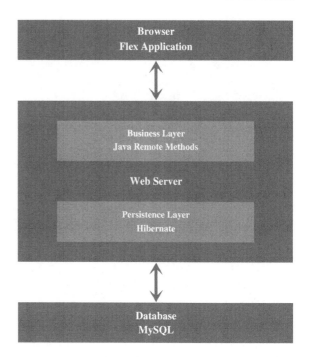

3.3 Flex Application Design

PureMVC is applied to creating the Flex application of the system. It is a light-weight framework with the goal of separating the application's coding concerns into three discrete tiers: Model, View, and Controller [5] so as to decrease coupling between different parts of the code and improve their reusability.

The framework of PureMVC has four tiers: the Model, the View, the Controller, and the Façade. The Model is made up of Proxies and Data Objects. Proxies are responsible for communicating with the remote server and managing Data Objects which store the retrieved data. Mediators and View Components comprise the View. A View Component caches references to one or several Flex UI components for a same or similar function. A Mediator manages a related View Component, manipulating the View Component's state, handling events of the View Component trigged by users, sending and receiving notifications to and from the rest of the application. The Controller is composed of Commands. Commands can retrieve and interact with Proxies and Mediators, send notifications to update View Components for instance, or execute other Commands. The application's business logic is implemented in Commands. The Façade initializes the other three tiers and registers all the notifications, providing the communication interface between each tier. All of the tiers are governed by Singletons. Typical procedures describing how applications communicate with J2EE remote servers based on BlazeDS can be concluded as follows:

Fig. 3 The structure of data
analysis module on flex side

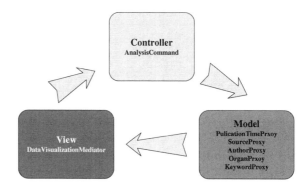

1. An event of a View Component is trigged by users.
2. The related Mediator handles this event and sends a notification.
3. The Command which listens for this notification starts to execute and invoke related Proxy's certain public method.
4. The Proxy invokes the remote methods located in the Web server through BlazeDS and waits for the requested data.
5. When the data are prepared, the Proxy sends a notification.
6. The Mediator receives this notification and updates its View Component's state.

In the light of the principles of PureMVC, the structure of Data Analysis Module on Flex side is designed as shown in Fig. 3.

In the view, Data Visualization Mediator manages a View Component consisting of five Flex UI components: a Button Bar object, a Pop Up Button, a Line Chart object, and two Data Grid objects. When clicked, the Button Bar object will correspondingly trigger an event which leads to a notification that requests for particular data. The other four Flex UI components are for the display of analysis results. In the Controller, Analysis Command receives the notification sent from Data Visualization Mediator and then invokes the method provided by one of the five Proxies to start communicating with the remote server. When the data are retrieved, the invoked Proxy will send a notification to Data Visualization Mediator with the data to be displayed.

The other two modules on Flex side, Data Management Module and User Management Module, are implemented in similar ways.

3.4 Server-Side Application Design

The primary role of the server-side application is responding to the Flex application's requests for retrieving data. According to different requests, the server-side application retrieves different data from the database, processes the data correspondingly, and then returns the data to the Flex application. In order to save time on interacting with the database, the data are preloaded in memory and stay unchanged unless users edit the literature information in Data Management Module.

The server-side application mainly consisted of three components: data access objects, data objects, and remote method objects. Data access objects are implemented based on Hibernate, providing the access to the database. Different data structures are implemented in data objects to store and organize different types of literature information such as author and keyword. Interacting with data access objects and data objects, remote method objects which can be invoked by the Flex Application accomplish different functions like uploading new metadata files to the database and analyzing data quantitatively.

4 Implementation of the Core Module

The core module of the system is Data Analysis Module, which has five sub-modules as shown below.

4.1 Distribution of Publication Time

This sub-module analyzes the publication time of literature, counting the amount of literature published per year. A HashMap is used to store the data.

4.2 Distribution of Source

The analysis of literature's source distribution is based on Bradford's Law [6]. Both Zone Analysis and Graphic Analysis [6] are implemented in this module. Zone Analysis contributes to identifying the core journals of the literature. Source Zone data objects and Source data objects are created to get the results of Zone Analysis. Their class diagrams are shown in Fig. 4.

The algorithm is shown as follows:

```
INPUT: List<Source> sourceList
int amount = the total amount of the input literature;
int processedAmount  = 0;
int boundryIndex = 0;
create    a    new    List<SourceZone>    object    named
sourceZoneList;
for (Source source : sourceList) {
processedAmount += source.getLiteratureAmount();
```

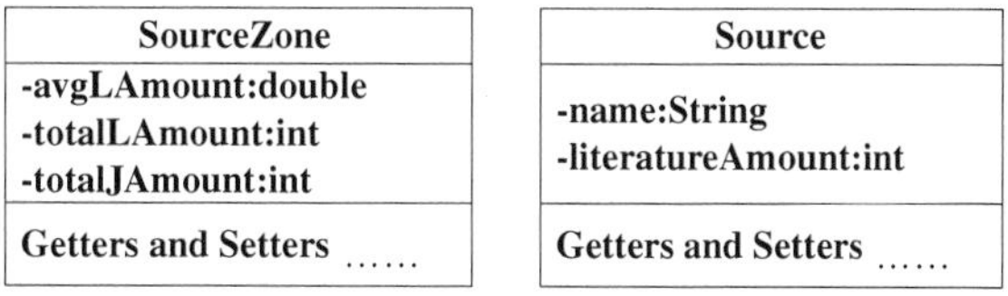

Fig. 4 Source zone and
source class diagrams

SourceZone
-avgLAmount:double -totalLAmount:int -totalJAmount:int
Getters and Setters ......

Source
-name:String -literatureAmount:int
Getters and Setters ......

```
int index = the index of source in sourceList;
if (processedAmount >= amount/3) {

create a new SourceZone object named sourceZone;
sourceZone.setTotalLAmount(processedAmount);
int totalJAmount = index + 1— boundryIndex;
sourceZone.setTotalJAmount(totalJAmount);
sourceZone.setAvgLAmount(processedAmount/
totalJAmount);
sourceZoneList.add(sourceZone);
boundryIndex = index + 1;
    processedAmount = 0;
    }
    }
    OUTPUT: List<SourceZone> sourceZoneList
```

Graphic Analysis is another way of using Bradford's Law. In Graphic Analysis, a distribution curve of source is plotted to find the core journals of the literature. The Drawing of the curve is based on the data of Zone Analysis and not complex. The discussion is omitted here.

4.3 Author Analysis

This sub-module analyzes authors of literature based on general Lotka's law [6] and Price's law [7] to describe the distribution of authors and discover the core group of authors in a particular field. Moreover, a co-occurrence matrix describing relationship between author collaboration is generated. Author data objects, shown in Fig. 5, are used to store the related data.

The algorithm of generating the co-occurrence matrix is shown as follows:

```
INPUT: List<Author> authorList
int size = authorList's size;
create a double[size][size] object named authorMatrix;
make the elements in the diagonal equal 1 and the others
equal 0;
for (Author author : authorList) {

int i = the index of author in authorList;
```

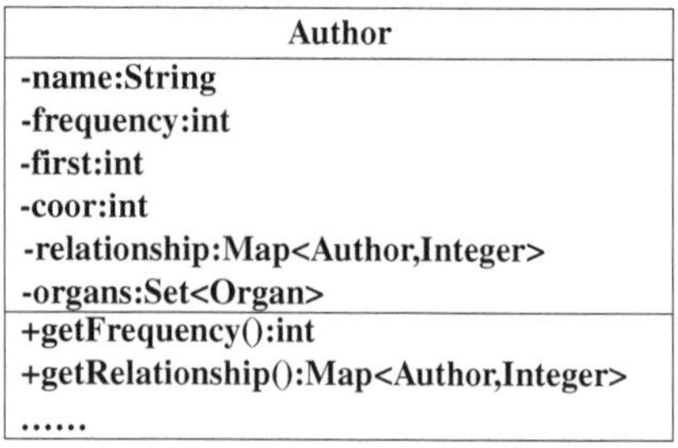

Fig. 5 Author class diagram

Author
-name:String
-frequency:int
-first:int
-coor:int
-relationship:Map<Author,Integer>
-organs:Set<Organ>
+getFrequency():int
+getRelationship():Map<Author,Integer>
......

```
int fi = author.getFrequency();
  for (Map.Entry<Author, Integer>entry : author.getRela-
  tionship.entrySet()) {

int j = the index of entry's author in authorList;
int fj = authorList.get(j).getFrequency;
authorMatrix[i][j] = entry.getValue() / (Math.sqrt(fi) *
Math.sqrt(fj))
  }
  }
  OUTPUT: doube[][] authorMatirx
```

4.4 Organs Analysis

This sub-module counts the amount of the literature published by per organ. A co-occurrence matrix describing relationship between organ collaboration is generated as well. Its implementation is similar to that of Author sub-module.

4.5 Keywords Analysis

Keywords represent main ideas of the literature. In this sub-module, the frequency of each keyword is calculated. And a co-word matrix is generated by the similar approach as discussed above.

4.6 Module Test Results

In order to test Data Analysis Module, the literature metadata of a certain file downloaded from CNKI (www.cnki.net) are uploaded into the system. Some of the

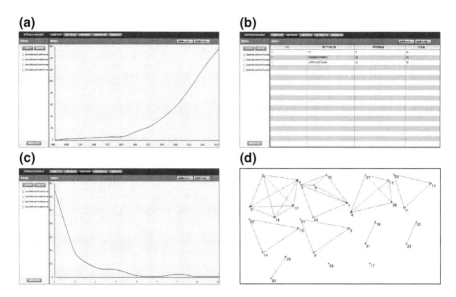

Fig. 6 Test results: **a** distribution curve of publication time, **b** results of zone analysis, **c** distribution curve of authors, and **d** collaboration network of authors

analysis results are shown in Fig. 6a–c. We also import the co-occurrence matrix of authors into Pajek, a complex network analysis tool, to generate the authors' collaboration network. The result is shown in Fig. 6d.

5 Conclusions

This paper discusses the design of Bibliometrics System and the implementation of its core module. The system is based on RIA which combines the advantages of both desktop applications and Web applications. Adobe Flex with PureMVC framework is used to create the browser-side application. J2EE is selected to create the server-side application. Because of B/S architecture, the system is easy to deploy, providing a useful tool for researcher of different fields to quantitatively analyze scientific and technological literature.

Acknowledgments This research is supported by the Fundamental Research Funds for the Central Universities.

References

1. Guojun, L., Entao, L.: Analysis and implementation of bibliometrics visualization software. Library Journal **30**, 72–78 (2011)
2. StatOwl. http://www.statowl.com/custom_ria_market_penetration.php
3. Peinan, Z., Kai, W.: The design and implementation of intelligent patrol web site based on J2EE. In: World Automation Congress 2012, pp. 1–4. IEEE Press, Puerto Vallarta (2012)
4. Hibernate. http://www.hibernate.org/
5. PureMVC. http://puremvc.org/component/option,com_wrapper/Itemid,31/
6. Junping, Q.: Informetrics. Wuhan University Press, Wuhan (2007)
7. Derek John de Solla Price: Little Science, Big Science. Columbia University Press, New York (1963)

SAR Image Filtering Based on Quantum-Inspired Estimation of Speckle Variance

Xiaowei Fu, Li Chen, Jing Tian, Xin Xu and Yi Wang

Abstract In this paper, a novel filtering method is proposed for synthetic aperture radar (SAR) images by applying the dual-tree complex wavelet transform (DTCWT) with an adaptive quantum-inspired estimation of speckle noise variance. Firstly, an improved Laplacian probability distribution function (PDF) with an adaptive smoothing factor is built up for the complex wavelet coefficients of log-transformed image signal. And then, considering the inter-scale dependency and inner-scale relativity, the quantum-inspired probability of noise is introduced in the estimation of speckle noise variance. Finally, according to the Bayesian maximum a posteriori (MAP) theory, an adaptive despeckling method is proposed for SAR images. Experiments demonstrate effectiveness of the proposed method, which can notably reduce speckle noise and effectively preserve details of SAR image at the same time.

Keywords Speckle noise · Quantum signal processing (QSP) · Dual-tree complex wavelet transform (DTCWT)

X. Fu (✉) · L. Chen · J. Tian · X. Xu · Y. Wang
College of Computer Science and Technology, Wuhan University of Science and Technology, Hubei Province Key Laboratory of Intelligent Information Processing and Real-time Industrial System, Wuhan, People's Republic of China
e-mail: fxw_wh0409@wust.edu.cn

L. Chen
e-mail: chenli@ieee.org

J. Tian
e-mail: jingtian@ieee.org

X. Xu
e-mail: xuxin@wust.edu.cn

Y. Wang
e-mail: erihppas@gmail.com

S. Patnaik and X. Li (eds.), *Proceedings of International Conference on Soft Computing Techniques and Engineering Application*, Advances in Intelligent Systems and Computing 250, DOI: 10.1007/978-81-322-1695-7_31, © Springer India 2014

1 Introduction

Synthetic aperture radar (SAR) is widely used in many fields nowadays, which has the advantage of all-weather, all-time, and high-resolution imaging, strong penetrability, and accurate determination of topographical features. However, SAR images are inherently suffered from multiplicative speckle noise, which is a signal-dependent noise. The inevitable presence of speckle noise degrades the SAR image quality and increases the difficulty of automatic scene analysis and understanding [1]. As a result, SAR image filtering is an important prerequisite in SAR image application.

Since speckles are multiplicative noise, a log-transformation is generally used to convert speckles to additive white noise in wavelet-based despeckling methods. OWTSURE is a successful wavelet-based method to reduce additive white noise [2]. However, the despeckling performance of Log-OWTSURE which denotes the OWTSURE operation after log-transformation is limited for SAR image despeckling. In the wavelet transform domain, despeckling techniques based on Bayesian estimation theory are effective, which imposes prior statistical probability distributions on image useful signal and speckle noise and then estimates the noise-free wavelet coefficients [3, 4]. In most wavelet-based methods, the results of despeckling depend on the selection of parameters, and the capability of speckle suppression is limited even with the optimized parameters. They sometimes cannot preserve details effectively with limited speckle reduction ability. With the advantages of approximate shift invariant and directional selectivity, the DTCWT is a valuable enhancement to the traditional discrete wavelet transform (DWT) [5]. Due to these advantages, DTCWT is employed in our method.

Borrowing from the principles of quantum mechanics and some of its interesting axioms and constraints, quantum signal processing (QSP) is a novel paradigm for signal processing [6]. The research of quantum-inspired image processing has just started recently [7, 8]. With the inspiration of QSP basic principle, a novel filtering method for SAR images is proposed in this paper. Experiments demonstrated the method can not only suppress speckle noise but also preserve edges and radiometric scatter points in SAR images effectively. This algorithm provides a feasible despeckling method to improve SAR images quality, which combines the basic quantum theory with the image processing technology.

The rest of this paper is organized as follows. The proposed method is presented in Sect. 2, experimental results are shown in Sect. 3, and finally, conclusions are discussed in Sect. 4.

2 Proposed Method

2.1 Statistical Models of SAR Images

The multiplicative speckle noise of SAR image can be converted to a signal-dependent additive noise by a logarithmic transformation. In DTCWT domain, since the DTCWT is a linear mathematic operation, the wavelet coefficients of the log-transformed SAR image can be expressed as

$$y = x + n \tag{1}$$

where y, x, and n are the complex wavelet coefficients of noisy SAR image, original SAR image, and speckle noise, respectively.

The existed literatures report that the distribution of the log-transformed speckle noise is close to a Gaussian distribution [9]. Therefore, in the proposed method, the statistical model of the log-transformed speckle noise is defined as

$$p_n(n) = \frac{1}{2\pi\sigma_n^2} \cdot \exp\left(-\frac{n_r^2 + n_i^2}{2\sigma_n^2}\right) \tag{2}$$

in DTCWT domain.

Since the distribution of noise-free image signal is changed with different images, an improved Laplacian statistical model is proposed for the noise-free log-transformed SAR image,

$$p_x(x) = \frac{1}{2\sigma^2} \cdot \exp\left(-\frac{\sqrt{2}}{\sigma}|x_r + ix_i| \cdot \exp(k)\right) \tag{3}$$

where the standard deviation σ and smoothing factor k are the parameters of the probability density function (PDF). Formula (3) is a traditional Laplacian statistical model when $k = 0$. Along with the scale increases, the noise of the corresponding wavelet coefficients attenuate quickly, while the wavelet coefficients of ideal image signal is relatively stable. In order to simplify parameter estimation, through some statistical experiments of SAR images, smoothing factor k is adaptively determined by the histogram distribution curve of the real DTCWT coefficients at level 2 with the 15° direction. The k is determined in the empirical range of $[-0.5, 3]$ on the supposition that $\sigma = \left(\sqrt{2P_{max}}\right)^{-1}$, where P_{max} is the maximum of the histogram distribution curve. Figure 1a, b shows a real SAR image and its log-transformed image, respectively, and c shows the fitting PDF curve with $k = 0.4$, which is determined adaptively. The red solid line is the corresponding (3) PDF curve with the fitting value $k = 0.4$. The blue dotted line is the corresponding (3) PDF curve when $k = 0$. The red solid line fits the real distribution much better than the blue dotted line adaptively, which can describe the real distribution of subband wavelet coefficients near zero much better.

2.2 Quantum-Inspired Speckle Noise Variance

With the dependency of coefficients, products of the coefficients and their parents are exploited to distinguish between signal and noise in the high-frequency subbands. With the elicitation of basic principle of QSP, according to the Ref. [8], the

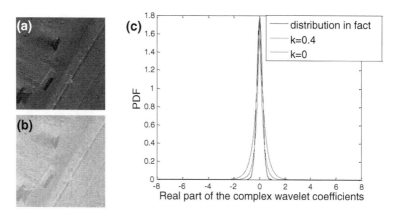

Fig. 1 **a** is the orginal image; **b** is the log-transformed image of (a); **c** draws the Fitting curves of smoothing factor k of (b)

state C_{sij} of the (i, j) wavelet coefficient at the sth scale level can be expressed in a superposition state of noise and signal in the high-frequency subbands. In our algorithm, a quantum-inspired speckle noise variance σ_n^2 is robustly estimated as follows

$$\hat{\sigma}_n^2(i,j) = \left[\frac{\text{median}\left(\left|y_{1r}^{45°}\right|\right)}{0.6745}\right]^2 \exp\left(\cos^2\left(\frac{\pi NC_{s\theta}^{ij}}{2}\right)\right) \tag{4}$$

where $y_{1r}^{45°}$ denotes the real component of wavelet coefficients at the first level in $45°$ direction subband, NC_{sij} denotes the normalized C_{sij}, and $NC_{sij} \in [0, 1]$ reflects the power of high-frequency signal in some degree.

Then, the standard deviation σ in (3) is calculated by

$$\hat{\sigma} = \sqrt{\left(\frac{1}{M}\sum_{(i,j)\in W}|y(s,i,j)|^2 - \hat{\sigma}_n^2\right)_+} \tag{5}$$

where $(f)_+ = \max(f, 0)$, $|y(s, i, j)|$ denotes the modulus of complex wavelet coefficient $y(i, j)$ at the sth scale, and M is the size of neighborhood W. Here, W is a 7×7 neighbor window. The speckle noise variance σ_n^2 is adaptively determined by the quantum-inspired probability of noise.

2.3 Proposed Method

According to Bayesian theorem, the proposed method gives a Bayesian MAP estimator with the quantum-inspired speckle noise variance.

The proposed estimator of x can be derived as

$$\hat{x} = \frac{\left(\sqrt{y_r^2 + y_i^2} - \sqrt{2}\exp(k)\sigma_n^2/\sigma\right)_+}{\sqrt{y_r^2 + y_i^2}} \cdot y \tag{6}$$

Equation (6) is the proposed soft threshold function, and the threshold T is defined as

$$T = \sqrt{2}\exp(k)\sigma_n^2/\sigma \tag{7}$$

However, soft threshold function shrinks all the wavelet coefficients, which may weaken the useful signal of SAR images. The wavelet coefficients have the characteristic of inner-scale relativity and convergence, which means that if current coefficient is small and its neighbor coefficients are also small, and vice versa. Therefore, a criterion that can distinguish signal from speckle noise is proposed in our method.

$$\begin{cases} \text{signal,} & \text{if} \quad \text{mean}[|y(s,i,j)|] \geq T \\ \text{noise,} & \text{else} \end{cases} \tag{8}$$

where $\text{mean}[|y(s,i,j)|]$ denotes the (i,j) mean value of the modulus coefficients at the sth scale in a 3×3 window. After judgment with the criterion, the proposed estimator of x is defined as

$$\hat{x} = \begin{cases} \dfrac{\left(\sqrt{y_r^2+y_i^2} - \sqrt{2}\exp(k)\sigma_n^2/\sigma\right)}{\sqrt{y_r^2+y_i^2}} \cdot y, & \text{signal} \\ 0, & \text{noise} \end{cases} \tag{9}$$

Our algorithm is summarized as follows.

Step 1: Take a log-transformation of the noisy SAR image.
Step 2: Decompose the log-transformed noisy image y by DTCWT.
Step 3: Calculate the quantum-inspired speckle noise variance σ_n^2 using (4).
Step 4: Calculate σ using (5) and estimate k mentioned in Sect. 2.1.
Step 5: Shrink complex wavelet coefficients based on the proposed estimator using (9)
Step 6: Perform the inverse DTCWT, followed by an exponential transformation, to obtain the despeckled SAR image.

3 Experiments

Our method is compared with various speckle suppression methods such as Lee filter [10], Log-OWTSURE method [2], homomorphic Wiener filter [11], Log-BI-DTCWT method [12], and Pizurica method [13]. Suppose the resolution of an image is $M \times N$, and $L = \text{round}(\log_2(\min(M,N)) - 4)$, where $\min(M,N)$ denotes

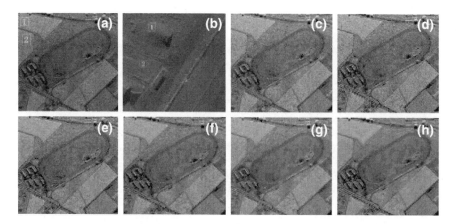

Fig. 2 Corresponding despeckled images using various methods. **a** Real SAR1. **b** Real SAR2.
c Lee. **d** Log-OWTSURE. **e** Log-BI-DTCWT. **f** Homomorphic Wiener. **g** Pizurica. **h** Proposed
method

Table 1 ENL and STD values for regions 1 and 2 in Fig. 1

Method	Region 1		Region 2	
	ENL	STD	ENL	STD
SAR1				
Before despeckling	23.33	37.85	20.59	37.85
Lee	95.78	30.97	71.17	30.97
Log-OWTSURE	35.93	36.20	29.84	36.20
Log-BI-DTCWT	216.71	33.18	139.99	33.18
Homomorphic Wiener	231.74	**29.90**	147.90	**29.90**
Pizurica	175.42	30.24	186.07	30.24
Proposed method $k = 0.3$	**654.61**	30.48	**555.73**	30.48
SAR2				
Before despeckling	8.94	25.44	7.70	25.44
Lee	29.28	17.95	23.02	17.95
Log-OWTSURE	8.94	25.44	7.70	25.44
Log-BI-DTCWT	44.21	15.90	27.23	15.90
Homomorphic Wiener	39.37	15.31	35.49	15.31
Pizurica	42.07	15.65	44.18	15.65
Proposed method $k = 0.4$	**67.07**	**12.70**	**153.56**	**12.70**

The proposed method can robustly obtain the largest ENL and the less STD in comparison with
the other methods

the minimum in M and N, and round () is a rounding function. L is the decomposing level of DTCWT. In homomorphic Wiener filter, a 5×5 window is used
to obtain the signal variance. Pizurica method employs the program supplied by its
authors with the optimal window size 5×5 and tuning factor $k = 3$. Three 1-m
resolution Ku-band SAR images from Sandia National Laboratory are presented in
this experiment, which are labeled as SAR1 and SAR2. They are shown in Fig. 2,

which are of 357×376, 811×811 pixels, respectively. In Fig. 2, two homo-geneous regions have been labeled in the two real SAR images, where the regions are made up of 41×29 and 31×31 pixels for SAR1 and 71×56 and 51×51 pixels for SAR2. ENL and STD obtained by the compared methods are shown in Table 1. The proposed method can robustly obtain the largest ENL and the less STD in comparison with the other methods.

The corresponding despeckled images using the various methods are also shown in Fig. 2. It is shown that the proposed method not only suppress much more noise than the others, but also preserves details and structures better.

4 Conclusions

In this paper, a quantum-inspired speckle noise variance and an improved La-placian statistical model have been introduced in the speckle suppression for SAR images. In DTCWT domain, the coefficients of the log-transformed noise-free signal are modeled as an improved Laplacian PDF with an adaptive smoothing factor k. Considering the inter-scale dependency and inner-scale relativity of coefficients, the proposed method gives a Bayesian MAP estimator with an adaptive quantum-inspired estimation of speckle noise variance. In our experi-ment, the proposed method shows good performance in reducing speckle noise and preserving edges and details for SAR images. In addition, this algorithm provides a feasible despeckling method to improve SAR images quality, which combines the basic quantum theory with the image processing technology.

Acknowledgments This work is supported by National Natural Science Foundation of China (Nos. 61201423, 61105010), Wuhan "Chen Guang" Project (No. 201150431095), the open fund project of Hubei Province Key Laboratory of Intelligent Information Processing and Real-time Industrial System (znss2013B016), and the Educational Commission of Hubei Province (Q20131108).

References

1. Bhuiyan, M., Ahmad, M.O., Swamy, M.: Spatially adaptive wavelet-based method using the Cauchy prior for denoising the SAR images. J. IEEE Trans. Circuits Syst. Video Technol. **17**(4), 500–507 (2007)
2. Luisier, F., Blu, T., Unser, M.: A new SURE approach to image denoising: interscale orthonormal wavelet thresholding. J. IEEE Trans. Image Process. **16**(3), 593–606 (2007)
3. Walessa, M., Datcu, M.: Model-based despeckling and information extraction from SAR images. J. IEEE Trans. Geosci. Remote Sens. **38**(5), 2258–2269 (2000)
4. Tian, J., Chen, L.: Image despeckling using a non-parametric statistical model of wavelet coefficients. J. Biomed. Signal Process. Control, 432–437 (2011)
5. Selesnick, I.W., Baraniuk, R.G., Kingsbury, N.: The dual-tree complex wavelet transform. J. IEEE Signal Process. Mag. **22**(6), 123–151 (2005)

6. Eldar, Y., Oppenheim, A.V.: Quantum signal processing. J. IEEE Signal Process. Mag. **19**(6), 12–32 (2002)
7. Xie, K.-F., Luo, A.: Research on quantum-inspired mathematical morphology. J. ACTA Electronica Sinica **33**(2), 284–287 (2005)
8. Fu, X.-W., Ding, M.-Y., Cai, C.: Despeckling of medical ultrasound images based on quantum-inspired adaptive threshold. J. Electron. Lett. **46**(13), 889–891 (2010)
9. Xie, H., Pierce, L., Ulaby, F.: Statistical properties of logarithmically transformed speckle. J. IEEE Trans. Geosci. Remote Sens. **40**(3), 721–727 (2002)
10. Lee, J.S.: Refined filtering of image noise using local statistics. J. Comput. Graphics Image Process. **15**(4), 380–389 (1981)
11. Jian, A.K.: Fundamentals of digital image processing. Prentice-Hall Inc., Englewood Cliffs (1989)
12. Sendur, L., Selesnick, I.W.: Bivariate shrinkage functions for wavelet based denoising exploiting interscale dependency. J. IEEE Trans. Signal Process., 2744–2756 (2002)
13. Pizurica, A., Philips, W., Lemahieu, I., et al.: Despeckling SAR images using wavelets and a new class of adaptive shrinkage estimators. J. Int. Conf. Image Process. **2**, 233–236 (2001)

Image Semantic Annotation Approach Based on the Feature Matching

Cong Jin and Jinglei Guo

Abstract There is a semantic gap between the low-level visual features of image and the high-level image semantic. In this paper, we deal with the annotation of image semantic from edge points of image and matching of feature points. According to the idea of particle swarm optimization (PSO), the feature point matching is proposed based on the image edge points. The image is transformed into its edge points set using edge extraction, and then, the feature ellipse of these edge points is also proposed. According to the center of the ellipse and reference triple, the matching of feature points can implement. The experimental results confirm that the proposed approach is very effectiveness. When we use proposed approach to Corel image library, the annotation results are satisfactory.

Keywords Image semantic · Semantic annotation · Feature matching · Edge extraction algorithm

1 Introduction

Increasing the number of images, it is difficult to automatically obtain useful information from these images. We know that different images have no semantic. Therefore, it is difficult that it is one-to-one correspondence between the low-level visual features and the high-level semantics. This is the famous "semantic gap". At present, the most common research framework of the image semantic annotation is to extract the low-level image features firstly and then use some technology mapping the low-level feature to high-level semantic of the image. The common mapping

C. Jin (✉) · J. Guo
School of Computer, Central China Normal University, Wuhan 430079,
People's Republic of China
e-mail: jincong@mail.ccnu.edu.cn

J. Guo
e-mail: guojinglei@mail.ccnu.edu.cn

S. Patnaik and X. Li (eds.), *Proceedings of International Conference on Soft Computing Techniques and Engineering Application*, Advances in Intelligent Systems and Computing 250, DOI: 10.1007/978-81-322-1695-7_32, © Springer India 2014

methods include the support vector machine (SVM), artificial neural network (ANN), and latent Dirichlet allocation (LDA). However, SVM and ANN are not suitable when the categories are very large. LDA is mainly used for natural language and text semantic processing, and it can generate text, theme, and the probability distribution of the words. When we applied LDA to image semantic, the two major problems must be faced, one is to ignore the spatial relationship of the objects in the image, and the second is the visual word of the image is not obvious.

2 Extraction Image Edge Points

At present, in the experimental environment, various image semantic methods have better performance and higher matching accuracy for image semantic annotation; however, we still cannot think that image semantic gap has completely been crossed. We believe that, when image visual features are mapped to the image semantic, the following principles should be satisfied: (1) image visual feature should be expressed using the least amount of data; (2) calculation should be as simple as possible; (3) image visual features can restore the image in a way. Based on these principles, an image semantic annotation approach based on feature matching will be proposed.

We know that edge of the image can maintain the most important information of the image. Thus, we do not deal directly with the image itself, but the edge of the image. We focus on Canny edge detection [1] algorithm, which is optimal for step edges corrupted by white noise and it is described as follows.

Algorithm 1 (*Canny edge detection*)
Step 1. Convolve an image I with a Gaussian of scale σ.
Step 2. Estimate local edge normal directions n using Eq. (1) for each pixel in the image.

$$n = \frac{\nabla(G * I)}{|\nabla(G * I)|} \tag{1}$$

Step 3. Find the location of the edges using Eq. (2)

$$\frac{\partial^2}{\partial n^2} G * I = 0 \tag{2}$$

Step 4. Compute the magnitude of the edge using Eq. (3)

$$|G_n * I| = |\nabla(G * I)| \tag{3}$$

Step 5. Threshold edges in the image with hysteresis to eliminate spurious responses.
Step 6. Repeat Step 1 through 5 for ascending values of the standard deviation σ.
Step 7. Aggregate the final information about edges at multiple scales using the feature synthesis approach.

(a)   **(b)**

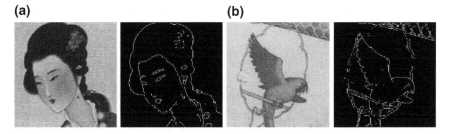

Fig. 1 (**a, b**) are the four original images and their edge detection results

Fig. 2 The center point and reference triple, where $T'_1(84,168)$, $T'_2(206, 90)$, and T'_3 (198, 232)

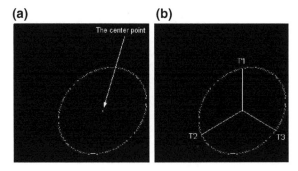

We denote edge point set as P. Experiment results are shown in Fig. 1a, b.

3 Reference Triple

By Zhang et al. [2], for edge feature point set $P = \{p_i = (x_i, y_i)^T | i = 1, 2, \ldots, F\}$, its feature ellipse is defined as

$$(x - c)^T E^{-1} (x - c) = 1/\alpha \tag{4}$$

where α is a positive integer, the size of the edge feature ellipse is determined by α, and $c = 1/F \sum_{i=1}^{F} p_i$ and $E = 1/F \sum_{i=1}^{F} (p_i - c)(p_i - c)^T$ are the center point and the second-order center moment of P, respectively. F is the number of edge feature point. Obviously, the center of the feature ellipse locates in the canter of the feature point set. Via adjusting the parameter α, the feature ellipse can be insured to lie into the convex hull of P. Draw three rays from the center point c of P, and each pair rays forms $120°$ angle. The rays intersect the feature ellipse at three points, which is the expected referenced triplet. Three intersect points compose a point set, denoted by P'. Obviously, P' locates in the local area around the center of P. For convenience, in this paper, we will ensure that one of three rays is vertical. For example, for image with 256×256 size, description of the center point and reference triple, please see Fig. 2.

4 Feature Matching

We obtain image's reference point set, denoted by Q, using the same method for tested image. Any three points of the reference point set compose a triple, denoted by Q'. Particle swarm optimization (PSO) [3] is used for features matching. Each particle adjusts its position in the search space from time to time according to the flying experience of its own and of its neighbors. It is initialized with a population of random potential solutions, and the algorithm searches for optima satisfying some performance. The potential solutions, called *particles*, are flown through a multidimensional search space. Each particle i has a position represented by a position vector X_i. A swarm of particles moves through a d-dimensional problem space with the velocity of each particle represented by a vector V_i. The particle velocity and position equations form are given by

$$V_i(t+1) = w \cdot V_i(t) + c_1 \text{rand}_1 (P_{i,\text{best}}(t) - X_i(t)) + c_2 \text{rand}_2 (P_{\text{global}}(t) - X_i(t)) \tag{5}$$

$$X_i(t+1) = X_i(t) + V_i(t+1) \tag{6}$$

where, t is current iteration number, w is inertia weight, c_1 and c_2 are positive constants, rand_1 and rand_2 are uniformly distributed random numbers in the range [0, 1]. $P_{i,\text{best}}$ and P_{global} are the best previously visited position of the particle i and the best value of all individual particle position values, respectively. Where $X_i(t) = (X_{i1}(t), X_{i2}(t), \ldots, X_{id}(t))$, and $V_i(t) = (V_{i1}(t), V_{i2}(t), \ldots, V_{id}(t))$. w, c_1 and c_2 are the predefined by the user. The fitness value of particle i, at iteration t, is

$$F_i(t) = \frac{1}{q} \sum_{k=1}^{q} \sum_{j=1}^{d} \left(X_{ij}^{(k)}(t) - \hat{X}_{ij}^{(k)}(t) \right)^2 \tag{7}$$

where $X_{ij}^{(k)}$ is the jth ideal component of particle i for the kth sample, and $\hat{X}_{ij}^{(k)}$ is the jth actual component of particle i for the kth sample. For measure the matching degree between two point sets P' and Q', our aim is the fitness function value as small as possible, namely the smaller the objective function value, the better the fitness value is. The best position $P_{i,\text{best}}$ of particle i is determined by the following equation

$$P_{i,\text{best}}(t+1) = \begin{cases} X_i(t), & \text{if} \quad F(X_i(t)) < F(P_{i,\text{best}}(t)) \\ P_{i,\text{best}}(t), & \text{if} \quad F(X_i(t)) \geq F(P_{i,\text{best}}(t)) \end{cases} \tag{8}$$

At each update step of PSO, the velocity of each particle is calculated according to (5) and the position is updated according to (6). When a particle finds a better position than the previously best position, it will be stored in the memory. The algorithm goes on until a satisfactory solution is found or the predefined number of iterations is met. Generally, the initial population is generated randomly, the generating range of particles is not arbitrary but limited to the local area around in

Table 1 Particle coding

x'_{j1}	y'_{j1}	x'_{j2}	y'_{j2}	x'_{j3}	y'_{j3}

the convex hull of Q. Each particle consists of a triple of points in the tested image. Its detailed construction is seriating positions of the three points and coding them into binary codes. As an illustration, let us consider the following example. Given the positions of three points $T'_1 = (x'_{j1}, y'_{j1})$, $T'_2 = (x'_{j2}, y'_{j2})$ and $T'_3 = (x'_{j3}, y'_{j3})$ in the test image. Table 1 shows the detailed construction of a particle consisting of the three points. Here, each x or y is represented by a binary code. Thus, the whole particle is a binary sequence.

Up to now, we can acquire image semantic annotation approach.

Algorithm 2 (*Image semantic annotations algorithm*)
Step 1. Given a template image I, calculate its referenced triplet P'.
Step 2. For test image, i.e., image I', calculate its feature point set Q.
Step 3. Evaluate P_{best} and P_{global} according to Eq. (8).
Step 4. Update P_{best} if the current particle's value P_{best} is better than P_{best}.
Step 5. Determine P_{global}. Choose the particle with the best value P_{best} of all.
Step 6. For each particle:
Step 6.1 Calculate particle's new velocity according to (5).
Step 6.2 Let the new velocity value be binary.
Step 6.3 Let the new position value be binary.
Step 6.4 Calculate particle's new position according to (6).
Step 7. While a sufficiently good P_{global} or stop condition is not yet attained.

5 Experiments and Results

5.1 Data Sets

In the experiments, we use an image library with 500 images from Corel. The image library is divided into 10 categories, and each category has 50 images. We arbitrarily selected 20 images as training samples in each category, and the remaining 30 image as the test sample. Thus, the total 200 training samples are obtained.

5.2 Parameters Initialization

In experiments, the training set is used to build the annotator and the testing set is used to evaluate the annotation performance of the annotator. The population size of PSO is selected as 60, and PSO was stopped after 150 iterations.

Approach				
[4]	horse leaf grass	flowers grass trees rose	grass water building	elephant leaf water grass
[5]	horse trees grass	flowers leaf grass rose	grass sky building	elephant water field grass
[6]	horse leaf grass	flowers grass leaf rose	grass water mountains	elephant leaf water grass
Proposed	horse trees grass	flowers leaf grass rose	grass water mountains	elephant sky field grass

Fig. 3 Annotation results of four images

5.3 Annotation Performance

In experiments, the annotation accuracies are measured as follows:

$$\text{Accuracy}(S'') = \frac{\sum_{i=1}^{|S''|} \text{Assess}(S_i)}{|S''|}, \quad S_i \in S'' \tag{9}$$

$$\text{Assess}(S_i) = \begin{cases} 1, & \text{if annotator } a(S_i) = a.c \\ 0, & \text{otherwise} \end{cases} \tag{10}$$

where $a.c$ is the actual class of sample S_i, and *annotator* $a(S_i)$ is return class of S_i by annotator a.

5.4 Results and Discussions

The Refs. [4–6] and our approach are used to annotate four different images, and the annotation results are listed in Fig. 3. The entire data set has 24 different keywords.

In [4], the subspace clustering algorithm was used for image annotation. In [5], the space mapping method based on SVM was also used, and it maps low-level features of the image to a certain level semantic concept in order to achieve an index. In [6], K-means clustering and Bayesian models were used for automatic semantic annotation. We notice that [4, 5]'s approaches annotate "sky" in elephant image into "water," and the proposed approach obtained the correct annotation result "sky," which shows that the proposed approach can get superior

Table 2 Results of annotation performance

Keyword	Average F_1 values					
	Wang et al. [4]	Lu and Ma [5]	Jeon et al. [6]	Li and Sun [7]	Ru and Ma [8]	Proposed
Flower	0.279	0.407	0.254	0.354	0.373	0.415
Horse	0.273	0.415	0.252	0.347	0.368	0.423
Building	0.252	0.371	0.226	0.317	0.349	0.390
People	0.264	0.402	0.242	0.325	0.357	0.411
Bus	0.285	0.423	0.267	0.361	0.382	0.433
Elephant	0.286	0.412	0.263	0.357	0.379	0.419
Mountain	0.258	0.373	0.232	0.316	0.354	0.388

performance. Then, we test more images than 200 and calculate their Precision, Recall, and F_1 according to Eqs. (11)–(13), respectively.

$$\text{Precision} = \frac{\text{Numble of Correct Images}}{\text{Numble of Retrieved Images}} \times 100\,\% \tag{11}$$

$$\text{Recall} = \frac{\text{Numble of Correct Images}}{\text{Numble of Class Images}} \times 100\,\% \tag{12}$$

$$F_1 = \frac{2 \cdot \text{Recall} \cdot \text{Precision}}{\text{Recall} \cdot \text{Precision}} \tag{13}$$

Recall reflects the approach's ability to retrieve relevant images, and Precision reflects the approach's ability to reject irrelevant images. F_1 value is the performance evaluation considering the Recall and Precision. According to keywords retrieving, the number of relating images in database is 50. When the number of retrieved images were 15, 30, 45, 60, 75, and 90, we let 7 kinds of images, i.e., flower, horse, building, people, bus, elephant, and mountain, be retrieved, respectively. The retrieval results are compared with the performances of [4–8] using average F_1 values, and the experiment results are shown in Table 2.

When the image semantic annotating, we consider the edge points set of the image, and which captures the kernel content of the image. From the annotating results, our approach not only reduces the semantic gap, but it also has good robustness.

6 Conclusions

In this paper, the edge points are extracted firstly. After obtaining the feature ellipse of these edge points, the center of the ellipse and reference triple also obtained. Using the PAO, the matching of feature points can be achieved. We have done a lot of experiments. From Table 2, we found that the average F1 value of the proposed approach is the highest, which indicates that the proposed image annotation approaches have good performance and the annotation results are satisfactory.

Acknowledgments This work was supported by Natural Social Science Foundation of China (Grant No. 13BTQ050), and social science foundation from Chinese Ministry of Education (Grant No. 11YJAZH040).

References

1. Canny, J.: A Computational approach to edge detection. IEEE Trans. Pattern Anal. Mach. Intell. **8**, 679 (1986)
2. Zhang, L.H., Xu, W.L., Chang, C.: Genetic algorithm for affine point pattern matching. Pattern Recogn. Lett. **24**, 9–19 (2003)
3. Kennedy, J., Eberhart, R.C., Shi, Y.: Swarm Intelligence. Morgan Kaufmann, San Francisco (2001)
4. Wang L., Liu L., Khan L..: Automatic image annotation and retrieval using subspace clustering algorithm. In: Proceedings of 2nd ACM International Workshop on Multimedia Databases, pp. 100–108. Washington, DC, USA (2004)
5. Lu, J., Ma, S.P.: Automatic image annotation based on concept indexing. J. Comput. Res. Dev. **44**, 452–459 (2007)
6. Jeon J., Lavrenko V., Manmatha R.: Automatic image annotation and retrieval using cross media relevance models. In: Proceedings of 26th Annual International ACM SIGIR Conference on Research and Development in Information Retrieval, pp. 119–126. Toronto, Canada (2003)
7. Li W., Sun M.S.: Automatic image annotation based on WordNet and hierarchical ensembles. In: Proceedings of 7th International Conference on Computational Linguistics and Intelligent Text Processing, pp. 417–428. Mexico (2006)
8. Ru, L.Y., Ma, S.P.: Boosting-based automatic linguistic indexing of pictures. J. Image Graph. **11**, 486–491 (2006)

Research on Transmission and Transformation Land Reclamation Based on BP Neural Network

Xi Wu, Hai-Ting Ming, Xue-Huan Qin and Wen-Jing Zhu

Abstract In this paper, the neural network structure, process flow, and error transmit are analyzed and the land reclamation suitability evaluation model based on BP neural network is built. The input layer was configured 7 according to soil parameters, the output layer was configured 4 according to soil degrees, and the hidden layer was configured 9 according to experience. The network was configured as 7-9-4 network structure and trained, tested, and validated in Levenberg-Marquardt algorithm. Experiments show that the soil degree recognized by BP neural network was equated with the actual soil degree. It proves the feasibility of BP neural network in land reclamation of transmission and transformation.

Keywords Land reclamation · Suitability evaluation · Neural network

1 Introduction

In transmission and transformation projects, the temporary land resources like construction plant may damage the nearby land to some extent. And the damaged land should be reclaimed to appropriate utilize according to soil texture and land reclamation suitability evaluation criteria.

X. Wu (✉)
Hubei Anyuan Safety and Environmental Protection Technology,
Wuhan 430040, Hubei, China
e-mail: wang_hust@qq.com

H.-T. Ming · X.-H. Qin
Hubei Electric Power Survey and Design Institute,
Wuhan 430040, Hubei, China

W.-J. Zhu
Hubei Institute of Geological Sciences,
Wuhan 430034, Hubei, China

S. Patnaik and X. Li (eds.), *Proceedings of International Conference on Soft Computing Techniques and Engineering Application*, Advances in Intelligent Systems and Computing 250, DOI: 10.1007/978-81-322-1695-7_33, © Springer India 2014

The main suitability evaluation criteria [1, 2] have limiting condition algorithm [3], exponential sum algorithm, and fuzzy evaluation algorithm [4, 5], which rely on human experience and influence the evaluation conclusion.

The essence of land reclamation suitability evaluation is a pattern recognition procedure which matches the evaluation parameter against different soil category.

The neural network is a supervise algorithm which reduces the output error of actual output relative to network output through iteration learning and network structure adjustment according to input data and avoid the impact of human factor. This paper talks about the feasibility and practicality of BP neural network in transmission and transformation land reclamation suitability evaluation.

2 BP Neural Network

The neural network is a data information process system, with the development of modern neurosciences, which simulates neurons' function in the brain. Figure 1 shows the neuron node's function, the neuron receives n input variables $p1$ to pn; each variable has different weight; the output of the neuron is comprehensive combination of input variable, weight, and active function [6].

BP network is a feed-forward network as shown in Fig. 2. The input layer receives variable x_n and output w_{ih} after process, the hidden layer receives w_{ih} as input and output h_{op} as input variable of output layer. The data flow spreads forward layer by layer from input layer to output layer. The neuron output y_{oq} is compared to expect output d_{oq}. If the error is less than the specified error ε, the iteration would end if the network has been adjusted to right state. Otherwise the error $\varDelta q$ would spread back-forward the network and adjust the network structure and parameters [7]. After several times iteration, the network error can approach the regular error. The iteration times depend on network's complexity [8].

The neural network need initiate before training. For initiate step, every variable weight is assigned a random value between -1 and 1, define error function e, computing precision ε and maximum study times M. Choose random k input sample and corresponding output as expression 1 and 2 [9].

$$x(k) = (x_1(k), x_2(k), x_3(k), \ldots, x_n(k)) \tag{1}$$

$$d_o(k) = (d_1(k), d_2(k), d_3(k), \ldots, d_q(k)) \tag{2}$$

The hidden layer input $hi_h(k)$ and output $ho_h(k)$ are defined as expression 3 and 4. w_{ih} is the weight of hidden layer. b_h is the threshold of hidden layer. P is hidden layer neuron number.

$$hi_h(k) = \sum_{i=1}^{n} w_{ih}x_i(k) - b_h \quad h = 1, 2, \ldots, p \tag{3}$$

Fig. 1 Neuron node function schematic diagram

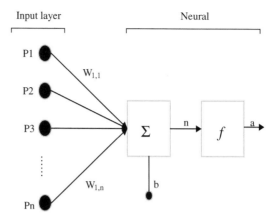

Fig. 2 Neural network schematic diagram

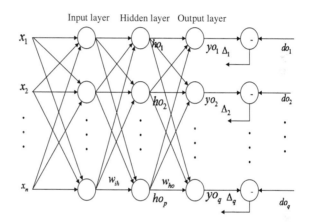

$$\mathrm{ho}_h(k) = f(\mathrm{hi}_h(k)) \quad h = 1, 2, \ldots, p \tag{4}$$

The output layer input $\mathrm{yi}_o(k)$ and output $\mathrm{yo}_o(k)$ are defined as expression 5 and 6. w_{ho} is the weight of output layer. b_o is the threshold of output layer. q is output layer neuron number.

$$\mathrm{yi}_o(k) = \sum_{h=1}^{p} w_{\mathrm{ho}} \mathrm{ho}_h(k) - b_o \quad o = 1, 2, \ldots, q \tag{5}$$

$$\mathrm{yo}_o(k) = f(\mathrm{yi}_o(k)) \quad o = 1, 2, \ldots, q \tag{6}$$

The error function is defined as expression 7 [10] and it is based on network output $d_o(k)$ and expects output $\mathrm{yo}_o(k)$. The partial derivative of error function e respect to output neuron $\delta_o(k)$ is computed as 8, 9, and 10.

$$e = \frac{1}{2}\sum_{o=1}^{q}(d_o(k) - \mathrm{yo}_o(k))^2 \tag{7}$$

$$\frac{\partial e}{\partial w_{\mathrm{ho}}} = \frac{\partial e}{\partial \mathrm{yi}_o}\frac{\partial \mathrm{yi}_o}{\partial w_{\mathrm{ho}}} \tag{8}$$

$$\begin{aligned}
\frac{\partial e}{\partial \mathrm{yi}_o} &= \frac{\partial\left(\frac{1}{2}\sum_{o=1}^{q}(d_o(k) - \mathrm{yo}_o(k))\right)^2}{\partial \mathrm{yi}_o} = -(d_o(k) - \mathrm{yo}_o(k))\mathrm{yo}'_o(k)\\
&= -(d_o(k) - \mathrm{yo}_o(k))f'(\mathrm{yi}_o(k))\\
&= -\delta_o(k)
\end{aligned} \tag{9}$$

$$\frac{\partial e}{\partial w_{\mathrm{ho}}} = \frac{\partial e}{\partial \mathrm{yi}_o}\frac{\partial \mathrm{yi}_o}{\partial w_{\mathrm{ho}}} = -\delta_o(k)\frac{\partial\left(\sum_{h=1}^{p}w_{\mathrm{ho}}\mathrm{ho}_h(k) - b_o\right)}{\partial w_{\mathrm{ho}}} = -\delta_o(k)\mathrm{ho}_h(k) \tag{10}$$

The partial derivative of error function e respects to hidden neuron $\delta_h(k)$ was computed as 11.

$$\frac{\partial e}{\partial w_{\mathrm{ih}}} = \frac{\partial e}{\partial \mathrm{hi}_h(k)}\frac{\partial \mathrm{hi}_h(k)}{\partial w_{\mathrm{ih}}} = -\delta_h(k)x_i(k) \tag{11}$$

$$\Delta w_{\mathrm{ho}}(k) = -\mu\frac{\partial e}{\partial w_{\mathrm{ho}}} = \mu\delta_o(k)\mathrm{ho}_h(k) \quad w_{\mathrm{ho}}^{N+1} = w_{\mathrm{ho}}^{N} + \eta\delta_o(k)\mathrm{ho}_h(k) \tag{12}$$

If the error e is greater than expect error, it means that the network has not been adjusted to proper state and need to be corrected. The weight $w_{\mathrm{ho}}(k)$ was corrected based on $\delta_o(k)$ and hidden neuron output $\mathrm{ho}_h(k)$. The weight $w_{\mathrm{ih}}(k)$ was corrected based on $\delta_h(k)$ and input neuron $x_i(k)$.

$$\begin{aligned}
\Delta w_{\mathrm{ih}}(k) &= -\mu\frac{\partial e}{\partial w_{\mathrm{ih}}} = -\mu\frac{\partial e}{\partial \mathrm{hi}_h(k)}\frac{\partial \mathrm{hi}_h(k)}{\partial w_{\mathrm{ih}}} = \delta_h(k)x_i(k)\\
w_{\mathrm{ih}}^{N+1} &= w_{\mathrm{ih}}^{N} + \eta\delta_h(k)x_i(k)
\end{aligned} \tag{13}$$

If the error of expect value is relative to actual output E less than regular precision ε, the iteration would be ended, Otherwise went into iteration until E approaches ε.

$$E = \frac{1}{2m}\sum_{k=1}^{m}\sum_{o=1}^{q}(d_o(k) - y_o(k))^2 \tag{14}$$

3 Land Reclamation Type

The land reclamation type has plow land, forest land, grassland, and garden plot. The soil characteristic lists are as follows (Table 1).

Table 1 Land reclamation classification

Type	Soil characteristic
Plow land	Slope <10°, organic matter content ≥3 %, available depth of soil >50 cm
Forest land	Slope <20°, available depth of soil >30 cm
Grassland	Slope <30°, available depth of soil >20 cm
Garden plot	Slope <30°, available depth of soil >40 cm

Table 2 Soil degree parameter range

	I	II	III	IV
Slope (°)	≤3	3–10	10–20	≥20
Thickness of soil (cm)	≥80	50–80	30–50	≤30
Total nitrogen (g/kg)	>2	1.5–2	1–1.5	<1
Organic content (g/kg)	>40	30–40	20–30	<20
Available nitrogen (mg/kg)	>150	120–150	90–120	<90
Available phosphorus (mg/kg)	>40	20–40	10–20	<10
Available kalium (mg/kg)	>200	150–200	100–150	<100

4 BP Neural Network Verification

This paper selected gradient, thickness of soil, total nitrogen, organic content, available nitrogen, available phosphorus, and available kalium seven soil parameters to classify the soil four degree. And the parameters are as follows (Table 2).

The network structure is configured as 7-9-4, the input layer neuron is 7, hidden neuron is 9, and output layer neuron 4. Every degree simulates 500 group data, which is in the limit of every parameter range. For the four degree 2,000 group data, 70 % is selected to train the network, 15 % is used to validate the network, and 15 % is used to test the network.

The neural network training, correct, and test results are displayed as conflict matrix as Fig. 3. Horizontal axis is the soil real degree, the vertical axis is the soil degree recognized by network.

Figures 3, 4, 5 and 6 are the training, correct, and test sample conflict matrixes. From the matrix, we can see that every sample can be recognized exactly as every sample was simulated. The error curve displays in Fig. 6, and we can see that the MSE is 10^{-5}, meets the requirements.

The above trained network can be used to recognize soil degree. The paper select 20 samples along the transmission and transformation project line, and the result is shown in Table 3.

From the result, we can see that every soil sample was recognized exactly, this proves the feasibility for BP neural network in transmission and transformation project.

Fig. 3 Network training
conflict matrix

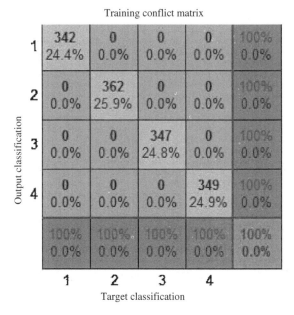

Fig. 4 Network correct
conflict matrix

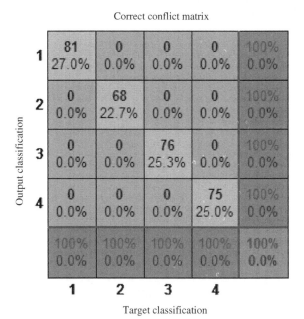

Fig. 5 Network test conflict
matrix

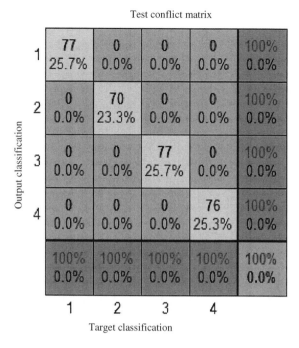

Fig. 6 Network training
error

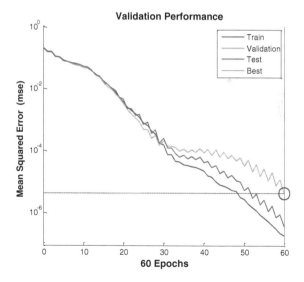

Table 3 Soil sample recognize result

Oder	Parameters							Result	
	Slope (°)	Soil thickness (cm)	Organic content (g/kg)	Available nitrogen (mg/kg)	Available phosphorus (mg/kg)	Available kalium (mg/kg)	Total nitrogen (g/kg)	Recognize type	Actual type
1	6.4	87.5	36.3	154.2	54.8	242.7	2.2	I	I
2	2.7	86.4	43.6	143.8	57.8	233.5	1.8	I	I
3	8.6	69.3	45.8	115.1	33.3	161.6	1.7	II	II
4	3.8	58.6	39.0	142.4	25.9	187.0	2.0	II	II
5	4.1	69.7	38.7	114.3	32.0	194.4	1.3	II	II
6	17.7	33.1	26.8	102.0	18.8	106.8	1.2	III	III
7	16.9	49.2	24.9	99.3	15.1	125.6	1.2	III	III
8	18.7	47.5	19.3	109.6	13.8	152.1	1.3	III	III
9	15.8	39.8	21.1	101.8	18.8	115.6	1.5	III	III
10	16.3	38.1	20.8	123.6	12.5	116.1	0.9	III	III
11	11.1	32.5	26.1	119.4	11.6	147.4	1.0	III	III
12	24.6	22.0	20.9	77.4	7.8	61.7	1.1	IV	IV
13	19.1	34.6	17.0	62.1	9.2	98.9	0.5	IV	IV
14	27.8	23.7	10.9	71.2	5.2	97.2	0.9	IV	IV
15	24.5	11.2	14.7	91.0	8.8	99.1	0.4	IV	IV
16	23.0	23.5	16.4	72.2	8.4	53.3	0.7	IV	IV
17	23.6	16.1	15.3	79.7	9.4	59.8	0.8	IV	IV
18	29.8	17.0	11.4	97.3	6.7	71.2	0.8	IV	IV
19	24.3	54.5	12.8	89.3	7.6	76.9	0.7	IV	IV
20	27.0	12.1	19.9	93.2	7.3	73.2	0.5	IV	IV

5 Conclusions

The paper established BP neural network model according to transmission and transformation project requirements. The input, hidden, and output neuron numbers were configured according to soil parameters. The network was trained using the sample, verified with practical soil sample. The results show that the network can recognize the soil degree exactly and prove the network feasibility.

References

1. Suguangquan, H.: Land resource suitability evaluation in ore district. Prog. Geogr. **17**(4), 39–46 (1998)
2. Wanghuan, W.: Commentary on methods for suitability evaluation of land reclamation. J Cent. South Univ. Forest. Technol. **30**(4), 154–158 (2010)
3. Wangshidong, G.: Study and application of suitability evaluation of land reclamation based on comprehensive extreme condition method. Sci. Surveying Mapp. (01) (2012)
4. Chengjianfei, L.: An integrated evaluation of land suitability based on fuzzy set theory. Resour. Sci. **21**(4), 71–74 (1999)
5. Caudill, M., Butler, C.: Understanding neural networks: Computer explorations, vols. 1 and 2. The MIT Press, Cambridge, MA (1992)
6. Charalambous, C.: Conjugate gradient algorithm for efficient training of artificial neural networks. IEEE Proceedings, vol. 139(3), pp. 301–310 (1992)
7. Wenzhan, DAI.: A method of multiobjective synthetic evaluation based on artificial neural networks and its applications. Syst. Eng. Theor. Pract. **19**(5), 30–33 (1999)
8. Irie, B., Miyake, S.: Capabilities of three-layered perceptrons. Proceedings of the IEEE International Conference on Neural Network, vol. 1, pp. 641–648 (1998)
9. Gorman, R.P., Sejnowski, T.J.: Analysis of hidden units in a layered network trained to classify sonar targets. Neural Netw. **11**(1), 75–89 (1988)
10. Pandafeng, L.: Study on the multi -index comprehensive evaluation method of artificial neural network. Syst. Sci. Compr. Stud. Agric. (02), 125–130 (1999)
11. Yangguodong, J.: BP neural network models for the reclaimed land suitability assessment of temporary land-use of highway. Syst. Eng. Theor. Pract **2**, 119–124 (2002)

TTP-ACE: A Trusted Third Party for Auditing in Cloud Environment

Songzhu Mei, Haihe Ba, Fang Tu, Jiangchun Ren and Zhiying Wang

Abstract Cloud computing offers an appealing business model, and it is tempting for companies to delegate their IT services, as well as data, to the cloud. But in cloud environment, lack of practical auditing party always put the users' data in danger. Users may suffer a serious data loss without any compensation for they have lost all their control on their data. We present in this paper a novel way to implement a trusted third party for auditing in cloud environment (TTP-ACE), a trusted and easy-to-use auditor for cloud environment. TTP-ACE enables the cloud service providers' accountability and protects the cloud users' benefits.

Keywords Cloud computing · Trusted computing · Cloud security · Audit

1 Introduction

Cloud computing has emerged as one of the most influential technologies in both the IT industry and academia. Cloud computing, as the Salesforce's definition, is a friendlier mode of business operation. It is rapidly revolutionizing the way IT

S. Mei (✉) · H. Ba · F. Tu · J. Ren · Z. Wang
College of Computer, National University of Defense Technology, Changsha, Hunan, People's Republic of China
e-mail: sz.mei@nudt.edu.cn

H. Ba
e-mail: haiheba@nudt.edu.cn

F. Tu
e-mail: fangtu@nudt.edu.cn

J. Ren
e-mail: jcren@nudt.edu.cn

Z. Wang
e-mail: zywang@nudt.edu.cn

S. Patnaik and X. Li (eds.), *Proceedings of International Conference on Soft Computing Techniques and Engineering Application*, Advances in Intelligent Systems and Computing 250, DOI: 10.1007/978-81-322-1695-7_34, © Springer India 2014

resources are managed and utilized [1]. Through combining a batch of technologies, such as virtualization, data center, utility computing, and service-oriented architecture (SOA), cloud computing endows itself a lot of characteristics as follows: on-demand self-service, ubiquitous network access, resource pooling, rapid elasticity, and measured service [2]. Among all the transformation cloud computing brings to IT, a most significant one is that more and more individuals, companies, and organizations begin to outsource their data into the cloud. From the user's perspective, on matter individuals and enterprises, storing data in the cloud in a flexible manner makes great sense to their benefits. Capital expenditure on hardware, software, and personnel maintenance can be reduced, some inherent difficulties in managing local storage systems can be relieved, and the access to their data will no longer be bounded by the geographical location [1].

While cloud computing makes so many significant advantages for IT enterprise, it also has some drawbacks and brings us new and challenging security threats. When we focus only on the data security, it is not difficult to find that data outsourcing forced the owners of data to relinquish the administrative right to the cloud provider. The loss of control inevitably causes some potential risk to users' data. The data's confidentiality, integrity, availability, and authentication (CIAA) may be threatened for following reasons. First, although the infrastructures under the cloud are much more powerful and stable, it still cannot be promised that CSPs would never fail. By contrary, the providers will have a boarder attack surface. They still face the large probability of intrusions and even Byzantine failures. Second, the CSPs themselves become new potential risk in the cloud environment. For some reason, such as storage capacity limit, CSPs may remove user's rarely accessed data. CSPs may dissemble some data loss silently to maintain their good reputation. In these situations, users' data and benefit will be harmed without any sense.

To protecting the users' benefits, cryptology has been introduced. But ciphers can only protect the confidentiality. The integrity of data is still exposed in danger. So a third-party auditor (TPA) should be imported to the cloud service system to audit and log the user and CSPs' behaviors. But in reality, there is no TPA being used in the cloud computing systems. The reasons may be as follows. (1) TPA may still untrustworthy, (2) traditional audit methods may cause performance issue, and (3) TPA can be attacked, and the logs can be tampered with.

In this paper, we propose some methods to build a trusted and practical TPA. The main contributions can be summarized as follows:

- Use the TPM-compatible USBKey for cloud users to enable remote attestation, so that cheating attacks can be avoided.
- Propose comprehensive methods to protect the TPA itself from malevolent tamper. Make sure the logs' security and the audit's authority and justice.
- Combine trusted computing technology into TPA. Make the TPA itself trustworthy and make sure the user and CSPs will not be cheated.

The rest of the paper is organized as follows. Section 2 introduces the system and threat model and our design goals. Then, we provide the main idea of our design in Sect. 3. Section 4 gives key technologies used in our design, followed by Sect. 5, which overviews the related work.

2 Problem Statement

2.1 Adversary Model

Wang et al. [3] describe a typical architecture for cloud storage system, which generally includes three main components: the cloud user (U), who has large amount of data files to be stored in the cloud; the cloud server (CS), which is managed by cloud service provider (CSP) to provide data storage service and has significant storage space and computation resources; and an optional TPA, who has expertise and capabilities that cloud users do not have and should be trusted to assess the cloud storage service security on behalf of the user upon request. According to Sect. 1, we explain that the TPA should be an indispensable part of the cloud storage system. But in practice, TPA still has little approaches to protect itself from various attacks. So nowadays, TPA is rarely introduced to the system.

In the open network environment, the TPA would face various threats from different subjects.

- CSP's misconfiguration of the cloud storage system makes some data loss, and CSPs want to conceal its mistake. So CSP hacked the TPA to distort the log files. The tampered log files may mislead arbitrators' evidence obtaining; thereby, the data owners may not get enough compensation;
- People's behaviors and habits, on the one hand, may be a good guide for enterprises' advertisements. On the other hand, these behaviors and habits are also considered as users' privacy. CSPs or some others who are interested in the users' customs or habits may hack the TPA to steal the log files for the analysis of users' customs;
- Traditional auditing methods require the auditor to be embedded into the system to log all the operations to the data. In cloud environment, a TPA cannot get the complete control and monitor over the CSP, and even if a TPA could embed into the CSP's infrastructures, the TPA would make a serious performance issue.
- Haeberlen [4] proposes a case for the accountable cloud. With an explicit audit instruction, user can challenge the CSP for the data integrity to make sure their data's correctness. The use of explicit audit command brings the users a new burden to use the cloud storage service conveniently.

2.2 Design Goals

To implement a practical TPA for the typical cloud storage system model, our TPA's design should achieve the following security and performance guarantee.

- Neither the CSPs nor the users can arbitrarily access and modify the TPA's log files, say nothing of the vicious subjects.
- With the TPA, user can implicitly query the data's integrity. If there are any invalid operations that violate the data's integrity, data owner should be notified.
- The TPA's auditing behaviors should have little impact on the service's performance.

3 Design of TTP-ACE

3.1 Requirements for TTP-ACE

There are some essential requirements for TTP-ACE to fulfill our design goals in cloud environment.

Trusted identification. We must make sure that the TTP-ACE knows which user or CSP is exactly audited. The TTP-ACE must avoid being cheated by man-in-the-middle attack to guarantee the log files' authority.

Strict storage security. TTP-ACE must strictly maintain its own storage space's security. It must use proper log updating mechanisms to ensure that there is only a unique log file for one user–CSP relationship, and only authorized users and CSPs can update the log file. Some other mechanisms also must be introduced to enhance the log files' security.

Active challenging and notifying. TTP-ACE must provide the users a well-designed mechanism to transparently raise a challenge to the CSP, and once some violation happens, this mechanism could detect it and notify the users for attention actively.

3.2 Architecture Designs

Figure 1 illustrates the architecture of TTP-ACE. It is divided into two domains, the public domain and the inner security domain. These two domains construct the TTP-ACE's perimeter, and all log files and information should never go out of this perimeter unless there are requests for evidence seeking.

1. *Public Domain*: The public domain processes basic audit operations. It contains two main components.

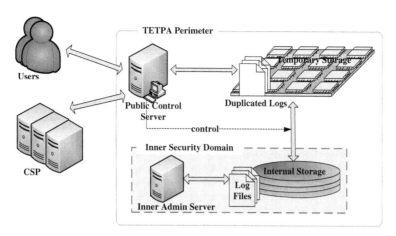

Fig. 1 Architecture of TTP-ACE

Public control server. This server is equipped with a trusted platform module (TPM) chip [5] or other security chips which implement the TPM specification for remote attestation. With TPM, users and CSPs can be protected from cheating attacks. Key and certification management mechanisms have been included in the public control server. These mechanisms ensure that the users' behavior information would not be intercepted, and the data integrity can also be guaranteed. The other important function of this public control server is to raise challenge to check the users' data integrity period, if CSPs' response for the challenge cannot match the specific rules and expectations. The public control server will notify the user who potentially suffered the data loss.

Temporary Storage. In this storage space, there exist duplications of the log files. These duplications are the data that can be accessed and modified from the outside of TTP-ACE's perimeter. The messengers and duplications are all encrypted and signed so the TPA itself cannot read and modify it neither. The temporary storage is also non-volatile media. We implement it with flash for performance and false-tolerant considerations which we will discuss in Sect. 4.

The audited information is organized as group of transactions, and each time the transaction is finished, the public control server will challenge for the users' and CSPs' identification. If the subjects are authorized, these duplications will be updated into the log files that exist in the inner security domain. Otherwise, they will be discarded and notify the CSPs or the users for potential violation.

2. *Inner Security Domain*: The inner security domain maintains the unique and effective log files. Log files are created according to the user–CSP relationships. Namely, when a new user contracts with CSPs for services and wants to use the TTP-ACE for benefits, TTP-ACE would create a new log file to record the audited information. After the log file is created, all the behaviors between user and CSP would use the ever-existed file. The inner security domain also consisted of two components.

Internal Storage. This storage space normally uses traditional hard disk array for cost and amount concerns. Log files and some metadata reside in this space. The access to this storage is strictly controlled by the public control server to ensure the log files' security and consistency.

Inner Admin Server. The main purpose of this server is to maintain the file stored in the internal storage. We use several methods, such as continuous data protection (CDP), and access and usage control [6], to enable the storage's availability and stability. We will discuss these methods in Sect. 4 for details.

We also enhance the users with TPM support. We assign the user a USBKey for identification attestation and system isolation, which maintains a trusted domain for users' keys and certifications and other important and sensitive data can be stored in this domain without worrying about data loss. And we consider that nowadays the CSPs' server should have TPM embedded in it.

3.3 Key Features of TTP-ACE

Bidirectional remote attestation. TTP-ACE and users, and TTP-ACE with CSPs can do attestation with the help of the TPM. So the cheating attacks could be avoided naturally.

Segment log file management. We use two-layer storage structure to support a better privacy-hidden mechanism. We only keep a newest segment in temporary storage; these segments compose the temp-duplicated log file. From these segments, a hacker can hardly distill the users' habits. We also strictly control the access and usage to the internal storage for protecting the complete user information.

High-performance auditing. We use flash as the storage media for temporary storage. It is well known that the flash has better performance than hard disk. With flash storage, we can avoid the TTP-ACE becoming the performance bottleneck.

Fault tolerance. We add various CDPs into the TTP-ACE to support fast error correctness. The flash is designed to work with the log-structured file system, so as to supporting block-level error correctness.

4 Key Technologies in TTP-ACE

4.1 Trusted-Computing-Enabled Strong Security

TCG has defined the specification for the widely implemented TPM [5]. The TPM is an international standard, hardware security component built into many computers and computer-based products. The TPM includes capabilities such as machine authentication, hardware encryption, signing, secure key storage, and

Fig. 2 In log-structured storage system, an update to an existed data block will be redirected to the end of the free space, and then, a link will be built between the two blocks to support data rolling back

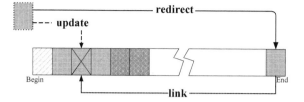

attestation. Encryption and signing are well-known techniques, but the TPM makes them stronger by storing keys in protected hardware storage. Machine authentication is a core principle that allows clouds to authenticate a known machine to provide this machine and user a higher level of service as the machine is known and authenticated.

4.2 Two-Layer Security Storage with Flash Memory

In the above sections, we have discussed why our two-layer storage can provide the users a more secure and reliable storage supports. Here, we will discuss the performance and availability benefits the flash memory brings to our system.

In performance aspect, flash storage systems such as solid-state drive (SSD) have shown a great superiority to traditional hard disk drive (HDD). All electronic structure endows the flash a better stability, too. SSD has been thought as a better choice to build a reliable system used in military and financial system.

In availability aspect, flash naturally supports a log-structured record manner [7], which is illustrated in Fig. 2. In this manner, newly written data will be append at the end of the free sequential storage space, and old version of a data block will not be covered until the free space is exhausted. Combining with the non-volatile feature of flash, once a disaster happens, the data can be roll back to its latest version easily.

4.3 Continuous Data Protection

CDP is a new paradigm in backup and recovery, where the history of writes to storage is continuously captured, thereby allowing the storage state to be potentially reverted to any previous point in time. Typically, the amount of history stored is limited by the operator, either in temporal terms using a CDP window (e.g., 2 days) or in terms of the amount of additional storage available for history data. CDP can be provided by different entities in the I/O path such as the host being protected. CDP has been widely used in various areas, e.g., database management, file system enhancement, and RAID system, to implement rapid data recovery.

In TTP-ACE, we synthesize block-level and file-level CDP into inner admin server to implement log files' backup and recovery. With the support of CDP, the availability and fault tolerance of the TTP-ACE can be significantly improved.

5 Related Works

Ateniese et al. [8] are the first to consider public auditability in their defined "provable data possession" (PDP) model for ensuring possession of data files on untrusted storages. Juels et al. [9] describe a "proof of retrievability" (PoR) model, where spot-checking and error-correcting codes are used to ensure both "possession" and "retrievability" of data files on remote archive service systems. Shacham and Waters [10] design an improved PoR scheme built from BLS signatures with full proofs of security in the security model defined in [9]. But in [8–10], the authors have not taken the trust issue in consideration, so they cannot be used in a complex open network environment like cloud.

Wu et al. [11] present a novel way to implement block-level CDP to implement a rapid data service recovery. Ma et al. [6] design a comprehensive system for data usage control, and this method can strictly limit the data usage extent to prevent sensitive data leakage. These two methods have been integrated into our TTP-ACE to implement inner security domain to improve the security of the log files.

Cheng et al. [12] design a TPM-specification-compatible USBKey, which is more flexible. The USBKey can be used as the users' TPM to support our system.

Acknowledgments This work was supported in part by the National Natural Science Foundation of China under Grant No. 60903204, No. 61272144, and No. 61070201.

References

1. Armbrust, M., Fox, A., Griffith, R., Joseph, A.D., Katz, R.H., Konwinski, A., Lee, G., Patterson, D.A., Rabkin, A., Stoica, I., Zaharia, M.: Above the clouds: A Berkeley view of cloud computing. University of California, Berkeley, Tech. Rep. UCB-EECS-2009-28, Feb 2009
2. Mell, P., Grance, T.: Draft NIST working definition of cloud computing. Referenced on 13 Nov 2010. Online at. http://csrc.nist.gov/groups/SNS/cloud-computing/index.html, (2009)
3. Wang, C., Wang, Q., Ren, K.: Ensuring data storage security in cloud computing. Charleston, SC, United States (2009)
4. Haeberlen, A.: A case for the accountable cloud. ACM SIGOPS Oper. Syst. Rev. **44**(2), 52–57 (2009)
5. Trusted Computing Group.: TPM main specification v1.2. TCG Press, Jul 2007
6. Ma, J., Ren, J., Wang, Z.: Implementing efficient management and security of removable storage by FVM. In: IEEE 2010 International Conference on Management Science and Information Engineering (ICMSIE 10), vol. 2, pp. 80–83. IEEE Press, Zhengzhou, China, Dec 2010

7. Douglis, F., Ousterhout, J.: Log-structured file systems. In: COMPCON Spring '89, pp. 124–129, Mar 1989
8. Ateniese, G., Burns, R., Curtmola, R., Herring, J., Kissner, L., Peterson, Z., Song, D.: Provable data possession at untrusted stores. In: Cryptology ePrint Archive, Report 2007/202. http://eprint.iacr.org/ (2007)
9. Juels, A., Burton, J., Kaliski, S.: PORS: Proofs of retrievability for large files. In: Proceedings of CCS'07, pp. 584–597. Alexandria, VA, Oct 2007
10. Shacham, H., Waters, B.: Compact proofs of retrievability. In: Proceedings of Asiacrypt 2008, vol. 5350, pp. 90–107, Dec 2008
11. Wu, J., Ma, J., Cheng, Y.: SSR-CDPS: A novel continuous data protection system supporting service recovery. In: IEEE 2010 International Conference on Management Science and Information Engineering (ICMSIE 10), vol. 2, pp. 88–92. IEEE Press, Zhengzhou, China, Dec 2010
12. Cheng, Y., Ma, J., Wu, J.: Research and implementation of a Trusted Removable USB Flash Device. In: IEEE 2010 International Conference on Management Science and Information Engineering (ICMSIE 10), Zhengzhou, China, vol. 2, pp. 84–87. IEEE Press, Dec 2010

Pattern Recognition Based on the Nonparametric Kernel Regression Method in A-share Market

Huaiyu Sun, Mi Zhu and Feng He

Abstract Fourteen kinds of technical pattern in A-share market have been automatically identified in this paper, using nonparametric kernel regression method and quantitative recognition rules that are constrained and improved to make the model more feasible and effective for investment practice. K–S test shows that the distribution of the stock yield following the recognized patterns statistically differs from that of random sampling. Clustering analysis also suggests that most of the identified technical patterns have distinguishable effects on the subsequent stock price movement. These findings provide useful guidance for the development of innovative investment models in high-frequency automatic stock trading.

Keywords Technical pattern recognition · Nonparametric kernel regression

1 Introduction

Technical analysis is done by means of data and figures, using prices and trading volume to predict stock price movement coming after. Pattern analysis is one type of the technical analyses, which is used to study the stock price chart and identify the meaningful chart patterns followed by specific price-moving trend.

H. Sun (✉)
Economics and Management School, Wuhan University, Wuhan, China
e-mail: sunhuaiyu@china-invs.cn

H. Sun · F. He
Research Department, China Investment Securities Co., Ltd, Shenzhen, China
e-mail: hefeng@china-invs.cn

M. Zhu
Product Design Department, UBS SDIC Fund Management Co., Ltd, Shenzhen, China
e-mail: Mi.Zhu@ubssdic.cn

S. Patnaik and X. Li (eds.), *Proceedings of International Conference on Soft Computing Techniques and Engineering Application*, Advances in Intelligent Systems and Computing 250, DOI: 10.1007/978-81-322-1695-7_35, © Springer India 2014

Traditional pattern analysis is first introduced by "Technical Analysis of Stock Trends", a book written by Edwards and Magee. The methods are widely used by technical analysts in various dimensions. However, most of the pattern identifications strongly rely on the subjective judgment of the observers, which can entirely be a matter of preference. Very few people have done the researches on pattern recognition through quantitative approach, among them, there were Levy [1], Chang and Osler [2], Lo et al. [3], Leigh et al. [4], etc., Chinese scholars such as Ouyang and Wang [5] and Chen [6] carried out relevant researches on the domestic stock market.

This paper uses an automated recognition system based on nonparametric kernel regression method, referring to Lo, Mamaysky, and Wang (LMW), to study and capture the technical pattern of China A-share market. In order to make the experiment closer to the real investment practice, the paper adopts more stringent recognition rules than those of LMW, improving the standardization of the pattern identity and controlling the frequency of the patterns' occurring. The number of detected patterns is also increased to fourteen, more than that of preceding experiments. Meanwhile, by separating the ascending triangle and the descending triangle from the general triangle category, the paper enhances the investment instructive effects on the technical pattern analysis. The study not only enriches the academic researches on technical analysis, but also provides more evidence and reference for the design and development of high-frequency quantitative trading on financial engineering innovations.

The structure of the paper is as follows: The first part is the introduction; the second part is the model design, including smoothing method based on nonparametric kernel regression and pattern recognition rules; the third part is the result of the experiment, from the K–S test and clustering analysis; and the last part is a summary.

2 Model Design

Technical patterns are composed of local maximum or minimum price points. Therefore, the key job of pattern identification is to find all of the local extreme points, with which some specific shapes are structured by proper rules. Finding local extreme points can be done by nonparametric kernel regression smoothing method, which is introduced by LMW. As for the pattern recognition, the paper adopts more stringent rules than those in LMW, making the patterns more representative and closer to the real investment practice.

Generally speaking, a local maximum value is the start point of a downward trend, while a minimum value is the starting point of an upward trend; the points captured in a slight concussion are hoped to be filtered out. It is difficult to give a valid judgment if rely solely on the discrete daily closing prices in the sequences.

Discrete points smoothing method can effectively solve the problem of judging the local extreme points. Let the stock price be $P_t = m(x) + \varepsilon_t$, x stands for time,

$m(x)$ stands for the smoothing price, and ε_t stands for the white noise. Suppose, we can get lots of P_t to approach $m(x)$, then ε_t can be eliminated by law of large numbers. Suppose, $\hat{m}(x)$ is satisfied by the relationship:

$$\hat{m}(x) = \frac{1}{T}\sum_{t=1}^{\tau}\omega_t(x)P_t \tag{1}$$

which means the smoothing value $\hat{m}(x)$ is the weighted average of various prices. The weight $\omega_t(x)$ is expected higher if t is closer, lower if t is farther. However, how much weight is appropriate? If too much weight allocates to the near points, smoothing curve can not be "smooth" enough; if too much weight to the far points, the curve could be too "smooth" to hardly find a local extreme point. Therefore, the determination for the weight becomes critical, it must be sufficient to filter out unwanted noise while retaining an appropriate volatility to identify the patterns.

The paper constructs $\omega_t(x)$ by kernel regression method. The Gaussian kernel is chosen as the weighted density function, and the parameter h (the bandwidth) is set for adjustment. Now, the smoothing prices can be represented by:

$$\hat{m}(x) = \frac{\sum\limits_{t=1}^{T}K_h(x-X_t)P_t}{\sum\limits_{t=1}^{T}K_h(x-X_t)} \tag{2}$$

The method of cross-validation is introduced to determine the value of h. By quadratic optimization, h is expected to satisfy:

$$\min\frac{1}{T}\sum_{t=1}^{T}\left(P_t-\hat{m}_{h,t}\right)^2, \quad \text{where } \hat{m}(x) = \frac{1}{T}\sum_{\tau\neq h}^{T}\omega_{\tau,h}(x)P_t \tag{3}$$

After obtaining an appropriate parameter h by optimization approach, the discrete stock closing prices can be smoothed reasonably.

Here comes the research framework: First, find a time window and smooth the discrete closing prices in the interval; second, recognize the local extreme points and check whether they conform to the specific pattern rules. If all rules are satisfied, then a technical pattern is recognized successfully from the price sequence.

In this paper, we set rules to define patterns according to the relative positions of five consecutive extreme points. Totally, fourteen different kinds of patterns are identified, including Head and Shoulders, Inverse Head and Shoulders, Broadening TOP, Broadening BOT, Symmetrical Triangle TOP, Symmetrical Triangle BOT, Ascending Triangle TOP (ATTOP), Ascending Triangle BOT, Descending Triangle TOP, Descending Triangle BOT, Ascending Wedge TOP, Ascending Wedge BOT, Descending Wedge TOP, and Descending Wedge BOT.

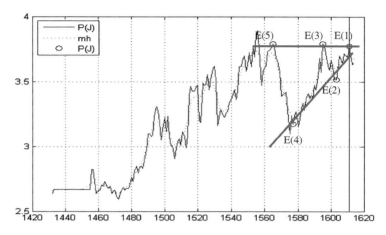

Fig. 1 Ascending triangle TOP (ATTOP)

Table 1 Recognition rules of ascending triangle TOP (ATTOP)

Pattern characteristics	Recognition rules
① Define three points on a horizontal line of top edge	E(1) E(3) E(5) fluctuate ±0.75 % from their mean
② Define the bottom edge	E(2) ≥ (1 + 0.75 %) * E(4)
③ Distinguish top and bottom edges	Min(E(1) E(3) E(5)) > Max(E(1) E(3) E(5))
④ Let top and bottom edges cross after E(1)	$\angle\theta_1$ is the angle between the line of E(1) and E(3) and the horizontal line
	$\angle\theta_2$ is the angle between the line of E(1) and E(2) and the horizontal line
	$\angle\theta_1 > \angle\theta_2$

Figure 1 and Table 1 show the example of how ATTOP is recognized. There are five consecutive extreme points in a stock price sequence. These five points satisfy four rules in Table 1, so the pattern is defined as an ATTOP.

The paper establishes a set of more stringent recognition rules applicable to the A-share market than LMW. For example, the collinearity criterion of the extreme points is tightening up from 1.5 to 0.75 % (like rules ① in Table 1), and the rules of angle are supplemented (like rules ② and ④ in Table 1). The improvement of the rules makes the patterns more recognizable, more reliable, and closer to the real investment practice.

Once, all of the patterns are identified, further statistical tests and chart analyses can be carried out.

Table 2 The proportion of passing K–S test from Day 1 to Day 30 after pattern recognition

HS	IHS	BTOP	BBOT	TTOP	TBOT
97 %	70 %	50 %	27 %	60 %	67 %
ATTOP	ATBOT	DTTOP	DTBOT	AWTOP	AWBOT
77 %	27 %	80 %	93 %	73 %	67 %
DWTOP	DWBOT				
63 %	87 %				

3 Experiment Results

The experimental results are interpreted from both the statistical angle and the practical investment angle. First, to prove the distribution of investment return following the recognized pattern is different from that of random sampling. Second, to calculate the proportion of positive yield in all the after-pattern returns; from a practical sense, a proportion deviating from 50 % suggests that the pattern have effect on the subsequent price movement.

The paper samples the closing price data of all A-share stocks from January 1, 2000 to October 31, 2011. Additional constraints include: Sample's trading days should cover 75 % of the market trading days; samples are divided into five groups based on market value, with 150 stocks from each group, 750 stocks in total.

The paper adopts the Kolmogorov–Smirnov test (K–S test) to do the hypothesis test, taking a confidence level at 95 %. The null hypothesis H_0 is: The distribution of return following the recognized pattern is the same as the random sampling. If the null hypothesis is rejected in the 95 % confidence level, it means the K–S test is passed. Through the test from Day 1 to Day 30 after the pattern identification, the impact of technical patterns on the stock prices can be verified.

Table 2 shows during Day 1–Day 30 after the pattern recognition, most of the null hypotheses are rejected, except the BTOP, BBOT, and ATBOT patterns whose unfavorable proportion could be related to the small number of patterns recognized from the samples. The results of the K–S test suggest that from the statistical aspect, most of the technical patterns do have obvious relevance to the following price movement, in the other word, the technical pattern analysis is effective for the A-share market and can be applied to the real investment practice.

From another aspect, the paper investigates the directivity of the pattern effect by the proportion of subsequent price rising for the first time. Clustering method is also used to observe the following price fluctuation. Compare the price of Day 30 to the price of Day 1 after the pattern recognition, and check whether the price is supported (upward) or suppressed (downward). Generally, a proportion higher than 55 % suggests price being supported while less than 45 % implies being suppressed. Falling in the range of (45, 55 %) is considered lack of directivity.

Referring to Fig. 2, all of the wedges and broadening categories fall in the range of (45, 55 %), according with their characteristic as kind of continuation pattern which has a sideway movement of stock price. There are seven categories located beneath 45 % level, showing that the prices' movements are suppressed by the patterns.

Fig. 2 The proportion of positive yield on Day 30 after the pattern recognition

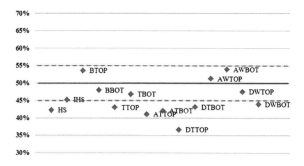

4 Conclusions

Using the nonparametric kernel regression method and establishing a set of more stringent identification rules, the paper detects and recognizes fourteen representative kinds of technical pattern in A-share market. Statistical analyses suggest most of these patterns have some distinguishable effects on the subsequent price movement, which means technical analysis is effective in A-share market.

Currently, the domestic technical analysis research is still at a tentative stage. The usage of innovative techniques such as data mining has profound theoretical and practical significance. The findings of this paper that the technical patterns have obvious impact on the subsequent stock price movement in A-share market provide important guidance for the development of quantitative investment model in high-frequency trading. Further empirical research is on-going to explore more features of the A-share technical patterns as well as to make the investment model more practical and effective.

References

1. Levy, R.A.: The predictive significance of five-point chart patterns. J. Bus. **44**(3), 316–323 (1971)
2. Chang, P.H.K., Osler, C.L.: Technical analysis and the irrationality of exchange-rate forecast. Econ. J. **109**, 636–661 (1999)
3. Lo, A.W., Mamaysky, H., Wang, J.: Foundations of technical analysis: computational algorithms, statistical inference, and empirical implementation. J. Finance **55**(4), 1705–1760 (2000)
4. Leigh, W., Paz, N., Purvis, R.: Market timing: a test of a charting heuristic. Econ. Lett. **77**, 55–63 (2002)
5. Ouyang, H., Wang, X.: Geometric technical trading rules: predictability and profitability. China J. Finance (1), 129–153 (2004)
6. Chen, Z., Song, F.: The information of 'charting'. Quant. Tech. Econ. **9**, 73–82 (2005)

The Research on the Detection and Defense Method of the Smurf-Type DDos Attack

Wantian Cao and Xingchuan Bao

Abstract Smurf attack is one of the major distributed denial of service (DDos) attack; it has a serious threat to network security of Internet. Combined with the attack principle and characteristics of Smurf attack, this paper analyzes the IP spoofing and ICMP reply attack. This paper presents the typical detection algorithm as Information entropy algorithm and congestion control strategy of RED algorithm. On the basis of specific algorithm introduction, it researches and analyzes the main guard way of Smurf attack systematically. Finally, it discusses the future direction in this field.

Keywords Smurf attack · ICMP · Attack detection · Attack prevention

1 Introduction

Smurf attack is one of the types of distributed denial of service (DDos) attacks. The DDos attack, is a denial of service attack based on the special form of Dos, is a distributed and collaborative large-scale attack. According to the protocol of TCP/IP, DDos attack can be divided into ICMP flood, IP flood, UDP flood, TCP flood types; Smurf attack is a kind of typical attack model based on ICMP. Then, it would analysis the principle of ICMP and Smurf attack.

W. Cao (✉) · X. Bao
China Electric Power Research Institute, Nanjing, China
e-mail: Caowantian@epri.sgcc.com.cn

X. Bao
e-mail: baoxingchuan@epri.sgcc.com.cn

S. Patnaik and X. Li (eds.), *Proceedings of International Conference on Soft Computing Techniques and Engineering Application*, Advances in Intelligent Systems and Computing 250, DOI: 10.1007/978-81-322-1695-7_36, © Springer India 2014

2 Introduction of the Smurf Attack Principle

Smurf attack takes advantage of the IP protocol loopholes, uses the software program to send a Ping packet to a plurality of network hosts on the Internet, so as to create the Reply packet, a lot of which leads to network traffic congestion. Then, it leads to a system crash or paralysis. Firstly, Smurf attack determined the attacked host and then use the two steps: At first, it used the host address as the source address, created a lot of ICMP response packet. Then, sent the packets out with the way of broadcast, it amplifies the ICMP response to respond more times packets' number, the packet would be sent back to the attack of network.

2.1 Introduction of ICMP Message

In order to better understand the process of Smurf attack, it needs to analyze the ICMP message; it is an "error detection and reporting mechanism," is generally used to detect network connection and ensures the accuracy of Internet line. Its main functions include the following four types:

- Detect remote host whether exist or not;
- Establish and maintain routing data;
- Redirect the transmission path of information; and
- Control the flow of information.

Each ICMP message can be represented with these four parts that include message type, type code, checksum, and specific type-code-related information.

Thus, the characteristic of ICMP protocol determines that it is very easy to be used for attack on the network routers and hosts; if host to send a large number of long-term and continuous ICMP data packet to the target, it may ultimately lead to paralysis of the system. Because a large number of ICMP data packet would form the ICMP flood, so that the target host consumes a large amount of CPU resources for processing, then CPU may be kept constantly on the run until the resources exhausted.

2.2 Smurf Attack Process

Smurf attack is mainly composed of three parts: the attacker, intermediate proxy site (hosts, routers, and other network equipment), and attacked target. First of all, we need to ensure a target host (usually some Web server), which is the attacked target. Then, looking as medium site, usually the attacker will select multiple intermediaries to amplify the attack, disguised himself. And then, the attacker sends an ICMP request packet with the IP deception method, not using its

Fig. 1 ICMP attack
principle

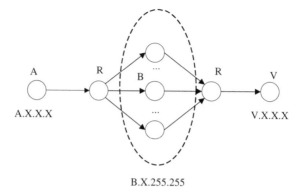

A

A.X.X.X

R

B

...

...

R

V

V.X.X.X

B.X.255.255

machine's IP as the source, and pseudo source IP by attacked object IP. Later, intermediate proxy site once receives the attack broadcast which is sent to no. 1 network would be responded to the source IP, sends an ICMP response packet to the attacked target. All network machine intermediate proxy site may also in response to ICMP request packet to the corresponding ICMP reply message, then the network congestion may occur at this time, or denial of service even system collapse. The attack principle is shown in Fig. 1.

In Fig. 1, R1, R2 express routers 1, 2. The attacker A camouflage their own IP to the attacked host V IP V.X.X.X, and their true IP is A.X.X.X. After selecting the medium network, the attacker sends the ICMP request packet, the target IP is B.X.255.255, once the middle agent site receives the ICMP request packet, it will request for the ICMP response packet, and the response object is the source IP address. In this way, the reply will be sent to the attacked host V. When the multiple agent site acts as a signal amplifier, one may also be attacked by Smurf and affect its own performance. If there are N number of sites in B network, ICMP request response after B subnet will be magnified n times to host V; bandwidth and CPU load of router R2 would increase, performance will decline. The object of attack would result in a denial of service, even to the collapse of the system [1]. And now, the various attack tools such as Shaft can change machine IP address and port number during attack, which makes the detection more difficult.

3 Detection Methods of Smurf Attack

For the detection of Smurf attack, it can use attack detection method based on statistics and controlling high-bandwidth aggregate test method. Here, we would introduce method to their typical strategy: Information entropy detection algorithm [2] and Congestion control strategy based on RED algorithm [3].

3.1 DDos Detection Algorithm Based on Information Entropy

Smurf attack has many features such as: network traffic increases suddenly, the network is blocking; there are a large number of identical packets, because the attack packet is due to the use of DDos tools that are automatically replicated; the same service requests. Information entropy will be displayed directly above characteristics.

In a network device using capture, software can pick the Ethernet data packet, then, analysis and classification statistics data's packet protocol head, then extract the corresponding feature information from it. It can be divided into the following steps: Firstly, analyze Ethernet protocol, pick the ones from all packets which the link layer type field to determine the value of IP is 0x0800. Secondly, analyze IP protocol head, gain the source IP address and destination IP address. When attack occurs, the data often have some changes compared with the normal situation. View the types of protocol type in eight protocol type field of IP data head, the value of 0x01 indicates the ICMP protocol. Thirdly, analyze the marker packet, the ICMP protocol judge message types through 8 types and 8 bit code, the ICMP request and response messages are marked in the type field. Detailed data packet structure is shown in Fig. 2.

Information entropy is defined as follows:

$$H(X) = -\sum_{i=1}^{N} P_i Log_2 P_i \tag{1}$$

An information source has N signal whose independent probability is p_i, X represents the state space source, there are N states, $X = (X_1, X_2,...,X_n)$. N represents different IP address number in a period, X_i shows each independent sample's logo (IP or port). X_i's appeared probability is P_i, P_i is equal to m/N, M is the number of no. i IP address occurrences $\sum_{i=1}^{N} P_i = 1$ [4].

Now, look at the calculation method of information entropy source based on IP address. Using the moving-window mechanism, we need to compute the probability of occurrence of each IP address in this custom window. If the window size is set to W, calculation steps are as follows:

1. Calculate each source IP address's probability of the occurrence of P_i and get the entropy H;
2. The first packet is marked as No. 1 package, P_1 represents the probability representation of the source IP address appears in all the W packages. Suppose, there is n packages in the W packages which is equal to no. 1 package's source data packet of IP, $P_1 = n/W$, we can get information entropy H_1 by the No. 1 formula.

4 bit version	4 bit heads	8 bit type of service	16 bit total length (in bytes)	
16 bit identification			3 bit identification	13 bit slice shift
8 bit life cycle (TTL)		8 bit protocol	16 bi first checksum	
32 bit IP source address				
32 bit destination address IP				
Options				
Data				

Fig. 2 IP packet structure diagram

3. When the first package is removed out, the No. $W + 1$ package is removed into the window, which can be concluded into the prior probability P_{W+1} and the P'_{W+1} that after entering, assuming there are W package slip into the window, the number of m packages' source IP is same as the No. $W + 1$ package, then $P_{W+1} = m/W$. After the No. $W + 1$ package entering, the source IP number increased to $m + 1$, then

$$P'_{w+1} = (m + 1)/W \tag{2}$$

4. By last step can be drawn $P'_1 = (n - 1)/W$, then calculate H'_1
5. Compare changes before and after sliding window, only P_1 and P'_{w+1} change, the other source of IP entropy does not change, so the new entropy is equal to:

$$H = H_1 - H_{w+1} + H'_1 + H'_{w+1} \tag{3}$$

6. Repeat the above steps 2–5, and obtain the entropy late.

With the above-mentioned steps, we can obtain a series of continuous data source address information entropy. The formula can be seen with the random distribution characteristics of source IP addresses: when entropy becomes bigger, it represents the source IP address distribution randomly, coverage area is wide; when entropy becomes smaller, it indicates that the source IP address range is small, some of which IP address appears greater probability.

In the normal situation, the network source IP distribution is relatively stable, entropy is also relatively concentrated, which only fluctuates in a certain range. When Smurf attack is initiated, the attacker can forge the attack packet source IP and then makes the attack intermediary to attack victims by the reflection. Thus, the number of the source IP address will increase suddenly; the entropy will fluctuate greatly and beyond the normal range, then it can detect the Smurf attacks.

Similarly, entropy model can be established based on the destination IP address, information the source port and destination port. Therefore, the information entropy can well reflect the variation characteristics of Smurf attacks, so as to afford the needs of the testing.

3.2 Congestion Control Strategy Based on RED Algorithm

Congestion control, as the name implies, through the data flow control in the network, the user's sending data would not submerge communication subnet, bottleneck resources would be rationally used. RED algorithm mainly monitors two independent unit index of router: ① the average length of the input queue and ② packet-loss probability, so it can be treated as two independent algorithms.

Algorithm I determines the router flow burst degree allowed. Algorithm II decides router-discarding packet frequency under the current load. The RED algorithm uses the exponentially weighted average algorithm to calculate the average queue length, which can be on congestion prediction basis to calculate the marking probability packet. Formula for calculating the average queue length is as follows:

$$\text{avg} = (1 - w_q) \times \text{avg} + w_q \times q \tag{4}$$

The q represents the queue length, w_q represents the queue length weighted coefficient, w_q meets $0 < w_q < 1$, avg represents the average queue length. RED algorithm uses the average queue length to determine whether congestion occurred, which can affect the balance of bursty traffic on the algorithm. The advantages of using average queue length to calculate the marking probability are the following: if the network traffic increases suddenly, avg changes would be very slow, it would not suddenly lead to a large number of packet discarded, the system can be adapted to the burst traffic increase.

Control threshold of RED algorithm $\min_{\text{th}}$ and $\max_{\text{th}}$ decides all arriving packets whether or not allowed to enter the queue. When avg $< \min_{\text{th}}$, all the arriving packet can enter the queue. When avg $> \max_{\text{th}}$, the arriving packet would be directly marked or dropped. When $\min_{\text{th}} < \text{avg} < \max_{\text{th}}$, mark the packet according to a certain probability, the formula is as follows:

$$P_b = \max_p \times (\text{avg} - \min_{\text{th}})/(\max_{\text{th}} - \min_{\text{th}}) \tag{5}$$

P_b represents the current packet-marking probability calculation results, $\max_p$ represents the marking probability, which is the preset. In actual use, the marking probability P_b needs to do the following correction calculation to spread evenly labeled packet and gets P_a as the marking probability.

$$P_a = P_b/(1 - \text{count} \times P_b) \tag{6}$$

In order to make the reach group number meeting the average distribution between twice packet discarding, which must meet the requirement of formula 4, it marks as X, which can smooth packet-marking process effectively. The formula is as follows:

$$\text{Prob}[X = n] = \frac{P_b}{1 - (n - 1) \times P_b}$$

$$\times \prod_{i=1}^{n-1} \left(1 - \frac{P_b}{1 - (i - 1) \times P_b} \right) = \begin{cases} P_b, \, 1 \le n \le 1/P_b \\ 0, \, n > 1/P_b \end{cases} \quad (7)$$

RED congestion control algorithm is not only a preventive strategy, but also a kind of recovery strategy. It can recover the network from the congestion state of Smurf attack.

4 Smurf Attack Prevention Method

In the detection of Smurf attack based on ICMP message, according to the characteristics of Smurf attacks and transmission mode, it introduces two kinds of Smurf attack defense mechanism:

- Attack source trace back;
- Detection and defense mechanism of puppet machine.

4.1 Attack Source Trace Back

Following-up the Smurf attack source, the ICMP request and response message source IP was found to be disguised; it cannot track, so we have to backtrack through other means.

In data packet transmission process, it needs to use the information-recording function in the routing process, but it will need a lot of storage space, so it is necessary to have the time of aging, to empty timeout information.

When the victim was found being attacked, the log will record the MAC address of ICMP packets, as shown in the figure; by the log information, the Smurf attack source can trace from the MAC address. All routers should store a part of the data packet information abstract that was being forwarded, such as a log information in the router log, which records the router information about IP, MAC address, and attack time that forwarding Smurf attack, from which we can read the link MAC to 00.0 a 00. a 52 d. Then, using the command show, ip arp can find out the upper level router's IP address which is 192.168.222.6.

After sending a request to a router R1, R2, R3, if the R1 router contains victim host providing IP address 192.168.222.6, it returns a positive answer; again query a level above R1, and so on, until the source of the attack is found out, see Fig. 3.

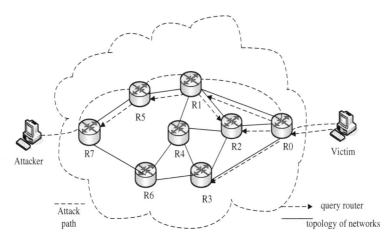

Fig. 3 Smurf attack topology

4.2 Detection and Defense Mechanism Based on Puppet Machine [5]

Puppet machine is the medium in Smurf attack, because the puppet machine is used as the implementation and dissemination of Smurf attack, so it is feasible and effective by Defense Technology in the puppet machine. The detection and defense mechanisms need to be defensive measures as a patch installed on each client, so that you can defense in these puppet computer through the puppet machine reflecting ICMP flood. It can make the defense burden and also can discard attack stream before they flow into the network, which would prevent the attack flow causing network congestion harm. If the client has been hacked, detection here can also spend reduced cost, the harm can be straight down to the lowest.

The detection principle is as follows, in Fig. 4: Before the client sends a data packet, analyze the data packet header, discard the source address which is not the machine address or data packet. Because ICMP is the network control data packets, general users do not frequently use ICMP request and response message, and it is generally used to diagnose logic error. In the normal situation, the general ICMP message is not more than 64–128 bytes. Therefore, through limiting the ICMP package in the unit time on the machine, it can effectively prevent DDos attack, and when receiving ICMP packets, only need to filter out the ICMP request packet that needs to reply, it can defend Smurf attacks. The following chart 5 can reflect the above process.

Fig. 4 ICMP packets
filtering process

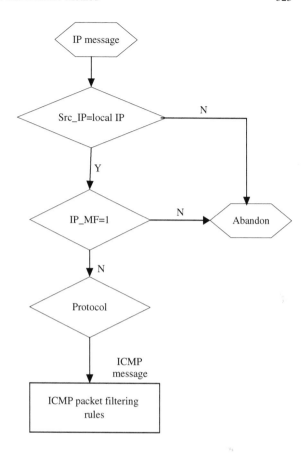

5 Conclusions

This paper describes the principle of the Smurf attack and puts forward the detection method. Based on the characteristics, defensive measures were presented. However, with the development of network technology, attack methods and attack software would be constantly updated, so network administrators need to dig out more and better methods on the existing basis and customized comprehensive security policy:

- Improve Smurf-testing efficiency of data transmission;
- Master Transmission condition of routers and other network controlled equipment to attack warning;
- Filter attack packet in the process of attack precautionary attack.

Above methods can more effectively ensure the safety and stability of the system and data.

References

1. Xu, Y., Zhang, K., Yang, Y., Liu, F.: Research on Smurf attack and its countermeasures. J. Nanjing Univ. Sci. Technol. **10** (2002)
2. Song, W., Zhu, S.: Information Theory and Application, Tsinghua University Publishing Company, pp. 15–25. Beijing (2005)
3. Floyd, S., Jacobson, V.: Random early detection gateways for congestion avoidance. IEEE/ACM Trans. Networking **1**(4), 397–413 (1993)
4. Shannon, C.E., Weaver, W.: The Mathematical Theory of Communication. University of Illinois Press, Illinois (1963)
5. Huang, Q.: Distributed DDoS Attack and Prevention Research Based on the Technology, p. 5. HeFei University of Technology, HeFei (2011)

A Preliminary Analysis of Web Usage Behaviors from Web Access Log Files

A Case Study of Prince of Songkla University, Thailand

Thakerng Wongsirichot, Sukgamon Sukpisit and Warakorn Hanghu

Abstract In our digital age, Internet becomes one of the most important factors in order to achieve information and knowledge. In educational institutes, Internet also plays important roles for people to promote comprehensive learning and teaching environments. However, there are always two sides of every coin. Overuse of Internet may lead to other problems. Our research objective is to investigate the Web usage behaviors from Web access log files. Data mining and statistical techniques have been employed to analyze for the purpose of descriptive and predictive aspects.

Keywords Web usage mining · Internet user behaviors · Web log mining

1 Introduction

Statistically, the world population in 2013 is reaching 7.1 billion, in which 77 % are in the developed countries and 31 % are in the developing countries. There is approximately 39 % of world population that has Internet access. In Thailand, a number of Internet users in 2012 are over 20 million, which has been increased from 2.3 million in 2000 [8]. The increase in the number of Internet users leads to in-depth research and development on Internet technologies such as network infrastructure and network security. On the other hand, research interests on

T. Wongsirichot (✉) · S. Sukpisit · W. Hanghu
Information and Communication Technology Programme, Prince of Songkla University,
Hat Yai, Songkhla, Thailand
e-mail: thakerng.w@psu.ac.th

S. Sukpisit
e-mail: sukgamon.s@psu.ac.th

W. Hanghu
e-mail: automaetic@hotmail.com

S. Patnaik and X. Li (eds.), *Proceedings of International Conference on Soft Computing Techniques and Engineering Application*, Advances in Intelligent Systems and Computing 250, DOI: 10.1007/978-81-322-1695-7_37, © Springer India 2014

Table 1 NetCache NetApp/6.1.1 data structure

Data field	Description	Sample
x-timestamp	Unix timestamp format (starting from January 1, 1970) [7]	1,226,077,502.049
Time-taken	Time in milliseconds	2,450
c-ip	Client's IP address	172.28.9.56
x-transaction	Connection status	TCP_MISS_PRIVATE_ON_STOP_LIST/200
Bytes	Size of object received or sent	14,222
cs-method	Type of request	GET
cs-uri	URL of destination	http://www.abc.com
x-username	Username of a client	user01
x-hiercode	Internal stored directory	SINGLE_PARENT/192.168.96.11
rs(content-type)	Content type	"text/html; charset=UTF-8"
x-note	Other predefined notes	Comments
x-smartfilter-categories	Internal data filter	mm
x-smartfilter-result	Results of data filtering	0

analyses of Internet users' behavior are also attracted. One of the indirect approaches to identify Internet users' behavior is to analyze Web logs by using data mining techniques. Web log files are stored information, each time a user requests for a Web site. The Web log files are located in Web servers, Web proxy servers, and client Web browsers [4]. The analyses of Web log files using data mining techniques are generally known as Web usage mining. Web usage mining techniques are approaches to understand users' behavior from Web access logs that are generated by Web servers [2, 3]. One of the main concrete objectives of the Web usage mining is to understand users' behavior on the Internet usages. However, it may expand to other aspects such as a formation of Internet usage policies in organizations and Internet usage quota management, etc.

2 Data Preprocessing

In our research study, we investigate Web log files of the Prince of Songkla University's proxy servers that are automatically generated by NetCache NetApp/ 6.1.1. The software is of Web cache product category. It generates and stores clients' Web log files, which include all of the students, lecturers, and other personnel in the university. The Web log files consist of data including x-time-stamp, time-taken, c-ip, x-transaction, Bytes, cs-method, cs-uri, x-username, x-hiercode, rs(content-type), x-note, x-smartfilter-categories, and x-smartfilter-result [6]. Table 1 represents descriptions of the data fields in the Web log files.

Data preprocessing of the Web log files is introduced. Our network management provides an authentication procedure for accessing to the Internet. Each user's identity is masked in order to maintain the users' privacy. In many similar research studies, the complications have been addressed and followed by practical solutions. Some research works proposed different thresholds for time-oriented heuristics based on empirical experiments. A number of researchers use a proposed reliable threshold at 25.5 min, which is confirmed to be a neutral value [1]. The full Web log files are extracted only for three main data fields including x-time-stamp, c-ip, and cs-uri for the purpose of our research study. Additionally, all of the Web access transactions in an identical c-ip are separated by using the threshold value mentioned above. In some circumstances, Web access transactions may be fragmented due to various reasons such as Web servers not responding and requested Web site not found. All of the incomplete records are instantly omitted.

3 Overview of the Proposed System

Fundamentally, our research work consists of four main processes.

- Import of Raw Data

The original Web log files are formatted according to the NetCache NetApp/6.1.1. In the import of raw data process, the Web log files are analyzed and a set of data fields are selected for further analysis. The subprocess, "Data Cleaning" in the import of raw data process, is described in the algorithm section.

- Web Site Classifiers

One of the core algorithms is to classify the Web access transactions into groups [5]. A set of simple Web site categories have been selected that include marketing, entertainment, education and training, sales, and publishing. In order to correctly classify a Web site category to a Web site, there are three main groups of techniques. Firstly, the techniques are designed to conduct analysis on parts of the Web access transactions' URI. Secondly, the techniques are designed to perform analysis on contents of selected Web sites. Lastly, some researchers perform hybrid analysis techniques by using a combination of the mentioned techniques.

- Subnet Managing

The scope of our research study is to perform analyses of students and personnel's Internet usage behavior in the Prince of Songkla University. With the size of the university, the Internet users are grouped by their physical locations, which can be identified by their initial IP address's subnets.

- Reporting System

Reporting system is designed to represent simple information regarding the Web access log analyses such as percentages of users accessed in each of the Web categories.

4 Proposed Algorithms and Techniques

4.1 Data Cleaning

According to the Web log file format, there are three main focused data fields that have been selected for further analysis. The selected data fields are x-timestamp, c-ip, and cs-uri. The data cleaning algorithm is presented below. Specifically, this procedure performs steps to extract only three selected data fields by processing each character on records in extracted Web log files. Each of the records contains 13 data fields separated by white spaces. A white space separator is presented in the algorithm. Additionally, only the records that contain "text/html" tags are selected due to the confirmation of data transmissions.

```
Procedure: Data Cleaning
Require: Rec is a record of an extracted web log file
Require: Rec' is a new formatted web log array
DataArray_{i,j} ← Rec
txt_{i,j} ← null
for all DataArray_{i,j}

  ifDataArray_{i,j} != SPACE

  txt_{i,j} =txt_{i,j} U DataArray_{i,j}
  end if
  j ++
end for
for all txt_{i,j}∈DataArray

  iftxt_{i,i+8}='text/html'

  iftxt_{i,i+8}∈ text _{i,9} AND !(``MISS'' ∈ text_{i,9})

    Rec'_{x-timestamp} ← txt_{i,0}
    Rec'_{c-ip} ← txt_{i,2}
    Rec'_{cs-uri} ← txt_{i,6}
  end if
  end if
end for
return Rec'
```

4.2 Web Site Classifier

Our proposed technique on the Web site classifier, a hybrid approach, has been selected. A set of predefined keywords have been selected and stored into a XML file with the following structure.

```
<categorydataset>

    <contentname>MSN Entertainment</contentname>
    <categoryname>Entertainment</categoryname>
    <contentkeyword>impaqmsn</contentkeyword>
</categorydataset>
```

The XML file contains three main data fields. The first data field, contentname, is a name of each category keyword, presumably as an identity of this data record. The second data field, categoryname, keeps the name of the Web site category. The third data field, contentkeyword, is a keyword that can be used to search on the URI of Web sites.

```
Procedure: Web Site Classifiers
Require: Trn is a URI of an unclassified web access transaction
Require: Dom is a set of predefined fragmented of URIs with
their associated web site categories
Require: Keyd is a set of predefined keywords with their
associated web site categories
for all Trn_i do
  Cat← null
  Curr_Trn_{c-ip}← Current IP of Trn_i
  Trn_t← Current Date/Time of Trn_i
  if (Curr_Trn_{c-ip} != Trn_{c-ip})

    t← 0
  else

    if t > 25.5

      if (Trn_{i,j}∈Dom_{i,j})

        Cat←Dom_j
        returnCat
      end if
      if (Trn_{i,j}∈Keyd_{i,j})

        Cat ←Dom_j
        return Cat
      end if
    else

      t = t +(Trn_t' - Trn_t)
    end if
  end if
Cat ← 'undefined'
return CAT
end for
```

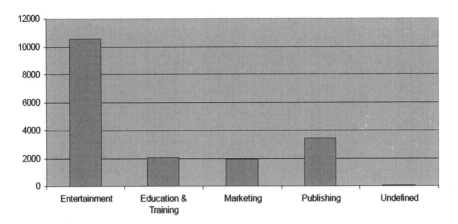

Fig. 1 A preliminary result of Web category analysis

5 Experimental Results and Interpretations

Two sample data sets were imported for in-depth analysis in our research work. Firstly, a sample data set is solely extracted from the Faculty of Science, Prince of Songkla University. Secondly, a sample data set is extracted from the overall Prince of Songkla University. A preliminary result that represents a majority category of the Internet usage is "Entertainment," followed by "Publishing," "Education and Training," and "Marketing," respectively. A bar chart is visualized in Fig 1.

According to the simple statistical representation in Fig 1, the "Entertainment" Web category is intentionally interested as it presents the highest frequency in the groups. Time series analysis techniques have been chosen to study the Web usage behaviors, especially in the "Entertainment" Web category. Two main time series analysis techniques have been employed, namely single moving average (SMA) and centered moving average (CMA). Table 2 represents MSE of predictive results of the SMA and CMA.

In order to conclude the above predictive results, MSE on each of the N must be evaluated. According to the result, the CMA is more accurate when compared to the SMA. It has the lowest MSE at $N = 2$, as shown in Fig. 2. Additionally, with extensive data, it is able to conclude that the users are surfing the Internet for the purpose of entertainment, especially during 12.00 to 14.00.

Table 2 MSE of predictive results of the SMA and CMA

N	MSE of SMA	MSE of CMA
1	2,136.53	–
2	2,136.53	903.83
3	5,331.98	1,825.47
4	8,922.42	2,466.14
5	12,831.43	3,236.33
6	1,686.18	4,135.13
7	20,741.92	5,136.91
8	24,355.63	6,097.84
9	27,677.19	7,312.34
10	30,238.40	8,513.92

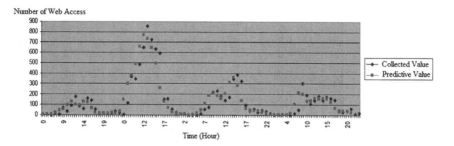

Fig. 2 A comparison between collected values and predictive values of CMA at $N = 2$

6 Conclusions

The analyses of Web log files are able to unveil many hidden truth behind digital walls, especially in terms of Internet users' behaviors. From our research work, the majority Web category that the Internet users accessed is the entertainment Web category, specifically during 12.00 to 14.00 h in weekdays. In the case study, some actions may require to be performed such as limiting speed for particular groups of users and quota allowance systems in order to achieve the university's objectives and goals. The result of our research works has been submitted for further analyses, and some of Internet usage regulations may be announced in order to balance between personal and organizational purposes.

References

1. Catledge, L.D., Pitkow, J.E.: Characterizing browsing strategies in the World-Wide Web. Comput. Netw. ISDN Syst. **27**(6), 1065–1073 (1995). Accessed 15 Aug 2013
2. Cooley, R., Mobasher, B., Srivastava, J.: Web mining: information and pattern discovery on the World Wide Web. In: Ninth IEEE International Conference on Tools with Artificial Intelligence (1997), pp. 558–567. Accessed 14 Aug 2013
3. Goel, N., Jha, C.K.: Analyzing users behavior from web access logs using automated log analyzer tool. Int. J. Comput. Appl. (0975–8887) **62**(2), 29–33 (2013). Accessed 15 Aug 2013
4. Grace, L.K., Maheswari, V., Nagamalai, D.: Analysis of Web Logs and Web User in Web Mining. Int. J. Netw. Secur. Appl. (IJNSA) 3, pp. 99–110 (2011). Accessed 12 Aug 2013. doi:10.5121/ijnsa.2011.3107
5. Kules, B., Kustanowitz, J., Shneiderman, B.: Categorizing web search results into meaningful and stable categories using fast-feature techniques. In: Joint Conference on Digital Libraries. JCDL'06, 11–15 June 2006
6. Patil, P., Patil, U.: Preprocessing of web server log file for web mining. World J Sci Technol **2**(3), 14–18 (2012). Accessed 15 August 2013
7. Sintay, B.: Unix timestamp. http://www.unixtimestamp.com/index.php. Accessed 12 Aug 2013
8. Worldometers: Worldometers: real time world statistics. http://www.worldometers.info/world-population. Accessed 21 Aug 2013

Assessment of BER Performance of a Power Line Communication System in the Presence of Transformer and Performance Improvement Using Diversity Reception

Munshi Mahbubur Rahman and S. P. Majumder

Abstract When a transformer is present in a power line communication (PLC) system and a high-frequency signal passes through it, the low-voltage distribution lines and customers' load characteristics will be changed due to high-frequency effects. When a transformer is present, the propagation of high-frequency signals in a PLC channel, the transmitted signals become a very strong function of frequency. In this paper, an analytical attempt is made to measure the attenuation of high-frequency signals across transformers in a PLC system considering the said effects. The BER performance analysis is carried out for a PLC system having a transformer on-load using the transfer function of the channel without and with impulsive noise and applying diversity reception. The performance results are evaluated for different number of orthogonal carriers and for several system bandwidths with and without impulsive noise and different number of diversity branches in the receiver. The results show that there is a significant improvement in system performance with the increase in the number of OFDM subcarriers; also, there is further improvement in receiver sensitivity with the increase in the amount of receiving antennas.

Keywords Maximal ratio combining · Broadband power line communication · Impulsive noise · Attenuation · Probability density function · Interference · Transfer function

M. M. Rahman (✉) · S. P. Majumder
Department of EEE, BUET, Dhaka 1000, Bangladesh
e-mail: mmr@eece.mist.ac.bd

S. P. Majumder
e-mail: spmajumder@eee.buet.ac.bd

S. Patnaik and X. Li (eds.), *Proceedings of International Conference on Soft Computing Techniques and Engineering Application*, Advances in Intelligent Systems and Computing 250, DOI: 10.1007/978-81-322-1695-7_38, © Springer India 2014

1 Introduction

The basic concept of PLC is to transmit information and electricity simultaneously along electricity lines as an alternative to constructing dedicated communication infrastructure [1, 2]. Generally, electricity lines are made from similar conductive materials as those used for telecommunications. The public power line network is designed for the transmission of AC power at typical frequencies of 50 or 60 Hz. It has only a limited capability to carry higher frequencies [3, 4]. Nowadays, higher-frequency signals are superimposed on the low-energy information signal of the power wave for communication purposes. In order to ensure that the power wave does not interfere with the data signal, the frequency range used for communication is very far from the one used for the power wave (50 Hz in Europe and 60 Hz in the United States). The frequency range used for PLC narrowband applications is 3–500 kHz and from 1 to 30 MHz for broadband applications [5].

One of the major causes of high-frequency attenuation in a power line communication channel (and by far most important) is the attenuation due to reflections from abrupt discontinuities and mismatched impedances (e.g., underground to overhead risers, taps, transformers, and capacitors) that occur along the power line. These reflections cause part of the signal to be diverted away from the receiver and absorbed in other parts of the system [6, 7].

Since users are supplied through a LV transformer, one needs to know the behavior of LV transformers, since they will have a significant effect on the PLC signals. If the PLC signals are to be generated at MV or HV substations, then the PLC signals must pass through the LV transformers, just like the ripple control signals do at present. If the PLC signals are coupled directly onto the LV lines, then the impedance of the transformer may be such that most of the PLC signals are lost in the transformer and do not reach the consumer [8, 9].

Diversity reception is a technique to improve the BER performance in this research. In diversity technique, each of the transmitted signals is received through a number of separate channels [10, 11]. When several numbers of individual channels carrying the same information are received over multiple channels that exhibits comparable strength [12, 13]. We employed different gains to each antenna to get better signal-to-noise ratio for the joint signals [14–16]. In this paper, we have presented an analytical approach to find out the effect of power line transfer function on the BER performance of a PLC system impaired by impulsive noise in the presence of transformer. We propose that in a PLC environment, firstly the BER performance can be improved by using OFDM, and secondly, performance is remarkably improved using diversity reception.

Fig. 1 PLC system block diagram

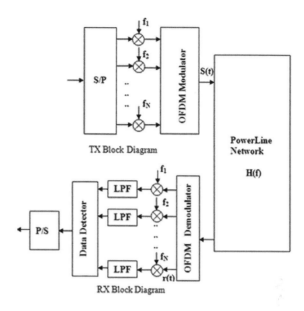

TX Block Diagram

RX Block Diagram

2 System Block Diagram

The block diagram of the PLC system with OFDM multicarrier modulation is shown in Fig. 1. In the transmitter, the signal is initially processed by a serial-to-parallel (S/P) converter, and then, OFDM is modulated to get $S(t)$. The $S(t)$ travels through power line network. In the receiver, the signal is demodulated first to get $r(t)$. The received signal $r(t)$ is then passed through low-pass filter (LPF) and detector. Finally, the signal is processed through a parallel-to-series (P/S) converter.

3 BER Analysis

3.1 BER in OFDM Scenario with Impulsive Noise

For BPL, the model for broadband power line in the frequency range of 1–30 MHz is given by [9],

$$H(f) = \sum_{i=1}^{N} g_i(t) \, e^{-(a_0 + a_1 f^k) d_i} e^{-2\pi j f \tau_i} \tag{1}$$

where g_i is the weighting factor for path i, a_0 and a_1 the attenuation parameters, f the frequency, k the exponent of the attenuation factor (0.5...1), d_i the length of the path and τ_i the delay. In OFDM, the BER under AWGN and impulsive noise is

$$BER = \frac{1}{N} \sum_{i=1}^{N} 0.5 \, \text{erfc} \left(\sqrt{\frac{E_b/N_0}{1 + \mu \lambda T \text{noise}}} |H(f)|^2 \right) \tag{2}$$

3.2 Formulation of Transfer Function and BER in OFDM Scenario Under Impulsive Noise in the Presence of Transformer

Now, the impulse response of the power line is given by $h(t) = F^{-1}[H(f)]$, so from Eq. (1), we get

$$H_1(f) = \left[\sum_{i=1}^{N} g_i \, e^{-(a_0 + a_1 f^k) d_i} e^{-2\pi f \tau_i} \right] \tag{3}$$

For an outdoor network, the power line model consisting of the transformer, the impulse function is given by [10]:

$$h_2(t) = \alpha \left[\text{Sin} \frac{4\pi}{T_{AC}} t + \beta \right] h_c(t)$$

$$= \alpha \, h_c(t) \, \text{Sin} \frac{4\pi}{T_{AC}} t + \alpha \beta \, h_c(t) \tag{4}$$

$$H_2(f) = \frac{\alpha \, h_c(t)}{\pi} \left[\frac{T_{AC}}{4 - T_{AC}^2 f_{ac}^2} - j \frac{\beta}{2f_{ac}} \right] \tag{5}$$

Now, the total impulse function of the outdoor channel is given by $h(t) = h_1(t) \otimes h_2(t)$; so, we get $H(f) = H_1(f) \cdot H_2(f)$ or,

$$H(f) = \left(\sum_{i=1}^{N} g_i e^{-(a_0 + a_1 f^k) d_i} e^{-2\pi f \tau_i} \right)$$

$$\times \left(\frac{\alpha \, h_c(t)}{\pi} \left[\frac{T_{AC}}{4 - T_{AC}^2 f_{ac}^2} - j \frac{\beta}{2f_{ac}} \right] \right) \tag{6}$$

Now,

$$BER = \frac{1}{N} \sum_{i=1}^{N} 0.5 \, \text{erfc} \left(\sqrt{m \frac{E_b}{N_0} |H(f)|^2} \right) \tag{7}$$

[as $m = \text{Rb/B}$].

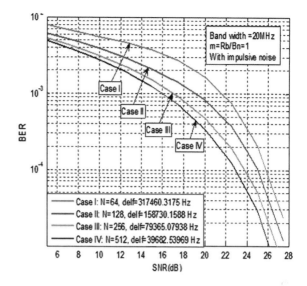

Fig. 2 BER performance in 20 MHz range with impulsive noise

3.3 BER in OFDM Scenario Under Impulsive Noise and Performance Improvement by Diversity Reception

For OFDM, in case of diversity repetition, let the number of receiver be L, then P_b is given by [17],

$$P_b = [0.5(1 - \mu)]^L \sum_{l=0}^{L-1} \binom{L - 1 + l}{l} [0.5(1 + \mu)]^l \qquad (8)$$

where

$$\mu = \sqrt{\frac{\Gamma_C}{1 + \Gamma_C}}, \quad \Gamma_C = 2\sigma_\alpha^2 \cdot \frac{E_b/N_0}{1 + \mu\lambda T \text{noise}} \quad \text{and} \quad \Gamma_C = 2\sigma_\alpha^2 \cdot \frac{m \cdot E_b/N_0 |H(f)|^2}{1 + \mu\lambda T \text{noise}}$$

(m = Rb/B = 1, 2, 3, ...).

4 Results and Discussion

In this paper, we have considered a suitable model for transfer function of broadband power line communication network in the presence of transformer. From the model, the BER performance (by MATLAB simulation) of broadband scenario with and without impulsive noise is depicted in Figs. 2, 3, 4, 5, 6, 7, 8, 9, 10, 11, 12, and 13. The BER performance with transformer in impulsive noise environment is shown in Fig. 7. Figure 8 shows the penalty with transformer in different channel bandwidths. In Figs. 9 and 10, the BER performance with

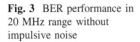

Fig. 3 BER performance in
20 MHz range without
impulsive noise

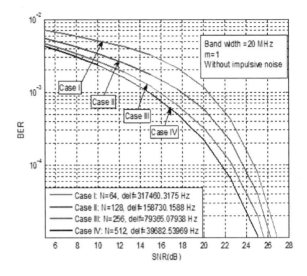

Fig. 4 Power penalty (SNR)
in 20 MHz range

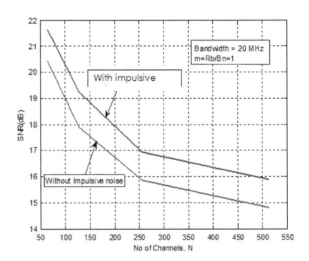

diversity reception in different noise conditions is shown. Figure 11 shows the
performance improvement in diversity versus number of antennas in different
noise conditions (schemes I and II). Figure 12 shows the performance improvement with and without transformer in diversity reception in impulsive noise
environment. Figure 13 shows the BER performance improvement with transformer in diversity. Finally, Fig. 13 shows the total performance improvement
(dB) with diversity in the presence of transformer.

These analyses show that there is a loss of BER at about 10–15 dB for transformer. The BER penalty due to transformer is shown in Fig. 8. All the simulation
results show that there is a significant improvement in BER performance by

Fig. 5 BER performance improvement in 20, 40, and 60 MHz range

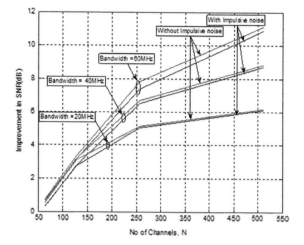

Fig. 6 Receiver sensitivity versus number of channels in 20, 40, and 60 MHz bandwidth

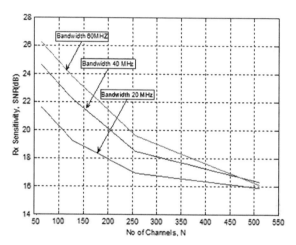

Fig. 7 BER performance in 20 MHz range in the presence of transformer with impulsive noise

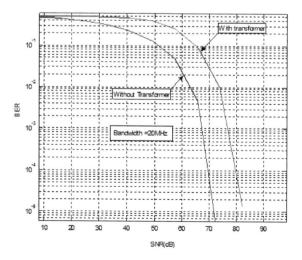

Fig. 8 BER penalty with transformer in different channel bandwidths

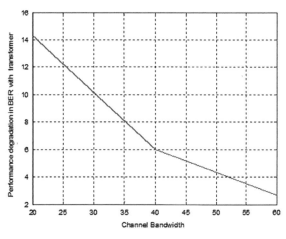

Fig. 9 BER performance in diversity reception. For Scheme I: Sigma^2 = 0.1, Tnoise = 0.0641, lamda = 1/0.0196. Scheme II: Sigma^2 = 0.5, Tnoise = 0.0607, lamda = 1/0.96. Scheme III: Sigma^2 = 1.0, Tnoise = 0.1107, lamda = 1/8.1967

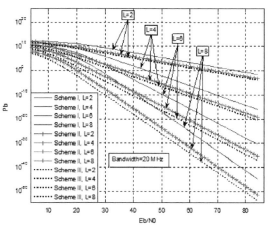

Fig. 10 BER performance in diversity reception

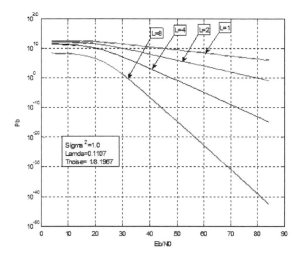

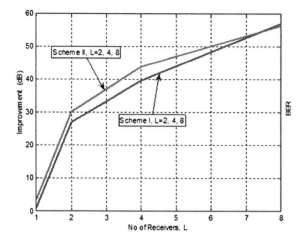

Fig. 11 Performance improvement (dB) in diversity reception versus number of antennas

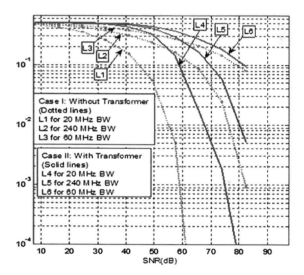

Fig. 12 BER performance in 20, 40, and 60 MHz range in impulsive noise environment with and without transformer in diversity reception

applying diversity reception. The performance degradation due to transformer can be well overcome by diversity, and furthermore, there is additional amount of improvement in the system performance by this proposed technique. However, the number of transformer and optimum number of receiving antennas is yet to be found out, which remains as a future work (Fig. 14).

Fig. 13 BER performance
improvement by diversity
reception ($L = 1, 2, 4,$ and 8)
with transformer in impulsive
noise

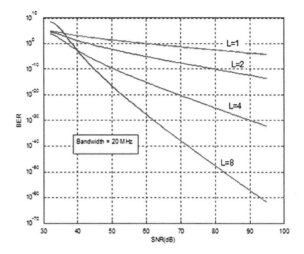

Fig. 14 Plot of performance
improvement (in dB) in
diversity reception with
transformer

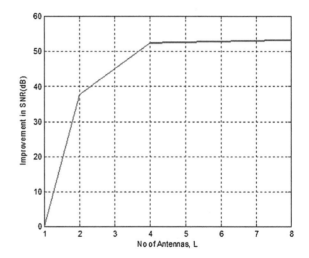

References

1. Tachikawa, S., Hokari, H., Marubayashi, G.: Power line data transmission. IEICE Technical Report, SSTA89-7, Mar 1989
2. Prasad, T.V., Srikanth, S., Krishnan, C.N., Ramakrishna, P.V.: Wideband characterization of low voltage outdoor powerline communication channels in India. AU-KBC Centre for Internet and Telecom Technologies Anna University, Chennai, 2001
3. Philipps, H.:Performance measurements of powerline channels at high frequencies. Research paper of Institute for Communications Technology, Braunschweig Technical University, Braunschweig, PLC'98 (1998)
4. Rennane, A., Konaté, C., Machmoum, M.: A simplified deterministic approach to accurate modeling of transfer function for the broadband power line communication. In: New, 2 Advanced Technologies Journal, Nantes of University, Nantes, www.intechopen.com. Aleksandar Lazinica, 2010

5. Danisman, B.: Analysis of conventional low voltage power line communication methods for automatic meter reading and the classification and experimental verification of noise types for low voltage power line communication network. Thesis for master of science in electrical and electronics engineering, Graduate School of Natural and Applied Sciences, Feb 2009
6. Çelebi, H.B.: Noise and multipath characteristics of power line communication channels. Thesis for master of science in Electrical Engineering, Department of Electrical Engineering College of Engineering, University of South Florida, 30 Mar 2010
7. Hooijenl, O.G., Han Vinck, A.J.: On the channel capacity of a european-style residential power circuit. Signal Communications, Institute for Experimental Mathematics, University of Essen, Essen, Germany
8. Sartenaer, T.: Multiuser communications over frequency selective wired channels and applications to the power line access network. Thesis presented for the Ph.D. degree in Applied Sciences, Université catholique de Louvain, Sept 2004
9. Zimmerman, M., Dostert, K.: A multipath model for the powerline channel. IEEE Trans. Commun. **50**(4), 553–559 (2002)
10. Nassar, M., Lin, J., Mortazavi, Y., Dabak, A., Kim, H., Evans, B.L.: Local utility powerline communications in the 3–500 kHz band: channel impairments, noise, and standards. IEEE Signal Process. Mag. **29**, 116–127 (2012)
11. CENELEC EN 50065.: Signaling on low voltage electrical installations in the frequency range 3 kHz to 148,5 kHz. Beuth-Verlag, Berlin, July 1993
12. Chan, M.H.L., Donaldson, R.W.: Amplitude, width, and interarrival distributions for noise impulses on intrabuilding power line communication networks. IEEE Trans. Electromagn. Compat. **31**(3), 320–323 (1989)
13. Janse van Rensburg, P.A.: Effective coupling for power line communications. D. Ing (Electrical and Electronics), University of Johannesburg, Jan 2008
14. Katayama, M.: Introduction to robust, reliable and high-speed power-line communication systems. IEICE Trans. Fundam. **E84-A**(12), 2958–2965 (2001)
15. Nosotti, J.: Power-line Communications (PLC) or Broadband-over-Power-line (BPL). CRC Powerline Proposal, Nov 2004
16. Zimmermann, M., Dostert, K.: Analysis and modeling of impulsive noise in broad-band powerline communications. IEEE Trans. Electromagn. Compat. **EMC-44**(1), 249–258 (2002)
17. Rahman, M.M., Majumder, S.P.: Analysis of a powerline communication system over a non-white additive Gaussian noise channel and performance improvement using diversity reception. Netw. Commun. Technol. J. **1**(1), 26 (2012)

A Multi-Constraint Anonymous Parameter Design Method Based on the Attribute Significance of Rough Set

Taorong Qiu, Lu Liu, Wenying Duan, Xiaoming Bai
and Zhongda Lin

Abstract To erase the imbalance phenomenon between the privacy protection and data availability caused by identifying all attributes have the same importance degree in traditional algorithm, a multi-constraint anonymous parameter design method based on the attribute significance of rough set is proposed taking into account the influence caused by various quasi-identifier attributes. In this algorithm, it is to carry out the dimension division automatically according to the quasi-identifier attributes significance and thereby actualizing designing multi-constraint anonymous parameters. After that, an anonymous operation is executing on the separate partition. Experimental results turn out that the method comes to a better balance between the privacy protection degree and data availability.

Keywords Privacy preservation · Anonymous parameters · Rough set · Attribute significance

1 Introduction

With the rapid development of information technology, it is available to collect and release relevant data for every walk of life, which may involve some confidential personal privacy information. In the literature [1], Sweeney proposed a K-anonymization rule to solve privacy leakage caused by link attack for the first time. Literature [2] advanced a new rule l-diversity to increase the diversity of sensitive attribute, thus reducing the risk of privacy leakage. Literature [3] introduces a t-closeness anonymous rule to resist the similar attack. Literature [4] and literature [5]

T. Qiu (✉) · L. Liu · W. Duan · X. Bai · Z. Lin
Department of Computer, Nanchang University, Nanchang 330031, China
e-mail: taorongqiu@163.com

S. Patnaik and X. Li (eds.), *Proceedings of International Conference on Soft Computing Techniques and Engineering Application*, Advances in Intelligent Systems and Computing 250, DOI: 10.1007/978-81-322-1695-7_39, © Springer India 2014

discuss the rule of (a, k)-anonymization and p-sensitive, respectively. All above can be implemented by generalization, and the list extends to (c, k)-safety [6], privacy skyline [7], and so on. However, the data precision and availability will be reduced because of the generalization. Literature [8] argues an exchange method to realize anatomy, a rule of high accuracy of data release which uses a lossy link method. In the meantime, (k, e)-anonymity [9] is also a typical anonymous model using the exchange method.

The traditional algorithms are presented by a single constraint to process K-partition, because they suppose all the quasi-identifier attributes have the same significance. But when they are applied to the high-dimensional data set, vast quantities of useful information will be lost. Literature [10] advances a multi-constraint rule to fit various constraint conditions. Though it is well to balance the privacy protection degree with data availability, how to set the constraint parameters for every constraint set is still not specified. The multi-constraint anonymous parameter design method based on the attribute significance of rough set proposed in the present paper is more concerned with the distinct influences of separate quasi-identifier attributes on sensitive attributes.

2 Relevant Concepts

Rough set [11] is a mathematical approach to study imprecise, uncertain information, which is mainly used to mine patterns and regulations from an incomplete data set.

Definition 1 Suppose $T = (U, A)$ be a decision table, $C = (c_1, \ldots, c_m)$ is the set of condition attributes, and $D = (d_1, \ldots, d_k)$ is the set of decision attributes. $U/(C \cup D) = \{Y_1, Y_2, \ldots, Y_k\}$ and $U/C = \{X_1, \ldots, X_l\}$ denote the partition of U on the attribute set C and $C \cup D$, respectively. The conditional information entropy of the decision attribute set D on the set of condition attributes C can explicitly be written as

$$I(D|C) = \left| \sum_{i=1}^{k} \frac{|Y_i|}{|U|} \log_2 |Y_i| - \sum_{i=1}^{l} \frac{|X_i|}{|U|} \log_2 |X_i| \right|$$

Definition 2 Let C and D be sets of condition and decision attributes respectively of the decision table $T = (U, A)$, the significance of attribute $c \in C$ can be normalized as

$$\text{NewSig}(c) = I(D/(C - \{c\})) - I(D/C) + I(D/\{c\})$$

3 Intelligent Selection of Multi-Constraint Anonymous Parameter Based on Attributes Significance

3.1 A Model of K Value Selection Based on the Significance

Definition 3 (*Relation between attribute significance and anonymous parameters*) Given two arbitrary attributes of a data table T, $\text{attr}_i(1 \leq i \leq m)$ and $\text{attr}_j(1 \leq j \leq m)$. k_{attr_i} and k_{attr_j} are attribute significances. Suppose K_i and K_j be the anonymous parameters of attr_i and attr_j; then, the relation between K_i and K_j can be expressed as: $K_i/K_j = k_{\text{attr}_j}/k_{\text{attr}_i}$

Definition 4 (*Average attributes signification of quasi-identifier*) Let T be a data table with m quasi-identifier attributes, and $k_{\text{set}} = \{k_{\text{attr}_1}, \ldots, k_{\text{attr}_m}\}$ is the set of attribute significances. The average attribute significance of quasi-identifiers can be calculated by:

$$\overline{K\text{Set}} = \frac{\sum_{i=1}^{m} K_{\text{attr}_i}}{m}$$

3.2 Intelligent Partition Algorithm of Constraint Set

Algorithm Description

Algorithm 1 Intelligent partition algorithm of constraint set
Suppose T is a data table to be released, QISet $= \{\text{Attr}_1, \ldots, \text{Attr}_m\}$ denotes the set of quasi-identifiers whose significance is K_{attr_i} $(1 \leq i \leq m)$.
Input: A data table T with n items and m quasi-identifiers
Output: Constraint set $C\text{Set} = \{C_1, C_2, \ldots, C_h\}$

Step 1: Compute the significance of every quasi-identifier in line with Definition 6 denoted as $K\text{Set} = \{K_{\text{attr}_1}, K_{\text{attr}_2}, \ldots, K_{\text{attr}_m}\}$;
Step 2: Set the threshold ε and divide the $K\text{Set}$. Get the significance subset $k\text{Set} = \{\text{sub}k\text{Set}_1, \text{sub}k\text{Set}_2, \ldots \text{sub}k\text{Set}_h\}$ and the partition set of quasi-identifier $\text{QISet}' = \{\text{subQISet}_1, \text{subQISet}_2, \ldots, \text{subQISet}_h\}$;
Step 3: Calculate the average attribute significance of every subset and put them into set $\overline{k\text{Set}} = \{\overline{\text{Sub}k\text{set}_1}, \overline{\text{Sub}k\text{set}_2}, \ldots, \overline{\text{Sub}k\text{set}_h}\}$
Step 4: Gain the set of anonymous parameters $\{K_1, K_2, \ldots, K_h\}$
Step 5: Output $C\text{Set}$

3.3 Anonymous Method on Multi-Constraint

On the basis of the above analysis, an anonymous method on multi-constraint is advanced, the process is as following:

Algorithm 2 Multi-constraint anonymous algorithm
    Input: A data table T to be released
    Output: A deliverable table T'

    Step 1: Get the constraint set in line with Algorithm 1
    Step 2: Divide table T according to the constraint set and obtain the subsets $T_1, T_2, \ldots T_h$
    Step 3: Perform the anonymization referring to $CSet = \{C_1, C_2, \ldots, C_h\}$
    Step 4: Combine the subset in accordance with ID to generate all tuples. Gain T'

4 Test and Comparison

4.1 Experiment Environment

The experiments are performed on a loaded intel Core i3M380 2.53 GHz, 2G of DDR3 Memory, Windows 7(32 bits), and Matlab 7.0.

4.2 Experimental Data Set Introduction

With the intention of testing our algorithm, an adult data set, the information from US census, generated by UCI is employed. We randomly take 1,000 samples from the total 45,222 items as test data.

4.3 Results Comparison

In our experiment, evaluations on the tuple link (DLD) are used for leakage analysis.

The risk evaluations with regarding to the proposed and MDVA algorithms are summarized in Table 1.

The MDVA method divides the tuples into several equivalence classes, while our approach is to partition on separate subset; therefore, the deliverable table does not satisfy K-anonymous condition, and at the same time the distortion is to a less degree.

Table 1 Risk evaluations of adult anonymous table

K	MDVA	The proposed algorithm
3	0.9920	0.9990
4	0.9620	0.9780
5	0.9410	0.9590
6	0.9140	0.9310
7	0.9380	0.9320
8	0.9090	0.9230
9	0.9150	0.9040
10	0.8850	0.8950

Table 2 Loss of the adult set

K	MDVA	The proposed algorithm
3	0.9509	0.7925
4	1.3323	0.8533
5	1.8666	1.3201
6	2.5334	1.8467
7	2.7182	2.0223
8	2.9507	2.5602
9	3.1022	2.6681
10	3.2308	2.9377

Because of the closer distance of every tuples, it increases the probability of linking to the original tuples.

Table 2 reveals the data availabilities of adult data set after adopting the two anonymous methods, respectively.

Experimental results exhibit that the proposed method reduces the dimension of primary data set that makes an equivalence partitioning toward every subset in a lower dimension. So then, it provides a better protection for the data veracity. By contrast, in traditional MDVA algorithm, anonymous costs increase along with the number of quasi-identifier (dimension raised). Selecting a larger K value is action taken to lower the chance of data leakage.

5 Conclusions

Different quasi-identifier attributes produce different influences on the sensitive attribute; therefore, it will generate unwanted information loss if we take the same anonymous parameters during the partition process. This paper aims to research and design a fresh anonymous rule for dimension division based on rough set. Owning to sum up the data on different levels, the rule presents in this paper produces diverse K-partition on the corresponding constraint subset rather than on the whole table. The results indicate that the advanced method can keep a better balance between privacy protection and data availability. In future studies, it is possible to perfect and optimize the algorithm, in the meantime come up with the model of appraising data availability aiming at various attribute types.

Acknowledgments This study was supported by National Natural Science Foundation of China (No. 61070139), Natural Science Foundation of Jiangxi Province (No. 20114BAB201039), and the Science and Technology Support Planning Project of Jiangxi Province (No. 20112BBG 70087).

References

1. Sweeney, L.: K-anonymity: a model for protecting privacy. Int. J. Uncertainty Fuzziness Knowl. Based Syst. **10**(5), 557–570 (2002)
2. Machanavajjhala, A., Gehrke, J., Kifer, D.: l-diversity: privacy beyond k-anonymity. In: Proceedings of the 22nd International Conference on Data Engineering (ICDE), pp. 24–35. Atlanta, Georgia, USA (2006)
3. Li, N., Li, T.: t-Closeness: privacy beyond k-anonymity and l-diversity. In: Proceedings of the 23rd International Conference on Data Engineering (ICDE), pp. 106–115. Istanbul, Turkey, (2007)
4. Wong, R.C.W., Li, J, Fu, A.W.C., et al.: (a, k)-Anonymity: an enhanced k-anonymity model for privacy preserving data publishing. In: Proceedings of the 12th ACM SIGKDD International Conference on Knowledge Discovery and Data Mining, pp. 754–759. ACM. Press, New York (2006)
5. Truta, T.M., Vinay, B.: Privacy protection: p-sensitive k-anonymity property. In: Proceedings of the 22nd International Conference on Data Engineering Workshops. IEEE Computer Society, Washington DC (2006)
6. Koudas, N., Srivastava, D., Yu T., et al.: Aggregate query answering on anonymized tables. In: Proceedings of the 23rd International Conference on Data Engineering (ICDE'7). IEEE Computer Society Press, Istanbul, Turkey (2007)
7. Chen, B.C., LeFevre, K., Ramakrishnan, R.: Privacy skyline: privacy with multidimensional adversarial knowledge. In: Proceedings of the 33rd International Conference on Very large Data Bases (VLDB'07), pp. 770–781. Vienna, Austria (2007)
8. Xiao, X., Tao, Y.: Anatomy: simple and effective privacy preservation. In: Proceedings of the 32nd Very Large Data Bases (VLDB) Conference, pp. 139–150. Seoul, Korea (2006)
9. Martin, D., Kifer, D., Machanavajjhala, A., et al.: Worst-case background knowledge in privacy. In: Proceedings of the 23rd International Conference on Data Engineering (ICDE'7). pp. 116–125. IEEE Computer Society Press, Istanbul, Turkey (2007)
10. Yang, X.C., Liu, X.Y., Wang, B., Yu, G.: K-anonymization approaches for supporting multiple constraints. China J. Software **17**(5), 1222–1231 (2006)
11. Liu, Q.: Rough Set and Rough Reasoning, 3rd edn, pp. 20–67. Science Press, Beijing (2005)

Design and Implementation of a Middleware for Service-Oriented Distributed Systems

Hong Xie, Donglin Su, Yijia Pan and Zhongfu Xu

Abstract Service-oriented distributed systems are internally loosely coupled in that the computation nodes are portable and reusable. They can be developed and maintained economically. Traditional middleware products enable the consistent operation on distributed objects and real-time communication. The middleware in a service-oriented distributed system provides more capabilities, such as analyzing data semantics, formulating data protocols, and defining event interfaces. This paper focuses on the design and implementation of a middleware for service-oriented distributed systems. The relationship between the distributed object and the object proxy is briefly introduced. The functional requirements on the middleware for service-oriented distributed systems are analyzed. The operational, system, and technical architecture design of the middleware is outlined. The programming technique for implementing the middleware is presented by focusing on the XML-based textualization of distributed object proxies. The application of the middleware is demonstrated by building a distributed simulation system.

Keywords Service-oriented · Distributed system · Middleware

1 Introduction

"Centralized" computer systems built upon a mainframe and a set of terminals can be regarded as the first-generation distributed system, in which the terminals provide the facilities for human–machine interaction and usually do not have

H. Xie · D. Su
School of Electronics and Information Engineering, Beihang University,
Beijing 100083, China

H. Xie · Y. Pan · Z. Xu (✉)
State Key Laboratory of Complex Electromagnetic Environment Effects
on Electronics and Information System, Luoyang 471003, China
e-mail: xuzf69@yahoo.com

S. Patnaik and X. Li (eds.), *Proceedings of International Conference on Soft Computing Techniques and Engineering Application*, Advances in Intelligent Systems and Computing 250, DOI: 10.1007/978-81-322-1695-7_40, © Springer India 2014

computing capability. This kind of distributed system usually shows low performance and requires high cost. As the computer network technology evolves, the distributed system with the client–server architecture emerges. In such a system, the client and the server both perform scientific computation in a network environment. Using the client–server architecture, the distributed system improves the overall performance, enlarges the deployment scale, and reduces the cost. The client–server architecture, however, does not enable the direct communication among clients, and as a result, impedes its application in a few domains, such as distributed interactive simulation. As the peer-to-peer (P2P) architecture comes into use, the publish-and-subscribe paradigm [1] and remote method invocation (RMI) on distributed objects are applied in the distributed system, supporting the direct communication and interoperation among computational nodes. The information exchanged among the nodes is carried out by distributed objects whose definitions reside on both sides. In such a way, however, the coupling of the distributed system structure is fastened, the portability and the reusability decline. Such shortcomings are overcome when the distributed system approaches the service-oriented paradigm [2], in which the self-describing Extensible Markup Language (XML) [3] facilitates interoperability at the "semantic" level and the nodes provide diverse services outwards depending on semantics of the information exchanged. The service-oriented distributed system structurally consists of a set of loosely coupled computational nodes whose interaction is determined solely by data and services. As a result, the interaction among the computational nodes in a service-oriented distributed system is promoted, and the portability and the reusability of computational resources in a service-oriented distributed system are enhanced. Meanwhile, the development and maintenance of a service-oriented distributed system become cheaper, especially when a computational node can be maintained without affecting the other nodes and the whole system.

This paper focuses on the design and implementation of a middleware for service-oriented distributed systems. Section 2 briefly introduces the relationship between the distributed object and the object proxy. Section 3 analyzes the functional requirements on the middleware for service-oriented distributed systems. Section 4 outlines the design of operational, system, and technical architecture of the middleware. Section 5 presents the programming technique for implementing the middleware. Section 6 demonstrates the application of the middleware in a distributed simulation system.

2 Distributed Object and Object Proxy

In the conventional P2P distributed system, information exchange among computational nodes is realized by applying the distributed object–object proxy paradigm which continues its application in the service-oriented distributed system. In the distributed object–object proxy paradigm, the information delivered outwards by a computational node is defined as the state attributes of distributed objects, and

Fig. 1 Distributed object
and object proxy

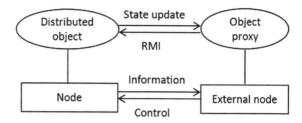

the control behavior on a computational node from the other nodes is defined as
the methods of distributed objects. As shown in Fig. 1, at run-time, distributed
objects update their state attributes to deliver information to external nodes, and
the external node invocates the remotely invocable methods of distributed objects
through the remote method interfaces provided by the object proxy to control the
behavior of the node, on which the distributed objects reside.

3 Functional Requirement of the Middleware

In some distributed systems, such as a conventional P2P distributed system,
a distributed object is owned solely by one computational node, and the other
nodes can access the distributed object only by means of the object proxy. In such
systems, the middleware support for consistent object operation emphasizes on the
state synchronization from the distributed object to the object proxy and the RMI
from the object proxy to the distributed object. In the service-oriented distributed
system, more capabilities are required for the middleware, e.g., to analyze data
semantics, formulate data protocols, define event interfaces, and generate data for
applying services.

The data protocol specifies the structure of data for applying services. Therefore, a data protocol is bound with a specific service. In a service-oriented distributed system, the services can be classified into two types, i.e., business logic
and underlying service. Business logics represent the system functionality to be
implemented by the application code. Underlying services form the common
infrastructure for business logic. They are implemented by the middleware and
disperse on all the computational nodes of the distributed system. The middleware
defines required data protocols and event interfaces for the underlying service. As
shown in Fig. 2, for example, *nodeA* declares subscription to *n* objects residing on
the other nodes. When *object1* changes its state, *nodeB* on which *object1* resides
sends to *NodeA* via network the request for updating the state of *object1*. In *nodeA*,
the middleware accepts the request and analyzes it and updates the state attribute
of the proxy of *object1*. Afterward, the middleware calls the event interface and
triggers the execution of the business logic related to the state update of *object1*.

Fig. 2 Underlying service
and business logic

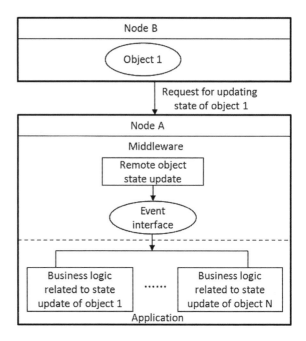

4 Design of the Middleware Architecture

Using the US DoD Architecture Framework [4] as reference, the middleware
architecture is designed, focusing on the operational, system, and technical aspects
of the middleware.

4.1 Operational Architecture

The middleware is a kind of soft "plugin." Its operational architecture describes
the tasks performed by the middleware, data sources external to the middleware,
and requirements of information exchange between the middleware and its
external data sources during the distributed computation.

At run-time of the distributed system, the external nodes, the local application,
and its data protocol files are the data sources external to the middleware. Figure 3
shows the information exchange between the middleware and its external data
sources at run-time.

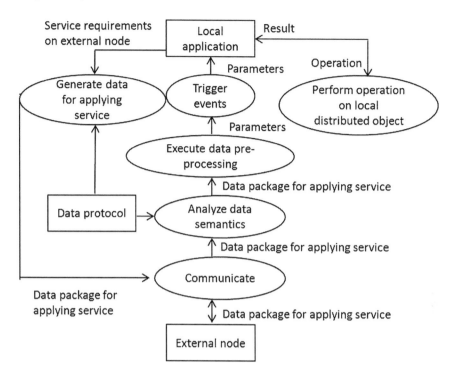

Fig. 3 Information exchange between the middleware and its external data sources at run-time

4.2 System Architecture

The system architecture describes the modules that constitute the middleware, their functionality (shown in Table 1), the collaboration among the modules and the data sources external to the middleware (shown in Fig. 4).

4.3 Technical Architecture

The technical architecture specifies the technology standards that regulate the design and implementation of the middleware. Table 2 lists the configurations of the technology standards.

Table 1 Functionality of the middleware modules

Module	Function description
Communication	Transforming texts to network data stream; sending and receiving data
Protocol analysis	Analyzing data protocols; invoking underlying service according to data semantics
Data preprocessing	Allocating memory space for received object proxy; updating the state of object proxy; processing subscription request; processing RMI; etc.
Data encapsulation	Generating data package for applying services available on external nodes
Event interface definition	Defining event interfaces
Local distributed object control	Performing operations on local distributed objects
Memory	Allocating memory space for local distributed objects, proxies of remote objects, subscriber information, etc.

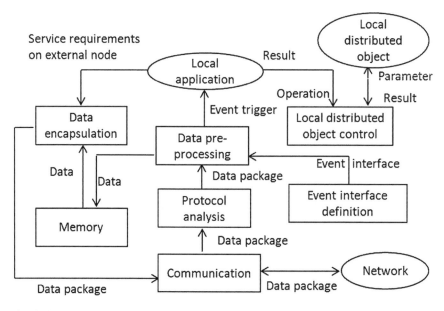

Fig. 4 Collaboration among the modules and the data sources external to the middleware

Table 2 Technology standards configurations of the middleware

Distributed object definition	Constrained by a meta-model
Communication protocol	TCP, UDP
Communication mode	Asynchronous communication
Threading	Thread pool
RMI	Reflecting
Data protocol description language	XML
Final product	Source code or DLL

5 Implementation of the Middleware

In Windows XP OS, a prototype of the middleware for the service-oriented distributed system is implemented using .NET and C# [5–7]. In the following, XML-based textualization of the object proxy is described in detail.

In the distributed system, the interaction between computational nodes is realized by associating the distributed object with its proxy. In some distributed system, such as the conventional P2P distributed system, the subscriber of the remote object gets (usually at compiling time) the remote object definition to create the object proxy. As a result, in such a system, the dependency among the computational nodes is strengthened. To deal with this issue, we use the XML-based textualization technique for the owner of the distributed object to create locally the textualized proxy of the distributed object and send the proxy via network to the subscriber of the distributed object. This way, the loosely coupled system structure is enabled in that an application residing on a computational node is independent of the other nodes.

XML and the reflecting technique are used to textualize the proxy of the distributed object. The state attributes of the distributed object are transformed into XML texts by the XmlSerializer. The operation definitions of the distributed object are transformed into XML texts using the reflecting technique. Additional efforts, however, are required to develop the algorithm for textualizing the operation parameters. In this work, an algorithm is developed, as shown in Fig. 5.

6 Example of Application

Using the middleware described above, a distributed training system is developed which is composed of six equipment applications, a routing server, an information display application, and a data management application, as shown in Fig. 6.

During the simulation, equipment applications send the equipment state information in real-time to the information display application and the data management application. The information display application controls the equipment behavior through RMI. This system shows two dominant features of a service-oriented distributed system: (1) The system is internally loosely coupled in that a single equipment application can join in or quit from the system execution without affecting the other applications. (2) The equipment application is reusable. When a new equipment application is to be developed, the code of the existing application may be reused without nontrivial modification.

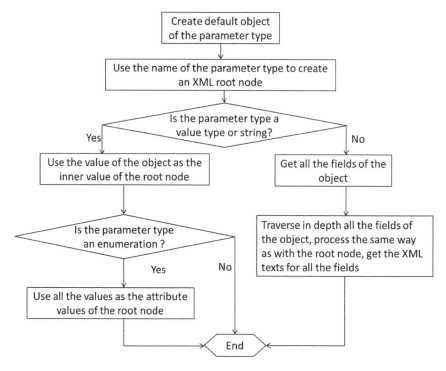

Fig. 5 Algorithm for XML-based textualization of operation parameters

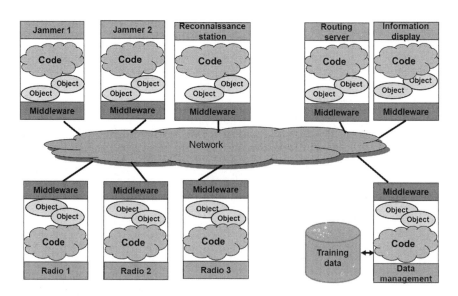

Fig. 6 Structure of a distributed training system

7 Conclusions

The service-oriented distributed system indicates the prospect of the distributed system. As an important aspect in a service-oriented distributed system, the middleware supports the consistent operations on distributed objects by providing a mechanism for coordinating the operations performed on a single object by different applications. With the networking facilities of the middleware, developers of the service-oriented distributed system can more concentrated on implementing the business logic.

References

1. Ma, J.G., Huang, T., Wang, J.L., Xu, G., Ye, D.: Underlying techniques for large-scale distributed computing oriented publish/subscribe system. J. Softw. **17**(1), 134–147 (2006)
2. Fawcett, J., Ayers, D., Quin, L.: Beginning XML, 5th edn. Wrox, NJ (2012)
3. Chen, X.L.: Research on service-oriented distributed operating system and the service composition technology. Dissertation, Press of University of Science and Technology of China (2007)
4. U.S. DoD: DoD architecture Framework 1.5. (2007)
5. Luo, J.Z., Li, B.T., Yang, M., et al.: TCP/IP Protocol and Network Programming Techniques. Tsinghua University Press, Beijing (2004)
6. Benz, B., Durant, J.: XML Programming Bible. Wiley, NJ (2003)
7. Watson, K., Nagel, C.: Beginning Microsoft Visual C# 2008. Wiley, NJ (2008)

Refactoring Structure Semantics Similar Clones Combining Standardization with Metrics

Xia Li, Xiaohong Su, Peijun Ma and Tiantian Wang

Abstract Eliminating code clones is good for improving the quality and maintainability of software. Structure semantics similar code clones are more difficult to be refactored than other types of clones. This paper presents an algorithm for refactoring structure semantics similar clones. Structure semantics similar clones are the clones that have similar program dependence graph (PDGs) after semantic equivalent standardization transformation. The graph-based standardization method can identify the semantically similar code clones effectively with a high computation complexity. The metrics-based method has a lower computation complexity but also has a lower accuracy of identification. To solve this problem, this paper first uses the metrics-based method to filter out most candidate clones not suitable for refactoring and then further confirms the structure semantics similar clones that are suitable for refactoring by code standardization, PDGs matching, and similarity comparison. Structure semantics we propose a new approach for refactoring structure semantics similar clones combining standardization with metrics. The experiments results in open source codes show that this method behaves well in refactoring structure semantics similar clones.

Keywords Program refactoring · Structure semantics similar clone · Code standardization · Metrics

1 Introduction

Code clones refer to the same or similar code fragment in source code file [1]. Studies have shown that the code clones account for about 7–23 % [2] in large software systems. In most cases, code clones are harmful to the system, and they increase the length of code, resulting in the software architecture bloated, easy to

X. Li (✉) · X. Su · P. Ma · T. Wang
School of Computer Science and Technology, Harbin Institute of Technology,
Harbin, China
e-mail: sxh@hit.edu.cn

S. Patnaik and X. Li (eds.), *Proceedings of International Conference on Soft Computing Techniques and Engineering Application*, Advances in Intelligent Systems and Computing 250, DOI: 10.1007/978-81-322-1695-7_41, © Springer India 2014

introduce defects [3, 4], and hard to maintain. Refactoring is to adjust the internal structure of the software, aiming to improve the understandability and reduce the modification costs without changing the observable behavior [5].

Roy [2] divided the code clone into four types: the first three are text-based similar code clones, and the last one is based on the functional equivalence.

At present, only specific types of code clones can be effectively detected. However, the actual software usually includes a variety of code clones types. Eliminating code clones is clearly good for improving the quality and maintainability of software. So this paper divides the code clones into the following three types, which are referred to as amorphous code clones.

Type 1: syntax similar code clone;
Type 2: structure semantics similar code clone;
Type 3: functionally equivalent code clone with different syntactically structure.

This paper proposes a method for refactoring structure semantics similar code clones. Metrics-based and graph-based combined approach (MGCA) [6] is proposed to detect structure semantics similar code clones effectively. The main ideas of the refactoring method are as follows: First of all, for the code to be refactored, standardized rules are established and then transformed the code into common form according to the rules. Secondly, the standardized code clone is detected and then determines the difference between the two code clone fragments. Thirdly, metrics-based differences between two code fragments are analyzed and then merged the differences of code fragments if the analysis result is suitable for refactoring. Preliminary experiments show that the method in this paper can implement the structure semantics similar clone refactoring correctly.

2 Related Work

In clone detection, it has already been proposed a number of methods and developed many detection tools, especially for the detection of syntax similar code clone [7–10]. Kim [11] presented a semantic similar clone detection method by comparing the program abstract memory. References [12, 13] put forward an algorithm of detecting functionally equivalent code clone by combining automatically dynamic testing and static program analysis. Kong [14] combined the K-nearest neighbor clustering, random assignment with dynamic testing to detect functionally equivalent code fragments. The method that Kong [14] proposed can reduce time complexity than references [12, 13], for the K-nearest neighbor clustering method can get the high cohesive and meaningful code fragments.

Balazinska [15] proposed to use 21 metrics as the refactoring guidance to help users determine whether the code clones are suitable for refactoring. Higo [16] developed ARIES tools, using the code detection tool CCFinder and environment

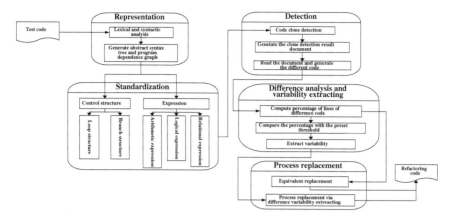

Fig. 1 Model to refactor structure semantics similar code clones

analysis tool Gemini to find the clone can be refactored. Schulze [17] proposed a new metric DIST (distance of the code clones).

In the refactoring of code clone, Feng [18] proposed a method using K-nearest neighbor clustering algorithm. Yu [19] raised an approach using abstract syntax tree and static analysis to refactor the code clones automatically.

3 The Approach of Structure Semantics Similar Clone Refactor Combining Standardization and Metrics

3.1 The Model to Refactor Structure Semantics Similar Code Clones

This paper discusses how to refactor structure semantics similar code clones accurately and efficiently. The structure semantics similar code clone refactoring involves five sub-processes: representation, standardization, detection, difference analysis and variability extracting, and procedure replacement, as shown in Fig. 1 and discussed in the sections that follow.

3.2 Code Standardization

The standardization includes two aspects: control structure standardization and expression standardization.

Loop structure standardization Loop structure standardization is mainly aimed at for and while statements. The abstract syntax tree representation is shown in Fig. 2.

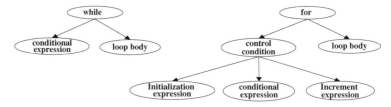

Fig. 2 Abstract syntax tree representation of while and for statements

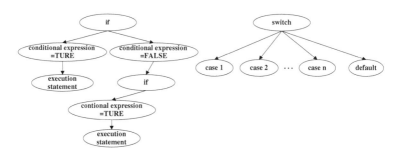

Fig. 3 Abstract syntax tree representation of if and switch statements

Branch structure standardization Branch structure standardization mainly includes if structure and switch structure. The abstract syntax tree of if structure and switch structure is shown in Fig. 3.

Arithmetic expression standardization Common arithmetic operators are +, −, *, /, and %, and the operand is mainly constants and variables. Here are the following 18 rules of standardized arithmetic expression conversion, E1, E2, and E3 mean the arithmetic expression. The standardization is achieved by repeated use of the rules. Standardization is achieved by traversing and conversion operation of the tree node. The rules of arithmetic expression standardization are shown in Table 1.

Logical expression standardization Common logical operators are &&, ||, and !, and the operand is mainly constants and variables. The rules of logical expression standardization are shown in Table 2.

Relational expression standardization Common relational operators are <, >, <=, >=, !=, and ==, and the operand is mainly constants and variables. The rules of relational expression standardization are shown in Table 3.

3.3 Difference Analysis and Variability Extracting

The percentage of different code lines is calculated. If the percentage is zero, the two code fragments are equivalent. Otherwise, the percentage and the preset

Table 1 Rules of arithmetic expression standardization

Rule number	Rule content
1	El + E2 = E2 + El
2	El × E2 = E2 × El
3	El + (E2 + E3) = (E1 + E2) + E3
4	El × (E2 × E3) = (El × E2) × E3
5	El × (E2 + E3) = El × E2 + El × E3
6	El × (E2 − E3) = El × E2 − El × E3
7	E1 + 0 = El
8	El − 0 = El
9	El − El = 0
10	El × 0 = 0
11	El × 1 = El
12	El/1 = El
13	E1/E1 = 1
14	E1 % 1 = 0
15	E1 % E1 = 0
16	E1 − (E2 + E3) = E1 − E2 − E3
17	E1 − E2 + E3 = E1 + E3 − E2
18	− (−E1) = El

Table 2 Rules of logical expression standardization

Rule number	Rule content
1	El&&E2 = E2&&El
2	El||E2 = E2||El
3	El&&(E2&&E3) = (E1&&E2)&&E3
4	El||(E2||E3) = (El||E2)||E3
5	El&&(E2||E3) = El&&E2||El&&E3
6	E1&&1 = E1
7	El&&0 = 0
8	El||0 = E1
9	El||1 = 1
10	E1&&!E1 = 0
11	E1||!E1 = 1
12	E1&&E1 = E1
13	E1||E1 = E1
14	!!E1 = E1

Table 3 Rules of relational expression standardization

Rule number	Rule content
1	El == E2 = E2 == El
2	El != E2 = E2 != El
3	El >= E2 = E2 <= El
4	El > E2 = E2 < El
5	El <= E2 = E2 >= El
6	El < E2 = E2 > El

Table 4 Refactoring experimental results based on different threshold

Threshold	Number of different clones		
	Equivalent	Suitable	Not suitable
0.1	8	8	14
0.3	8	12	10
0.5	8	14	8

threshold are compared to determine whether the code is suitable for refactoring or not. Finally, extract the variability of the suitable code clones using "if–else if" statement and control variables.

4 Experimental Results and Analysis

We choose open source program Siemens to do experiment. In order to obtain better contrasting experimental results, this paper uses the manner of manually injecting and modifying code into Siemens to verify the method proposed in this paper.

Firstly, standardize the code and do the clone detection. This paper can obtain 30 groups of structure semantics similar code clones. The next step is the code clones analysis. Reference [19] pointed that the difference threshold needs to take specific circumstances into account, it set the threshold = 0.5. In this paper, the threshold is, respectively, set to different values. The refactoring experiment results of different threshold are shown in Table 4. Table 4 shows the experiment results when the threshold values are set to 0.1, 0.3, and 0.5, the 30 groups contain three classes: (1) equivalent, (2) suitable for refactoring, and (3) not suitable for refactoring. When the threshold decreases, the number of code suitable for refactoring will become less and less. With the threshold value increases, the number of code suitable for refactoring will become more and more.

5 Conclusions

A method of refactoring structure semantics similar clones combining standardization with metrics is proposed in this paper. Firstly, the code blocks are standardized using the standardization rules that we designed. Secondly, the code difference is analyzed based on the percentage of different code lines. Finally, the code which is suitable for refactoring is merged and the procedure replacement via variability extracting is developed. The refactored codes are obtained through the procedure replacement and procedure calling. Experimental results show that this method can refactor structure semantics similar code clones correctly and effectively.

Acknowledgments This research is supported by the National Natural Science Foundation of China (Grant No. 61173021) and the Research Fund for the Doctoral Program of Higher Education of China (Grant Nos. 20112302120052 and 20092302110040).

References

1. Kamiya, T., Kusumoto, S., Inoue, K.: CCFinder: a multilinguistic token-based code clone detection system for large scale source code. IEEE Trans. Softw. Eng. **28**(7), 654–670 (2002)
2. Roy, C.K., Cordy, J.R., Koschke, R.: Comparison and evaluation of code clone detection techniques and tools: a qualitative approach. Sci. Comput. Program. **74**, 470–495 (2009)
3. Wang, Q.: C Code Clone and Related Software Defect Detection Based on Sequence Mining, pp. 1–47. Harbin Institute of Technology Computer Science and Technology School, Harbin (2009)
4. Chou, A., Yang, J., Chelf, B., Hallem, S., Engler, D.: An empirical study of operating systems errors. In: Proceedings of the Eighteenth ACM Symposium on Operating systems principles, SOSP'01, pp. 73–88 (2001)
5. Fowler, M.: Refactoring: Improving the Design of Existing Code. Addison Wesley, Reading (1999)
6. Wang, T.: The study to identify structure semantics similar programs. Doctoral Dissertation, Harbin Institute of Technology (2009)
7. Baker, B.S.: On finding duplication and near-duplication in large software systems. In: Working Conference on Reverse Engineering (WCRE), pp. 86–95 (1995)
8. Baxter, D., Yahin, A., Moura, L., Sant'Anna, M., Bier, L.: Clone detection using abstract syntax trees. In: ICSM, pp. 368–377 (1998)
9. Jiang, L., Misherghi, G., Su, Z., Glondu, S.: Deckard: scalable and accurate tree-based detection of code clones. In: ICSE, pp. 96–105 (2007)
10. Kamiya, T., Kusumoto, S., Inoue, K.: CCFinder: a multilinguistic token-based code clone detection system for large scale source code. TSE **28**(7), 654–670 (2002)
11. Kim, H., Jung, Y., Kim, S., Yi, K.: MeCC: memory comparison-based clone detector. In: 33rd International Conference on Software Engineering, ACM, pp. 301–310 (2011)
12. Jiang, L.: Scalable detection of similar code. Techniques and Applications. Computer 82–119 (2009)
13. Jiang, L., Su, Z.: Automatic mining of functionally equivalent code fragments via random testing. In: ISSTA '09 Proceedings of the Eighteenth International Symposium on Software Testing and Analysis, pp. 81–91 (2009)
14. Kong, D., Su, X., Wu, S., et al.: Detect functionally equivalent code fragments via K-nearest neighbor algorithm. IEEE 5th International Conference on Advanced Computational Intelligence(ICACI), Oct. **18**(20), pp. 94–98. Nanjing, Jiangsu, China (2012)
15. Balazinska, M., Merlo, E., Dagenais, M., et al.: Advanced clone-analysis to support object-oriented system refactoring In: Proceedings. Seventh Working Conference on Reverse Engineering, pp. 98–107. IEEE (2000)
16. Higo, Y., Kusumoto, S., Inoue, K.: A metric-based approach to identifying refactoring opportunities for merging code clones in a Java software system. J. Softw. Maintenance Evol. Res. Pract. **20**(6), 435–461 (2008)
17. Schulze, S., Kuhlemann, M., Rosenmüller, M.: Towards a refactoring guideline using code clone classification. In: Proceedings of the 2nd Workshop on Refactoring Tools. ACM, p. 6 (2008)
18. Feng, J.: Code clone restructuring of c programs via K-nearest neighbor algorithm. Harbin Institute of TechnologyComputer Science and Technology School, Harbin, pp. 1–57 (2011)
19. Yu, D., Peng, X., Zhao, W.: Automatic refactoring method of cloned code using abstract syntax tree and static analysis. J. Chin. Comput. Syst. **30**(9), 1752–1760 (2009)

A Retail Outlet Classification Model Based on AdaBoost

Kai Liu, Bing Wang, Xinshi Lin, Yeyun Ma and Jianqiang Xing

Abstract This paper proposes a framework to get a stable classification rule under unsupervised learning, and the term "stable" means that the rule remains unchanged when the sample set increases. This framework initially makes use of clustering analysis and then use the result of clustering analysis as a reference-studying sample. Secondly, AdaBoost integrated several classification methods is used to classify the samples and get a stable classification rule. To prove the method feasible, this paper shows an empirical study of classifying retail outlets of a tobacco market in a city of China. In this practice, k-means is used to make clustering analysis, and AdaBoost integrated RBF neural network, CART, and SVM is used in classification. In the empirical study, this method successfully divides retail outlets into different classes based on the sales ability.

Keywords Retail outlet classification · AdaBoost · Unsupervised learning

1 Introduction

1.1 Background

It is a problem deserving to be discussed for manufacturers that how to decide the amount of goods for each retail outlet on consignment sales or in the planned economy pattern. If the amount of supply is larger than that of demand, the cost for retail outlets to store goods will rise. On the other hand, the period for manufacturers to take back their money will extend. Therefore, an appropriate amount of supply can make work between production and sales more efficient. An approach which can help to decide the amount is predicting present and future sale

K. Liu (✉) · B. Wang · X. Lin · Y. Ma · J. Xing
Xiamen University, Siming S. Rd. 422, Xiamen, China
e-mail: liu.kai@stu.xmu.edu.cn

S. Patnaik and X. Li (eds.), *Proceedings of International Conference on Soft Computing Techniques and Engineering Application*, Advances in Intelligent Systems and Computing 250, DOI: 10.1007/978-81-322-1695-7_42, © Springer India 2014

condition according to the past sale condition in each retail outlet. However, it is difficult to do such thing for millions of retail outlets.

This paper puts forward a solution. The first step is classifying retail outlets into different kinds according to sales ability of each. Subsequently, specialists work out the amount of goods for each kind of retail outlets through the market analysis. After that, retail outlets that belong to the same kind can receive the same amount of goods. Classification is based on commercial scale and historical sale, which reflect sailing condition. Once the amount has been decided, product launching can be put into practice.

Consequently, this paper emphasizes the problem that how to classify retail outlets and work out an approach to get a classification rule which is stable while the size of samples changes. That way, once there is a new sample, it can be classified into a certain kind according to its features. Since the classification rule remains the same, it can provide a beneficial way for specialists to decide the amount for each kind through market analysis.

1.2 Previous Study

Harald Hruschkaa did a research on market segmentation and compared the performance of self-organizing neural network with k-means clustering in 1999 [1]. And the research comes to a conclusion that self-organizing neural network makes applicability more extensive. However, both of these classification rules are changing, while the number of samples is getting larger.

Jih-Jeng Huanga does a research on market segmentation with support vector machine in 2007 [2]. The article gets a classification rule which will not change with the number of samples becoming increasingly large. But the disadvantage is that the classification rule only takes a single classifier, which may not suitable for certain cases.

Data mining algorithms such as decision tree, SVM, k-means, neural network have been widely applied to classification and clustering problems, and previous studies [3, 4] have shown that these algorithms are very efficient. In particular, some hybrid models such as bagging and AdaBoost can improve the classification accuracy while comparing to individual model [5, 6]. Therefore, this paper focuses on these algorithms on retail outlets' classification problem.

2 Research Methods

2.1 Overview of the Framework

Because the relative retail outlets which have never been defined into a certain kind before, the classifier has no reference sample to learn from. Clustering can divide samples into different kinds according to some similar features without the reference-studying sample. Thus, this paper will perform clustering first to get some reference. After clustering, different kinds show the distinction in the ability to sell and the classification result obtained through clustering can be the standard for classifiers as their reference.

Before clustering, what to do beforehand is picking up attributes related to sales ability (otherwise, it can be the wrong classification features with elements which are not related to sales ability). Also, the dimension of sample properties should be reduced to make sure that the number of the attributes of samples is not too much (classifying high-dimensional data lessens the accuracy of the result). At the same time, it assumes that the change in sales ability with the change in the value of selected attributes is monotonous (for example, monotonic increase). It can be expressed in mathematical language. If there is a sample X with some attributes $\{x_1, x_2, \ldots, x_n\}$, function $G(x)$ represents sales ability (though it is not sure that function $G(x)$ can be found in reality), and it is monotonically increasing; thus, it has the following quality (1):

$$\frac{\partial G}{\partial x_i} > 0 \qquad \forall i \in \{1, 2, \ldots, n\} \tag{1}$$

On the other hand, to eliminate dimensional differences, it is necessary to standardize the data, that is,

$$x_i \in [0, 1] \qquad \forall i \in \{1, 2, \ldots, n\} \tag{2}$$

Therefore, sample X can be a point in the space $[0, 1]^n$.

After all these above things have been done, cluster analysis can be carried out with the rectified data. Hopefully, some categories can be worked out, that is, $A = \{C_1, C_2, \ldots, C_k\}$. Each category means different sales ability (or the expected amount of goods to be put into the market). If new samples are added into or old samples are replaced, the result remains the same. To get a classification rule that does not change while the number of samples increased. An approach, which takes clustering first then makes classification based on the clustering result, is considerable. This paper mainly uses AdaBoost algorithm that integrated various kinds of weak classifiers to do ensemble learning so that to get a strong classifier. It takes advantages of different kinds of classifications and makes sure the result to a maximum level of accuracy.

2.2 Cluster Analysis

The main idea is to divide the retail outlets into k sorts to make it convenient for experts to calculate the volume of supply of each retail outlet. Therefore, the cluster method that is suited for this research should generate exactly k clusters as we want. Also, since each retail outlet belongs to a sort, the cluster method should not leave outliers. Thirdly, the distribution of sample set of each cluster should have a significant difference in space (rather than one encircle another). Thus, the clustering can reflect the difference in sales abilities, which means the supply to the retail outlets we should assign.

Based on the three considerations mentioned above, our research chose k-means method to conduct the cluster analysis.

2.2.1 *K*-means Algorithm

K-means clustering is a typical clustering algorithm based on distance. It takes distance as evaluation of similarity.

Given a problem that partition n samples into k clusters $C_1, C_2,..., C_k$, and the mean of points in each cluster is $u_1, u_2, ..., u_k$, $dis(x, y)$ is the distance between sample x and sample y under a certain measure.

K-means solves this problem by optimizing the following:

$$\text{minimize} \sum_{i=1}^{k} \sum_{x_j \in C_i} \text{dis}(x_j, u_i) \tag{3}$$

This algorithm uses the idea of greedy algorithm. It repeatedly updates the location of mean points in each cluster. Finally, $\sum_{i=1}^{k} \sum_{x_j \in C_i} \text{dis}(x_j, u_i)$ will converge.

The performance of this algorithm depends on the initial mean of points. We can get a better performance by using genetic algorithm or simulated annealing to determine the initial mean of points.

2.3 *AdaBoost*

AdaBoost is a machine learning algorithm invented by Freund and Schapire [7]. It can composite other machine learning algorithms and improve their performance. AdaBoost algorithm trains the basic classifier (weak classifiers) using multiple training sets and add the result up by their performance to get a stronger final classifier (strong classifier). The theory proved that if each weak classifier performs better than random guesses, the error rate of the strong classifier will converse to zero as the number of weak classifiers goes to infinity [7].

The algorithm generates a different training set by adjusting distribution of weights of each sample, which indicates the importance of each sample in the set for the classification. Let $X = \{x_1, x_2, ..., x_n\}$ be the set for the classification and $Y = \{y_1, y_2, ..., y_n\}$ be the corresponding class of each sample. $Y = \{y_1, y_2, ..., y_n\}$ is a repeated permutation of 1, 2, ..., k if samples are divided into k classes. In this paper, we use a vector to represent the distribution of weights (in training round i), say $D_i = \{d_{i,1}, d_{i,2}, ..., d_{i,n}\}$, satisfying $\sum_{j=1}^{n} d_{i,j} = 1$.

Suppose we want m samples in training round i, then the training set is S_i, and let the probability be $P(x_j \in S_i) = d_{i,j} \times m$. Therefore, the sample size n_i of training set S_i has the expected value $E(n_i) = m \times \sum d_{i,j} = m$.

At first, each sample has the same weight $d_{0,j} = 1/n$, $j = 1, 2, \ldots, n$. We generate a training set through this distribution of weights and train a weak classifier named H_1 under this training set.

On this round, the weights of incorrectly classified samples under H_1 are increased, and at the same time, the weights of correctly classified samples are decreased. By doing so, the algorithm gets a new distribution for the weak classifier H_2 in the next round. Thus, H_2 will focus on the samples that H_1 cannot classify. H_1 also gets a weight α_1 based on the error rate it makes, so does H_2. The lower the error rate, the higher the weight. After T training rounds, AdaBoost algorithm produces T weak classifiers and T weights. Finally, the strong classifier is produced by adding T weak classifier up by the corresponding weights.

The algorithm is shown with pseudo-code (Algorithm 1).

---

1 initialize $d_{0,j} = 1/n$, $j = 1, 2, \ldots, n$

2 for $i = \{1, 2, \ldots, T\}$ do

3   Generate S_i from distribution D_i

4   Use the training set S_i to train weak classifier H_i

5   Use H_i to do the classify work with sample set X, and calculate the error rate $\epsilon_i = \sum_{H_i(x_j) \neq y_j} d_{i,j}$

6   if $\epsilon_i > 1/2$ then

7 $i = i - 1$

8   continue

9   end

10   Give H_i weight $\alpha_i = \frac{1}{2} \log \frac{1 - \epsilon_i}{\epsilon_i}$

11   Determine the distribution weight for next training round

$$d_{i+1,j} = \frac{d_{i,j}}{Z_i} \begin{cases} e^{-\alpha_i} & H_i(x_j) = y_j \\ e^{\alpha_i} & H_i(x_j) \neq y_j \end{cases}, \quad Z_i \text{ is the normalizing factor makes}$$

$\sum d_{i+1,j} = 1$

12 end

13 The strong classifier H is produced by add $H_{1,2,\cdots T}$ up by $\alpha_{1,2,\cdots T}$,

$H_{final}(x_j) = a$ if

$\sum_i I(H_i(x_j) = a)\alpha_i = max\{\sum_i I(H_i(x_j) = a)\alpha_i | a = 1, 2 \cdots k\}$

Algorithm 1: AdaBoost

2.4 Several Classification Methods

In our research, we take three classification methods, support vector machines (SVM), decision trees, and neural networks (NN), as weak classifier to complete AdaBoost's learning, respectively. These methods have the function to do linear classification, decision tree classifier, and nonlinear classification.

Linear SVM is used to classify the linear classifiable part of our data. As linear SVM can only divide the data into two parts, it needs to do some change while using linear SVM to do the classification work. We use stepwise classification method, solving this problem by classifying 1, 2, ..., $k - 1$ classes and k class, and then 1, 2, ..., $k - 2$ classes and $k - 1$ class, and so on.

2.4.1 Linear SVM

Linear support vector machine (linear SVM) is a supervised learning model, which classifies the linear classifiable part of our data [8].

The linear SVM uses two hyperplanes $w \cdot x - b = 1$ and $w \cdot x - b = -1$ to separate samples into two parts. And for any x_i, satisfying $(w \cdot x_i - b \leq -1) \vee (w \cdot x_i - b \geq 1)$.

The linear SVM algorithm aims at maximizing the distance between two hyperplanes. By using geometry, we have distance $= 2/|w|$. Thus, the algorithm essentially aims at minimizing $|w|$ [9].

2.4.2 Classification and Regression Tree

The classification and regression tree (CART) is used for our decision tree classification. Each inside node conducts a test on an attribute and has two branches. Each leaf node indicates a class. The samples go through the nodes and fall into exactly one leaf node. The nodes describe the rules which are used to classify the samples [10].

2.4.3 RBF Neural Networks

Radical basis function (RBF) neural networks can solve the nonlinear classification part of our research. RBF neural network is a feedforward neural network that contains the input layer, hidden layer, and output layer. A RBF network is determined by center of radial functions, variance of radial functions, and weights [11].

The training of RBF network maintains these parameters to minimize the error of output layer.

3 Empirical Study

In this paper, we apply the above-mentioned technique to proceed with a precision marketing classification study on a tobacco market of city A, which is located in Fujian Province, China. In the following text, we name it as market A.

3.1 Background

Known as the most important market in Fujian Province, marketing methods of market A include direct sale and consignment sale. However, no matter which method for a company to carry out, the more precise the prediction of the expected tobacco supply volume is, the more efficient the business operations will be.

In the following text, we classify tobacco retail outlets in city A by classifying their sales volume in order to analyze and predict the expected tobacco supply volume of each retail outlet.

The research is based on a database contains most tobacco retail outlets of A, which includes more than 10 thousands samples and upwards of 50 attributes.[1]

3.2 Data Processing

We select five attributes which are mostly related to sales ability among 67 attributes: (1) historical total supply volume, (2) historical total order volume, (3) historical count of making orders, (4) area of counter, and (5) number of employees.

While these attributes are chosen, other attributes such as license number, company code, and name of business corporation are abandoned. And then we normalize these five attributes by mapping (dividing the numbers by their maximum values) vectors from R^5 (which represent attributes of each sample) to $[0, 1]^5$.

It is obvious to find that attribute 1 (historical total supply volume), attribute 2 (historical total order volume), and attribute 3 (historical count of making orders) are of similar index, which reflect the historical sales ability, while area of counter and number of employees are of "hardware" index, which reflect potential sales ability. Here, we use entropy weight method [12, 13] to reduce historical total supply volume, historical total order volume, and historical count of making orders to one index and area of counter and number of employees to another index. And we use these two indexes to proceed with a cluster analysis (however, we still use five attributes to proceed with an AdaBoost classification).

By applying entropy weight method, sample attributes are weighted in accordance with the amount of information of the sample properties and make weight of attributes with a great amount of information. Let sample be x_i, the attribute of sample be x_{ij},

$$p_{ij} = \frac{x_{ij}}{\sum_{i=1}^{n} x_{ij}} \tag{4}$$

entropy will be

---

[1]. Please contact me for the data if anybody wants to continue research on this problem.

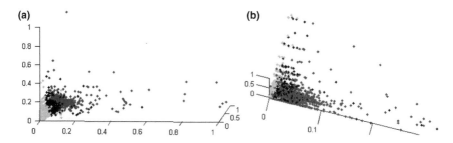

Fig. 1 The result of cluster analysis. **a** The distribution of attributes 1, 3, and 4. **b** The result of cluster analysis 2, 4, and 5

$$e_j = -\frac{1}{\ln(n)} \sum_{i=1}^{n} p_{ij} \ln(p_{ij}) \tag{5}$$

so weight of entropy will be

$$w_j = \frac{1 - e_j}{\sum_k 1 - e_k} \tag{6}$$

Attributes are weighted in accordance with these weights so that we can reduce attributes by mapping vectors from $[0, 1]^5$ (which represent attributes) to $[0, 1]^2$. The economic meanings of these two dimensions are historical sales ability and sales conditions.

3.3 Cluster Analysis

Samples are classified into three categories by using k-means algorithm to make a cluster analysis.

Diagrams show distinction of the data. As there are five attributes for the data, so we can only select three attributes to show each time and plot them in the three-dimensional Cartesian coordinate system. We map values of attribute 1 to x-axis, attribute 3 to y-axis, and attribute 4 to z-axis in Fig. 1a and attribute 2 to x-axis, attribute 4 to y-axis, and attribute 5 to z-axis in Fig. 1b. Clusters are separated by different colors.

Samples are well separated into three categories as shown in Fig. 1.

3.4 Classification through AdaBoost

First, we generate several classifiers which are required in AdaBoost learning including classification and regression tree, RBF neural network, and linear SVM

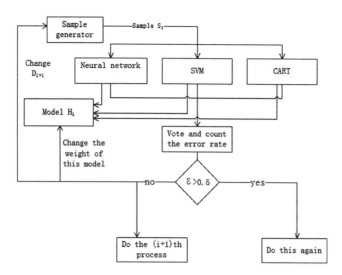

Fig. 2 The *i*th process

(the source code is from the SVM_light, an open source project [14]. Classification and regression tree is compiled as "CART," RBF neural network is compiled as "RBF," and linear SVM is compiled as "SVM." A program that generates the samples according to the distribution of weight is also included and compiled as "sample_generator." A program that calculates the output by voting each sub-output of three classifiers is compiled as "vote."

For loop *i*, "sample_generator" generates a set of data in accordance with D_i. And then, the algorithm calls "SVM," "CART," and "RBF". Each algorithm yields two files as the outputs SVM_data_i.model, SVM_data_i.out, RBF_data_i.model, RBF_data_i.out, CART_data_i.model, CART_data_i.out. The files with suffix ".model" represent classifiers and with ".output" represent the performance of each. Program "vote" takes an equal-weighted vote on SVM_data_i.out, RBF_data_i.out, and CART_data_i.out, which will generate a comprehensive result. The next step is checking the error rate of this classifier generated by "vote"; if error rate is less than 0.5, then we accept this loop of learning. Thereafter, "vote" gives feedback to "sample_generator" and sets the weight for classifiers through the *i*th study. Procedures above are shown in Fig. 2.

We take 11,000 samples out from 11,737 samples for AdaBoost ensemble learning and 737 samples as test samples. Each time we take 5,500 (expected) samples for learning. We conduct a total of 10 times learning. The accuracies[2] of each are shown in Table 1.

---

[2] Here accuracy rate is defined as 1- error rate. As the sum of d_{ij} is less than than 100 %, the accuracy is slightly larger than the actual accuracy

Table 1 Accuracy

1	2	3	4	5	6	7	8	9	10
99.18 %	99.18 %	99.98 %	98.69 %	100 %	98.96 %	99.87 %	99.81 %	99.83 %	99.04 %

Table 2 Weights of the weak classifiers

1	2	3	4	5	6	7	8	9	10
2.7463	2.7463	4.6051	2.5109	4.6051	2.6269	3.6689	3.4790	3.5347	2.6672

We treat zero error rate as 0.001 % in case of zero division error, which will affect the weighted result. The weights of ten weak classifiers obtained by calculation are shown in Table 2.

At last we classify 11,000 samples used for AdaBoost learning, and the accuracy is 100 %. And then, we classify 737 samples not involved in AdaBoost learning and the accuracy is 99.86 %.The results indicate that AdaBoost integrated learning has a good performance on this study.

4 Conclusion

Actually, our research proposes a framework to solve similar problem, that is, the classification problem with no reference-studying samples and expecting to get a classification rule which will not change over the enlargement of data set. This framework works by taking cluster analysis first, then using the result as reference-studying samples for classifier to learn. Adaboost helps a lot to improve the performance. Empirical study indicates that this method gets a very good result. However, there are many aspects that can be improved under the framework of this paper, and the clustering method and the weak classifier can be changed to fit specific cases.

Acknowledgments We are very grateful to our project mentor Prof. Defu Zhang for his great support on algorithms in the data mining field. This work has been partially supported by the National University Student Innovation Program of China and the National Natural Science Foundation of China (Grant No. 61272003).

References

1. Hruschkaa, H., Natter, M.: Comparing performance of feedforward neural nets and k-means for cluster-based market segmentation. Eur. J. Oper. Res. **114**, 346–353 (1999)
2. Huanga, J.J., Tzeng, G.H., Onga, C.S.: Marketing segmentation using support vector clustering. Expert Syst. Appl. **32**, 313–317 (2007)
3. Zhang, D.F., Chen, Q.S., Wei, L.J.: Building behavior scoring model using genetic algorithm and support vector machines. In: Lecture Notes in Computer Science, vol. 4488, pp. 482–485 (2007)

4. Zhang, D., Leung, S.C.H., Ye, Z.: A decision tree scoring model based on genetic algorithm and k-means algorithm. In: Third International Conference on Convergence and Hybrid Information Technology, vol. 1, pp. 1043–1047 (2008)
5. Zhang, D., Zhou, X., Leung, S.C.H., Zheng, J.: Vertical bagging decision trees model for credit scoring. Expert Syst. Appl. 37(12), 7838–7843 (2010)
6. Zhang, D., Huang, H., Chen, Q., Jiang, Y.: A comparison study of credit scoring models. In: Third International Conference on Natural Computation, vol. 1, pp. 15–18 (2007)
7. Freund, Y., Schapire, R.E.: A decision-theoretic generalization of on-line learning and an application to boosting. In: CiteSeerX: 10.1.1.56.9855 (1995)
8. Tan, P., Steinbach, M., Kumar, V.: Introduction to Data Mining. Addison Wesley Press, Reading (2006)
9. Cortes, C., Vapnik, V.: Support-vector networks. Mach. Learning 20, 273–297 (1995)
10. Breiman, L., Friedman, J.H., Olshen, R.A., Stone, C.J.: Classification and Regression Trees. Wadsworth and Brooks/Cole Advanced Books and Software, Monterey (1984)
11. Broomhead, D.S., Lowe, D.: Radial basis functions, multi-variable functional interpolation and adaptive networks. Technical report, p. 4148 (1988)
12. Aczel, J., Daroczy, Z.: On Measures of Information and their Characterizations. Academic Press, New York (1975)
13. Lin, J.: Divergence measures based on the Shannon entropy. IEEE Trans. Inf. Theor. 37(1), 145–151 (1991)
14. Joachims, T: SVM light-support vector machine, http://svmlight.joachims.org/ (2012). Accessed 10 Aug 2012

Extensions of Statecharts with Time of Transition, Time Delay of Message Transmitting, and Arrival Probability of Message

Junqiao Li, Jun Tang and Shuang Wan

Abstract In this paper, we focus on new extensions to the current statecharts to support performance modeling of complex reactive systems. The extensions include time description of transition, time delay description of message transmitting, arrival probability (reliability) description of message. The new extensions not only need to extend the syntaxes of current statecharts, but also need to modify the current semantics to support the executability. We describe the new extended statecharts at syntactic level and discuss them at semantic level.

Keywords Statecharts · Performance modeling · Timing description · Arrival probability

1 Introduction

Statecharts, as a graphics-oriented modeling language, pioneered by David Harel in 1987, are used to specify the dynamic behavior of real-time reactive systems. For the features of hierarchy, concurrency, and broadcast mechanism, modeling method of statecharts is widely used in the system modeling, and has been incorporated into the fundamental behavior views of UML since version 1.0.

J. Li (✉) · J. Tang · S. Wan
Department of Electronic Engineering, Tsinghua University,
Room 10105 Rohm Building, Beijing, China
e-mail: li-jq11@mails.tsinghua.edu.cn

J. Tang
e-mail: tangj_ee@mail.tsing-hua.edu.cn

S. Wan
e-mail: wanshuang@mail.tsinghua.edu.cn

S. Patnaik and X. Li (eds.), *Proceedings of International Conference on Soft Computing Techniques and Engineering Application*, Advances in Intelligent Systems and Computing 250, DOI: 10.1007/978-81-322-1695-7_43, © Springer India 2014

The statecharts, which aim at specifying the logical behavior of reactive systems, have been widely used in software field from its outset. Meanwhile, people have been trying to make use of statecharts in system performance analysis. However, people quickly discovered their insufficiencies, especially in specifying the stochastic and timing features of actual systems. Naturally, extensions to the statecharts have been made in these two aspects.

There are two outstanding representatives of the researchers extending the statecharts; one is Vijaykumar, and the other one is Jansen. In [1–3], Vijaykumar first proposed the probability node to the statecharts to describe non-deterministic state transitions that widely exist in the actual systems and proposed the description method of stochastic events with exponential distribution that widely exist in the actual environment. In [4, 5], Jansen made formal extensions of real time, probabilistic choice, and stochastic timing to statecharts and proposed the corresponding method of model checking to the extended statecharts.

To further expand the application of statecharts in the performance analysis of actual systems, we propose several new extensions to the current extended statecharts. First, we propose to add time parameter to the actions of state transition to specify the executing time of action. Second, we propose to add time parameter to the triggered events to specify the delay time of event message transmitting. Finally, we propose to add probability parameter to the triggered events to specify the arrival probability of event message.

Remaining of the paper is organized as follows. Section 2 briefly introduces the current statecharts, including original statecharts and extended statecharts. The semantics is also introduced. Section 3 introduces our new extensions to current statecharts. Section 4 discusses the executable semantics of the new extended statecharts. Finally, conclusions are drawn, and directions of future work are elucidated.

2 Current Statecharts

2.1 Basic Statecharts [6]

For clarity, take Fig. 1 (supplied by [7]) as an example of conventional statecharts. From a top–down view of the statechart in Fig. 1, state $s0$, as a root, is refined into three substates, $s1$, $s2$, and $s3$. State $s1$ is a concurrent state that contains two concurrent substates, $s4$ and $s5$. State $s4$ is further refined into two states, namely $s6$ and $s7$. State $s5$ is further refined into two states, namely $s8$ and $s9$. States like $s0$, $s1$, $s4$, and $s5$ are called composite states, and states like $s2$, and $s3$ are called basic states.

A transition in a statechart is labeled by "$m[c]/a$," where m is the **message** triggering the transition, c is a **condition** that guards the transition from being enabled unless it is true when m occurs, and a is an **action** that is carried out when the transition is taken. It is noted that all of these parts are optional.

Fig. 1 An example of
conventional statecharts

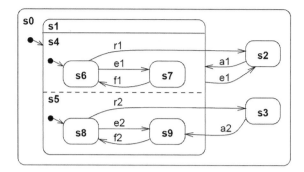

Fig. 2 An example of
stochastic statecharts

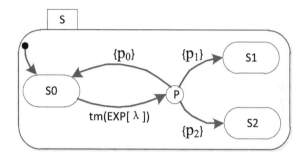

The active configuration is a maximal set of states the system can be in simultaneously. An example of active configuration of the statecharts in Fig. 1 is: $\{s0, s1, s4, s5, s6, s8\}$, which can be simplified as $\{s6, s8\}$.

Due to the nature of randomness, real-reactive systems for which performance is to be analyzed are, in general, represented by stochastic models. So, extension of randomness has to be made to the statecharts [1–5]. The first extension is to include stochastic description to the timing and external stochastic events. The second extension is to include probabilities to the non-deterministic internal state transitions.

Figure 2 is an example of stochastic statecharts. Assuming the system is initially in state $S0$, when the external stochastic event $tm(EXP[\lambda])$ is triggered, the possible destinations of transition are $S1$, $S2$, and $S0$ (self-transition), whose probability is specified by the discrete probability distribution $P = \{p_0, p_1, p_2\}$, such that (a) $0 \leq p_i \leq 1$, $\forall i = 0, 1, 2$; and (b) $p_0 + p_1 + p_2 = 1$.

2.2 Basic Semantics of UML Statecharts

There is a large number of semantics for statecharts, each adapting to a specific situation. An overview of some can be seen in [8, 9]. In this paper, we only discuss the popular UML statecharts semantics [7–12], which is object-oriented and

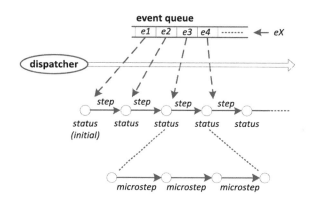

Fig. 3 Basic semantics of UML statecharts

implementation-level, opposed to the requirement-level statecharts semantics in STATEMATE [13–15].

Figure 3 illustrates the executable semantics of UML statecharts. In UML statecharts, there is an event queue for all the triggered events, whether internal or external. A dispatcher is to select the top event in the event queue and broadcast it when the system is in a stable status after responding to the previous event by RTC principle. So, the events in the queue are processed by first-in-first-out (FIFO) principle, and only one event can be selected at one time. It is worth noting that a step may take time not zero. The time of step depends on the executing time of its corresponding actions, including exit/entry actions and actions triggered by transitions.

A conflict may happen when several transitions are enabled simultaneously. Priority is used to resolve the conflicts. Simply speaking, a transition with a higher-level source state has priority to the other transitions. Take Fig. 1 as an example. Assuming that the system's active configuration is $\{s0, s1, s4, s5, s6, s8\}$, and event $e1$ is triggered, the transition from $s6$ to $s7$ and the transition from $s1$ to $s2$ are both matching and are in conflict. According to the above priority rule, transition from $s6$ to $s7$ is fired, and transition from $s1$ to $s2$ is neglected.

3 New Extended Statecharts

3.1 Time of Transition

In the current UML statecharts semantics, a state transition may take time not zero, which depends on the total executing time of actions. However, the executing time of action seems not be specified but be dependent on the computer power or the coding efficiency. Apparently, this uncontrollable method is not suitable for objective and accurate analysis of system performance, especially when the system is time-sensitive. So, we propose to add time parameter to the time-consuming

Fig. 4 Example of a
statechart with parametric
time of actions

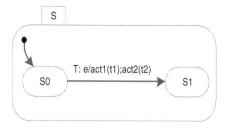

actions. In the simulation time mode, when the action is executed, the system time
will advance by the time parameter.

Figure 4 is an example of the newly extended statecharts. Here, we assume
states $S0$ and $S1$ have no entry/exit actions. Transition T is accompanied by two
actions of $act1$ and $act2$, whose consuming time is, respectively, $t1$ and $t2$, so the
transition T takes $t1 + t2$ time units.

3.2 Time Delay of Message Transmitting

Communication and network have penetrated into many complex systems. Though
statecharts take broadcast mechanism for event message transmitting, there is no
parametric description of the characteristics of such a broadcast mechanism.
Instead, the ideal hypothesis of infinite bandwidth and infinite speed is made. This
hypothesis may be rational in such systems whose time delay of message transmitting is much shorter than the response time of message, so that the time delay
of message transmitting can be ignored. However, in most cases, the time delay of
message transmitting cannot be ignored, as the time delay is closely related to the
systems' performance.

It is difficult and even impossible to describe all the features of the communication network in the statecharts. But, it is possible for us to abstract and simplify certain critical features, such as time delay of message transmitting. Our new
method is shown in Fig. 5, where the broadcast mechanism between objects is
changed to the peer communication mechanism and the broadcast mechanism is
only taken within certain area. So, in Fig. 5, delay time is $t1$ for transmitting event
f from $S0$ to $S1$, and $t2$ for transmitting event f from $S0$ to $S2$.

3.3 Arrival Probability of Message

In practical systems, the transmission of message is not always reliable. Just as in
the communication system, transmission reliability is described by bit error rate; in
statecharts, using probability to describe the reachability of the event message is

Fig. 5 Method to describe the time delay and arrival probability of event message

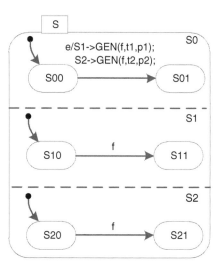

reasonable. The probability extension is based on the extension of 3.2, which just adds the probability parameter to the triggered event. As shown in Fig. 5, $p1$ and $p2$ are, respectively, the arrival probabilities of event f from $S0$ to $S1$ and $S2$.

4 Executable Semantics of New Extended Statecharts

To make our newly extended statecharts executable, the simulated mode, under which time advances according to the scheduled tasks, is taken. Our executable semantics is compatible with the UML statecharts semantics and takes extensions in the following aspects:

- **Time control of transition**: The events and actions are both treated as tasks, and the system time will advance to the next task time only when the current task is done. So, in Fig. 4, the system time advances to $t1$ when $act1$ is done and advances to $t1 + t2$ when $act2$ is done. So, consumption time of transition is controlled to $t1 + t2$.
- **Time delay control of message transmitting**: As shown in Fig. 6, the events with time parameter in the queue are like being deferred. The event in the queue will be dispatched only if its time parameter is the smallest one and the current task has been done. If several events' time parameter is the same value, the dispatcher will select the event by FIFO principle. Accordingly, the system time will advance by the selected event time.
- **Arrival probability control of message**: As shown in Fig. 6, when the current task is done, event with the smallest time parameter will be selected first, then a number will be generated by the Boolean random number

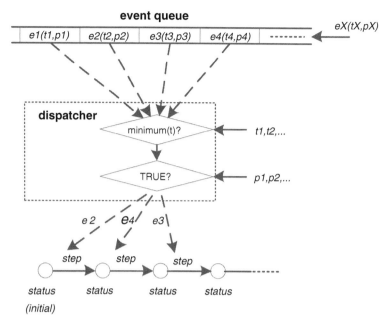

Fig. 6 Executable semantics of the new extended statecharts

generator with its probability parameter p. When the generated number is TRUE, the event will be dispatched successfully, or the event will be ignored. It is worth noting that this probability processing mechanism is quite similar to the processing mechanism of probabilistic transition in the stochastic statecharts.

5 Conclusions and Future Work

To expand the application of statecharts in actual system performance modeling, many extensions have been made to the statecharts proposed by David Harel. In this paper, we further extend the statecharts with time of transition, time delay of message transmitting, and arrival probability of message at the levels of syntax and executable semantics, respectively.

Future research directions should be directed to:

- Make extension of communication bandwidth to the statecharts. As we all know, the time delay and bandwidth are two key parameters of the communication-related systems.
- Apply our newly extended statecharts to system modeling and simulation based on which performance can be analyzed.

References

1. Vijaykumar, N.L.: Statecharts: Their use in specifying and dealing with performance models. Ph.D. thesis, Informatics and IEC-ITA and São José dos Campos, Brazil (1999)
2. Vijaykumar, N.L., Carvalho, S.V., Abdurahiman, V.: On proposing statecharts to specify performance models. Int. Trans. Oper. Res. (ITOR) 9(3), 321–336 (2002)
3. Vijaykumar, N.L., Carvalho, S.V., Andrade, V.M.B.: Introducing probabilities in statecharts to specify reactive systems for performance analysis. Comput. Oper. Res. 33(8), 2369–2386 (2006)
4. Jansen, D.N., Hermanns, H., Katoen, J.P.: A Probabilistic Extension of UML Statecharts: Specification and Verification. Formal Techniques in Real-Time and Fault-Tolerant Systems, vol. 2469, 7th edn, pp. 355–374. Springer, Berlin (2002)
5. Jansen, D.N.: Extensions of Statecharts with Probability, Time, and Stochastic Timing. Ph.D. thesis, Twente University, Bern, Switzerland (2003)
6. Harel, D.: Statecharts: a visual formalism for complex systems. Sci. Comput. Program 8(3), 231–274 (1987)
7. Latella, D., Majzik, I., Massink, M.: Towards a formal operational semantics of UML statechart diagrams. Proceedings of the IFIP TC6/WG6.1 Third International Conference on Formal Methods for Open Object-Based Distributed Systems (FMOODS), pp. 465–481. Deventer, The Netherlands (1999)
8. von der Beeck, M.: A comparison of statecharts variants. In: Formal Techniques in Real-Time and Fault-Tolerant Systems: Third International Symposium, vol. 863, pp. 128–148. Springer (1994)
9. Crane, M.L., Dingel, J.: UML vs. classical vs. Rhapsody statecharts: not all models are created equal. Proceedings of the 8th International Conference on Model Driven Engineering Languages and Systems, pp. 2–7, Oct 2005
10. Börger, E., Cavarra, R., Riccobene, E.: Modeling the dynamics of UML state machines. In: Abstract State Machines, Theory and Applications, International Workshop, ASM Proceedings 2000, vol. 1912, pp. 223–241. Springer, Berlin (2000)
11. Jürjens, J.: A UML statecharts semantics with message-passing. Proceedings of the 2002 ACM Symposium on Applied computing, pp. 1009–1013 (2002)
12. Harel, D., Kugler, H.: The RHAPSODY semantics of statecharts (or, on the executable core of the UML), vol. 3147, pp. 325–354. Integration of Software Specification Techniques for Application in Engineering, Lecture Notes in Computer Science. Springer, Berlin (2004)
13. Harel, D., Pnueli, A., Schmidt, J.P., Sherman, R.: On the formal semantics of statecharts, vol. 2, pp. 54–64. Annual IEEE Symposium on Logic in Computer Science, Ithaca, USA (1987)
14. Harel, D., Naamad, A.: The STATEMATE semantics of statecharts. ACM Trans. Softw. Eng. Methodol. 5(5), 293–333 (1996)
15. Harel, D., Gery, E.: Executable object modeling with statecharts. IEEE Comput. 30(7), 31–42 (1996)

The Optimization of Hadoop Scheduling Algorithms on Distributed System for Processing Traffic Information

Weizhen Sun and Xiujin Wang

Abstract Traffic information retrieval and data mining are not only the hotspots and key techniques in the intelligent transportation, but also the research issue of massive data's distributed processing. With the development of urban traffic acquisition technology, the traffic data have increased to PB level. In order to manage these traffic data effectively and serve for intelligent transportation, we need to use efficient algorithm to process them in the distributed environment. In a distributed platform, this paper optimizes the Hadoop schedule algorithm that is used in processing traffic data and makes up the shortcomings of real-time traditional algorithms. The results of experiments show that the optimized scheduling algorithm used in a distributed environment, whether it is compute-intensive or I/O-intensive, has the most minimum calculation time, the best performance, better capacity of processing the traffic data, and better real time.

Keywords Hadoop · Mapreduce · Intelligent transportation · Scheduling optimization

1 Introduction

With the development of traffic information collection technology, the urban traffic data grow rapidly. After many year accumulations of traffic data from Olympic Games, Beijing Fair, Fair Park, and other large events and the collection of data

W. Sun (✉) · X. Wang
College of Information Engineering, Capital Normal University,
105 West Third Ring Road, Beijing, China
e-mail: sunweizhen@cnu.edu.cn

X. Wang
e-mail: wxj839437532@163.com

S. Patnaik and X. Li (eds.), *Proceedings of International Conference on Soft Computing Techniques and Engineering Application*, Advances in Intelligent Systems and Computing 250, DOI: 10.1007/978-81-322-1695-7_44, © Springer India 2014

from Taxi industry and special industry floating car, the traffic data are large, complex and can be over PB level in magnitude. Thus, data mining and real-time processing of these traffic data must use distributed computing platform [1]. Distributed scheduling algorithm is the key to coordinate the work of each node on distributed computing platforms. Among them, Hadoop is not only open source but also suitable for deploying all kinds of resources and data on low-cost machines [2], which is in distributed storage and distributed management. Since Hadoop platform released, it has obtained wide application and promotion. Hadoop is a distributed infrastructure, which is composed by Hadoop distributed file system (HDFS) and MapReduce calculation mode model. HDFS is mainly responsible for massive data stream storage, and MapReduce is mainly responsible for the massive amounts of data on a distributed computing.

Urban intelligent transportation systems (ITS) need to deploy massive traffic information on processing platform rapidly and to schedule tasks, providing the effective traffic information and meeting the real-time requirements for users. Hadoop is a massive data parallel processing system run [3], which runs on a large cluster and schedules thousands of tasks. To efficiently enable the system to complete jobs submitted by users, we need to optimize the schedule algorithm of the Hadoop platform, so that we will obtain better operation efficiency [4]. In this paper, based on the performance of the Hadoop open-source framework and intelligent traffic data-processing algorithm and features for urban traffic data, we propose an Improved Capacity Scheduling algorithm, which effectively improves the massive data-processing efficiency in intelligent transportation.

2 Distributed Scheduling Algorithm

In distributed systems, job scheduling, computing power balance, and efficiency of every node are the key indicators to weigh the pros and cons of an algorithm. Capacity scheduling algorithm is developed by Yahoo, which is to simulate the independent computing capacity with the specified cluster resources for different user's job to provide services. The scheduling algorithm is divided into two levels: The top one is scheduling among queues, and the below one is scheduling in a queue [5, 6]. Capacity scheduling algorithm does not take full account of the diversity of users, scheduling strategy is still FIFO in the single queue, real time, and urgency of traffic information that decides the strong interaction in the algorithm when processing traffic information. What is more the real time of an algorithm is closely related with the proportion of local calculation.

For the particularity of traffic information data, scheduling queue should have better matching strategy in allocating tasks [5, 7]. More complex the type of task is, the stronger dependency among the tasks have; it is also bound to increase the delay in calculation [8]. Thus, it is necessary to design scheduling algorithm in a distributed system that is more suitable for the characteristics of urban traffic to improve resource utilization and computational efficiency of platform.

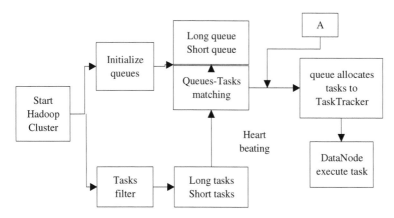

Fig. 1 Matching between queues and tasks

3 Improved Scheduling Algorithm

According to the above research on distributed job scheduling algorithms and analysis of urban traffic information status, we add some strategies on the original algorithm. Such as FIFO and short operating strategy, emergency steals strategy, matching job and queue strategy and delay-scheduling strategy. The improved algorithm inherits good stability, scalability, and parallelism superiority from classical Capacity scheduling algorithm with multi-queue job scheduling model.

Capacity scheduling algorithm supports multiple queues to ensure its computing capacity. Taking the diversity of users that asks for service from transport platform into account, the queues would be divided into long queue and short queue. The long queue handles long task, as well as short queue. The approach for processing tasks is targeted strongly and can improve handling efficiency. In order to classify the queue, additional job filter is necessary, which is responsible for the job's division and classification. Figure 1 is the matching between queue and tasks.

In the initialization, the queues are divided into long-task queue and short-task queue on the demands. Job scheduler allocates different tasks to different queues and optimizes configurations of queues' property at the same time. For example, short queue for the short tasks owns a smaller scale, but the quantity is large and the type is complex, so that minMap and minReduce can be set up bigger. The majority of traffic data own such character, so majority resource was allocated for the short tasks. By the improvement, the execution of different types of tasks becomes more ordering and more efficient and acquires better interactivity.

Within the scheduling queue, FIFO algorithm is taken to choose tasks by default, so adding the short operating strategy in FIFO algorithm means that performing the FIFO algorithm first and then sorting the tasks by short operating

Fig. 2 Added short
operating strategy

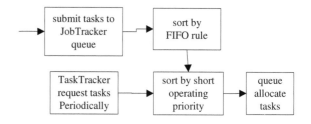

strategy, increasing system throughput. Figure 2 will be added to the image above at dot A.

Within a single job queue, if a user submits a job in a variety of different types, such as strong data-dependent task and weakly dependent task. According to a scheduling algorithm (e.g., short operating strategy), small tasks often appear on different nodes of the same queue. Then, the strategy will cause a large network data congestion. So, the smart way is to join delay-scheduling strategy and then put the above three strategies together:

```
Initial the queue of tasks JobList
if node n send heart beat to master then
if node n release the free pool then
for task J=JobList.get(i),0=<i<JobList.size(),i++
if a node n who has a local task on task J then
execute the task now
elseif network load>=M then
call delay function delaysometime()
if task J.wait<T then
execute long task or strong data-dependent task or strong
local task
else
stop delay function
endif
endif
endfor
endif
endif
```

In this scheduling strategy, if the network load exceeds a certain value M, it will automatically trigger delay-scheduling function and then during the delay-scheduling computer, resources will turn to long tasks, or strong data-dependent operations, or strong local jobs. Meanwhile, scheduled delay time has a maximum value T, and it can be changed according to network load conditions adjusted in real time. Delay strategy is favorable for strengthening data locality and reducing pressure on the network data transmission between nodes and then improving system throughput.

Fig. 3 Improved state
transition diagram

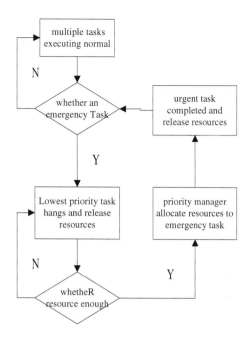

There should have very strong interaction in algorithm, which processes traffic information in real time and urgently. As for scheduling algorithm, how to choose the next appropriate task run is the core of the problem. All queues are initialized to a predetermined amount of resources, if the queue is idle; its resources is to be allocated to the busiest queue. This is the characteristics that Capacity Scheduling algorithm has. But, if all the queues in the queue are in the case of no idle resources, then there is also a task need to be executed urgently. Low-priority task would not be suspended until the urgency task completed, releasing the resource and according to real-time priority to judge whether to continue execute-suspended task. Figure 3 is the improved state transition diagram.

4 Implementation and Experiments

The object of the improved algorithm is still queue, job, and task. The three kinds of granularity, which maintains some of tasks information by itself. The improved algorithm is based on the Capacity Scheduling algorithm, so that it basically maintains the structure of the original scheduling algorithm. Code is mainly composed of the five java files, and Fig. 4 is the improved algorithm structure class diagram.

As it can be seen from the figure above, the core class of algorithm is Ecapacity Task Scheduler, which associated with core class are file configuration class Ecapacity Scheduler Conf, initialized class EJobInitialzation Poller, queue

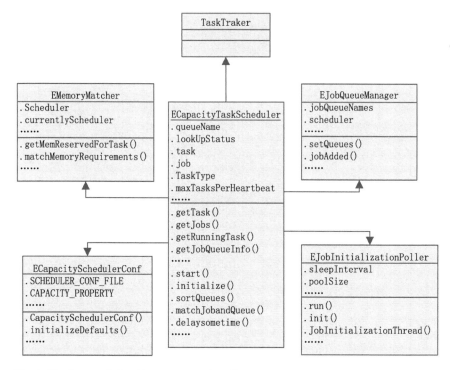

Fig. 4 The improved algorithm structure class diagram

management class EJob Queue Manager, and memory capacity matching class EMemory Matcher. Major point of the improvement is adding some variables and functions, such as setting and acquiring priority functions (set Job Priority ()/get Job Priority ()), judges short-task functions is Short Job(), Rewritten sort Queue() method, etc.,. In addition, do some parameter configuration and modification for file setting in the configuration file conf, including the maximum and minimum number of map and reduce, percentage of queue and its allocated capacity, maximum load capacity etc.,.

The whole cluster consists of four computers PC, three data nodes, and a master node. In order to verify the effectiveness of the improved algorithm, this experiment includes two type of test procedure: I/O-intensive and compute-intensive, which individually counts the average running time of different algorithm and different data size tasks in multiuser and multitask submit mode.

For I/O-intensive test, we use the word frequency statistics WordCount benchmark sample, because the data structure of distributed discrete in intelligent transportation is similar to the word frequency statistics. Figure 5 shows the results.

Figure 5 shows that, under the condition of multiuser and multitask, the effect on FIFO algorithm is significantly smaller than the other three kinds of algorithms. So, choosing the FIFO algorithm is unwise. What is more with the increment in

Fig. 5 Wordcount
benchmark sample

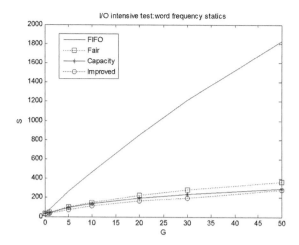

Fig. 6 Compute-intensive
test

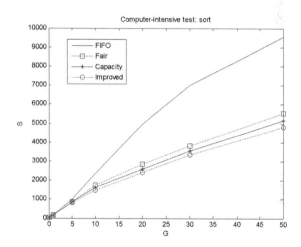

scale of data, Capacity Scheduling algorithm and the improved algorithm will be selected, and improved algorithm is a little better than original algorithm.

For compute-intensive test, the data are a week of GPS information from a taxi company, extracting the route and the average velocity. We have tested the data-processing performance with the classical and improved algorithm. The result is shown in Fig. 6.

From the figure above, if the data is below 5G, the execution time of the four algorithms is not far off. When the amount of data exceeds 5G, the improved algorithm is the best one and the FIFO algorithm is the worst one.

5 Conclusions

According to the characteristics of urban traffic data, traditional scheduling algorithm and improved Capacity Schedule algorithm have been optimized and implemented, making up the shortcomings of original algorithm and making the improved algorithm adapt to urban intelligent transportation data processing very well. Experiments show that improved scheduling algorithm has certain advantages either from the interaction, or from the data-processing performance, on processing distributed large urban traffic data. Urban transport will be one of the future focuses of the study of huge amount data. The next step will be to use the algorithm of urban traffic data for a deeper treatment and explore more practical applications to meet the needs of users.

References

1. Liu, X., Lu, F., Zhang, H., et al.: Estimating Beijing's travel delays at intersections with floating car data. In: Proceedings of the 5th ACM SIGSPATIAL International Workshop on Computational Transportation Science, ACM, pp. 14–19 (2012)
2. Xu, X., Wu, J., Yang, G.: Mass data processing system based on large scale low cost computing platform. Appl. Res. Comput. **29**(2), 049 (2012)
3. Edwards, M., Rambani, A., Zhu, Y., et al.: Design of Hadoop-based framework for analytics of large synchrophasor datasets. Procedia. Comput. Sci. **12**, 254–258 (2012)
4. Fischer, M.J., Su, X., Yin, Y.: Assigning tasks for efficiency in Hadoop. In: Proceedings of the 22nd ACM Symposium on Parallelism in Algorithms and Architectures, ACM, pp. 30–39 (2010)
5. Deng, C., Fan, T., Gao, F.: Resource scheduler algorithm based on statistical optimization under Hadoop. Appl. Res. Comput. **30**(2), 417–419 (2013)
6. Wang, F.: Scheduling algorithm of Hadoop cluster jobs. Programmer **12**, 1–19 (2009)
7. Dean, J., Ghemawat, S.: Mapreduce: simplified data processing on large clusters. Commun. ACM **51**(1), 107–113 (2008)
8. You, H.H., Yang, C.C., Huang, J.L.: A load-aware scheduler for MapReduce framework in heterogeneous cloud environments. In: Proceedings of the ACM Symposium on Applied Computing, ACM, pp. 127–132 (2011)

Understanding the Capacity Scaling of Personal Communications Services

Zheng Wang

Abstract Mobility management consumes a significant portion of resources for personal communications services (PCS) network. Modeling this overhead of mobility management and evaluating how it scales with the service area and user density is the key to planning the PCS systems. This study provides the model of capacity scaling of PCS networks in the context of a constant user arrival rate and presents numerical results and analysis on how the capacity of PCS networks scales with the size of the service area or the number of cells.

Keywords Capacity scaling · Random walk · Personal communications services network

1 Introduction

Personal communications services (PCS) network provides communication services to mobile users via their mobile stations (MS) or handset within its service area. The radio coverage of a base station (BS) is called a cell. The service area of PCS networks usually consists of many cells, which are configured as a hexagonal network given in Fig. 1. In our performance study of PCS networks, MS moves from one cell to another by a random walk until it leaves the service area after sufficient steps. Akyildiz et al. proposed a two-dimensional random walk model to reduce the computational complexity of state transition [1]. There were some

Z. Wang (✉)
Computer Network Information Center, Chinese Academy of Sciences, Beijing 100190, China
e-mail: wangzheng@conac.cn

Z. Wang
China Organizational Name Administration Center, Beijing 100028, China

S. Patnaik and X. Li (eds.), *Proceedings of International Conference on Soft Computing Techniques and Engineering Application*, Advances in Intelligent Systems and Computing 250, DOI: 10.1007/978-81-322-1695-7_45, © Springer India 2014

Fig. 1 An illustration of a
5-subarea service area for
PCS networks

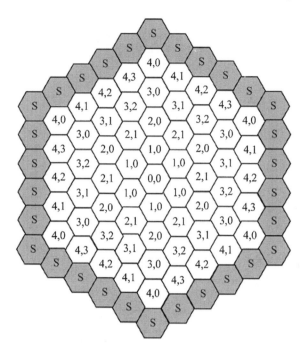

efforts on designing mobile tracking scheme [2], dynamical adaptive registration
areas [3], and minimum bandwidth location management [4, 5] for efficient
mobility management.

A central problem in PCS systems is to optimize resource usage, while
providing quality-of-service (QoS) guarantees to mobile users. Network mobility
management consumes a significant portion of resources, which is a necessary
overhead for supporting mobile users. We focus our efforts on modeling this
overhead of mobility management.

Based on the random walk model of user mobility, our study models the
capacity scaling of PCS networks in the context of a constant user arrival rate. It
also provides numerical results and analysis on how the capacity of PCS networks
scales with the size of the service area or the number of cells.

2 Residence Steps of MS Under Random Walk

The capacity of PCS networks is dependent on the residence steps of MS in the
service area. The more steps MS takes to leave the service area, the more resources
of PCS networks are required for the mobility management and communication
service of the MN.

The capacity of PCS networks is dependent on the residence steps of MS in the service area. The more steps MS takes to leave the service area, the more resources of PCS networks are required for the mobility management and communication service of the MN.

To derive the movement states of MN, each cell in the service area of PCS networks is marked as (x, y) where x represents the hop count distance from the center cell and $y \geq 0$ represents the $y + 1$ st type of the cell illustrated in Fig. 1. Some cells with identical (x, y) label are considered equivalent because they have the identical set of neighbors in the clockwise direction. For example, cells labeled with $(3, 0)$ has common neighbors of $(2, 0)$, $(3, 2)$, $(4, 3)$, $(4, 0)$, $(4, 1)$, and $(3, 1)$ in the clockwise direction. Specifically, state S is an absorbing state indicating that MN has left the service area of the PCS network. Refer to the classification algorithm proposed by Akyildiz et al. to label the cells [1].

Based on the cell classification, we can write the transition matrix. The elements in each column and row in the transition matrix are listed in the following (x, y) order: $(0, 0)$, $(1, 0)$, $(2, 0)$, $(2, 1)$, $(3, 0)$, $(3, 1)$, ..., $(x - 1, 0)$, $(x - 1, 1)$, ..., $(x - 1, x - 2)$, $(x, 0)$, $(x, 1)$, ..., (x, y), ..., S. The row number of state (x, y) (where $1 \leq x \leq n - 1$) state can be written as

$$R_{(x,y)} = 1 + 1 + 2 + 3, \ldots, x - 1, y + 1$$
$$= \frac{(x - 1)x}{2} + y + 2 \tag{1}$$

Let $S(n)$ be the total number of states for a n-subarea random walk. Then, $S(1) = 2$ and if $n > 1$

$$S(n) = R_{(n-1,n-2)} + 1$$
$$= \frac{(n - 1)(n - 2)}{2} + n + 1 \tag{2}$$

The expected number of transitions $W_{(x,y)}$ made by a MN starting in state (x, y) before leaving can be derived based on the transition matrix. $W_{(x,y)}$ is also the residence steps of MS in the service area of PCS networks.

3 Capacity Scaling of PCS Networks

To simplify our analysis, we assume that the time axis is discretized as time slots with the same lengths. Each step taken by all MNs roaming in the service area happens at the beginning of every time slot. We consider a user arrival model that each cell of the PCS networks has the same user arrival rate. To integrate into the random walk model, the user arrival is also assumed to take place at the beginning of every time slot.

Fig. 2 The maximum
expected residence steps of
MN for a fixed user arrival
rate per cell

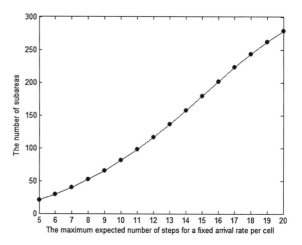

For a fixed user arrival rate per cell, we first illustrated how the maximum expected residence steps of MN scales with the number of subareas. Obviously, the (0, 0) cell always exhibits the longest distance from the boundary of the service area and therefore has the maximum expected residence steps. In other words, we have (3) for a n-subarea random walk ($n > 1$)

$$W_{(0,0)} = \arg \max_{\substack{1 \leq x \leq n-1 \\ 0 \leq y \leq x-1}} W_{(x,y)} \tag{3}$$

Figure 2 shows how $W(0, 0)$ scales with the number of subareas. We can see that the maximum expected residence steps of MN grow rapidly with the size of serve areas.

To evaluate the overall capacity requirements, we consider the PCS networks as a queuing system where MNs arrive at each cell of the network at the rate of λ and on average, MNs initially entering cell (x, y) stay in the system for $W(x, y)$ time units. By Little's formula, the expected number of MNs in the n-subarea PCS networks is given as

$$L_n = \left(W_{(0,0)} + 6W_{(1,0)} + 6W_{(2,0)} + \ldots, +6W_{(n-1,n-2)} \right)\lambda \tag{4}$$

Figure 3 shows how L_n scales with the number of subareas. We can see that the expected total residence steps of MN grows more rapidly with the size of serve areas compared with the maximum expected residence steps of MN in Fig. 2. This is due to fact that besides the growth of average residence steps of MN, the increase in the number of cells also contributes to the overall results. As the overall capacity requirements for PCS networks scale up dramatically with the service area diameter, a large PCS network should be depreciated in terms of efficiency consideration.

From another perspective, we then investigate the overall capacity requirements if the total arrival rate of MNs is fixed for PCS networks with different sizes. For this case, $W(x, y)$ remains the same as (4), while the MN arrival rate λ has an

Fig. 3 The expected total residence steps of MN for a fixed user arrival rate per cell

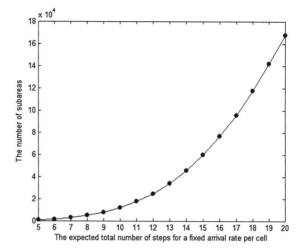

Fig. 4 The expected total residence steps of MN for a fixed overall arrival rate

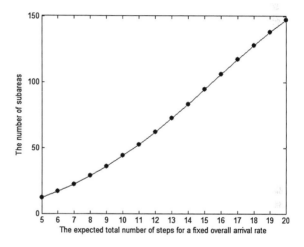

inverse relationship with the number of cells in n-subarea PCS networks. Thus, we can modify (4) as

$$L'_n = \frac{c \times \left(W_{(0,0)} + 6W_{(1,0)} + 6W_{(2,0)} + \ldots, + 6W_{(n-1,n-2)} \right) \lambda}{3n^2 - 3n + 1} \quad (5)$$

where c is a constant.

Figure 4 shows how the overall capacity requirements for a fixed overall arrival rate L'_n scales with the number of subareas. We can see that due to the diminish effect of the MN arrival rate per cell, the growth speed of the overall capacity requirements is significantly suppressed. However, another conclusion drawn from Fig. 4 is that densely distributed MNs require much lower resources than sparsely distributed ones for PCS networks.

4 Conclusions

This study provides the model of capacity scaling of PCS networks in the context of a constant user arrival rate and presents numerical results and analysis on how the capacity of PCS networks scales with the size of the service area or the number of cells.

Acknowledgments This work was supported by the National Key Technology R&D Program of China (No. 2012BAH16B00) and the National Science Foundation of China (No. 61003239).

References

1. Akyildiz, I.F., Lin, Y.B., Lai, W.R., Chen, R.J.: A new random walk model for PCS networks. IEEE J. Sel. Areas Commun. **18**(7), 1254–1260 (2000)
2. Liang, B., Haas, Z.J.: Predictive distance-based mobility management for multidimensional PCS networks. IEEE/ACM Trans. Netw. **11**(5), 718–732 (2003)
3. Varsamopoulos, G., Gupta, S.K.S.: Dynamically adapting registration areas to user mobility and call patterns for efficient location management in PCS networks. IEEE/ACM Trans. Netw. **12**(5), 837–850 (2004)
4. Bhattacharjee, P.S., Saha, D., Mukherjee, A.: An approach for location area planning in a personal communication services network (PCSN). IEEE Trans. Wireless Commun. **3**(4), 1176–1187 (2004)
5. Bejerano, Y., Immorlica, N., Naor, J., Smith, M.: Efficient location area planning for personal communication systems. In: Proceedings of the 9th Annual International Conference on Mobile Computing and Networking, vol. 3(4), pp. 109-121 (2003)

VHDL Implementation of Complex Number Multiplier Using Vedic Mathematics

Laxman P. Thakare, A. Y. Deshmukh and Gopichand D. Khandale

Abstract The fundamental and the core of all the digital signal processors (DSPs) are its multipliers, and the speed of the DSPs is mainly determined by the speed of its multiplier. This paper presents a design of efficient complex number multiplier using the Vedic sutra "Urdhva Tiryakbhyam" from ancient Indian Vedic mathematics. The Urdhva Tiryakbhyam sutra (method) was selected for implementation since it is applicable to all cases of multiplication. Multiplication using Urdhva Tiryakbhyam sutra is performed by vertically and crosswise. The most significant aspect of the proposed method is the development of a multiplier architecture based on vertical and crosswise structure of Ancient Indian Vedic Mathematics. On account of these formulas, the partial products and sums are generated in one step, which reduces the carry propagation from LSB to MSB. The implementation of the Vedic sutras and their application to the complex multiplier ensure substantial reduction of propagation delay.

Keywords VHDL implementation · Signal processing algorithms · Vedic multiplier · Complex number multiplier · Vedas · Urdhva Tiryakbhyam

1 Introduction

Complex number operations are the backbone of many digital signal processing algorithms, which mostly depend on extensive number of multiplication. A systems performance is generally determined by the performance of the multiplier, since the multiplier is generally the slowest element in the system [1]. Complex number multiplication involves four real number multiplication and two additions/

L. P. Thakare (✉) · A. Y. Deshmukh · G. D. Khandale
Department of Electronics Engineering, G. H. Raison College of Engineering, Nagpur, India
e-mail: laxman.thakre@gmail.com

S. Patnaik and X. Li (eds.), *Proceedings of International Conference on Soft Computing Techniques and Engineering Application*, Advances in Intelligent Systems and Computing 250, DOI: 10.1007/978-81-322-1695-7_46, © Springer India 2014

subtractions [2]. While doing real number multiplication, carry needs to be propagated from the least significant bit (LSB) to most significant bit (MSB) when binary partial products are added. The overall speed is drop down by the addition and subtraction after binary multiplication [3, 4].

Vedic Mathematics is an ancient mathematics, which is based on 16 sutras and 16 sub-sutras invented by Jagadguru Shankaracharya Bharati Krishna Teerthaji Maharaja (1884–1960) [5]. Mainly multiplication in Vedic mathematics is carried out using three sutras Nikhilam Navatascaraman Dasatah, Ekadhikena Purvena, and Urdhva Tiryakbhyam [5]. Urdhva Tiryakbhyam sutra is the targeted Vedic sutra (algorithm) as it is suitable for all cases of multiplication. The most common multiplication algorithms used are array multiplication and Booth's algorithm. The computational time in case of array multipliers are comparatively less since the partial results are calculated in parallel. Multiplication using Booth's algorithms takes comparable computational time [6]. This paper presents multiplication algorithm that may be useful in the efficient implementation of signal processing algorithms. The framework of this multiplication algorithm is based on Urdhva Tiryakbhyam sutra of Vedic mathematics.

This paper is organized as follows. In Sect. 2, the implementation of complex number multiplier using Gauss's multiplication equations, and Urdhva Tiryakbhyam sutra of Vedic mathematics is explained with example. The proposed methodology for complex multiplication is discussed in Sect. 3. The results obtained for 4-bit, 8-bit, and 16-bit complex multiplications are presented in Sect. 4 along with the conclusion in final section.

2 Implementation

2.1 Complex Number Multiplication

While implementing complex number multiplication, the multiplication system can be divided into two main components giving the two separate results known as real part (R) and imaginary part (I).

$$R + jI = (A + jB)(C + j) \tag{1}$$

Gauss's algorithm for complex number multiplication gives two separate equations to calculate real and imaginary part of the final result. From Eq. (1), the real part of the output can be given by $(AC - BD)$ and the imaginary part of the result can be computed using $(BC + AD)$. Thus, four separate multiplications are required to produce the real as well as imaginary output [6–8].

2.2 Vedic Multiplication Method

Multiplication process is the critical part for any complex number multiplier design. There are three major steps involved for multiplication. Partial products are generated in first step. In second step, partial product reduction to one row of final sums and carries is done. Third and final stages add the final sums and carries to give the result.

The proposed complex number multiplier is based on the Vedic multiplication formulae (sutras). These sutras have been traditionally used for the multiplication of two numbers in decimal number system. In this work, the same ideas are applied to binary number system, to make the proposed algorithm compatible with the digital hardware. The multiplication based on Urdhva Tiryakbhyam sutra is discussed below.

2.3 Urdhva Tiryakbhyam Sutra

Urdhva Tiryakbhyam sutra is suitable for all cases of multiplication. It literally means "Vertically and crosswise"; vertically means straight above multiplication, and crosswise means diagonal multiplication and taking their sum [9]. The feature of this method is any multi-bit multiplication can be reduced down to single-bit multiplication and addition [5]. The multiplication is illustrated below using an example. The crosswise and vertical multiplication be implemented starting either from right-hand side or from left-hand side [9, 10, 11, 12].

Example 1: Multiplication of 42 and 13

Step 1: Starting at the left multiply two left hands most significant digits vertically and set down results underneath as the left hand most significant part of the answer.

$$
\begin{array}{cc}
4 & 2 \\
\underline{1} & \underline{3} \\
4 & ((4 * 1) = 4)
\end{array}
$$

Step 2: Next multiple crosswise and add these partial results. Set down the result of addition as illustrate below.

$$
\begin{array}{cc}
4 & 2 \\
\underline{1} & \underline{3} \\
4 & 4 \quad (((4 * 3) + (1* 2)) \\
= 14) & \\
1 &
\end{array}
$$

Step 3: Multiply two rights hand least significant digits vertically and set down results underneath as the right hand least significant part of the answer.

```
4      2
1      3
4      4      6 ((2 * 3) =
6)
1
```

Step 4: Finally, add the digits vertically as illustrated

```
4      2
1      3
4      4      6
1
5      4      6
```

Result of multiplication 42 * 13 = 546.

Thus, the above method is equally applicable for binary multiplication. Figure 1 shows the general multiplication procedure of the 4×4 multiplication. The multiplication of two 4-bit binary numbers $a_3a_2a_1a_0$ and $b_3b_2b_1b_0$ is performed starting from left-hand side. Every step in Fig. 1 has a corresponding expression as follows:

$$r0 = a0b0 \tag{1}$$

$$c1r1 = a1b0 + a0b1 \tag{2}$$

$$c2r2 = c1 + a2b0 + a1b1 + a0b2 \tag{3}$$

$$c3r3 = c2 + a3b0 + a2b1 + a1b2 + a0b3 \tag{4}$$

$$c4r4 = c3 + a3b1 + a2b2 + a1b \tag{5}$$

$$c5r5 = c4 + a3b2 + a2b3 \tag{6}$$

$$c6r6 = c5 + a3b3 \tag{7}$$

With c6r6r5r4r3r2r1r0 being the final product [2, 3, 9].

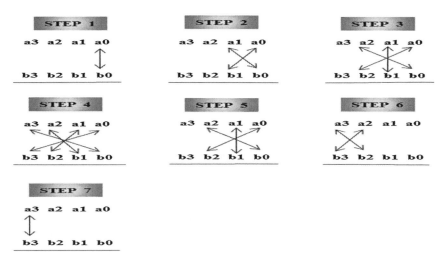

Fig. 1 Vertically crosswise multiplication of four-bit binary number [2]

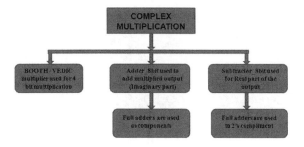

Fig. 2 Proposed methodology for complex number multiplication

3 Discussion

The complex number multiplier using Urdhva Tiryakbhyam sutra is implemented using VHDL. The methodology used to implement the complex multiplier using Gauss's multiplication equations are described with the help of block diagram and is shown in Fig. 2. As discussed earlier, complex multiplication requires four multiplications and an addition and subtraction units.

Using the proposed algorithm, 4-bit, 8-bit, and 16-bit complex multiplications are achieved. Functional verification of the code through simulation is carried out using Xilinx ISE simulator. The complete code is synthesized using Xilinx synthesis tool (XST). Table 1 indicates the device utilization summary of the Vedic complex multiplier for 4-bit, 8-bit, and 16-bit multiplications.

Table 1 Device utilization summary

Vedic complex algorithm	No. of slices	No. of 4 input LUTs	No. of IOs	No. of bonded IOBs	Delay in nS.
4-bit	84	147	33	33	18.419
8-bit	385	674	64	64	30.900
16-bit	1166	2038	128	128	40.250

Fig. 3 RTL schematic (partial) of 16-bit complex multiplier (Vedic multiplier)

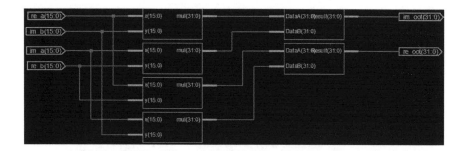

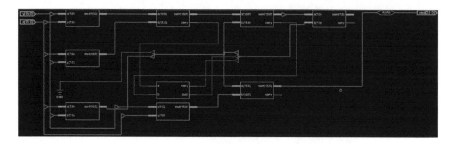

Fig. 5 RTL schematic (partial) of 4-bit complex multiplier (Vedic multiplier)

Figure 3 shows the device utilization, and Figs. 4, 5, and 6 show the RTL schematic of 16-bit complex multiplier using Vedic algorithm. The methodology used for the implementation of complex multiplier in each case is same and is implemented using VHDL language.

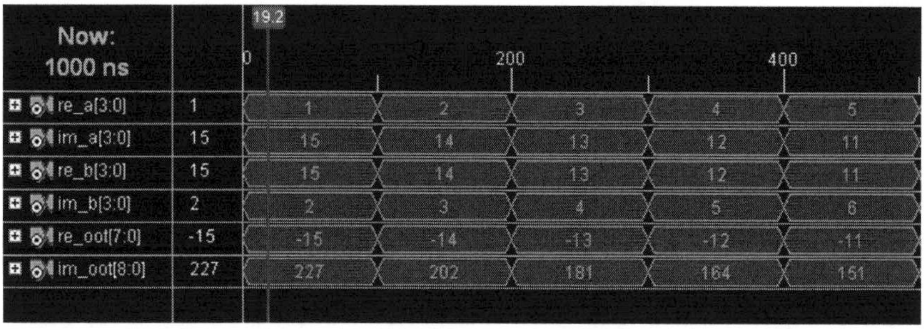

Fig. 6 Simulation waveforms for 4-bit complex multiplication

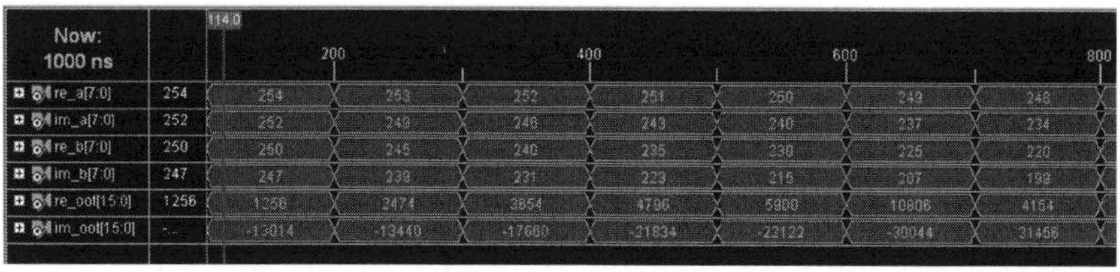

Fig. 7 Simulation waveforms for 8-bit complex multiplication

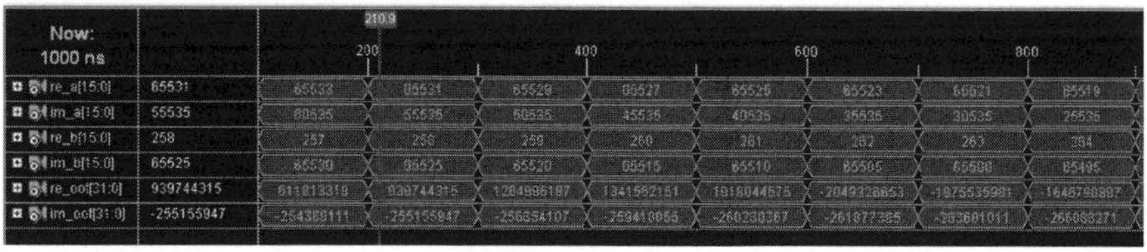

Fig. 8 Simulation waveforms for 16-bit complex multiplication

4 Results

The work presented in this paper was implemented using VHDL, logic simulation was done using Xilinx ISE simulator, and synthesis was done using Xilinx project navigator. The design was synthesized for Spartan3 (xc3s200-5-ft256) device. The obtained results are presented in Table 1, and waveforms for 4-bit, 8-bit, and 16-bit complex multiplications using Urdhva Tiryakbhyam are shown in Figs. 6, 7, and 8, respectively.

5 Conclusions

In many real-time DSP applications, a number of complex multiplications are involved, in which high performance is a prime target. However, achieving this may be done at the expense of area, power dissipation, and accuracy. So efforts have to be made to decrease the number of multipliers and to increase their speed. A high-speed complex number multiplier design using Vedic Mathematics (Urdhva Tiryakbhyam sutra) is implemented using VHDL. This sutra is applicable to all cases of multiplication. The results show that Urdhva Tiryakbhyam sutra with less number of bits may be used to implement high-speed complex multiplier efficiently in digital signal processing algorithms.

References

1. Sriraman, L., Prabakar, T.N.: Design and implementation of two variable multiplier using KCM and Vedic mathematics. In: First International Conference on Recent Advances in Information Technology, RAIT (2012)
2. Sandesh, S.S., Banakar, R.M., Saroja, S.: High speed signed multiplier for digital signal processing applications. In: Proceedings of IEEE (2012)
3. Kerur, S.S., Prakash Narchi, J.C.N., Kittur, H.M.: Implementation of Vedic multiplier for digital signal processing. Int. J. Comput. Appl. **16**, 1–5 (2011)
4. Jaina, D., Sethi, K., Panda, R.: Vedic mathematics based multiply accumulate unit. International Conference on Computational Intelligence and Communication System, pp. 754–757 (2011)
5. Jagadguru Swami Sri Bharati Krishna Teerthaji Maharaja.: Vedic Mathematics. Motilal Banarsidas Publishers Pvt. Ltd, Delhi (2001)
6. Kong, M.Y., Langlois, J.M.P., Al-Khalili, D.: Efficient FPGA implementation of complex multipliers using the logarithmic number system. In: Proceedings of IEEE, pp. 3154–3157 (2008)
7. Dhillon, H.S., et al.: A reduced bit multiplication algorithm for digital arithmetic. Int. J. Comput. Math. Sci. **2**, 64–69 (2008)
8. Ismail, R.C., Hussin, R.: High performance complex number multiplier using Booth-Wallace algorithm. In: Proceedings of ICSE2006, pp. 786–790. Kuala Lumpur, Malaysia (2006)
9. Thakre, L.P., et al.: Performance evaluation and synthesis of multiplier used in FFT operation using conventional and Vedic algorithms. Third International Conference on Emerging Trends in Engineering and Technology, ICETET, pp. 614–619 (2010)
10. Rudagi, J.M., et al.: Design and implementation of efficient multiplier using Vedic mathematics. International Conference on Advances in Recent Technologies in Communication and Computing, pp. 162–166 (2011)
11. Jayaprakasan, V., Vjayakumar, S., Kanchana Bhaaskaran, V.S.: Evaluation of the conventional vs. ancient computation methodology for energy efficient arithmetic architecture. In: Proceedings of IEEE (2011)
12. Dayalan, D., Priya, S.D.: High speed energy efficient ALU design using Vedic multiplication techniques. In: Proceedings of IEEE, ACTEA, pp. 600–603. Zouk Mosbeh, Lebanon, 15–17 July 2009

Key Security Technologies of Cloud Computing Platforms

Liang Junjie

Abstract The information systems built on the basis of cloud computing is the current development trend. How to utilize the cloud technology to enhance the security and defense ability is a challenge facing the cloud users. It is also crucial for maintaining the healthy and sustainable development of cloud computing technology. This article first introduces the cloud computing security model. Then, the key security technologies of cloud platforms are briefly illustrated from four aspects: infrastructure security, virtualization security, service security, and data security. The security problems unique to cloud platform are the emphasis of this article. Moreover, the significance of virtualization security technologies, as well as the current status, realization principle, and development trend are also the main concerns.

Keywords Cloud computing · Cloud computing security model · Virtual machine · Virtualization security

1 Introduction

Cloud computing represents a major breakthrough in the information technology domain. But with its rapid development and applications in various fields, the security incidents frequently occur in recent years. In the early morning of April 21, 2011, the cloud computing platform of Amazon in north Virginia broke down. The cloud services of Amazon were suspended for nearly 4 days, including the answer service Quora, news service Reddit, Hootsuite, and position tracking service Four Square. In 2009, Google "accidentally" disclosed the private information of the customers. In March 2011, Google mail experienced another massive

L. Junjie (✉)
Faculty of Mathematics and Computer Science, Hubei University, Wuhan, China
e-mail: Ljjhubu@163.com

S. Patnaik and X. Li (eds.), *Proceedings of International Conference on Soft Computing Techniques and Engineering Application*, Advances in Intelligent Systems and Computing 250, DOI: 10.1007/978-81-322-1695-7_47, © Springer India 2014

user data leakage incident. About 150 thousand Gmail users woke up in the morning to find that all their mails and chat logs were deleted. Some other users found that their accounts had been reset.

Such a high frequency of security incidents of cloud computing inevitably reduces the user experience. In the meantime, the reliability and stability problems of cloud computing have been recognized by many as the challenges that should be coped with. The privacy, confidentiality, and integrity of user data should be safeguarded by taking effective measures.

As an emerging commercialized computing model, cloud computing has some intrinsic security problems stemming from its model design. Firstly, the traditional security boundary has disappeared in the unification of network architecture under cloud computing environment. With the high integration of storage and computing resources, cloud computing is constantly extending the boundary. A new security model has to be developed for the cloud computing. Secondly, as the key technology of computing technology, virtualization is also being challenged by security problems. To realize the network infrastructure of cloud computing center and the virtualized delivery of data storage and application services, there is a higher technical demand that should be met in relation to the design, construction, and deployment of security equipments. Thirdly, the security threats brought about by data centralization are affecting the user data storage, processing, and network transmission in a cloud environment. The avoidance of potential risks associated with multi-user coexistence and the establishment of security mechanism that consists of the identity authentication of cloud service, authentication management of access control, and the security auditing are among the major security challenges for the cloud computing environment. Fourthly, the stability and reliability problems of cloud computing arise. The user data and business application procedures in a cloud environment depend on the virtualization services enabled by cloud computing. It is not difficult to conceive that a higher requirement has to be met on the stability, security strategy deployment, disaster recovery ability, and event processing auditing for cloud computing.

2 Cloud Computing Security Model

Cloud computing provides service on three layers, IaaS, PaaS, and SaaS. The hardware, software, and data resources are distributed on the basis of user demand. The diversity of combinations of service mode and distribution mode in cloud computing will directly result in the various extents of cloud computing resources monitoring and customization. Thus, a higher requirement is placed on the security model of cloud computing. Reliable security strategies should be deployed for diverse cloud computing modes.

Literature [1] proposes a security model of cloud computing, as shown in Fig. 1. This security model covers IaaS, PaaS, and SaaS platforms. It incorporates the basic equipment security, system security, and network security technologies

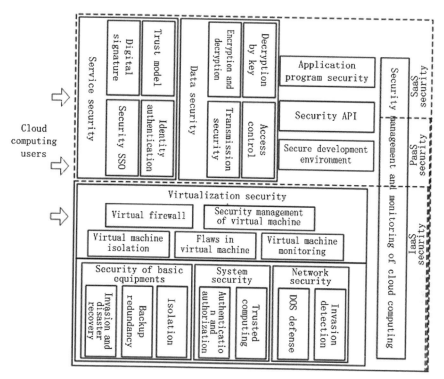

Fig. 1 A possible security model of cloud computing

in infrastructure security; virtual firewall, virtual machine security, virtual machine isolation, flaws in virtual machine, and virtual machine monitoring in virtualization security; encryption and decryption, key management, transmission security, and access control technology in data security; security SSO, digital signature, trust model and identity authentication in service security.

3 Security Technology Analysis of Cloud Computing

3.1 Infrastructure Security

The infrastructure of cloud computing includes the storage, computing, and network resources of cloud computing. The security of infrastructure is a guarantee for the upper-layer applications of cloud computing. Infrastructure security is divided into basic equipment security, network security, and system security.

3.2 Virtualization Security

The basic unit of cloud computing platform is virtual machine, the security of which is the fundamental requirement of the cloud computing platform. The threat to virtualization security is unique to cloud computing and is the main subject of this study.

Virtualization technology refers to the virtualization of the lower-layer hardware, including the servers, storage, and network equipments. By means of virtualization technology, a demand-based resources sharing, distribution, and control platform can be constructed. A variety of isolation applications are possible according to the upper-layer data and business patterns. Thus, a service-oriented, extendable infrastructure is built. The cloud computing services such as the lending of IT infrastructure resources are provided to the users. The providers of cloud infrastructure based on virtualization technologies have to make security and isolation guarantees to the users.

Malicious Code Detection Technologies Based on Virtual Machine. Intel launched the Intel VT-x, the hardware virtualization technology targeted at personal computer in 2005 [2]. The matching processors were released in 2006. It is a hardware virtualization platform for X86-based PC. In 2006, AMD also released AMD-V virtualization technology for hardware, which is based on the development code named Pacifica. In the same year, the matching processors were released [3].

Many software manufacturers developed the software that supports the hardware virtualization. Among them, VMware Workstation 5.5 and the later versions and the processors by VMware [4] are compatible with Intel VT. Microsoft released Hyper-V, the virtualization platform based on Hypervisor, which manages the system's processor in early 2008 [5, 6]. Hyper-V manages and schedules the construction and operation of the virtual machine, in addition to the virtualization of the hardware resources.

Currently, many software producers are engaged in the developing the software that supports the hardware virtualization technologies. However, due to the immaturity of hardware virtualization technologies and the involvement of the lower layer of computer system, a proper coordination is needed between hardware virtual machine, lower-layer hardware, and the resources of upper-layer operation system. There are very few mature software in this field. Most of them fulfill only limited functions, which are basically the management and optimization of hardware resources. However, no software has been known to cope with the security issues.

Among the applications of virtualization technologies in the analysis and detection of malicious code, the representatives are Paladin [7], Anubis [8], Azure [9], and Hypersight Rootkit Detector [10].

Paladin runs on VMW are Workstation, in Linux environment. Its detection principle is that the system calls related to files and processes in the users' operating system are first intercepted. Then, this relevant information of system calls is queried and compared against the data in trusted database.

Anubis is developed based on QEMU virtual machine [11], with main function of analyzing the malicious code. The precursor is TT Analyze proposed in Ulrich Bayer's master's degree thesis [12]. TT Analyze is a tool for analyzing the behavior of Windows PE-executables.

Azure, based on Intel VT, is a malicious code analyzer of transparency using KVM virtual machine monitor. Azure can analyze the Rootkit in Linux environment and implement the automatic unpacking of malicious codes, such as Armadillo and Aspack, via decompilation technology. However, it is currently at the stage of theoretical prototype.

Hypersight Rootkit Detector, based on hardware virtualization technology, is an experimental program that detects the malicious codes. It is developed by North Security Laboratory of Russia. Monitoring the behavior of the target operating system, Hypersight Rootkit Detector intercepts any operations that may damage the operating system.

With the constant improvement on the hardware virtualization technologies and the support of matching software, the detection of malicious codes based on hardware virtualization technology will become the major trend of computer security research.

Partitioning Technologies of Trusted Virtual Domain. Cloud computing is pushing forward the resources integration, which is associated with the disappearance of the boundaries among the systems. Using the partitioning technology of trusted virtual domain, a system boundary can be delineated for the cloud computing on the virtualization layer. Through the implementation of local security strategies, the virtual machine isolation and security of user data can be realized.

Trusted virtual domain is composed of a series of distributed virtual machines. The partitioning of the trusted virtual domain refers to the dividing of the distributed virtual machines into trusted virtual domains in accordance with the security strategies. The trusted virtual domains mark the security boundary of these virtual machines. Within the same virtual domain, the secure communications can be expected between the virtual machines. Meanwhile, they are also isolated from the virtual machines outside this domain (including the virtual machines in other virtual domains).

Isolation is defined as the non-communication of information between non-trusted virtual machines located in different trusted virtual domains. For the same trusted virtual domain, the virtual machines share one security level. They follow a given security strategy when interacting with the virtual machines outside this trusted virtual domain. Before being included as a member of the trusted virtual domain, the virtual machine has to pass the security authentication. Once admitted, the virtual machines will acquire a unified security configuration of the virtual domain. The security strategies defined by each trusted virtual domain include the access control of the resources and the inter-domain and intra-domain information stream interaction.

Information Stream Control between the Virtual Machines. The information stream control between the virtual machines relies on the network control and monitoring of the virtual machines. Any abnormality of information stream between the virtual machines can be detected and reported. Virtual machine monitoring enables the monitoring of the behavior of virtual machines on the virtualized platform. The virtual machines with their lower layer invaded will be excluded to prevent further damage to the platform on which they are situated. At present, hypervisor technology is available to monitor a diversity of virtual machines.

Virtual Machine Isolation. After one physical machine is virtualized into multiple virtual machines, the virtual machine isolation technology will become necessary to prevent the information leak among the virtual machines and the independence of each virtual machine. Besides the virtual machine isolation, isolation also occurs between memories, shared storage space, and shared data.

Other Virtualization Security Technologies. The trusted cloud computing environment is built based on trusted computing theory. In virtualization security technologies, the virtual server equipped with TPM security module is used to construct the trusted virtual platform.

Virtual firewall is set up between the applications systems of virtual machine for security defense and access control. The flow between the application systems is monitored and regulated, while a centralized management is adopted for users' firewall.

3.3 Service Security

Cloud computing provides resources to the users through services. Thus, the service security is the most important security issue. In a complex environment such as cloud environment, there arises a need for user authentication due to user diversity, including identity authentication, access control and auditing and tracking of user behavior.

3.4 Data Security

On cloud platform, users entrust the cloud service supplier with the management of database. Providing the data storage and query services, the server is no longer the center of trust. More considerations should be given to the security protection of user data. The technologies involved are data encryption technology, cipher text retrieval technology, data integrity authentication technology, data isolation and database recovery and backup technology, data protection technology, secure transmission technology and digital authentication technology.

4 Summary

Cloud computing is a convenient tool for providing services, but the information security risks deserve serious considerations. As a security problem unique to cloud computing, virtualization security can be better dealt with the emerging technologies and research achievements. The relevant security technologies include the malicious code detection technology based on virtual machine, partitioning technology of trusted virtual domain, information stream control technology between the virtual machines and virtual machine isolation technology. All these technological breakthroughs offer important guarantee for the secure and reliable applications of cloud platform as well as its healthy, sustainable development.

References

1. Lin, Z., Fu, X., Wang, R., Han, Z.: Research on security challenges in cloud computing. Electron. Eng. (2009)
2. Intel Corporation.: Leading virtualization 2.0[EB/OL]. White Paper: Intel Virtualization Technology (Intel VT), March 28,2008
3. Advanced Micro Devices Inc.: Live Migration with AMD-V Extended Migration Technology[EB\OL]. April 2008
4. VMware Inc.: VMware Workstation 7[EB/OL],2009
5. Armadillo[EB/OL].: 29 March, (2010). http://www.siliconrealms.com
6. PEiDSO EB/OL. 4, November 2007
7. Baliga, A., Chen, X., Iftode, L.: Paladin: Automated detection and containment of rootkit attacks[D]. Dept Comput. Sci. Rutgers University. (2006)
8. Bayer, U., Milani, P., Hlauschek, C., et al.: Scalable, Behavior-Based Malware Clustering[C]. 16th Annual Network and Distributed System Security Symposium (NDSS 2009), San Diego, February,2009
9. Royal, P.: Alternative Medicine: The Malware Analyst's Blue Pill. Black Hat, USA (2008)
10. North Security Labs.: Hypersight Rootkit Detector A New-Generation Rootkit detector EB/OL, 27 January, (2008)
11. Anthony Liguori.: QEMU Emulator User Documentation EB/OL. 29 January (2010)
12. Bayer, U.: TT Analyze A Tool for Analyzing Malware. Vienna University of Technology, Vienna (2005)

The Measurement and Analysis of Software Engineering Risk Based on Information Entropy

Ming Yang, Hongzhi Liao, Rong Jiang and Junhui Liu

Abstract In software engineering, risk is objective existence; identifying and analyzing the risk can improve the quality of software engineering to reduce risk, this makes the engineering benefit achieve maximization. But the analysis of software engineering risk cannot completely copy the traditional risk theory. This paper, on the basis of traditional risk analysis method, combined with mathematical method, proposes measurement and analysis method of software engineering risk based on the information entropy, makes up the defect of the traditional method that risk is difficult to measure, and analyzes seriously the occurrence probability and consequences of risk with the instance, proposes the concepts of risk entropy and benefit entropy, gives the specific quantitative method of software capability maturity.

Keywords Information entropy · Software engineering · Risk analysis · Risk measurement · Software capability maturity

1 Introduction

The development of software engineering involves the support of software and hardware environment, the storage of information, technical requirements, and system management. In these activities, there are human factors, human activities are an uncertainty, it means in the process of software development the risk is inevitable [1].

Risk is a potential uncertain event [2], it contains two meanings, one is the uncertainty of the occurrence of a risk, another is the uncertainty of risk

---

M. Yang (✉) · H. Liao · R. Jiang · J. Liu
School of Software, YunNan University, Kunming, China
e-mail: httx133@126.com

S. Patnaik and X. Li (eds.), *Proceedings of International Conference on Soft Computing Techniques and Engineering Application*, Advances in Intelligent Systems and Computing 250, DOI: 10.1007/978-81-322-1695-7_48, © Springer India 2014

consequence, so it can be said that risk is the combination of risk possibility and risk consequences. In the process of software engineering development, the risk is closely related to the activity. It is inversely proportional to the quality of software engineering, the greater the risk is, the benefit is smaller. On the contrary, the lower the risk is, the higher the quality of the software engineering development is, the project can obtain more economic benefits. Software risk analysis usually includes the following: risk identification, risk estimate, risk evaluation, and risk management. The main purpose is to reduce the uncertainty of the risk, to fully understand the existent risk, to make risk prevention, and in addition, through the reasonable measures, to minimize the loss caused by risks.

However, the traditional risk analysis methods mostly stays on the basis of qualitative analysis, the probability of risk uncertainty and the consequences size have not a quantitative index. In process of software project development, experts can only rely on experience to judge the possibility of risk, through test and observation, and to reduce the consequences of risk, such a project often exists the problems that cycle is too long and benefits shrink. Therefore, this paper proposes the analysis methods of software engineering risk based on information entropy, expect to provide reliable data for risk identification and risk control, it has realistic value.

2 Analysis of Software Engineering Based on Information Entropy

2.1 The Concept of Information Entropy

Information is different from the material and energy; it is a concept that is difficult to measure. Until 1948, Shannon combined with thermodynamic entropy, first propose the concept of information entropy, the information can be measured. Information refers to reduce degree of things uncertainty for people.

Assume that there is a certain event in a system, its probability of occurrence is $P(A)$, thus, the calculation formula of information contained in event is:

$$I(A) = \log P(A). \tag{1}$$

As shown in formula (1), the bigger probability of occurrence of event A is, the greater the information it contains. Suppose that a system has N possible events X_i, $i = 1, 2, \ldots, n$, their probability of occurrence is $P_1, P_2, \ldots, P_n$, $\Sigma P_i = 1$, thus the average information contained in the system is:

$$S = -\sum_{i=1}^{n} P_i \ln P_i. \tag{2}$$

This value is called the information entropy, its unit is bit, and its value is inversely proportional to the amount of information contained in the system. The defined entropy is generalized entropy, when the system and event have different definitions, entropy has a different meaning.

2.2 The Definition of Risk Entropy and Benefit entropy

According to the above-proposed generalized entropy, assuming that in a software project the probability of occurrence of risk i is P_i, the probability of loss caused by it is C_i. So, this project's risk entropy and benefits entropy are shown as follows:

$$H_p = -\sum_{i=1}^{n} P_i \ln P_i, \quad H_c = -\sum_{i=1}^{n} C_i \ln C_i. \tag{3}$$

According to the above definition, it can be known that the greater the value of H_p is, the more risk that is difficult to control the project has; on the contrary, the smaller the risk entropy is, the less risk of the project has, and the risk is easy to control. Similarly, the benefits entropy H_c is inversely proportional to the benefits that project could obtain.

2.3 The Definition of Software Maturity Based on the Information Entropy

As known to all, the actual project risk entropy must be between maximum entropy and minimum entropy, so that through the following formula it can normalize risk entropy and benefits entropy:

$$E_p = H_p/H_{p\max} = H_p/\ln N. \tag{4}$$

$$E_c = H_c/H_{c\max} = H_p/\ln N. \tag{5}$$

So the calculation formula of entire project maturity is as follows:

$$E = 1 - E_p E_c. \tag{6}$$

According to the definition of the capability maturity model (CMM), the project maturity entropy can be divided into five levels {0.1, 0.3, 0.1, 0.3, and 0.9}, it is corresponding to {initial stage, repeated level, defined level, management level, and optimizing level}, respectively. When project maturity E is higher, then the more mature the project ability is. Conversely, the project maturity is lower.

3 The Mathematical Description of Software Engineering Risk

The mathematical description of software engineering risk is the description for probability of occurrence of risk and uncertainty of consequence. This paper divides the risk into management risk of project, technology risk, e- and business risk, and puts forward the evaluation standards for probability of occurrence of risk and uncertainty of consequence, the standards are divided into {small, small, average, big, and big} five levels, they are corresponding to $U_i = \{0.1, 0.3, 0.5, 0.7,$ and $0.9\}$ and $V_i = \{0.1, 0.3, 0.5, 0.7,$ and $0.9\}$ as shown in Table 1:

The standard puts forward a solution for how to quantify probability of occurrence of risk and uncertainty of consequence, but one of the main problems of this method is subjective factors that is bigger, it is not defined from the dialectical perspective. Therefore, this paper, on the basis of this standard puts forward a comprehensive evaluation matrix λ for assignment of the risk weights; its expression is as follows:

$$\lambda_i = \begin{bmatrix} \lambda_{11} & \lambda_{12} & \cdots & \lambda_{15} \\ \lambda_{21} & \lambda_{21} & \cdots & \lambda_{25} \\ \lambda_{31} & \lambda_{32} & \cdots & \lambda_{35} \end{bmatrix}$$

The value of λ_{ij} expresses the expert proportion supporting that the probability of occurrence of risk i is j, then according to the following formula:

$$\omega_i' = \{\omega_1', \omega_2', \omega_3'\} = U_i \lambda^T, \quad q_i' = \{q_1', q_2', q_3'\} = V_i \lambda^T \tag{7}$$

Again through the formulas $\omega_i = \omega_i' / \sum \omega_i'$ and $q_i = q_i' / \sum q_i'$, the data will be normalized, the obtained ω_i is the comprehensive evaluation weight about probability of occurrence of risk, thus the q_i is the comprehensive evaluation weight about uncertainty of consequence. The obtained data will be more persuasive and representative than the directly assigned data, put them into formula (3) can get the risk entropy and benefits entropy of the whole project, so that to measure the maturity of the whole project.

4 Case Studies

If software professionals have carried on the comprehensive evaluation for the risk probability and the probability of losses about an office system project, then it makes the number of experts to be normalized, it will get the comprehensive evaluation matrix as shown in the following:

Table 1 Standards for probability of occurrence of risk and uncertainty of consequence

Risk probability	0.1	0.3	0.5	0.7	0.9
Privacy risk	Management optimization	Quantitative management	Documented management	Set up a management system	Management confusion
Technology risk	Advanced technology	Technical good	Technology standardization	Technology is general	Technology is bad
Business risk	Guaranteed investment	Have a legal contract	Signed the agreement	Oral agreement	No agreement
Risk loss Probability	0.1	0.3	0.5	0.7	0.9
Information disclosure	Least data leakage	Less data leakage	Half data leakage	More data leakage	Most data leakage
Cycle is too long	Delay within 1 month	Delay within half year	Delay within 1 year	Delay within 2 years	Delay without 2 years
Investment increase	Least over budget	Over budget 5 %	Over budget 5–20 %	Over budget 20–50 %	Over budget 50 %

$$\lambda_1 = \begin{bmatrix} 0.818 & 0.182 & 0 & 0 & 0 \\ 0 & 0.83 & 0.833 & 0.083 & 0 \\ 0.769 & 0.154 & 0.077 & 0 & 0 \end{bmatrix}$$

$$\lambda_2 = \begin{bmatrix} 0.8 & 0.2 & 0 & 0 & 0 \\ 0 & 0.077 & 0.769 & 0.154 & 0 \\ 0.909 & 0.091 & 0 & 0 & 0 \end{bmatrix}$$

According to the method proposed in this paper, put the data in matrix into formula (7), it can get the weight of risk probability and loss probability ω and q, its calculation process is as follows:

$$\omega_i' = \left(\omega_1', \omega_2', \omega_3' \right) = U_i \lambda_1^T = (0.1, 0.3, 0.5, 0.7, 0.9) \lambda_1^T$$

$$q_i' = \left\{ q_1', q_2', q_3' \right\} = V_i \lambda_2^T = (0.1, 0.3, 0.5, 0.7, 0.9) \lambda_2^T$$

$$\omega_i = \frac{\omega_i'}{\sum \omega_i'} = (0.171, 0.627, 0.202)$$

$$q_i = \frac{q_i'}{\sum q_i'} = (0.181, 0.666, 0.153)$$

Next, calculated by formula (3), project risk entropy and the benefits entropy are shown as follows:

$$H_p = 0.918, \quad H_c = 0.867$$

Take the above entropy into formulae (4) and (5), it can get the following results:

$$E_p = H_p / H_{p\max} = 0.836, \quad E_c = H_c / H_{c\max} = 0.789$$
$$E = 1 - E_p E_C = 0.341$$

- To observe the above results, it can be found that the project maturity entropy value is 0.659; it lies between repeated levels and defined level. Compared risk entropy with benefits entropy, it can be seen that risk entropy is significantly higher than the benefits entropy, it shows high risk entropy is the main factor that results the decrease in project maturity.
- According to the further decomposition analysis of risk entropy, it can get the privacy risk entropy, technology risk entropy, and business risk entropy as 0.302, 0.293, and 0.323 in turn, privacy risk entropy > technology risk entropy > business risk entropy. It means that business risk has the greatest impact on risk entropy, the second is privacy risk, and technology risk is the weakest; it

also can find that the business risk is the most difficult to control, and technical risk is under the control range.

- Through the above quantitative analysis, it indicates to improve the ability of the whole project maturity, it needs to pay more attentions on the business risk and privacy risk, and to do risk prevention and risk control seriously, as far as possible according to the standard to reduce their impact on the overall project risk entropy.

References

1. Lu, H.: The application study of structure model of software engineering system. Comput. CD Softw. Appl. **18**, 200–201 (2012)
2. Wu, H., Han, R.: Study on the measurement of software engineering supervising risk. J. Hebei Polytech. Univ. **6**(4), 108–110 (2006)

GRACE: A Gradient Distance-Based Peer-to-Peer Network Supporting Efficient Content-Based Retrieval

Jianming Lv, Can Yang and Kaidong Liang

Abstract Beyond the simple file name-based search supported in peer-to-peer (P2P) file sharing networks, content-based search aims to precisely locate the files containing the desired contents. Several existing P2P systems based on structured overlays provide content-based search by indexing shared contents to numeric keys. However, these systems are usually costly when publishing and maintaining the index of contents in dynamic network churn. In this paper, we propose a novel P2P network, GRACE, which probably constructs connections among peers according to the *gradient distance* of their shared contents. GRACE can achieve high efficiency of the content-based search while gaining a significantly lower maintenance cost than those of previous efforts.

Keywords Peer-to-peer · Overlay network · Gradient distance · Content-based search

1 Introduction

Content-based search is one of the most important functions provided in peer-to-peer (P2P) content management systems and aims to precisely locate the files containing the desired contents. While P2P systems are usually large-scale,

J. Lv (✉) · C. Yang · K. Liang
South China University of Technology, Guangzhou 510006, China
e-mail: jmlv@scut.edu.cn

C. Yang
e-mail: cscyang@scut.edu.cn

K. Liang
e-mail: liangkaidong@gmail.com

S. Patnaik and X. Li (eds.), *Proceedings of International Conference on Soft Computing Techniques and Engineering Application*, Advances in Intelligent Systems and Computing 250, DOI: 10.1007/978-81-322-1695-7_49, © Springer India 2014

427

heterogeneous, highly distributed, and dynamically changed, it is challenging to perform efficient and robust content-based search.

Recently, some content-based search algorithms [1–3] over structured P2P overlays are proposed by mapping the shared contents to low-dimensional keys. The mapping operations adopted in above systems inflict high cost to publish and maintain the key-based indexes, especially in the dynamic networks where peers join and depart frequently.

In this paper, we present a probably constructed P2P network, GRACE, to support efficient, resilient content-based search with low maintenance cost. In the GRACE network, peers build the neighboring connections probably according to the *gradient distance* between their shared documents. Compared with traditional P2P content-based search algorithms, GRACE can support efficient content-based search of text documents with relatively low maintenance cost. GRACE needs no overhead of publishing and maintenance of key-based indexes, which is necessary in traditional search algorithms [1–3] based on structured overlays. The maintenance cost of GRACE is less than 1 ‰ of those.

The rest of this paper is organized as follows. Section 2 introduces related works. Section 3 presents the model of the GRACE network. The experimental results are discussed in Sect. 4. Finally, Sect. 5 summarizes the paper.

2 Related Work

Structured overlay networks [4–6] have a kind of distributed infrastructures to support O (log N) key-based search, which are usually named as the distributed hash table (DHT). The topology of the network is constructed based on the global unique ID assigned to each peer. Given any numeric key, the structured overlay can locate the peer having the closest ID to the key within O (log N) hops.

Several recent research efforts [1–3, 7] build content-based search algorithms over DHT. Specifically, the researches [1, 2] map each keyword of a shared text document into a unique numeric key by a uniform hash function to support P2P full-text search. Muller et al. [3] propose a content-based image retrieval algorithm by mapping the color histogram vector of each image into a fixed set of numeric keys. Tang et al. [7] map the high-dimensional document vectors into low-dimensional numeric keys by using the latent semantic index (LSI) method. All of them need to publish index items constructed as <*key, element*> in the networks to support efficient search, which can cause the following latent problems:

- High communication cost is required to maintain the consistence of the index in the dynamic network.
- Uneven distribution of the shared data elements can cause load imbalance in these systems. In full-text search systems [1, 2], the Zipf-distributed keyword popularity can make the load of the peers maintaining the index items of hot keywords extremely heavy.

3 The GRACE Network

3.1 Basic Model

The GRACE model builds up connection probably between any pair of peers according to the *gradient distance* between their shared documents. The *gradient distance* is a kind of document distance with the following *gradient property*: Given any document v in a document dataset, the number of remaining documents within x distance of v is proportional to x^β, $x \in (1, +\infty)$, where β is a constant bigger than zero.

In the GRACE model, the *gradient distance* between any pair of documents is defined as follows:

$$d(v_i, v_j) = 1 / \left(\frac{v_i \cdot v_j}{|v_i||v_j|} + \theta \right) \quad (0 < \theta < 1) \tag{1}$$

Here, θ is a constant to prevent the denominator to be zero. v_i and v_j are the content vectors of the two documents. For a text document, the content vector is defined as the term vector [8], whose elements indicate the frequencies of the terms appearing in the document.

Based on the gradient distance defined above, any peer P_i in the GRACE network selects another peer P_j as its neighbor with the following probability:

$$l(P_i, P_j) = \alpha |H(P_i)|^{-1} \sum_{v_m \in H(P_i)} \sum_{v_n \in H(P_j)} d(v_m, v_n)^{-\beta} f(v_n)^{-1} \tag{2}$$

where $H(P_i)$ and $H(P_j)$ denote the document sets shared by P_i and P_j. $d(v_m, v_n)$ is the gradient distance between any two documents v_m and v_n shared by P_i and P_j, respectively. $f(v_n)$ means the ratio of peers in the network sharing the document v_n, and the parameter β is the constant defined in the gradient property of the distance. $\alpha \in (0, +\infty)$ is a constant used to normalize the probability.

3.2 Construction of the GRACE Network

The main task of constructing a GRACE network is to build up the neighborship of online peers. A simple solution is to select the neighbors of each peer according to the connection probability with all existing peers in the network calculated as Eq. (2). However, this global selection strategy is impractical for its huge communication cost. A more efficient two-phase selection mechanism is adopted in GRACE as follows:

Phase I: Each peer selects several peers randomly from the network to form its candidate set.

Phase II: The peer selects peers from the candidate set as its neighbors. The probability for a candidate to be selected is calculated according to Eq. (2).

To implement the above phases in a distributed manner, each peer maintains four tables: *Host Table, Random Seed Table, Candidate Table,* and *Neighbor Table*.

The Host Table is used to store the contact information of the hosts appearing in the peer's interaction history. For a peer joining the GRACE network for the first time, the Host Table is initialized to contain some stable online peers like what Emule [9] do.

The Random Seed Table consists of peers randomly selected from the Host Table and used for candidate selection in Phase I. Each host in the table is named as a *random seed*. The Candidate Table records the selected candidate peers.

The procedure to construct the Neighbor Table is illustrated in Algorithm 1. Step 1 and step 2 of Algorithm 1 are executed to construct the Random Seed Table from the Host Table. Steps 3–9 illustrate how to construct the Candidate Table by interacting with the random seeds. The candidates selected here can be viewed as a random sample of all peers in the network. Step 10 and step 11 are to select the GRACE neighbors from the candidates according to the probability defined in Eq. (2). The document collections $H(P_i)$ transferred in step 4 and step 5 are presented as content vectors and can be compressed to a small size in practical systems.

Algorithm 1: The joining procedure of a peer

Input: P_0: The peer to join the network.
Output: The Neighbor Table of P_0.
Notations:
Hosts (P_i): The Host Table of the peer P_i.
Random_Seeds (P_i): The Random Seed Table of P_i.
Candidates (P_i): The Candidate Table of P_i.
Neighbors (P_i): The Neighbor Table of P_i.
Method:
//construct P_0's Random Seed Table.
1. **for** $\forall (P_i) \in Hosts\ (P_0)$ **do**
2. P_0 inserts P_i to *Random_Seeds* (P_0) with probability c.
//construct P_0's Candidate Table.
3. **for** $\forall (P_i) \in Random\_Seeds(P_0)$ **do**
//exchange the shared document vectors
4. P_0 sends $H(P_0)$ to P_i
5. P_i sends $H(P_i)$ to P_0
//select all Random Seeds of P_i as candidates
6. **for** $\forall (P_j) \in Random\_Seeds\ (P_i)$ **do**
7. $Pr_j \leftarrow l(P_0, P_j)$ //Eq. (2)
8. P_i send (P_j, Pr_j) to P_0
9. P_0 insert (P_j, Pr_j) into *candidates*(P_0)
//select the GRACE neighbors form candidates
10. **for** $(P_j, Pr_j) \in candidates(P_0)$ **do**
11. P_0 insert P_j to *Neighbors* (P_0) with probability Pr_j

---

3.3 Content-Based Search in GRACE

A simple greedy algorithm is adopted to perform content-based search in GRACE. While a peer forwards a query in the network, it selects another peer from its Neighbor Table as the next hop, which shares the document with minimal distance to the query. Each searched peer independently checks its local stored documents and returns back the records of the top k documents closest to the query. Then, the initiator of the query gathers all returned records, ranks the documents according to the distance to the query, and composes the search result as the top k documents.

4 Experiments

4.1 Data Collection in Experiments

We evaluate the experimental systems on the dataset TREC WT10G [11], which contains 1,692,096 Web documents from 11,680 Web sites. We deem each Web site to be a peer and associate the documents on a site to the corresponding peer as shared contents. We also construct the query set containing 100 queries by randomly sampling as [12]. Each query is obtained by first randomly choosing a document from the dataset and then randomly choosing 5 terms from the document as the searched keywords of the query. The search task is to search the top k documents most related to the query.

4.2 Comparison with Other Overlays

We compare GRACE with two algorithms described as follows:

- *K-Chord* [1, 2]. K-Chord is based on Chord [5] and supports keyword-based search.
- *Rand.* Connections are randomly built among peers [10] and random-walker mechanism is adopted to route search requests.

4.3 Search Latency Analysis

To understand how the performance of GRACE is affected by the network scale, we randomly select peers from the whole datasets to form the subsets with different scales. Then, we build the P2P networks over each subset and test the search efficiency.

Fig. 1 Search latency in
different networks

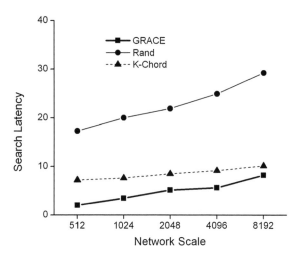

Fig. 2 Communication cost
of network maintenance.
Here, *Join* means the cost for
a new peer to join the
overlay. *PTM* means
periodical topology
maintenance cost. *PIM* means
periodical index maintenance
cost. *Leave* means the cost for
a peer to leave the overlay

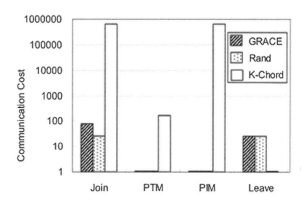

Figure 1 shows the average search latency of GRACE, Rand, and K-Chord under different network scales in the WT10G dataset. We can see that GRACE is more efficient than K-Chord and Rand to locate the top *k* documents most similar to the query.

4.4 Communication Cost

Figure 2 illustrates the communication cost of different systems run on the WT10G dataset. The *y*-axis of the figure is the communication cost measured by the number of messages. We can see that the joining cost of K-Chord is much higher than that of GRACE and Rand. Moreover, in K-Chord, the cost for joining and indexing maintenance makes up the largest percentage. In fact, this overhead is required in most of structured overlay-based systems for publishing and maintaining data indexes. Compared with these systems, the advantage of GRACE is to support efficient search with no need of index publishing, so its maintenance cost can be less than 1 ‰ of K-Chord.

5 Conclusions

The GRACE presented in this paper probably constructs connections among peers according to the gradient distance of their shared contents. GRACE provides efficient content-based search according to the similarity of shared contents without any additional key-based index. Experiments validate the search performance of GRACE and show that its maintenance cost is less than 1 ‰ of structured overlay-based search systems.

Acknowledgments The work described in this paper was supported by the grants from National Natural Science Foundation of China (No. 61300221), the Comprehensive Strategic Cooperation Project of Guangdong Province and Chinese Academy of Sciences (No. 2012B090400016), the Technology Planning Project of Guangdong Province (No. 2012A011100005), and the Fundamental Research Funds for the Central Universities (Project No. 2011ZM0069).

References

1. Reynolds, P., Vahdat, A.: Efficient peer-to-peer keyword searching. In: Proceedings of 4th ACM/IFIP/USENIX International Middleware Conference, pp. 21–40. (2003)
2. Yang, Y., Dunlap, R., Rexroad, M., Cooper, B.F.: Performance of full text search in structured and unstructured peer-to-peer systems. In: Proceedings of 25th IEEE International Conference on Computer Communication, pp. 1–12. (2006)
3. Muller, W., Boykin, P.O., Sarshar, N., Roychowdhury, V.P.: Comparison of image similarity queries in P2P systems. In: Proceedings of 6th IEEE Conference on Peer-to-Peer Computing, pp. 98–105. (2006)
4. Ratnasamy, S., Francis, P., Handley, M., Karp, R., Shenker, S.: A scalable content-addressable network. In: Proceedings of ACM SIGCOMM, pp. 161–172. (2001)
5. Stoica, I., Morris, R., Karger, D., Kaashoek, F., Balakrishnan, H.: Chord: a scalable peer-to-peer lookup service for internet applications. In: Proceedings of ACM SIGCOMM, pp. 149–160. (2001)
6. Zhao, B.Y., Huang, L., Stribling, J., Rhea, S.C., Joseph, A.D., Kubiatowicz, J.D.: Tapestry: a resilient global-scale overlay for service deployment. IEEE J. Sel. Areas Commun. **22**(1), 41–53 (2004). doi:10.1109/JSAC.2003.818784
7. Tang, C., Xu, Z., Dwarkadas, S.: Peer-to-peer information retrieval using self-organizing semantic overlay networks. In: Proceedings of ACM SIGCOMM, pp. 175–186. (2003)
8. Baeza-Yates, R., Ribeiro-Neto, B.: Modern Information Retrieval, pp. 41–44. ACM Press, New York (1999)
9. Emule.: Official emule site. http://www.emule.org/

10. Lv, C., Cao, P., Cohen, E., Li, K., Shenker, S.: Search and replication in unstructured peer-to-peer networks. In: Proceedings of 16th ACM International Conference on Supercomputing, pp. 84–95. (2002)
11. TREC.: TREC web dataset: WT10g. http://ir.dcs.gla.ac.uk/test_collections/wt10g.html. (2003)
12. Bawa, M., Manku G.S., Raghavan, P.: SETS: search enhanced by topic-segmentation. In: Proceedings of ACM SIGIR, pp. 306–313. (2003)

Image Denoising Using Discrete Orthonormal S-Transform

Feng-rong Sun, Paul Babyn, Yu-huan Luan, Shang-ling Song and Gui-hua Yao

Abstract S-transform is an effective time-frequency analysis technique, which can provide simultaneous time and frequency distribution information similar to the wavelet transform (WT). Discrete orthonormal S-transform (DOST) can reduce the redundancy of S-transform further. We introduce the ideas of wavelet transform-based image denoising into DOST domain and propose the soft-thresholding-based image denoising method using DOST. Simulations and the application in myocardial contrast echocardiography (MCE) image denoising illustrate the good performance of the proposed method and its application prospects.

Keywords Discrete orthonormal S-transform · Image denoising · Myocardial contrast echocardiography (MCE) · Wavelet transform

1 Introduction

S-transform is a time-frequency analysis method proposed by R. G. Stockwell in 1996 [1]. S-transform inherits and develops the localization concept of short-time Fourier transform (STFT) and continuous wavelet transform (CWT). S-transform improves upon the time-frequency resolution of STFT and can be regarded as an

F. Sun (✉) · Y. Luan
School of Information Science and Engineering, Shandong University, Jinan, China
e-mail: sunfr.cn@gmail.com

P. Babyn
Department of Medical Imaging, University of Saskatchewan, Saskatoon, Canada

S. Song
The Second Hospital, Shandong University, Jinan, China

G. Yao
Qilu Hospital, Shandong University, Jinan, China

S. Patnaik and X. Li (eds.), *Proceedings of International Conference on Soft Computing Techniques and Engineering Application*, Advances in Intelligent Systems and Computing 250, DOI: 10.1007/978-81-322-1695-7_50, © Springer India 2014

extension of wavelet transform (WT) in multi-resolution analysis domain. However, a large amount of computation time and storage are needed for a signal of moderate size since its S-transform is highly redundant. R. G. Stockwell proposed the discrete orthonormal S-transform (DOST) in 2007 [2] to combat this redundancy, which partitions the time-frequency domain into a quantity of regions, and each region is represented by one coefficient. This makes DOST much more convenient and practical than conventional S-transform.

Because of the favorable performances of S-transform and DOST, many scholars have been studying its applications in recent years. For example, S-transform was applied to laser Doppler flowmetry to analyze hyperemia phenomena [3] and applied to magnetic resonance imaging [4], including removal of artifacts from functional magnetic resonance imaging (fMRI) data [5]. DOST also shows strong potential in image compression [6], image texture characterization [7], and image restoration [8]. This paper focuses on the application of DOST to image denoising.

It is well known that WT has been applied to image denoising successfully. As an extension or a reforming of WT [9], S-transform can be expected to possess much better performance over WT in image denoising field. So we introduced the ideas of thresholding-based image denoising of WT [10] into S-transform and proposed the thresholding-based image denoising method using DOST. Simulations validated our method, and the implementation of our method in myocardial contrast echocardiography (MCE) image denoising demonstrated its application prospects.

2 Theory

2.1 S-transform

The relationship between the S-transform of function $h(t)$ and its CWT is defined as:

$$S(\tau, f) = e^{-i2\pi f\tau} W(\tau, d) \tag{1}$$

where $W(\tau, d)$ is the generating function (i.e., the mother wavelet) of CWT and has to satisfy the admissibility condition. Equation (1) indicates that S-transform can be regarded as a phase correction version of CWT, but it does not need to satisfy the admissibility condition.

S-transform maintains a direct relationship with Fourier spectrum and has more straightforward physical significance and engineering usability. When averaged over time, S-transform becomes the FT of the original signal:

$$\int_{-\infty}^{+\infty} S(\tau, f) d\tau = H(f) \tag{2}$$

where $H(f)$ stands for the FT of signals. Utilizing this relationship between S-transform and FT, we can get the original signal easily:

$$h(t) = \int\limits_{-\infty}^{+\infty} \left\{ \int\limits_{-\infty}^{+\infty} S(\tau,f)d\tau \right\} e^{i2\pi ft} df \tag{3}$$

2.2 Discrete Orthonormal S-transform

The efficient representation of S-transform can be defined as the inner product between a time series $h[k]$ and the basis function $D[k]$, with the parameters υ, β, τ.

$$\begin{aligned}
S_{[\upsilon,\beta,\tau]} &= \left\langle D[k]_{[\upsilon,\beta,\tau]}, h[k] \right\rangle \\
&= \frac{1}{\sqrt{\beta}} \sum_{k=0}^{N-1} \sum_{f=\upsilon-\beta/2}^{\upsilon+\beta/2-1} \exp\left(-i2\pi\frac{k}{N}f\right) \exp\left(i2\pi\frac{\tau}{\beta}f\right) \exp(-i\pi\tau) h[k]
\end{aligned} \tag{4}$$

where υ specifies the center of each frequency band, β is the bandwidth, and τ specifies the location in time. To make the family of basis vectors in (4) orthogonal, the three parameters υ, β, and τ have to be chosen suitably. R. G. Stockwell defined the DOST basis vectors by introducing an integer p to index the frequency bands:

$$\text{if } p = 0, \quad D[k]_{[\upsilon,\beta,\tau]} = 1 \tag{5}$$

$$\text{if } p = 1, \quad D[k]_{[\upsilon,\beta,\tau]} = \exp(-i2k\pi/N) \tag{6}$$

$$\text{for } p = 2, 3, \ldots, \log_2 N - 1, \quad \begin{cases} \upsilon = 2^{p-1} + 2^{p-2} \\ \beta = 2^{p-1} \\ \tau = 0, \ldots, \beta - 1 \end{cases} \tag{7}$$

For an $N \times N$ image, the number of its S-transform coefficients is $N^2 \times N^2$, but that of DOST coefficients still is $N \times N$. So DOST reduces the storage requirement greatly. Taking advantage of fast Fourier transform (FFT), we can achieve the fast calculation of DOST, which can be depicted as follows:

$$\begin{aligned}
S_{[\upsilon,\beta,\tau]} &= \left\langle D[k]_{[\upsilon,\beta,\tau]}, h[k] \right\rangle \\
&= \frac{1}{\sqrt{\beta}} \sum_{f=\upsilon-\beta/2}^{\upsilon+\beta/2-1} \exp(-i\pi\tau) \exp\left(i2\pi\frac{\tau}{\beta}f\right) H[f]
\end{aligned} \tag{8}$$

3 Thresholding-Based Image Denoising Method Using DOST

3.1 Basic Ideas

S-transform is WT in essence [11], and the latter has been applied to image denoising widely and successfully. It inspires us to introduce the ideas of WT-based image denoising into S-transform domain. The WT of a noisy signal is the linear combination of the noise's transform and the original signal's; an appropriate threshold can restrain the noise while retain the useful signal features [12]. Generally, the signal's WT coefficients are larger than the noise's. We can consider that those larger WT coefficients are related to the signal, but those smaller are mainly due to the noise. Appropriate thresholds can be selected to keep the signal's WT coefficients while reducing most of the noise's. Here we introduce these ideas into DOST and propose the thresholding-based image denoising method using DOST. The proposed method first chooses an appropriate threshold T, then maps all those S-transform coefficients less than T to zero, and shrinks other coefficients using the soft-thresholding-based image denoising strategy.

3.2 Method Description

The proposed soft-thresholding-based image denoising method using DOST is as follows:

1. Transform the noisy image to get its DOST coefficients $S(i, j)$.
2. Process $S(i, j)$ using the soft-thresholding strategy. It includes selecting an appropriate threshold T, letting those DOST coefficients that meet $abs(S(i, j)) < T$ be zero, and shrinking all other coefficients. We can depict the applied soft-thresholding strategy by the following equation:

$$S(i, j) = \begin{cases} 0 & |S(i, j)| < T \\ \frac{|S(i,j)| - T}{|S(i,j)|} S(i, j) & \text{others} \end{cases} \tag{9}$$

Threshold T in Eq. (9) is determined as that in WT-based image denoising [10]:

$$T = a * \sigma \sqrt{2 \log(N)} \tag{10}$$

where a is a constant greater than zero and is called threshold coefficient in this paper, σ is the variance of noise, and N is the image size.

3. Conduct inverse DOST to get the denoised image.

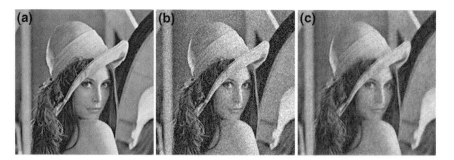

Fig. 1 An experiment: **a** original image, **b** noisy image, and **c** denoised image

4 Simulation Experiments

4.1 Lena Image

White Gaussian noise with variance $\sigma^2 = 0.01$ was added to the 256×256 Lena image, and then, the noisy Lena image was processed utilizing our soft-thresholding-based image denoising method. Signal-to-noise ratio (SNR) and mean-square error (MSE) criteria were calculated to evaluate the performance of our method. Figure 1 illustrates an experiment, where (a) is the original Lena image, (b) is the noisy Lena image, and (c) is the denoised image by our method. Their SNR and MSE indexes are given as follows: MSE = 621.19, SNR = 33.987 (for the noisy image); MSE = 236.91, SNR = 43.627 (for the denoised image).

4.2 Edge Preservation Performance Analysis

Since the capability for an image denoising method to preserve edge information is important for applications, we used figure of merit (FOM) [13] to measure the edge preservation performance of our method. The FOM index is defined as:

$$\text{FOM} = \frac{1}{\max(N_I, N_q)} \sum_{i=1}^{N_q} \frac{1}{1 + \alpha d_i^2} \tag{11}$$

where N_I and N_q, respectively, are edge pixel numbers of the original image and the denoised/noisy image, α is a constant typically set to 1/9, and d_i is the Euclidean distance between the ith edge pixel of the denoised/noisy image and the nearest edge pixel of the original image. The quantity of FOM ranges between 0 and 1.0 and reaches its maximum value 1.0 when two images have fully identical edges.

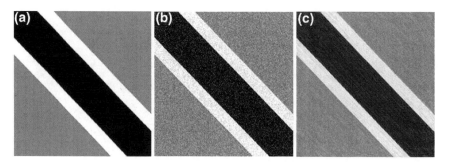

Fig. 2 Edge preservation performance analysis: **a** original image, **b** noisy image, and **c** denoised image

We first generated a 256×256 synthetic image as shown in Fig. 2a, and b shows its noisy version with additive white Gaussian noise ($\sigma^2 = 0.01$). Figure 2c shows the denoised image by our method. Their SNR and FOM criteria are as follows: SNR = 36.98, FOM = 0.10981 (for the noisy image); SNR = 47.424, FOM = 0.64367 (for the denoised image).

The simulation experiments presented above indicate the soft-thresholding-based image denoising using DOST not only is capable of increasing SNR of noisy images effectively, but also has beneficial edge preservation performance.

5 Myocardial Contrast Echocardiography Image Denoising

MCE is an echocardiography technology with important diagnostic and prognostic values. Clinicians can assess the myocardial perfusion of patients noninvasively by means of the quantitative analysis of MCE images [14]. And the accurateness of this analysis depends on MCE images quality to a great degree. But MCE images usually have heavy speckle noise. The noise brings about the spatial and contrast resolutions being reduced and covers the statistical characteristics of MCE images, so that it throws large negative to the accurateness of the quantitative analysis of MCE images. So suppressing the speckle noise in MCE images effectively is of great clinical significance and is beneficial to promoting the clinical status of MCE.

As a matter of fact, speckle noise is ubiquitous in medical ultrasound image and has complicated formation mechanism and statistical characteristics. General speaking, restraining the speckle noise in ultrasound image effectively is challenging. The previous methods for medical ultrasound image denoising include homomorphism wiener filter, self-adaptive median filter, anisotropic diffusion filter, and so on. As a whole, these works have certain scientific significance, but their clinical values are not satisfactory.

Fig. 3 MCE image denoising: **a** original image and **b** denoised image

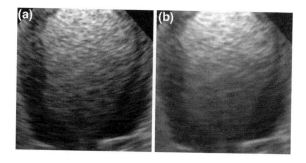

We applied the soft-thresholding-based image denoising method to MCE images and obtained good results. Figure 3 demonstrates the image denoising results, where (a) is the original MCE image and (b) is the denoised MCE image. Their SNRs are as follows: 1.1769 (for the original image); 2.2309 (for the denoised image). It can be seen that our method is able to suppress the speckle noise of MCE images effectively. This guarantees the accurateness of the quantitative analysis of MCE images.

6 Conclusions

S-transform integrates the advantages of STFT and WT, and DOST further reduces the redundancy of S-transform to lessen the storage requirement and to achieve fast computation. Image denoising is a very important procedure in image processing; its purpose is to increase SNR and highlight expected image features. In this presentation, a soft-thresholding-based image denoising method using DOST is established. Simulations and the application in MCE image denoising show the good performance of the proposed method and its bright application prospects.

Acknowledgments This work was supported by the National Nature Science Foundation of China under the Grant 61071053 and the Nature Science Foundation of Shandong Province under the Grant ZR2010FM012.

References

1. Stockwell, R.G., Mansinha, L., Lowe, R.P.: Localization of the complex spectrum: the S transform. IEEE Trans. Signal Process. **44**, 998–1001 (1996)
2. Stockwell, R.G.: A basis for efficient representation of the S-transform. Digital Signal Process. **17**, 371–393 (2007)
3. Assous, S., Humeau, A., Tartas, M., Abraham, P., L'Huillier, J.-P.: S-Transform applied to laser Doppler flowmetry reactive hyperemia signals. IEEE Trans. Biomed. Eng. **53**, 1032–1037 (2006)

4. Drabycz, S.: Efficient S-Transform Techniques for Magnetic Resonance Imaging, pp. 160–176. Library and Archives Canada, Canada (2009)

5. Goodyear, B.G., Zhu, H., Brown, R.A., Mitchell, J.R.: Removal of phase artifacts from fMRI data using a Stockwell transform filter improves brain activity detection. Magn. Reson. Med. **51**, 16–21 (2004)

6. Wang, Y., Orchard, J.: Fast discrete orthonormal Stockwell transform. Soc. Ind. Appl Math. **31**, 4000–4012 (2009)

7. Drabycz, S., Stockwell, R.G., Mitchell, J.R.: Image texture characterization using the discrete orthonormal S-transform. J. Digit. Imaging **22**, 696–708 (2009)

8. Wang, Y., Orchard, J.: The discrete orthonormal Stockwell transform for image restoration. ICIP, pp. 2761–2764 (2009)

9. Ventosa, S., Simon, C., Schimmel, M., Danobeitia, J.J., Manuel, A.: The S-Transform from a wavelet point of view. IEEE Trans. Signal Process. **56**, 2771–2780 (2008)

10. Donoho, D.L.: Denoising by soft-thresholding. IEEE Trans. Inf. Technol. **41**, 613–627 (1995)

11. Peter, C.G., Michael, P.L., Gary, F.M.: Letter to the editor: Stockwell and wavelet transforms. J. Fourier Anal. Appl. **12**, 713–719 (2006)

12. Donoho, D.L., Johnstone, I.M.: Adapting to unknown smoothness via wavelet shrinkage. J. Am. Stat. Assoc. **90**, 1200–1224 (1995)

13. Yongjian, Yu., Scott, T.A.: Speckle reducing anisotropic diffusion. IEEE Trans. Med. Imaging **11**, 1260–1270 (2002)

14. Sun, F.-R., Zhang, M.-Q., Jia, X.-B., et al.: Numerical methods and workstation for the quantitative analysis of real-time myocardial contrast echocardiography. IEEE Trans. Inf Technol. Biomed. **14**, 1204–1210 (2010)

Analyzing Services Composition Using Petri Nets

Jiajun Xu and Shuzhen Yao

Abstract In this paper, a Petri Net-based method is proposed to model and analyze service composition process. The service composition is related to structure and interaction association. This paper first studies structure association service composition description language BPEL; while to the service interaction, it models the interactive features by applying Petri Net. It gives the definition of Composited Service Stochastic Petri Net and gives the mapping method, containing structure and interaction associated service composition modeling and simplification method, and analysis of the model. The approach is validated through a case, and further research directions are pointed out.

Keywords Service composition · Petri Net · Modeling · Analysis

1 Introduction

Web services are emerging as a major technology for deploying automated interactions between distributed and heterogeneous applications [1, 2]. Web service composition is a distributed model to construct new Web services on top of existing primitive or other composite Web services, through a certain association. The association generally includes two categories: structural association and interaction association. Structural associated composition includes sequence, switch, loop, and parallel logical relationship; interaction is involved in the service composition between requester/provider. There are questions like the completion or correctness of service composition that are left to be answered. However,

J. Xu (✉) · S. Yao
School of Computer Science and Engineering, Beihang University, Beijing, China
e-mail: xujiajun@cse.buaa.edu.cn

S. Yao
e-mail: szyao@buaa.edu.cn

S. Patnaik and X. Li (eds.), *Proceedings of International Conference on Soft Computing Techniques and Engineering Application*, Advances in Intelligent Systems and Computing 250, DOI: 10.1007/978-81-322-1695-7_51, © Springer India 2014

current researches are with little effort being dedicated to the verification of the explicit process model from two aspects.

Existing powerful Petri Net-based techniques could be used to verify the service composition by answering the aforementioned questions, provided that it is possible to successfully analyze the BPEL process model by using Petri Net. This paper will focus on the structural association and interaction association, introduce the current mapping method, and construct composited service model, to perform the analysis.

2 Related Work

To capture the behavior of service composition in some formal way, a variety of analysis techniques have been proposed. Among which, Petri Net with the characters of concurrency, asynchronous, and uncertainty is a strong dynamic modeling and analysis tool. It can not only be used to describe the local behavior, but also the combination characteristics, for modeling and analyzing of the system.

BPEL has become the main description language that is related to the structure. Comparing to the interaction association, there are several major typical method for modeling of the interaction: process algebra [3], state transition diagrams (STD) [4], finite automata (FA) [2], π-calculus [5], Petri Net [6], fuzzy Petri Net [7], colored Petri Net (CPN) [8], and some other techniques and methods.

There has been a lot of work focused on the modeling and analysis of BPEL based on Petri Net. Hinz et al. [9] proposes a tool BPEL2PN to transform BPEL to Petri Net. The tool could support a variety of transformation operations (including standard, and exception handling-related behavior) and could be used for understanding the BPEL inner semantics. Lohmann et al. [10] proposes a transformation tool called BPEL2oWFN which is mainly used for analysis of control flow-related properties of BPEL. Ouyang et al. [11] describes another tool called BPEL2PNML which focus on the control structure of BPEL; the Petri Net model can be used after transformation for testing unreachable and conflict. The work proposed in this paper provides valuable reference for our Petri Net based on modeling and the analytical methods.

In the following, the paper would focus on service-composited Petri Net model from viewpoint of the structural association and interaction association and apply on Petri Nets theory to perform analysis of the models and then with examples to verify the usability and correctness of the method.

3 CPN-Based Service Composition Model

3.1 CPN-Based Structure Model

BPEL is used to describe the behavior and composition of Web services, and its statute contains basic activities and structured activities. Each defined activity in its statute can be abstracted into a Petri Net model. In this paper, the CPN-based

Table 1 Petir net-based basic BPEL activities

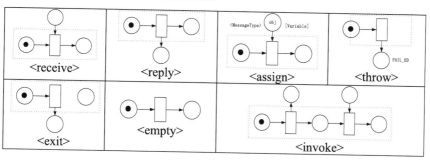

formal representation of the BPEL process is presented, to express the control flows of the BPEL process and the message interaction.

Definition 1 (*CPN-based formal representation of the BPEL*)

The formal representation of the BPEL process is a CPN: $N = (P, T, F, C, W, M_0, \lambda)$,

1. P is a finite set of places. $P = P_I \cup P_m$, and $P_I \cap P_m = \varnothing$, P_I is the set of internal places. P_m is the set of message places. (Internal places P_I represent the internal control logic, and message places P_m represent the messages exchanged between services.)
2. T is a finite set of transitions. $F = F_1 \cup F_m$, and $F_1 \cap F_m = \varnothing$, F_1 is the normal arc where $F_1 \subseteq (P_1 \times T) \cup (T \times P_1)$, and F_m is the arc from P_m, $F_m \subseteq F_m^R \cup F_m^S$. All the arc is associated with a color set C. $F_m^R \subseteq (P_m \times T)$ means arc of receiving message, and $F_m^S \subseteq (T \times P_m)$ means arc of sending message $(P \cap T = \varnothing)$.
3. C is a color set, $C : P \cup T \to \Sigma C$, it links each Place P to a color set $C(p)$ and links each transition to a color set $C(t)$, which means all the tokens in P and T belong to a color type and stands for message; each color in C stands for a message.
4. $W : F \to (\text{ID}, \text{CDS})$ means the function on the arc, ID means the identity of the Arc, and CDS means the data structure or object of service interface.
5. M_0 is the initial marking of CPN.
6. λ is the firing rate set on the transitions, $\lambda_i \in \lambda$ is the firing rate of transition t_i.

In BPEL, a process is built up by plugging language constructs together. Therefore, we translate each construct of the language into a Petri Net. Such a net forms a pattern of the respective BPEL construct. Each pattern has an interface for joining it with other patterns as it is done with BPEL constructs.

The BPEL activities can be divided into two categories: basic activities and structured activities. The seven kinds of basic activities are defined and they describe the basic behavior of Web services, as shown in Table 1.

Table 2 Petir net-based structured BPEL activities

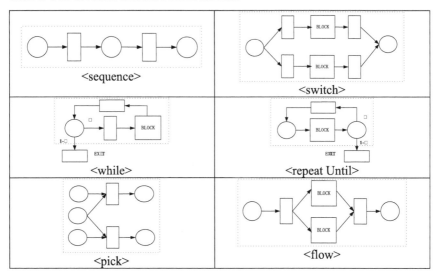

The six kinds of structured activities are defined, as shown in Table 2. Through connecting all of the basic activities and specify their running path, structured activities construct complex service processes. A structured activity embeds at least one activity. Thus, only the interface of an embedded pattern is visualized and all other information of the pattern is hidden.

3.2 CPN-Based Interaction Model

To interaction model, the service could be modeled as simple input and output, and this paper uses Petri Nets to model loosely coupled Web service and to perform the analysis of service interactive mode.

The Web services composition includes services interaction between the available services and user interaction; former will build a number of available services as a whole, and the latter is used to satisfy the service requester. From the viewpoint of interaction protocols, the two can be unified as the service requester and service by establishing session. Session is a set of associated meta message m. The session $(m_1, m_2, \ldots, m_n)$ represents a set of meta message $m_1, m_2, \ldots, m_n$ exchanging.

Places, transitions, and tokens in the traditional Petri Net are lack of semantic description; data processing cannot be supported. Moreover, the state space explosion problem in the Petri Net model greatly limits its usage for large, complex systems modeling and analysis capabilities. To describe the session of service-oriented systems, the following definition is introduced.

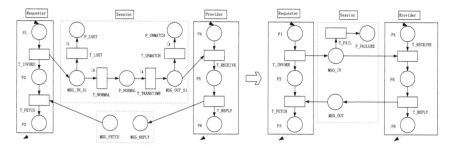

Fig. 1 Single service requester/provider session model

Definition 2 (*Composited Service Stochastic Petri Net—CSSPN*)

Given a CPN of requester model *Requester* = $(P_1, T_1, F_1, C_1, W_1, M_{1,0}, \lambda_1)$ and a CPN of provider model *Provider* = $(P_2, T_2, F_2, C_2, W_2, M_{2,0}, \lambda_2)$. After the commutation of *Requester* and *Provider*, the Petri Net after interaction CSSPN = <*Requester*, *Provider*, *Session*>.

In CSSPN, it is based on the concept of CPN and stochastic Petri Net. In the Places of CSSPN, data type of token is defined, to stand for complex data structures or objects or service message and reduce the complexity of modeling. We define token to be message, and the arc means that the message-passing mechanism. A CSSPN model could be linked with the other CSSPN models through the arc. Transition is on behalf of the message processing unit. CSSPN is very suitable to describe the distributed, parallel system, and it not only can describe the static model of service interaction protocols, but also dynamic trusted interaction.

Different types of service requests reflect the differences in cooperation, thereby forming kinds of cooperative interaction. The composite service is obtained by combining the new service by a plurality of basic services in accordance with certain logical structural relationship. However, because actual deployed service applications interact through the session, which makes the changes in session state leading to a greater difference in performance. In this case, if the service composition modeling does not consider the impact of session state, the results of the analysis of the model will result in a large deviation.

Figure 1 represents a single service requester/provider session model, in which places mean the state, and transitions mean processing of the message. In the session of request–response protocol, when the service requester makes a request to invoke a service, the session begins; the service provider receives the message from the requester and makes a reply, after the processing is completed. The transition T_LOST represents that interactive messages are lost in the network environment, and it goes to place P_LOST, after the termination. The transition T_NORMAL represents that interactive messages are transmitted normally, and the transition T_UNMATCH represents that the detected message exchange sequence is a departure from the defined standard message exchange sequence. The firing rate of T_LOST, T_UNMATCH, T_NORMAL, and T_TRANSTIME are λ_1, λ_2, λ_3, and λ_4, respectively.

In order to reduce the scale of the model, we propose a simplified session model. The places MSG_IN_S1, P_NORMAL, and MSG_OUT_S1, and transitions T_NORMAL and T_TRANSTIME are merged to place MSG_IN. Transitions T_LOST and T_UNMATCH are merged to transition T_FAIL. P_LOST and P_UNMATCH are merged to place P_FAILURE.

4 Case Study and Analysis

In order to better show the usability and correctness of the method, we use the case of the tourism services system (TSS). As shown in Fig. 2, it gives BPEL model of TSS. First, the user sends a request. TSS, including airfare, hotel, and tour services, receives the request and starts travel planning process. The service provider TSS provides three concurrent tasks, which would interact with hotel center service, ticket center service, and tour center service (three services can be executed in parallel). Then, the system receives the results of three concurrent services. After the join of three services, it invokes another service provider bank credit service, standing for the credit card payment process and then receives the results of the bank credit information. After confirmed by the end user, the process ends.

We use tool BPEL2PNML [12] to transform BPEL model to PNML and use tool PIPE [13] taking PNML as input, to perform the analysis of the model. Given BPEL of TSS, we first created CSSPN including system structure and interaction association, as shown in Fig. 2. It models the user requests, three concurrent services Hotel, Air ticket and Tour, and BANK service. The dotted line boxes INVOKE_HOTEL, INVOKE_TICKET, INVOKE_TOUR and INVOKE_BANK represent the invoking of the hotel services, ticketing services, tour services, and bank credit card services. Firing identifies the message on the arc, receiving message indicates the start of this behavior, and replying the message indicates the completion.

In order to avoid the state explosion, it is necessary to simplify the primitive model. In the CSSPN original model, the Petri Net elements transition T_B_RE-CEIVE, place BANK_UP, transition T_B_REPLY, and place MSG_REPLY in the dashed boxes are merged to one arc. Similar simplification would be done in the dashed boxes: INVOKE_HOTEL, INVOKE_TICKET, and INVOKE_TOUR.

We define the initial marking as $M_0 = [1,0,0,0,0,0,0,0,0,0,0,0,0,0,0,0,0,0,0,0,0,0,0,0,0,0,1,0,0,0,0,0]$, that is, only p_1 and P_START have an initial token. $M_0(p_1) = 1$, $M_0(P\_START) = 1$. Txhe weight functions of arc are 1.

The reachability (coverability) graph of the TSS model is shown in Fig. 3, this is a little complex Petri Net. There are 74 markings in the reachability graph including 65 tangible states and 9 vanishing states. There is a finial state $S_{73} = [0,0,0,0,0,0,0,0,0,0,1,0,0,0,0,0,0,0,0,0,0,0,0,0,0,0,1,0,0,0,0,0]$, that is S_{73} $(P\_END) = 1$. When the marking could reach the final state, the BPEL could reach its final service from its starting, which proves the completeness and correctness of the BPEL model.

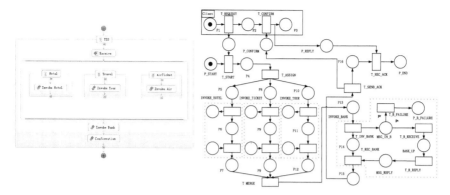

Fig. 2 BPEL model of TSS

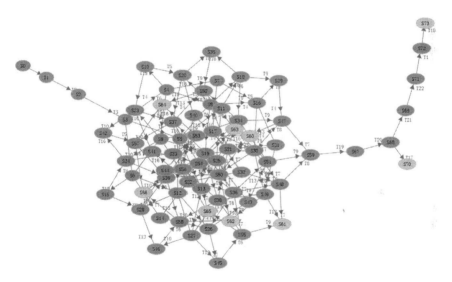

Fig. 3 The reachability graph of the TSS model

5 Conclusions and Future Work

The modeling and analysis of service structures and interactions modeling is a feasible and effective method for verifying the completeness and correctness of service composition. It is a topic worthy of further study.

In this paper, a Petri Net-based services method could easily be used for modeling service composition, and the model is analyzed by using reachability (coverability) graph. All these will be the basis of our follow-up work.

References

1. Fan, W., Geerts, F., Gelade, W., et al.: Complexity and composition of synthesized web services. In: Lenzerini, M., Lembothe, D. (eds.) Proceedings of 27th ACM SIGMOD-SIGACT-SIGART Symposium on Principles of Database Systems, pp. 231–240. ACM, Vancouver BC (2008)
2. Deng, T., Huai, J.P., et al.: Complexity of synthesis of composite service with correctness guarantee. Sci. Sinica Inf. **42**(7), 789–802 (2012)
3. Ferrara, A.: Web services: a process algebra approach. In: Proceedings of the 2nd International Conference on Service Oriented Computing (ICSOC'04), pp. 242–251. IEEE, Piscataway, NJ (2004)
4. Ru, F., Zou, Z.L., Stratan, C., Fong, L.: Dynamic support for BPEL process instance adaptation. In: IEEE International Conference on Services Computing (SCC'08), vol. 1, pp. 327–334 (2008)
5. Lucchi, R., Mazzara, M.: A pi-calculus based semantics for WS-BPEL. J. Logic Algebraic Program. **70**(1), 96–118 (2007)
6. Zhang, J., Chung, J.Y., Chang, C.K., Kim, S.: WS2Net: a Petri Net based specification model for web services. In : Proceedings of t he 2nd IEEE International Conference on Web Services, San Diego, California, USA, pp. 420–427 (2004)
7. Fu, R., Dong, W., Yang, G., Mei, Y., Dong, X.: Fuzzy Petri Net-based optimized semantic web service composition. In: Seventh International Conference on Grid and Cooperative Computing GCC'08, pp. 496,502, 24–26 Oct 2008
8. Zhu, J., Guo, C., et al.: A runtime monitoring web services interaction behaviors method based on CPN. J. Comput. Res. Dev. **48**(12), 2277–2289 (2011)
9. Hinz, S., Schmidt, K., Stahl, C.: Transforming BPEL to Petri Nets. In: Proceedings of the 3rd International Conference on Business Process Management (BPM'05), pp. 220–235. IEEE, Piscataway, NJ (2005)
10. Lohmann, N., Massuthe, P., et al.: Analyzing interacting WS-BPEL processes using flexible model generation. Data Knowl. Eng. **64**(1), 38–54 (2008)
11. Ouyang, C., Verhaek, E., van der Aalst, W., et al.: Formal semantics and analysis of control flow in WS-BPEL. Sci. Comput. Program. **67**(2/5), 162–198 (2007)
12. BPEL 2PNML is available at www.bpm.fit.qut.edu.au/projects/babel/tools
13. Platform Independent Petri Net Editor 2 is available at www.pipe2.sourceforge.net

FPGA-Based Image Processing for Seamless Tiled Display System

Mingyu Wang, Yan Han, Rui Wang, Xiaopeng Liu and Yuji Qian

Abstract In this paper, we present an image processing method for seamless tiled display system. An ordinary personal computer (PC) is employed to preprocess the system, i.e., to generate geometry calibration and intensity blending parameters, with software. And a hardware image processor based on field programmable gate array (FPGA) is employed to divide the source video image into two sub-images according to the parameters mentioned above. The sub-images projected on the screen splice seamlessly, presenting a visually comfortable high-resolution image to the audience. Taking advantages of software and hardware, this method could process high-resolution video image with low cost and power consumption. Experimental results show that the proposed method is effective.

Keywords Multi-projection · Geometry calibration · Intensity blending · FPGA

M. Wang · Y. Han (✉) · X. Liu · Y. Qian
Institute of Microelectronics and Photoelectronics, Zhejiang University, Zhejiang,
310027 Hangzhou, People's Republic of China
e-mail: hany@zju.edu.cn

M. Wang
e-mail: wmy9066@zju.edu.cn

X. Liu
e-mail: lxp@zju.edu.cn

Y. Qian
e-mail: qianyuji@zju.edu.cn

R. Wang
State Key Laboratory of CAD&CG, Zhejiang University, Zhejiang,
310058 Hangzhou, People's Republic of China
e-mail: rwang@cad.zju.edu.cn

S. Patnaik and X. Li (eds.), *Proceedings of International Conference on Soft Computing Techniques and Engineering Application*, Advances in Intelligent Systems and Computing 250, DOI: 10.1007/978-81-322-1695-7_52, © Springer India 2014

1 Introduction

Large screen display system is widely used as a display interface of multimedia applications. Kinds of tiled display solutions have been proposed, as introduced in [1]. And the solution of multi-projector seamless tiled display system becomes more and more popular. It mainly consists of three parts: a large screen, a video image processing equipment, and an array of projectors. Introduction of building the system can be found in [2]. PC cluster and complex hardware processor (usually consists of GPUs or FPGA arrays) are usually used as the image processing equipment, which increase the cost and power consumption significantly.

FPGA is a kind of programmable chip which is applied in the field of data computing frequently. And the computing capability of a single FPGA chip is proved to be high enough for real-time image processing by [3, 4].

In this paper, we propose a real-time image processing method which is mainly based on FPGA. This method could achieve high flexibility, low cost and low power consumption at the same time. In the next section, a dual-projection display system is taken as an example to present the mechanism of the proposed method, and the hardware framework of the image processor is also proposed. In Sect. 3, an experiment is presented to illustrate our approach. Finally, the conclusion is drawn at the end of this paper.

2 Method

To build a dual-projection seamless tiled display system, a large screen and two projectors should be placed steadily first, making sure that the projection regions of the projectors are overlapped slightly. Geometry calibration and intensity blending (or edge blending) are indispensable to splice the projected images seamlessly [5–8].

In our work, we defined a geometry calibration parameter α and an intensity blending parameter β for each pixel of the images accepted by the projectors. Parameter α represents the corresponding pixel's coordinate in the source image when the image on the screen is displayed without distortion. And parameter β represents the blending coefficient to reduce the brightness of the overlapping region.

An ordinary PC is firstly employed to export two reference images to the projectors. The PC analyzes the image on the screen and calculates out the geometry calibration and intensity blending parameters, which are then stored into a SD card in the form of a text file. Once the parameters are gotten, the preprocessing step is finished, and the PC becomes useless and should be removed from the system in daily projection. If the relative positions of the screen and the projectors are changed someday, we only need to preprocess the system again and update the parameters in SD card.

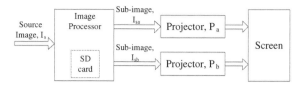

Fig. 1 Schematic representation of image processing step

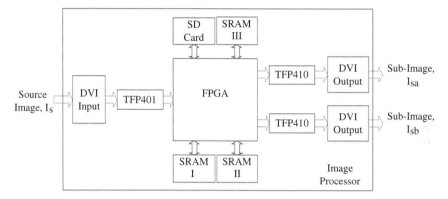

Fig. 2 Framework of the image processor

Once the preprocessing step is finished, the parameters could be used for image processing. As shown in Fig. 1, a hardware image processor, with the SD card mentioned above on its board, is connected with the projectors. The processor can receive a video image (source image, I_s) from any device, such as a computer, a camera. According to the parameters stored in the SD card, the processor divides and processes the source image into two sub-images (I_{sa} and I_{sb}), which are then transported separately to the projectors. The images projected on the screen are spliced seamlessly and a visually comfortable high-resolution image is displayed on the screen.

The key components of the image processor are shown in Fig. 2. A Xilinx Spartan-6 FPGA is employed as the core device of the processor. The Texas instruments TFP401 chip between the DVI input interface and the FPGA is used to decode the DVI format source image data. Two pieces of Samsung SRAMs (SRAM I and II) are connected with the FPGA, working as the buffers for caching the decoded image data. Since the SD card is able to store data without power and convenient for carry, a Kingston SD card is used to store the parameters mentioned above. However, the working frequency of SD card is not high enough for the real-time image processing. So a Samsung SRAM (SRAM III), whose highest working frequency could reach 250 MHz, is employed to temporarily cache the parameters stored in SD card. Two DVI-encoding chips (TFP410) are connected with the FPGA and the DVI output interfaces, to encode the sub-images into standard DVI video signals.

When powered on, the parameters in the SD card are read out by the FPGA and stored into SRAM III for real-time processing. At the same time, the image data decoded by TFP401 are stored into SRAM I and SRAM II in a Ping-Pong mode. Then, the FPGA starts to generate the sub-images pixel by pixel. For each pixel, FPGA reads out the relevant parameters from SRAM III firstly. According to the geometry calibration parameter α, FPGA reads out the corresponding pixel's color value from the source image stored in SRAM I and II, and then, the color value read out is multiplied by intensity blending parameter β. The result of the multiplication is just the pixel's color value of the sub-images we want. Finally, the sub-images are encoded by TFP410 chips and exported to the projectors through DVI cables.

3 Experiment Results

To verify the proposed method, an experimental system is established. Two BenQ MP771 projectors are employed to project images toward the large screen. The image processor is realized on a PCB, with a length of 15.0 cm and a width of 10.0 cm. The power consumption of the processor is 7.2 W (12 V, 0.6 A), whereas the power consumption of a PC or a complex hardware processor is at least tens of watts. The function of the image processor is mainly determined by the design of FPGA. Since the pixel frequency of the source video image is very high, pipelining and fan-out controlling techniques are widely used to improve the performance in the design of FPGA.

In the experiment, a computer is used to export a source video image (24-bit, 1600×720 resolution, 60 Hz) to the processor. When powered on, the FPGA-based processor could export two sub-images (24-bit, 1024×768 resolution, 60 Hz) in 25 s. The processor is firstly employed to divide the source image half-and-half, without geometry calibration or intensity blending. As shown in Fig. 3a, the sub-images projected on the screen are distorted and partly exceed the scope of the screen. And the overlapping region of the screen is much brighter than non-overlap regions. Then, the processor is employed to divide the source image according to the calibration and blending parameters. As shown in Fig. 3b, the sub-images are properly projected within the scope of the screen, and they are seamlessly spliced together without distortion. The brightness of the overlapping region is normalized. Thus, a visually comfortable high-resolution image is displayed on the screen.

Fig. 3 a Image projection without geometry calibration or intensity blending; **b** Image projection with geometry calibration and intensity blending

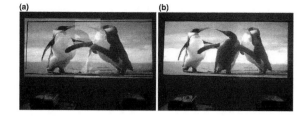

4 Conclusions

In this paper, we demonstrate a software–hardware cooperative approach for multi-projector seamless tiled display system. Two steps are included in this approach: PC-based preprocessing and FPGA-based image processing.

Experiment of a dual-projection implantation shows that a high-resolution video image is divided into two low-resolution sub-images by the proposed image processor, and the sub-images are projected onto a large screen, presenting a visually comfortable high-resolution image to the audience. The mount of projectors could be extended if a higher performance FPGA is used on the processor.

The proposed method is based on software–hardware cooperation, which results in the increase in the system's cost performance. Benefiting from the various available software blending and calibration algorithms, a high-resolution seamless tiled display system with flexible display performance can be implemented. Comparing to the existing methods based on PC cluster [8] or complex hardware processor [9], this method based on a simple FPGA-based processor could reduce the size, cost, and power consumption of the system in daily use, which results in a good application in multi-projection market.

References

1. Ni, T., et al.: A survey of large high-resolution display technologies, techniques, and applications. IEEE Virtual Reality Conference (2006)
2. Hereld, M., Judson, I.R., Stevens, R.L.: Introduction to building projection-based tiled display systems. IEEE Comput. Graphics Appl. **20**(4), 22–28 (2000)
3. Antoniewicz, A.: FPGA Implementation of Decomposition Methods for Real-Time Image Fusion. Image Processing and Communications Challenges 4, pp. 163–170. Springer, Berlin (2013)
4. Han, D.: Real-Time digital image warping for display distortion correction. 2nd International Conference on Image Analysis and Recognition (ICIAR), pp. 1258–1265. Springer, Berlin (2005)
5. Raskar, R., Welch, G., Fuchs, H.: Seamless projection overlaps using image warping and intensity blending. Fourth International Conference on Virtual Systems and Multimedia. Gifu, Japan (1998)

6. Sun, Yi., et al.: Computer vision based geometric calibration in curved multi-projector displays. 3rd IEEE International Conference on Computer Science and Information Technology (ICCSIT), vol. 6. (2010)
7. Sajadi, B., et al.: Color seamlessness in multi-projector displays using constrained gamut morphing. IEEE Transactions on Visualization and Computer Graphics, pp. 1317–1326. (2009)
8. Liu, X., et al.: Construct low-cost multi-projector tiled display system for marine simulator. 16th IEEE International Conference on Artificial Reality and Telexistence–Workshops, ICAT'06. (2006)
9. 3D Perception AS, http://www.3d-perception.com/

2D Simulation of Static Interface States in GaN HEMT with AlN/GaN Super-Lattice as Barrier Layer

Imtiaz Alamgir and Aminur Rahman

Abstract In this paper, two-dimensional simulation of interface charge effects in GaN-based high-electron-mobility transistors (HEMT) with AlN/GaN super-lattice (SL) device is performed. Charges of different polarity and magnitude are introduced in the interface, and their relative modification of drain current is studied. We have found that drain current has very different response to positive and negative charges of equal magnitude. Free hole accumulation at the interface is considered to give raise to different sensitivity for positive and negative interface charges. We have also investigated the drain current response for changed SL period thickness and equivalent Al composition of the quasi-alloy with different interface charge. The implication of our study in current collapse and related dispersion effects is discussed.

Keywords GaN HEMT · AlN/GaN super-lattice barrier · Interface states · Current collapse

1 Introduction

As a promising candidate for high-frequency and high-power applications, the high-electron-mobility transistors (HEMT) based on AlGaN/GaN hetero-structure have been attracting a lot of attention in the last decade. However, there are still some complexities that remain in production of GaN-based devices. The presence

I. Alamgir (✉) · A. Rahman
Department of Electrical and Electronic Engineering,
Ahsanullah University of Science and Technology, 141-142 Love Road,
Tejgaon Industrial Area, Dhaka 1208, Bangladesh
e-mail: alamgir_imtiaz@yahoo.com

A. Rahman
e-mail: aminur.rahman.eee@gmail.com

S. Patnaik and X. Li (eds.), *Proceedings of International Conference on Soft Computing Techniques and Engineering Application*, Advances in Intelligent Systems and Computing 250, DOI: 10.1007/978-81-322-1695-7_53, © Springer India 2014

of dispersion effects, which is a major reason for device degradation, has limited the initial expectation. Dispersion effects are not generally observed under DC test conditions, because they take place in primarily in dynamic situations, when a transient signal is applied to device terminals. Some of the observed effects are threshold voltage shift [1, 2], light sensitivity [3–6], transconductance frequency dispersion [7], and limited microwave power output [8, 9]. One of the major dispersion effects is the current collapse—defined as a transient and recoverable reduction in drain current response. A great effort has been dedicated to understand this negative effect and their possible elimination or, at least, minimization.

In present, several studies have shown that introduction of nanostructure into HEMT can enhance device performance. For example, recent reports have confirmed that by replacing conventional AlGaN barrier layers with AlN/GaN super-lattices (we call it quasi-AlGaN) composed of several nanometer-thick binary alloys, electric field induced by macroscopic polarization becomes much stronger, and the sheet carrier density is higher [10]. The AlN/GaN SLs have strong polarization effect and large conduction offset than the hetero-structure of AlGaN/GaN, and the SLs have deeper triangular well [11]. The carriers in the hetero-interface are confined in two dimensionally and have strong quantum effects which make more density of 2-DEG and higher mobility. These advantages have stimulated the interest to study the growth of HEMT with SL barrier and calculate the sheet carrier density in detail [12]. But, to the best of our knowledge, there is no literature focusing on the current collapse of quasi-AlGaN-based HEMT. However, for conventional HEMT, there exists extensive research on current collapse mechanism, which concludes that the presence of trapping centers related to surface, material, and/or interface is majorly responsible for this effect [9, 13–17]. Filling and emptying of these traps would change the density of charge and influence the recombination statistics. These donor-like traps states may be present at the surface and cause device degradation due to interaction with the free holes [18]. At the same time, acceptor-like traps in the interface of the device may cause the variation trends of drain currents [19]. All of these literatures provide more or less clear impression about the effect of surface and interface trap states in the performance of conventional AlGaN/GaN devices only, yet these type of comprehensive study is absent for quasi-AlGaN. But the parasitic effects have also strong influence on different parameters including sheet carrier density and mobility in quasi-AlGaN. Hence, it is very important to include them in physical models in order to fit to the operation of a real AlN/GaN SL HEMT device.

The aim of this work is to gain insights about the current collapse mechanism due to the static interface trap states of GaN HEMT with AlN/GaN SL as barrier layer. A previous essential step to understand these transient trapping phenomena is to study from an electrostatic point of view so that their effects can be studied and understood in an independent manner, providing a clear conception about the influence of these traps in modulating the electrical response. Thus, it will provide a very practical base intuition of inner mechanisms of these responses regardless of their actual time dependence.

Fig. 1 Schematic diagram of
GaN/AlN HEMT

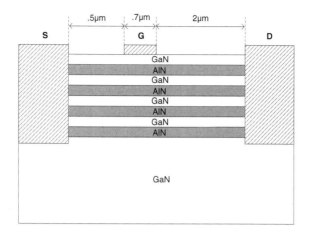

2 Method of Analysis and Physical Models

The two-dimensional device simulator used in this work was Silvaco Atlas. We used a drift–diffusion mode-space approach where the solution is decoupled into Schrödinger's equation in transverse direction that takes into account the quantum confinement effects and 1D classical drift–diffusion equation to account for carrier transport in each subband. Although theses equations are solved in steady-state conditions, our study considers a wide range of interface charges representing partially ionized donors or acceptors.

The super-lattice-based GaN HEMT device taken to illustrate our results is shown in Fig. 1. Substrate is assumed to be ideally semi-insulating with acceptor density (N_a) of 1×10^{15}/cm^3. The gate length was $L_g = 0.7$ µm, the separation between the gate and drain was $L_{GD} = 2$ µm, and the separation between gate and source was $L_{GS} = 0.5$ µm. AlN/GaN super-lattice was used as barrier layer where the topmost layer was GaN. The equivalent Al composition of the quasi-AlGaN was changed by adjusting the thickness of AlN and GaN layers. The equivalent Al composition is defined as:

$$Al\% = \frac{d_{AlN}}{d_{AlN} + d_{GaN}} \times 100\% \tag{1}$$

where, d_{AlN} = Thickness of AlN layer. d_{GaN} = Thickness of GaN layer.

The GaN mobility was taken as 1,300 and 30 cm^2/V s for electrons and holes, respectively, and mobility of 100 and 5 cm^2/V s was used in AlN for electrons and holes, respectively [18]. Saturation velocities for electron and holes in GaN were taken as 1.9×10^7 cm/s and 1×10^6 cm/s, respectively, according to Monte Carlo fits [20, 21]. The drain and source contacts are n-doped with a density of 5×10^{18}/cm^3. The built-in fields due to spontaneous and piezoelectric polarization are defined as fixed sheet charges at the top of AlN layer and AlN/GaN interface. A positive sheet charge $+\sigma_{pol}$ with density $+2 \times 10^{13}$/cm^2 was defined at the

Fig. 2 Drain current in the HEMT structure. Analysis with $V_{GS} = 0$ V for the following cases in Fig. 3

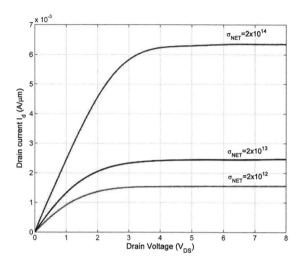

interface, and equivalent negative sheet charge of $-\sigma_{pol}$ was defined at the AlN surface. Newton numerical method was used for calculation and temperature of 300 K was used.

3 Result and Discussion

The interface trap states were considered as fixed charge of density σ_T—both positive and negative, uniformly distributed in the interface between AlN and GaN. This charge is added to the space charge term in the Poisson's equation. Although additional sources of charge, due to traps in locations other than the interface (such as bulk GaN or surface) may be present and have significant effect, the scope of our work is limited to the interface and presence of various types of charge in it.

As mentioned earlier, fixed sheet charges should be defined in the AlN surface and AlN/GaN interface in order to account for piezoelectric and spontaneous polarization. This is an intrinsic factor and should be considered prior to HEMT modeling. But if static interface states are also modeled, the original charge dipoles at the interface are modulated, as it gets added or subtracted with the static charges, and hence, the net modified interface charge density is expressed as, $\sigma_{net} = \sigma_{pol} \pm \sigma_T$.

Figure 2 shows the I_d–V_{DS} characteristic of the HEMT device whose structure was detailed in Sect. 2. Three cases are considered: (a) presence of original polarization charges are considered in the model, $-\sigma_{pol}$ at the AlN surface and $+\sigma_{pol}$ at the AlN/GaN interface. (b) A net charge of $+10\sigma_{pol}$ is considered at the AlN/GaN interface corresponding to the modeling of a positive charge density (PCD) at the interface. This results in $\sigma_{net} = +\sigma_{pol} + \sigma_T = 10\sigma_{pol}$. (c) A net

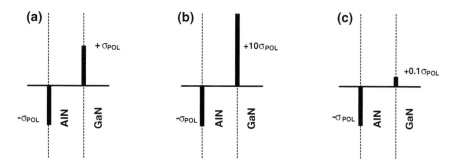

Fig. 3 **a** Original polarization charges are considered in the model, $-\sigma_{pol}$ at the AlN surface and $+\sigma_{pol}$ at the AlN/GaN interface. **b** A net charge of $+10\sigma_{pol}$ is considered at the AlN/GaN interface. **c** A net charge of $+0.1\sigma_{pol}$ is considered at the interface

charge of $+0.1\sigma_{pol}$ is considered at the interface, due to modeling of negative charge density (NCD) at the interface, which results in $\sigma_{net} = +\sigma_{pol} + \sigma_T = +0.1\sigma_{pol}$. All these three cases are shown in Fig. 3. The final effect in drain current can be observed for all three cased in Fig. 2. This figure clearly illustrates that there is a significant increment in drain current if the interface positive charge is increased by a certain magnitude. Similarly, a decrease in the drain current response is obtained when a decrease in positive interface charge is used in the model relative to the original dipole. It is evident that when negative charge $-\sigma_T$ is added to the original charge, $+\sigma_{pol}$, at the interface, the drain current does not give substantially different results with respect to case (a). On the other hand, sensitivity to the positive charges is comparatively higher.

At this point of discussion, it is necessary to mention the properties of donor and acceptor traps. A donor-like trap can be either positive or neutral like the donor, and an acceptor-like trap can be either negative or neutral like the acceptor. A donor-like trap is positively charged (ionized) when empty and neutral when filled (with an electron). An acceptor-like trap is neutral when empty and negatively charged (ionized) when filled (with an electron). There are three types of charges in the AlN/GaN interface, which are the positive polarization charges, the negative charged electron, and ionized trap charge, either acceptor or donor. If the interface charge is due to donor traps, which are partially ionized, a positive step in gate voltage will further ionize the donor traps and increase the positive-ionized donor-like traps. If the interface charge is due to partially ionized acceptor traps, a positive step in gate voltage will would decrease the occupation of acceptors, decreasing the negative-ionized acceptor-like traps in the interface. For both cases, because of the restriction of charge neutrality condition in the interface, this increment in the positive charge is followed by the increment in the electron concentration, leading to a significant increase in drain current. According to the trends in Fig. 2, this in only possible if the positive charge is added to the interface, thus increasing the positive part of the dipole. By similar argument, current reduction measured from DC conditions is due to either addition in NCD or to the

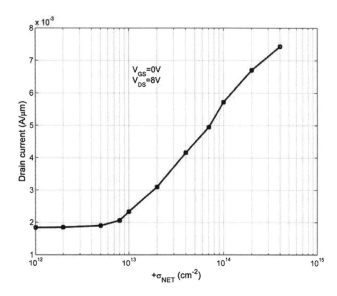

Fig. 4 Sensitivity in HEMT to variations in the net charge density. The drain current response for $V_{GS} = 0$ versus the net interface charge density

reduction in PCD—which is only possible if negative charge is added to the interface reducing the positive part of the dipole (Fig. 2). Hence, our steady-state analysis provides a compatible explanation for current collapse due to interface donor and acceptor traps in AlN/GaN super-lattice HEMT device, and this view is consistent with the experimental observations seen in case of conventional AlGaN/GaN HEMT, in which an increase in the gate-pulsed voltage produces a drain lag with an increase in the drain current [18, 19, 22, 23]. However, our result indicates that the sensitivity in the drain current is mainly for the increased PCD, not for increased NCD as shown in Fig. 4.

Figures 5 and 6 show the effect of SL period thickness and equivalent Al composition on drain current for positive and negative interface charge. The reason behind this modification is due to the dependence of 2-DEG on the thickness of AlN barrier [12], given by

$$n_s = \frac{\sigma}{e} \frac{\varepsilon_2 d_1}{\varepsilon_2 d_1 + \varepsilon_1 d_2} \left(1 - \frac{(E_D + E_F)\varepsilon_1\varepsilon_2}{\sigma e \varepsilon_2 d_1}\right)$$

or

$$n_s = \frac{\sigma}{e} \frac{\varepsilon_2 d_1}{\varepsilon_2 d_1 + \varepsilon_1 d_2} \left(1 - \frac{(e\varphi_B + E_F)\varepsilon_1\varepsilon_2}{\sigma e \varepsilon_2 d_1}\right) \tag{2}$$

where n_s is the 2-DEG density, d_1 and d_2 are the thickness of AlN barrier and GaN layer, respectively, E_F is the Fermi-level position with respect to the GaN

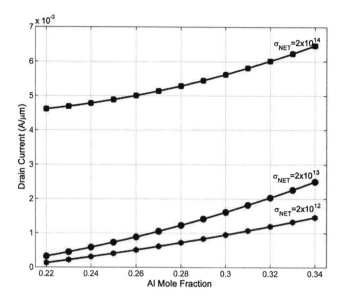

Fig. 5 Relationship of drain current and Al mole fraction

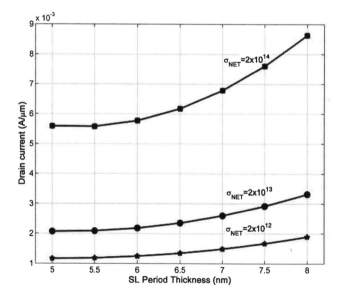

Fig. 6 Relationship of drain current and SL period thickness

conduction-band edge, ε_1 and ε_2 are the dielectric constants of AlN and GaN, respectively. According to the equation above, the 2-DEG density increases with d_1 and the increment of d_1 indicates the increment of both Al mole fraction and of

SL period thickness as shown in Figs. 5 and 6. Also, less sensitivity to negative interface charge can be observed for both cases.

At this point, it is interesting to analyze why the 2-DEG channel is less sensitive to the presence of an additional negative charge at the interface (Fig. 4). If vertical cuts are considered between the contacts and charge concentration is analyzed, a determinant factor is observed: Free holes accumulate at the interface and increase in such a way that they compensate for the added negative interface charge. This neutralization of negative charge is the reason of the great difference in sensitivity of the drain current to the positive and negative interface charge. An increment of the negative charge at the interface does not produce a significant reduction in the drain current response because is neutralized by holes.

4 Conclusions

In this work, the impact of static interface charges in AlN/GaN super-lattice HEMT device has been simulated and presented. The structure has been analyzed by means of a 2D device simulator, Silvaco. The mechanism of drain current changes has been discussed. It has been obtained that the sensitivity of the drain current is crucially dependent on the polarity of the interface charge. Hole accumulation next to the interface has been identified as the main cause of this sensitivity. The presence of positive charges at the interface has turned out to be the main cause of current collapse. Our results allow us to identify the nature of the traps introducing the interface charge and collapse effect.

References

1. Meneghesso, G., Chini, A., Zanoni, E., Manfredi, M., Pavesi, M., Boudart, B., Gaquiere, C.: Diagnosis of trapping phenomena in GaN MESFETs. In: IEDM Technical Digest, pp. 389–392 (2000)
2. Hasumi, Y., Kodera, H.: Simulation of the surface trap effect on the gate lag in GaAs MESFETs. Electron. Commun. Jpn. 85(2), 18–26 (2002)
3. Tirado, J.M., Sanchez-Rojas, J.L., Izpura, J.I.: 2-D simulation of static surface states in AlGaN/GaN HEMT and GaN MESFET devices. Semicond. Sci. Technol. 20(8), 864–869 (2005)
4. Trassaert, S., Boudart, B., Gaquiére, C., Théron, Y., Crosnier, Y., Huet, F., Poisson, M.A.: Trap effects studies in GaN MESFETs by pulsed measurements. Electron. Lett. 35(16), 1386–1388 (1999)
5. Klein, P.B., Freitas, J.A., Binari, S.C., Wickenden, A.E.: Observation of deep traps responsible for current collapse in GaN metal–semiconductor field-effect transistors. Appl. Phys. Lett. 75(25), 4016–4018 (1999)
6. Vetury, R., Zhang, N.Q., Keller, S., Mishra, U.K.: The impact of surface states on the DC and RF characteristics of AlGaN/GaN HFETs. IEEE Trans. Electron Devices 48(3), 560–566 (2001)

7. Kruppa, W., Binari, S.C., Dovespike, K.: Low-frequency dispersion characteristics of GaN HFETs. Electron. Lett. **31**(22), 1951–1952 (1995)

8. Binari, S.C., Klein, P.B., Kazior, T.E.: Trapping effects in GaN and SiC microwave FETs. Proc. IEEE **90**(6), 1048–1058 (2002)

9. Binari, S.C., Ikossi, K., Roussos, J.A., Kruppa, W., Park, D., Dietrich, H.B., Koleske, D.D., Wickenden, A.E., Henry, R.L.: Trapping effects and microwave power performance in AlGaN/GaN HEMTs. IEEE Trans. Electron Devices **48**(3), 465–471 (2001)

10. Smorchkova, I.P., Chen, L., Mates, T., Shen, L., Heikman, S., Moran, B., Keller, S., DenBaars, S.P., Speck, J.S., Mishra, U.K.: AlN/GaN and (Al, Ga) N/AlN/GaN two dimensional electron gas structures grown by plasma-assisted molecular-beam epitaxy. J. Appl. Phys. **90**, 5196–5201 (2001)

11. Kawakami, Y., Nakajima, A., Shen, X.Q., Piao, G., Shimizu, M.: Improved electrical properties in AlGaN/GaN heterostructures using AlN/GaN superlattice as a quasi-AlGaN barrier. Appl. Phys. Lett. **90**, 242112 (2007)

12. Xu, P., Jiang, Y., Chen, Y., Ma, Z., Wang, X., Deng, Z., Li, Y., Jia, H., Wang, W., Chen, H.: Analyses of 2-DEG characteristics in GaN HEMT with AlN/GaN super-lattice as barrier layer grown by MOCVD. Nanoscale Res. Lett. **7**, 141 (2012)

13. Khan, M.A., Shur, M.S., Chen, Q.C., Kuznia, J.N.: Current–voltage characteristic collapse in AlGaN/GaN heterostructure insulated gate field effect transistors at high drain bias. Electron. Lett. **30**(25), 2175–2176 (1994)

14. Klein, P.B., Binari, S.C., Ikossi-Anastasiou, K., Wickenden, A.E., Koleske, D.D., Henry, R.L., Katzer, D.S.: Investigation of traps producing current collapse in AlGaN/GaN high electron mobility transistors. Electron. Lett. **37**(10), 661–662 (2001)

15. Klein, P.B., Binari, S.C., Ikossi, K., Wickenden, A.E., Koleske, D.D., Henry, R.L.: Current collapse and the role of carbon in AlGaN/GaN high electron mobility transistors grown by metalorganic vapor-phase epitaxy. Appl. Phys. Lett. **79**(21), 3527–3529 (2001)

16. Ibbetson, J.P., Fini, P.T., Ness, K.D., DenBaars, S.P., Speck, J.S., Mishra, U.K.: Polarization effects, surface states, and the source of electrons in AlGaN/GaN heterostructure field effect transistors. Appl. Phys. Lett. **77**(2), 250–252 (2000)

17. Verzellesi, G., Pierobon, R., Rampazzo, F., Meneghesso, G., Chini, A., Mishra, U.K., Canali, C., Zanoni, E.: Experimental/numerical investigation on current collapse in AlGaN/GaN HEMT's. In: IEDM Technical Digest, pp. 689–692 (2002)

18. Tirado, J.M., Rojas, J.L.S., Izpura, J.I.: Trapping effects in the transient response of AlGaN/GaN HEMT devices. IEEE Trans. Electron Devices **54**(3), 410–417 (2007)

19. Zhang, W., Zhang, Y., Mao, W., Ma, X., Zhang, J., Hao, Y.: Influence of the interface acceptor-like traps on the transient response of AlGaN/GaN HEMTs. IEEE Electron Device Lett. **34**(1), 45–47 (2013)

20. Device Simulator Atlas Ver. 5.10.0.R. Atlas User's Manual, Silvaco Int., Santa Clara, CA (2005)

21. Farahmand, M., Garetto, C., Bellotti, E., Brennan, K.F., Goano, M., Ghillino, E., Ghione, G., Albrecht, J.D., Ruden, P.P.: Monte Carlo simulation of electron transport in the III-nitride wurtzite phase materials system: binaries and ternaries. IEEE Trans. Electron Devices **48**(3), 535–542 (2001)

22. Chini, A., Lecce, V.D., Esposto, M., Meneghesso, G.: Evaluation and numerical simulation of GaN HEMTs electrical degradation. IEEE Elec. Device Lett. **30**(10), 1021–1023 (2009)

23. Tirado, J.M., Rojas, J.L.S., Izpura, J.I.: Simulation of surface state effects in the transient response of AlGaN/GaN HEMT and GaN MESFET devices. Semicond. Sci. Technol. **21**, 1150–1159 (2006)

Study on Model and Platform Architecture of Cloud Manufacturing for Aerospace Conglomerate

Jihong Liu, Hongfei Zhan and Wenting Xu

Abstract With rapid development of emerging information technologies such as cloud computing, the Internet of Things, cloud manufacturing, which emphasizes manufacturing resources virtualization and service-oriented manufacturing capacity, has become an emerging network-based manufacturing. Based on the characteristics and needs of aerospace conglomerate, the conglomerate cloud manufacturing model that puts emphasis on management and control of core resources is proposed. And the architecture of the conglomerate cloud manufacturing platform is proposed based on the conglomerate cloud manufacturing model. Four core services are detailed. Meanwhile, the key technologies of conglomerate cloud manufacturing are discussed.

Keywords Cloud manufacturing · Cloud computing · Cloud service · Conglomerate cloud manufacturing centralized management and control platform

1 Introduction

As a national strategic high-tech enterprise, aerospace conglomerate, which promotes the depth of integration of industrialization and information technology, has an important role to accelerate the transformation and upgrading of China's manufacturing. Aerospace conglomerate has core resources that are the company's

J. Liu (✉) · H. Zhan · W. Xu
School of Mechanical Engineering and Automation, Beihang University, Beijing, China
e-mail: ryukeiko@buaa.edu.cn

H. Zhan
e-mail: feihong1225@126.com

W. Xu
e-mail: xuwenting8911@126.com

S. Patnaik and X. Li (eds.), *Proceedings of International Conference on Soft Computing Techniques and Engineering Application*, Advances in Intelligent Systems and Computing 250, DOI: 10.1007/978-81-322-1695-7_54, © Springer India 2014

most important strategic resources and drives the sustainable development of corporate economy. These core resources have gradually become the key factor of the core competitiveness of enterprises and product innovation capability. Aerospace conglomerate product development processes with lots of characteristics, such as many participators in research with a wide geographical distribution, a large amount of data with frequenting interaction, large research tasks with short development cycle, so fully sharing of resources and business in close collaboration is urgently needed. However, aerospace product development still faces some problem:

1. Large-scale high-performance computing hardware and software resources configuring are distributed with low utilization. Large-scale high-performance computing hardware and software that are required for design, analysis, simulation, and testing are distributed in each unit.
2. The conglomerate's centralized management and control capabilities and management efficiency need to be improved. Existing operation and management of the conglomerate are more extensive with the dispersion of core resources and fragmented subordinate units.
3. Conglomerate's internal-product development knowledge storage is dispersed and has lack of effective management. The conglomerate carries out all the business around product development. A variety of knowledge and data resources that are produced in the development process are currently limited to the various units within the department.

Cloud manufacturing model proposition has provided a feasible way for solving the above problems [1–3]. Cloud manufacturing model draws on the concept of cloud computing [4], and uses the cloud computing technology, the Internet of Things [5] technology to virtualize all kinds of manufacturing resources and make the manufacturing capacity service-oriented [6].The core of cloud manufacturing is to build the cloud manufacturing services platform to integrate a variety of hardware and software manufacturing resources that are required for manufacturing enterprises and provide 4A (active, agile, aggregative, and all-aspects) services to manufacturing enterprises.

The current studies of cloud manufacturing focus on model and architecture. Zhang [7] has analyzed characteristics and differences of cloud manufacturing and related advance manufacturing models. Zhan [8] has proposed a cloud manufacturing for manufacturing and management of conglomerate. Most of the current study are the versatility of cloud manufacturing model and architecture and cannot fully meet the different manufacturing environments of aerospace conglomerate. The conglomerate requires a different cloud manufacturing model corresponding to its own characteristics. It also needs a different architecture to guide the build of its cloud manufacturing system. Cloud manufacturing model for conglomerate, which is proposed, does not fully reflect the characteristics and needs of the conglomerate, and it is to be further investigated.

2 Aerospace Conglomerate Cloud Manufacturing Model

Aerospace conglomerate cloud manufacturing model contains four types of roles: Conglomerate cloud manufacturing centralized management and control platform (CCM–CMCP), the subsidiary of the conglomerate under the cloud, the suppliers of the conglomerate, and the conglomerate vendor. The specific model is shown in Fig. 1.

Conglomerate realizes the centralized management and control of the subsidiaries of the Conglomerate through CCM–CMCP. The subsidiaries take core resources into CCM–CMCP through the Internet of Things, Industrial Ethernet, and Internet. CCM–CMCP takes the use of virtualization technology to build a virtual resource pool and packages the form of the service-oriented virtual resource. Conglomerate realizes unified handling orders, integration of resources, and decomposition of the business to blind to its subsidiaries as appropriate. In the implementation of the specific business process, Conglomerate takes real-time traffic scheduling and provides the appropriate resources. CCM–CMCP fully shares and realizes centralized management and control of supply chain resources, and realizes uniform distribution of equipment, raw materials, parts, and products. Conglomerate cloud manufacturing model has four typical characteristics:

Efficient sharing of hardware and software resources. Overall amount of investment in large-scale equipment and scarce software is great, but the subsidiaries of the Conglomerate cannot cover everything since resources relatively scattered and the subsidiaries unable to focus on the advantages to solve the problem of large-scale engineering development, then these hinder the level of the design and the design ability of the Conglomerate to further enhance. Conglomerate efficiently shares various subsidiaries of computing power, large-scale and high-precision equipment through CCM–CMCP, and CCM–CMCP greatly enhances equipment resources sharing rate and reduces repeat purchase, efficient managing of a large number of devices.

Sharing and centralized management and control of supply chain resources. Suppliers and logistics resources are the key link in the supply chain. Supplier resources and logistics resources cannot be adequately shared leading to high cost of material procurement and logistics resources, taking vehicle logistics in an automobile group as an example. When there is vehicle shipping, logistics vehicles are basically full. And it often led to the postponement of the delivery due to inadequate logistics capabilities, affecting the entire chain of the development and production; when logistics vehicles return, it always occur the phenomenon of no-load. CCM–CMCP integrates conglomerate-wide supply chain resources, realizes centralized management and control of logistics system, improves the utilization of logistics resources, enhances their capacity peak, and reduces procurement and logistics costs.

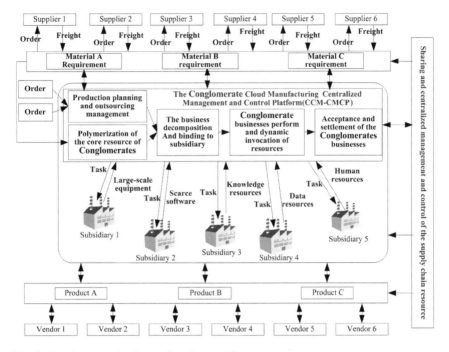

Fig. 1 Conglomerate cloud manufacturing model

Integration of the knowledge resources of the whole conglomerate, the effective realization of creation, representation, sharing, application and inheritance of knowledge resources. Knowledge resources are important to guarantee for the sustainable development of enterprises and the enhancement of corporate innovation capability. Knowledge resources include domestic and foreign patents, standards, technical papers, typical design/process/manufacturing processes, and experience skills, etc. CCM–CMCP realizes unified management of information and knowledge, standardizes the form of various information resources, facilitates information sharing and knowledge accumulation, and accelerates the value of the use of knowledge and technological achievements into the business.

Centralized management and control of the whole conglomerate. Conglomerate realizes centralized management and control of the whole conglomerate subsidiaries through CCM–CMCP. When the subsidiaries conduct business, the CCM–CMCP provides all the available resources in the conglomerate to support them completing the work efficiently. After the completion of the business, the conglomerate also achieves acceptance and settlement of the subsidiaries' business through CCM–CMCP.

3 Architecture of Aerospace Conglomerate Cloud Manufacturing

As it is shown in Fig. 2, this paper has put forward a new architecture of aerospace conglomerate cloud manufacturing based on the aerospace conglomerate cloud manufacturing model. This architecture consists of three layers: resource management layer, core business layer, and the platform portal layer.

Resource management layer. This layer mainly provides the production of hardware resources access, such as high-end precision CNC machine tools and other processing equipment; software resources access, such as UG software. And it also completes the resource registration, monitoring, resource access security authentication, access authorization control and management, etc.

Core business layer. The core business layer mainly includes four parts: the platform management center (data center, shared resource service center, information systems operation and maintenance management center), research and production control services, collaborative product development process and data management services, and research and production services.

The platform portal layer. This layer includes portal layout design and management, the information content of the polymerization of the portal, user organizations and user management tools, portal unified authentication and personalization management.

The core business layer is in charge of the flow directions and combinations of cloud manufacturing services to integration of the knowledge resources of the whole conglomerate. And the research and production services are the core services to support models for product development that includes concurrent design and simulation verification service, manufacturing process and resource management service, testing the trial resource service and maintenance service.

Concurrent design and simulation verification service. The concurrent design and simulation verification service that is design task-oriented, decomposes multi-disciplinary design tasks into a plurality of single disciplines, and achieves the concurrent design of the product and simulation, analysis and optimization of the key components through the coordination and connection between single-disciplines design tasks.

Manufacturing process and resource management service. Manufacturing process and resource management service that matches the needed process initiatively according to the users' demand, and the constraint condition provides the planned optimum processing services chain and realizes the dynamic matching of the processing capacity simultaneously. To ensure coordinated implementation of the production planning and production scheduling, it also provides dynamic scheduling driven by production emergencies and constraint scheduling.

Testing the trial resource service. Testing the trial resource service is of users' demand-oriented, this service transforms the diversified test trial demands of users into unified description form and achieves the required testing trial devices through matching the similarity of the function or performance of testing trial devices with

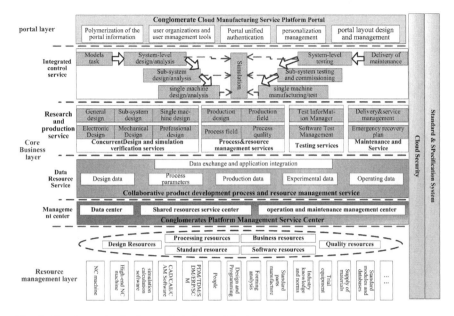

Fig. 2 Conglomerate cloud manufacturing architecture

demands. In addition, the evaluation results of the testing trial for the service and the handling mechanism to the change of the users' demand should be given.

Maintenance and service. Maintenance and service provides customized delivery, maintenance and service management to users. When encounter emergency, this service will give matched contingency plans according to situation description, so that it can provide high-quality service to users.

4 Key Technologies of the Cloud Manufacturing for Aerospace Conglomerate

The virtualization of diversified soft- and hard-manufacturing resources and cloud adaptation access technology. Compared to the involved resources of cloud computing, the type of soft- and hard-manufacturing resources are more complex, which involve in various types of manufacturing information application software (CAD, CAE, etc.), business management software (PDM/PLM, ERP, etc.), and numerous processing, experiment, test equipment devices, in addition, the different conditions of resources virtualization. Even more, the technologies of cloud computing, the Internet of Things, embedded software are still developing and can reference with low maturity. Therefore, the virtualization of diversified heterogeneous soft- and hard-manufacturing resources and cloud adaptation access technology are the key point for cloud manufacturing that must be solved firstly.

The technology of agile dynamic manufacturing services for Multi-Agent. The character of manufacturing resources service for conglomerate's cloud manufacturing service is the service demand related with large-scale tasks, at the same time a lot of services are required to provide which constraint conditions are stricter. Therefore, at this condition, the demand for manufacturing resource services are possibly numerous, concurrent, and personalized, while the supply of manufacturing resource services are with ranged capacity and dynamic changes, how to get the optimal or better service scheme through quick match and efficient portfolio are the key technologies.

The modeling technology and method for business collaboration services for complex conglomerate cloud manufacturing environment. Conglomerate cloud manufacturing platform application faces a global, complex, and open competitive environment, which leads to the selection of key index system and model building for business collaboration and management decision making to be very complex. Business collaboration of conglomerate involves not only organizational structure, management way, but also technical capacity and task character of conglomerate. Therefore, facing to complex cloud manufacturing environment, the modeling technology and method of scientific efficient business collaboration services and decision-implementing process for conglomerate are the key technologies.

5 Conclusions

Being a sort of new network-based manufacturing in the frontier of science, there is no successful experience for this new model, inevitably, the services model and architecture combined with the specific characteristics and needs of the aerospace conglomerate determine the capability of cloud manufacturing platform. Thus, the conglomerate cloud manufacturing model that puts emphasis on management and control of core resources is proposed,and based on the model this paper gives the three-tier architecture of the conglomerate cloud manufacturing platform. The conglomerate cloud manufacturing platform realizes unified order processing, unified resources allocation and business dynamic matching; so it can enhance the conglomerate's overall business and innovative ability to reduce operating costs. Finally, the three key technologies of the cloud manufacturing for aerospace conglomerates are discussed that are to lay the foundation for further research.

References

1. Li, B., Zhang, L., Ren, L., Chai, X., Tao, F., Luo, Y., Wang, Y., Yin, C., Huang, G., Zhao, X.: Further discussion on cloud manufacturing. Comput. Integr. Manuf. Syst. **17**(3), 449–457 (2011). (in Chinese)
2. Li, B., Zhang, L., Zhang, S.: Cloud manufacturing: a new service-oriented networked manufacturing model. Comput. Integr. Manuf. Syst. 16(1), 1–7, 16 (2010) (in Chinese)

3. Xu, X.: From cloud computing to cloud manufacturing. Robot. Comput.-Integr. Manuf. **28**(1), 75–86 (2012)
4. Rings, T., Caryer, G., Gallop, J., Grabowski, J., Kovacikova, T., Schulz, S., Stokes-Rees, I.: Grid and cloud computing: opportunities for integration with the next generation network. J. Grid Comput. **7**(3), 375–393 (2009)
5. Bandyopadhyay, D., Sen, J.: Internet of things: applications and challenges in technology and standardization. Wireless Pers. Commun. **58**(1), 46–69 (2011)
6. Ren, L., Zhang, L., Zhang, Y.B., Tao, F., Luo, Y.L.: Resource virtualization in cloud manufacturing. Comput. Integr. Manuf. Syst. **17**(3), 511–518 (2011) (in Chinese)
7. Zhang, L., Luo, Y., Fan, W., Tao, F., Ren, L.: Analyses of cloud manufacturing and related advanced manufacturing models. Comput. Integr. Manuf. Syst. **17**(3), 458–465 (2011) (in Chinese)
8. Zhan, D., Zhao, X., Wang, S., Cheng, Z., Zhou, X., Nie, L., Xu, X.: Cloud manufacturing service platform for group enterprises oriented to manufacturing and management. Comput. Integr. Manuf. Syst. **17**(3), 487–494 (2011) (in Chinese)

Service Composition Algorithm for Vehicle Network Based on Multiple Ontology

Yamei Xia and Chen Liu

Abstract In order to increase the efficiency and satisfaction degree of vehicle network's service composition, this paper put forward a service composition algorithm for vehicle network using multiple ontology. This algorithm builds not only vehicle network-domain ontology, object ontology, and service ontology, but also social relation ontology and preference ontology. On the basis of the above ontology, we generate relevant ontology instances according to historical data, which enable better application of social relation preference and user preference when making vehicle network service composition. At the same time, as the vehicle network has its peculiarities, we adopt a lightweight service organizing, managing, and issuing method to save the time spent in selecting services. The above algorithm enables a significant increase in the efficiency and satisfaction degree of the service composition.

Keywords Vehicle network · Social relation · Ontology · Service composition · Preference

1 Introduction

With the rapid development of the Internet of things (IoT) and wireless communication technologies, the vehicle network, one of the model applications of IoT, has become a new focus in this field [1]. How to construct a vehicle network, how

Y. Xia (✉)
Software Institute, Beijing University of Posts and Telecommunications, Beijing, China
e-mail: ymxia@bupt.edu.cn

C. Liu
School of Computer Science, Beijing University of Posts and Telecommunications, Beijing, China
e-mail: lchen@bupt.edu.cn

S. Patnaik and X. Li (eds.), *Proceedings of International Conference on Soft Computing Techniques and Engineering Application*, Advances in Intelligent Systems and Computing 250, DOI: 10.1007/978-81-322-1695-7_55, © Springer India 2014

to maximize its functions, and how to deliver beneficial services on this platform have become important problems needed to be solved. Compared with traditional Internet technologies, the vehicle network, because of its real-time changing network nodes and typology, has such characteristics as dynamics, lightweight, mobility, self-organization, domain, and socialization [2]. This paper, based on the above characteristics, makes an in-depth research on the related service composition algorithm in the field of mobile vehicle network. The successful implementation of this project will provide useful suggestion and reference for the construction of the vehicle network and will be of great significance to the development of the IoT.

2 Related Researches

To date, many technologies and approaches have been used by researchers to analyze Web service composition. BPEL [3] is an XML-based language for Web services composition. A BPEL process provides a Web service interface; behind the scenes, it calls other Web services to do the actual work. WS-BPEL 2.0, the latest version of BPEL, is an OASIS Committee specification as of January 2007.

The Web service modeling ontology [4] (WSMO) is one of these initiators' related technologies. WSMO is based on the Web service modeling framework (WSMF). The core concepts are ontology, goals, Web services, and mediators.

OWL-S [5] specifies a set of ontology based on OWL, which are used to describe the different aspects of a semantic Web service. The primary aim of OWL-S is to support automatic service discovery, invocation, composition, and interoperation. The set of ontology to describe an OWL-S service includes an upper service ontology, service profile, service model, and service grounding.

WSMO and OWL-S are the main specifications for describing semantic Web services. Both of these descriptions aim at providing automatic discovery, composition, invocation, and interoperation of Web services.

Details for other technologies, such as UML modeling and rule reasoning, are left out here. There are some service composition approaches applying the above technologies of BPEL, OWL-S, and WSMO: semantics matching approach based on interface, service composition approach based on model driving, service composition approach based on services ontology, etc. Recently, many other new Web service composition approaches are put forward. Lin Lin and Ping Lin have presented an overview of BPEL as an orchestration language for Web services and CCXML and SCXML as orchestration languages for call control and other similar contexts. BPEL, CCXML, and SCXML can be complementary tools in an orchestration toolkit, and this can be used to solve real-time communications, including telecom applications and converged telecom data applications [6]. Timm and Gannod [7] present a model-driven architecture-based approach for specifying semantic Web services composition through the use of a UML profile that extends class and activity diagrams. In addition, some scholars are also researching social relation attributes [8], which can also be used in the generation of services composition.

In addition, there are some researchers introducing preference into service composition [9]. Some researchers take the QoS into consideration in the analysis of service composition, which shows the needs for users' preference in service composition [10, 11]. These service compositions, filtering services by adding preferences and QoS parameters, reduce the scope of selected services and increase the satisfaction of composite services.

The first and second parts of this paper are introduction and related researches, the third part puts forward the model of service composition, and the fourth part gives a conclusion and further research.

3 Algorithm

3.1 General Structure for Vehicle Network Services Platform

Figure 1 is the basic framework of this study with the focus on service composition methods. In order to express the integrity of this study, Fig. 1 also gives the platforms of vehicle network, social relation network, and vehicle network. What follows is the analysis of the above contents.

- Service composition method for self-organized vehicle network

The service composition algorithm for self-organized vehicle network has different features from the service composition issues on the Internet, and it has higher requirements on dynamics and real-time performance. The dynamics feature refers to that the outer environment around the service composition is in constant change and needs to be paid close attention to. The real-time performance feature refers to that the information of all kinds of changes must be acquired in real time and be used in service composition; otherwise, the composed services will be of no use.

- Services for vehicle network

The key services for the field of vehicle network have a clear feature of domain. Some key services for vehicle network proposed in this paper include navigating service for road locations, journey plan service, reservation service, notification service, etc.

- Social relation network for vehicle network service composition

Some of the relations that the vehicle network show have very close connections with the social relation attributes. In this paper, we will build a mapping mechanism between the vehicle network relation and the social relation network and develop related domain ontology and preference ontology using social relation

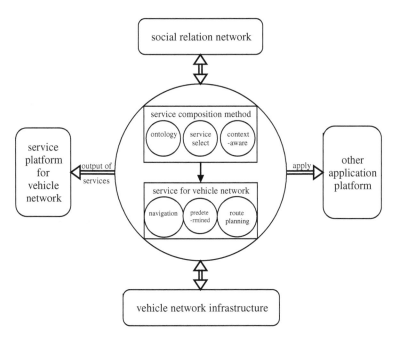

Fig. 1 Service composition structure for vehicle network

attributes to make service composition for vehicle network to increase the service quality of the composed services in the field of vehicle network and the individualized experience of the users.

3.2 Ontology Building

In this paper, there are domain ontology, object ontology, services ontology, social relationship ontology, and preference ontology for service composition. The domain ontology, object ontology, social relationship ontology, and preference ontology are built with OWL, and the services ontology is built with OWL-S.

Social relation ontology mainly describes such subject elements as human being, community, organization, unit, etc., analyzes the characteristics and attributes of each subject element, and accordingly excavates all the related attributes of each subject element and builds social relation network (Fig. 2).

In service composition system, a service has multiple attributes, and each person has a different preference weight on each attribute. In order to express user's preference more accurately and really, the possible preference of a service is decomposed into many attributes, and the preference values of each attribute will be obtained respectively, after which the preferences of multiple attributes will be further compounded to express users' preference to this service. Different attribute

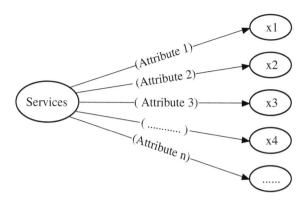

Fig. 2 Decomposing attributes of the services

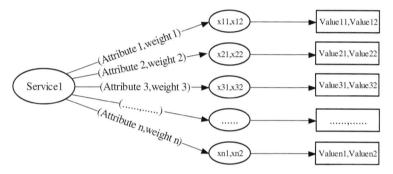

Fig. 3 Decomposing the service preference into many attributes

preferences have different attribute types and operation definitions which are defined in the preference ontology. In this paper, ontology is introduced to express user's preference, in which the service is expressed by the concept of Class in OWL, the attribute decomposition is expressed by object attribute type, and the domain of each service attribute preference is directed at the description source of this attribute preference through the mark which is defined in OWL specification. Details for the building process and approach of preference ontology are left out here (Fig. 3).

Among them, *attribute* denotes the attribute decomposition of the service concept, and *weight* denotes the weight that the attribute takes in the concept decomposition. Just as subject–attribute decomposition, the sub-subject preference of the decomposition in subject–subject decomposition also has a weight to correspond. In addition, every concept in the preference ontology is divided into the must-be-met preference and choose-to-meet preference. The preference parameters in the preference ontology can be acquired by many preference extracting algorithms.

3.3 Algorithm Realization

The service composition for the Internet generally uses universal description, discovery, and integration (UDDI) to register, issue, and seek services, but for the vehicle network services, the UDDI has some inherent disadvantages including that it has a much higher operation cost, there are problems in the real-time updating of the information, there is no universal UDDI, etc. In addition, the searching of the service is realized by seeking in a great amount of services using searching engine with a heavy cost, which decreases the efficiency of the service searching in some degree. In this study, we adopt an easier service organizing method using vector mapping to deal with the service set of the vehicle network to decrease the scope of service searching and increase the system's performance.

In addition, because of the importance of "geographic location" of the vehicle network service, we organize the vectors according to geographic locations and use multi-chained lists in the implementation: The services with the same geographic location and the same kind are organized together as the same sector, and the first node of each kind of service has a pointer to point to other kind of services to denote these services in the same location. In this way, the searching efficiency will be significantly increased. Additionally, although the services in the neighboring locations are located in different vectors, there are links between those vectors to facilitate the searching of the location. What follows is the service organizing method (Table 1).

3.4 The Complexity Analysis

In order to find the service with the highest degree of satisfaction, the service composition needs to make ergodic computation of all the service on the service links on which the candidate services are. We set the amount of the location as n, the service number that each location includes as m_i ($i = 1 \ldots n$), and then the algorithm complexity of finding the service on the i location of the first item is $o(n) + o(m_i)$. In this regard, by using such organization mode of service chained list, the algorithm complexity is decreased. If we do not use this kind of organization mode, the algorithm complexity of finding a service is $o(\sum_{i=1}^{n} m_i)$, and $\sum_{i=1}^{n} m_i$ is the service amount in the service ontology.

To compute the preference of each service, we need to compute all the attributes of the service. Set the service w_{ij} at location i, which includes t_{ij} attributes, so we get the algorithm complexity of the service with the highest satisfaction degree at the location i as $o(\sum_{j=1}^{m_i} t_{ij})$.

Table 1 Service composition algorithm based on ontology

Input parameter: object requirements, domain ontology, social relation ontology, users'
   preference ontology, etc.

Output parameter: services composition list SCL

1. For the object requirements, use task decomposition module to decompose it automatically into
   series($WS_R, WS_1, WS_2, \ldots, WS_n, WS_P$); // WS_R denotes the user's object requirement input, and
   WS_P denotes the user's object output

2. Initialize queue `SCL = NULL`;

3. `SL1 = Get Map Next Services (WSR)`; // to acquire the service list that matches WS_R
   interface

`If SL1 == NULL`

`Return NULL`; //services composition fails

`Else`

`{`

`WS1 = GetMaxSatisfyingServices(SL1)`; //use social relation ontology and user's
   preference ontology to acquire services with the biggest user's preference value in the found
   service list

`SCL = EnQueue (WS1)`; //put the service with the biggest user's satisfy value into services
   composition list SCL

`{`

`SL`$_i$ `= GetMapNextServices(WS`$_i$`)`;

`If SL`$_i$ `==NULL`

`Empty(SCL)`;//services composition fails, and clear services composition list

`Return NULL`;

`Else`

`WS`$_{i+1}$`=GetMaxSatisfyingServices(SL`$_i$`)`;

`SCL = EnQueue (WS`$_{i+1}$`)`;

  `i++`;

`}until (WS`$_P$ `== GetServicesOut (WS`$_{i+1}$`))`//the service composition finishes when
   finding the final output meet the user's object output

`}`

4. return to services composition list SCL, and give its to engine for implementation

4 Conclusions

This paper puts forward a vehicle network service composition algorithm based on
multiple ontology for the vehicle network application, which applies domain
ontology, object ontology, service ontology, social relation ontology, and preference to make service composition. Because of the limitation of the paper's length,
we have not made in-depth analysis of the social relation and preference attributes,
and as a result, the generated ontology instances have limited depth and scope,
which will be improved in our future work.

References

1. Mitra, P., Poellabauer, C.: Efficient group communications in location aware mobile ad-hoc networks. Pervasive Mob. Comput. **8**(2), 229–248 (2012)
2. Wei, D., Wang, Y., Wang, J., et al.: A survey on mobility models of vehicular adhoc networks. Chin. J. Comput. **36**(4), 677–700 (2013)
3. Web Services Business Process Execution Language v.2.0, OASIS Committee Specification, 31 Jan 2007
4. Lauren, H., Roman, D., Keller, U.: Web services modeling ontology-standard (WSMO Standard). http://wsmo.org/2004/d2/v0.2/2004
5. OWL-S: semantic markup for web services, W3C member submission 22 Nov 2004. http://www.w3.org/Submission/OWL-S (2004)
6. Lin, L., Lin, P.: Orchestration in web services and real-time communications. Web Serv. Telecommun. IEEE Commun. Mag. **45**(7), 44–50 (2007)
7. Timm, J.T.E., Gannod, G.C.: Specifying semantic web service compositions using UML and OCL. In: IEEE International Conference on Web Services (2007)
8. Jahanbakhsh, K., King, V., Shoja, G.C.: Predicting missing contacts in mobile social networks. Pervasive Mob. Comput. 698–716 (2012)
9. Tsinaraki, C., Christodoulakis, S.: A user preference model and a query language that allow semantic retrieval and filtering of multimedia content. In: IEEE Semantic Media Adaptation and Personalization (SMAP'06) (2006)
10. Xia, Y., Cheng, B., Chen, J., et al.: Optimizing services composition based on improved ant colony algorithm. Chin. J. Comput. **35**(2), 270–281 (2012)
11. Zhao, X., Song, B., Huang, P., Wen, Z., Weng, J., Fan, Y.: An improved discrete immune optimization algorithm based on PSO for QoS-driven web service composition. Appl. Soft Comput. **12**(8), 2208–2216 (2012)

Design and Application of Virtual Laboratory for Photography

Xuefei Shi

Abstract This paper discusses design and development method of virtual experiment teaching system based on VRML, the function framework of virtual experiment system, and the interface design of VRML and Java language through the "photographic laboratory" case and also discusses some problems in designing, development, and application of virtual experimental system in teaching.

Keywords Language (VRML) · Virtual experiment system · Simulation laboratory for photography · Design and application

1 Introduction

Virtual laboratory can effectively reduce the experimental cost and enhance the experimental efficiency. It can conducive to the realization of distance learning and resources sharing. Therefore, this experimental teaching mode is becoming an important aspect of educational information construction in our country. This paper makes a preliminary analysis on the design, development, and application issues surrounding the virtual experiment system based on VRML (Virtual Reality Modeling Language), through the design, development, and application of "photographic simulation laboratory."

2 Design and Analysis of Virtual Experimental System

The virtual experiment system is often referred to as virtual laboratory. It is mainly to create software and hardware environment and experimental platform in a computer system that can assist in experiment teaching, simulation experiment,

X. Shi (✉)
Shenyang Normal University, Shenyang 110034, China
e-mail: xuefeishi@yahoo.com.cn

S. Patnaik and X. Li (eds.), *Proceedings of International Conference on Soft Computing Techniques and Engineering Application*, Advances in Intelligent Systems and Computing 250, DOI: 10.1007/978-81-322-1695-7_56, © Springer India 2014

and even some alternative or fully substitute traditional experimental processes and links, by means of multimedia and virtual reality technology. Experimental tools and subjects of the real experiment process are the physical form, and virtual experiment is to operate various experimental artifacts in a virtual scene. In the virtual experimental environment, the experimenter can also complete a variety of predetermined experimental projects like in a real environment, and the learning or training effect is similar to or even better than the experimental results obtained in real environment.

The virtual experiment is established in a virtual experimental platform. It pays attention to the interaction of experimental manipulation and simulation of the experimental process and the experimental effect. Generally, a typical virtual experiment teaching system is operated by the user interface, experiment management system, data management system, and management and control system in integration. The user interface is interactive window. The database includes all kinds of database and knowledge base, method base, rule base, model base, and case base. Integrated management system is a control section of virtual experiment system, and it calls database information according to the experimental management system command, at the same time the data resource protection and coordinates the work between each function module cohesion and process.

Laboratory management is the core of the virtual experiment system [1]. It constructs the experimental model according to the user's input information that performs the experimental task and processes the data according to the model. The experimental process can directly call the database, and comprehensively analysis, and treat to the user operation. Experimental teaching is to make students understanding the related knowledge, concept, or principle, being familiar with experimental apparatus with the correct use method, mastering the experimental process operation links, feeling and understanding the experimental results. Therefore, as the experimental management system, design of experiment, experimental apparatus, experimental scene rules, experimental results, and related support system will become the core and key of the virtual experiment teaching system design.

3 Functional Architecture of Simulation Experimental System of Photography

3.1 Target Analysis and Technical Scheme

Simulation of photographic laboratory is auxiliary teaching system of "photography" curriculum experiment. It is designed in order to deepen the students' understanding and application of the camera, and to provide students with a photographic skills training in the virtual scene, ultimately enhance the student photography skills practice. The design target of the system is mainly reflected in: first, students can understand the main components and function of the camera, in the virtual scene to observe the 3D camera structure model; second, the students can be trained to randomly shoot, using simulate camera to adjust the focal length, aperture, shutter control regulation, and scene; third, the system can be run on the network, and its interactivity is better; in fourth, the operation process is real time, and the experiments can reflect synchronized photo effect; in fifth, the simulation degree is high, and the results can reflect the specific experimental parameters of the camera operation.

According to the design requirements of the simulation system of photographic experiments, we selected VRML and Java to be the main technical support. VRML is a kind of virtual reality modeling language that can be suitable for the network and three-dimensional and interactive environment, and it also has the advantages of interactive and platform independence. As the mainstream Web programming tools, Java has the features of being object oriented, being safe and reliable, and being to transplant and cross platform. Based on interactive technology of Java and VRML, we can realize the design function on the complex arithmetic, logic processing, network communication, and precise control of the scene, so as to effectively compensate for the inadequacy of VRML itself and can enhance the animations and interactive ability of VRML.

3.2 Functional Architecture of Virtual Experimental System

Photography of the virtual experiment system is an integral part of the curriculum teaching system [2]. It integrated content, equipment, teachers guide, students' operation, and related support tools for teaching. It is a bridge between theory study and practice in the photography course. According to the objectives and requirements of photography teaching, function structure of simulation photography experiment system is divided into system management module, virtual experiment module, and related support module.

According to the teaching characteristics of photography, we have designed different types of experimental items, such as type of observation, experiment, simulation training, and comprehensive creation. Simulation laboratory needs to construct some realistic experimental scene (Fig. 1), so that the user can conduct various kinds of simulation experimental practices and skills training. Students can watch the main components and function of camera, and the system alerts for some operations that should be banned or error and can truly reproduce the experimental results generated by improper operation. By adjusting the lens focal length, the user can intuitively feel the change about the depth of field of the subject, and by adjusting the aperture and shutter, the user can set the exposure parameters of camera. Through the adjustment of indoor lighting, the user can feel directly and feel the light distribution effect on a variety of portrait photography. The users can roam and find a view in the virtual campus, they also can achieve photo shooting through the operation of the shutter, and they can get the pictures effects with the corresponding camera parameters.

Fig. 1 The scene of experiment on photo shoot outdoor roaming

4 The Key Techniques in the System Development

A simple scene on the virtual experiment system can be built directly through the VRML editor, and the more complex scene modeling needs to be first created in 3Dmax and then guided into VRML format file. In VRML, with the script, routing, and Java technology to realize the interaction, the user can conduct experiments training through the browser containing plug-in. The key technology in the system mainly includes the interface problems, camera control (including the focal length of the lens changes and exposure parameters) and real-time synchronization technology on the effect of imaging and real-time image generation and synchronization of picture display technology in VRML and Java.

4.1 Scheme of Interface of VRML and Java

Being used in the traditional method of interaction of SAI and EAI, JDK version restrictions is below JDK1.4, and we cannot make full use of the new function gradually enhanced in JAVA. In order to solve this problem, we adopt interactive scheme based on the Webpage script: Webpage scripting language implemented in the client, so it can enhance dynamic interaction on the Webpage. It can interact with the VRML scene, and it also can facilitate to call each other with Java program. In the mechanism of interaction based on Webpage script, VRML and Java applet are not directly communicated and the Webpage script in JavaScript language (JS for short) is their bridge. That is, when the VRML needs to call the Java program to perform a task, first, it calls the JS function in the same Webpage, and the JS function calls Java applet program in this Webpage to perform specific functions. When the parameter is modified in the Java applet region, it calls the JS function in the same Webpage and then updates the attribute values of an object in a scene by the JS function, so as to achieve the purpose of interaction between VRML and Java.

4.2 Synchronous Technology Between the Regulating Focal Distance and Imaging Effects

Zoom is the important performance index of the lens. Adjusting the focal length has a direct influence on the perspective, depth of field, scene, and space effect of perspective of the photo image. When the user makes focusing or zooming operation through the click of a mouse, this system will not only update focal length parameter values, and the image displayed on the viewfinder frame must also correspondingly synchronous change. This process is controlled mainly by the VRML script and routing design, in which the key problem is how to achieve the effect of the focal length of the lens changes on the photo view and screen depth.

In order to reflect the changes of the view with focal length regulation, the frame is defined in view coordinate, it changes with viewpoint and the sight changes. When the focal length is larger (or smaller), the view will move along the line of sight direction, the relationship is $s = k * D$ on the mobile distance (s) and amplitude of focal length (d). Among them, K is a transform coefficient, it can set different values according to different range of focal length, when the K value is 2, and the change in camera angle is quite obvious in the system. The original coordinate on the viewpoint of the coordinate origin is (x_0, Y_0, Z_0), and our view is always negative in the direction of Z axis in view coordinate; then, the new coordinate (x, y, z) of the viewpoint of coordinate system is into $x = x_0$, $y = Y_0$, $z = z_0 + s$, and the domain value of translation of the point coordinates will update by the values of (x, y, z) by means of a routing.

In order to reflect the influence of the depth of field with the lens focal length change, according to the regularity of the focal length of the lens and the depth of field, when the viewer moves back and forth along the line of sight, it simultaneously enlarges or reduces its visible domain value of visibility range using Fog of VRML. The relationship between the domain value of the visibility range and the amplitude of focal length changes is $V = a * d + v_0$. Among them, V is used to represent the new value of visibility range. A can be set to different values according to different focal length range, and the atomization effect is more obvious when the value of A is—11 in this system; D is the variation amplitude of focal length; V_0 is the current value of visibility range. The value of visibility range is smaller, the atomization effect produced by a scene is stronger, and the fuzzy state of the scene filmed before and behind the object focus is more obvious. Strictly, the comparison between image clarity and ambiguity in this design does not fully meet the depth effect changes in reality, but it can let the experimenter to realize the corresponding changes of the scene of crisp and fuzzy with the change in focal length in the visual perception, and its experimental effect is as shown in Fig. 2.

Focal length is 100mm   **Focal length is 200mm**

Fig. 2 Effect diagram on focal length adjusting

4.3 Synchronous Technology of Adjusting Exposure and Imaging Effects

EV(Exposure Values) is the characterization of technical parameters of photography exposure; at the same time, it also reflects the relationship between lens aperture and shutter speed. In the present experimental system, the minimum value of EV of exposure is 2; the maximum is 18, in a total of 17 grades. In order to corresponding normal light conditions in reality, we take EV10 as the reference value. In experiment operation, if the exposure value is EV10, then the effect of photograph generated is same to the effect of image captured by screenshots directly. When the value of EV decreased a grade, the brightness on the photo effect will reduce by 15, that is three color quantization values of R (red), G (green), B (blue) were decreased by 15; When the value of EV increased a grade, the brightness on the photo effect will increased by 15, that is three color quantization values of R (red), G (green), B (blue) were, respectively, increased by 15. According to the relationship between EV value and aperture coefficient (A) and the shutter speed (T), exposure effects by camera operator and changes of image depth can be shown in the experimental results on synchronization.

4.4 Real-Time Generation of Photo and Synchronous Display of Images

There are two main types of method for generating photograph that is preprocessed images and drawing according to the viewpoint and the information of scene file. The former is easy to cause the deviation of the test results so that it reduces the system simulation. However, although the latter can truly reflect the effect of actual shooting, but it involves a large number of operations, to occupy more resources, high requirements on hardware, it will affect the system's real-time if running in the general client. Combining of thought based on image, we presented the pictures generation technique based on scene graph, namely first to capture the scene area about the view dependent as the initial image and then combined with the camera parameters form a final photograph to the initial image by using a certain algorithm. Since VRML itself does not have the capture function, if the photograph needs to be displayed in a certain area, we need to use Java technology to assist. In a Webpage using Java applet and scenes, photographs displayed in the Java applet area in real time, and function of the setting of camera parameters is provided. Scene frame is equivalent to a rectangle that moves in synchronism with perspective. In the premise of the certain screen resolution, the location of visual frame on the screen is fixed. To obtain the absolute coordinates of visual frame in screen coordinates through the test, we can capture the initial image and store the picture pixel information. In the experiment of photographic simulation, the ultimate effect of photograph is closely associated with the camera parameters. After obtaining the initial image information, we need to process the initial picture according to the camera parameters, in order to make the effect of image display correspond to the experimental parameters. Based on the algorithm of camera parameters on initial photos, We set A as the aperture value, and T as the shutter speed, so the algorithm of the initial photograph processing can be described as follows: (1) to calculate the corresponding exposure value EV according to A and T; (2) to calculate the value of d of the brightness variation according to the formula $d = D$ (ev $- 10$) $\times 15$; (3) to calculated for each pixel of the initial image, you can access the new image data that are consistent with the experimental parameters, then the new images can be generated in accordance with experimental parameters, and they display on the screen synchronously.

5 Problems on Teaching Application of Virtual Experimental System

In the virtual photography laboratory environment, users can self-plan and self-control to complete the experiment project and evaluate the experimental results, according to their own learning needs. In addition, the virtual experiment system can also serve as tutors, it can create a realistic experimental situation in order to arouse study interest, and it also can provide rich information resources and good technical support and service. Development and application of the virtual experiment system will produce great influence on the reform of experimental teaching. To the student users, the virtual experiment is still belongs to a new way of learning; therefore, it needs appropriate training to users in use. At the same time, fully help content in system such as the description of the method used, download function of the system plug, and answers to common questions is provided in support module, so that the user can use the virtual experiment system to learn. The teaching practice proves that virtual experiment system has important educational value to improve the level of education informatization, to improve the experiment teaching and training environment, optimizing teaching process, enhance the learning efficiency, enhance the learning interest of the students, practical ability, and innovation ability, etc.

References

1. Ying, J.: Cognition of virtual laboratory constriction by virtual reality technology. Lab. Sci. **2**(1), 102 (2006)
2. Li, JZ., Sun, S.L., Jiang, P., et al.: Construction of virtual laboratory teaching platform based on network. College Days, 125–126 (2006)

The Research of Travel-Time Tomography Based on Forward Calculation and Inversion

Yaping Li and Suping Yu

Abstract The technology of travel-time tomography, widely used in the fields such as resource exploration and quality testing of construction, is characterized by high resolution, wide detection range, and visual imaging. The technology includes two parts—the estimation of ray tracing and travel time and the rebuilding of images. This paper, based on ray theory, provides forward calculation by using shortest path ray tracing; it also presents fast inversion imaging based on SIRT by comparing different inversions; finally, it provides the stimulation results, and the comparison as well as the analysis of the results proves the validity and practicability of forward calculation and inversion.

Keywords Travel-time tomography · Forward calculation · Inversion · Tracing · SIRT

1 Introduction

Tomography technique (CT) is a method of the inversion of internal objects based on the projection data. Proposed by the Austrian mathematician J. Radon in 1917, the travel-time tomography technology [1] was further developed by the research of A. M. Cormack, Langan, Bishop, Chiu, and others. With the features of high resolution, wide detection range, and intuitive imaging results, in recent decades, tomography detection technology has been widely used in areas such as

Y. Li (✉) · S. Yu
College of Computer and Information Engineering,
Luoyang Institute of Science and Technology,
Luoyang 471000, China
e-mail: jsjxexam@163.com

S. Yu
e-mail: badkid@126.com

S. Patnaik and X. Li (eds.), *Proceedings of International Conference on Soft Computing Techniques and Engineering Application*, Advances in Intelligent Systems and Computing 250, DOI: 10.1007/978-81-322-1695-7_57, © Springer India 2014

engineering exploration, geological structure exploration, resource exploration and development, seismic detection, and engineering quality inspection.

Tomography projection method can be divided into travel-time tomography, amplitude tomography, and waveform tomography, among which the travel-time tomography is the most commonly used in engineering. This paper conducted two studies on travel-time tomography: the forward method for the estimation of the ray path and the travel time and the inversion method for image reconstruction.

2 Forward Calculation

The forward algorithm of the tomography can be divided into two categories [2]: one is the numerical simulation of the wave field and the other is the ray tracing method based on ray theory. At present, the tomography method based on ray theory is more mature and is the most widely used in engineering.

Take cross-hole measurement as an example, the transmitting antenna and receiving antenna are installed on both sides of the test region (two boreholes). Fix the transmitting antenna, and then move the receiving antenna along the drill regularly and receive once every corresponding distance; you can get a series of rays; then, move the transmitting antenna and receive the other group until the rays cover the whole region and achieve a certain density. The process of the CT measurement is shown in Fig. 1 [3].

The travel-time tomography includes velocity tomography and attenuation chromatography. Take the velocity tomography as an example. Its computation formula is [4]

$$L \cdot S = T. \tag{1}$$

The L is the distance array, the S is the slowness (reciprocal of the speed) array, the T is the travel-time matrix of the ray propagation.

Discrete the formula (1) and we can get

$$
\begin{bmatrix}
l_{11} & l_{12} & \cdots & l_{1M} \\
l_{21} & l_{22} & \cdots & l_{2M} \\
\cdots & \cdots & \cdots & \cdots \\
l_{N1} & l_{N2} & \cdots & l_{NM}
\end{bmatrix}
\cdot
\begin{bmatrix}
S_1 \\
S_2 \\
\cdots \\
S_M
\end{bmatrix}
=
\begin{bmatrix}
T_1 \\
T_2 \\
\cdots \\
T_N
\end{bmatrix}
\tag{2}
$$

l_{ij} is the path length of ray i in Unit j; $S_i = 1/V_j$ is the average slowness degree value (reciprocal of the speed value); T_i is the measured walk of ray i.

The forward calculation of travel-time tomography is to accurately obtain the travel time T of the ray. In formula (1), as S is entered as a known quantity, the exact calculation of the ray travel time depends on the COLA coupled with the travel time from the wave source to the subprime source, is the travel time from the wave source to the grid point. Conduct tracking comparison successively, and we can get a more accurate ray path (Figs. 2, 3).

Fig. 1 Ray CT diagram

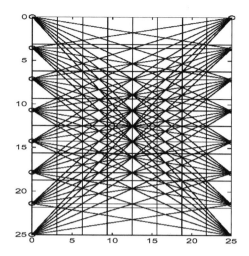

Fig. 2 The CT diagram of the shortest ray path

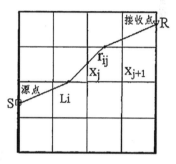

Fig. 3 The diagram of source points and subprime source points

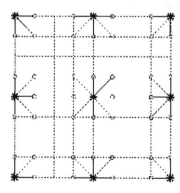

In the specific implementation, you can increase several nodes in the cell boundaries of the discretization grid, so that the true path of ray can be achieved more accurately.

3 Inversion Algorithm

Tomography inversion algorithm mostly adopted the iterative type of algorithms, including gradient method and projection iterative method. Based on the gradient, the most optimized iterative method is a method which moves the initial model along the gradient direction, gradually approaching the true model in the model space. It mainly includes the singular value decomposition (SVD), the conjugate gradient algorithm (CG), the least square QR decomposition (LSQR), etc.; projection iterative method maps the travel-time anomalies step by step along the path to the slowness anomalies after several iterations, including algebraic reconstruction technique (ART) and simultaneous iterative reconstruction method (SIRT). In practical engineering applications, the internal structure of the test body is extremely complex, increasing the instability of the inversion algorithm greatly, and this time, ART and SIRT method combined with the projection BP algorithm will be exceptionally advantageous, being able to find the real solution space faster and more accurately.

ART method distributes the travel-time residuals to each grid, in accordance with the proportion of length of ray passing each grid length of the whole length of ray. The SIRT method averages travel-time residuals and then allocates it to each grid. These shortcomings such as distortion caused by the individual data errors and concentration of errors caused by uneven distribution of ray can be overcome. Therefore, compared with ART algorithm, SIRT has better convergence and higher imaging accuracy [5]. For this reason, this paper conducted the inversion imaging in the SIRT method combined with the projection BP algorithm.

3.1 Backward Projection Method

Reconstructing image in the backward projection (BP) method is a basic method of CT imaging. It reverses the collected travel-time information along the ray path to each grid point, getting the speed value of each grid point. It is a coarse discrete image reconstruction method, and specific procedure is as follows [6]:

Assign projection data to each pixel along the ray. In allocation, make the proportion of length of ray i in the pixel to the total length of ray $\sum_{j=1}^{J} l_{ij}$ as a weight, and then add the weighed travel time of all the ray passing pixel j, and divide the total ray length in the unit to obtain the slowness of the medium of the unit $\widehat{S_j}$.

$$\widehat{S_j} = \sum_{j=1}^{J} l_{ij} \left(T_j \Big/ \sum_{j=1}^{J} l_{ij} \right) \Big/ \sum_{j=1}^{J} l_{ij}. \qquad (3)$$

In the formula l_{ij} is i rays is the length of path of ray I in the jth unit, T_i is the measured walk time value of the ith ray, J is total number of pixels in the inversion region.

BP imaging results are of low resolution, but is simple and fast in calculation. This paper defines the initial iteration model using the BP method.

3.2 Simultaneous Iterative Reconstruction Method

Simultaneous iterative reconstruction method (SIRT) is one of the most representative method of the back projection methods. Its implementation steps are as follows: Assume that n rays pass through a certain unit, and ray m is one of them, whose measured travel time is T_m. Using the direct line or curved ray tracing algorithm, we can get the length l_m of ray m passing through this unit, as well as the total ray length L_m and ray travel time T_m^c, and the travel-time error assigned to the unit is

$$\varepsilon_m = \left(T_m - T_m^c\right) \cdot l_m / L_m. \tag{4}$$

The total travel-time error of n rays passing through the unit is $\sum_{m=1}^{n} \varepsilon_m$, the total ray length is $\sum_{m=1}^{n} l_m$, and unit slowness S is corrected with the following formula:

$$S^{(k+1)} = S^k + \sum_{m=1}^{n} \varepsilon_m \Big/ \sum_{m=1}^{n} l_m. \tag{5}$$

If the velocity variation range in the survey area is known, $S_{\min} < S^{(k+1)} < S_{\max}$ can be used as constraints. Using the constraints and adding a suitable damping factor in calculation can improve the convergence effect and imaging accuracy.

4 Numerical Example

Assume that the forward model is a 4×6 m high-speed transmission area, in which a 1.5×1.5 m square low-velocity anomaly zone exists, as is shown in Fig. 4. Take the horizontal and vertical step lengths as 0.5 m, i.e., dividing the forward area into 20×30 grids. Assume that the transmitting antenna is launched vertically in each grid point for a total of 30 times. Corresponding to each launch, the receiving antenna on the other side of the calculation region receives it along the grid successively, getting 30 travel-time data. Then, 30 launches collect a total of 900 travel-time value. Make all the travel-time value into the formation of a vector, and we can get the travel-time value curve as is shown in Fig. 5.

Fig. 4 The forward model

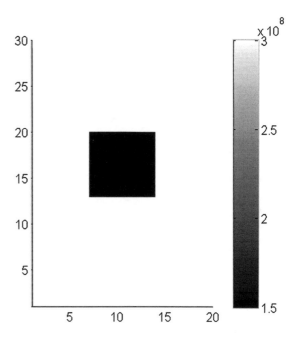

Fig. 5 Forward calculation
of travel-time value

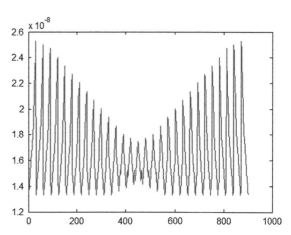

After the obtainment of the travel-time value, we can conduct the inverse imaging. In order to avoid the use of the same array L in the simulation process of forward and inversion, the construction of array L in inversion algorithm in this paper uses the direct ray CT method. Substitute the travel-time value obtained through the use of ray tracing method into the right side of formula (1) and then array L constructed by the direct ray CT method into the left end of formula (1). Conducted the rapid imaging in the BP method described in above chapters and the imaging results are shown in Fig. 6a.

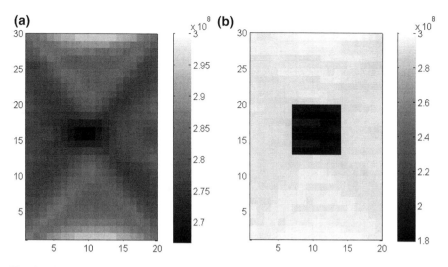

Fig. 6 The imaging results **a** BP imaging and **b** SIRT imaging

On the basis of BP imaging, take its value as the initial value and then conduct accurate imaging in SIRT method, and the imaging results are shown in Fig. 6b.

The target shape can be vaguely seen in the imaging results of the BP method, but there are many indistinct areas in the image, which clearly retains the ray trace, and the imaging results are rough. However, taking it as the initial model, the imaging results of the SIRT inversion clearly and correctly reflect the characteristics of the target's location, shape, size, etc., proving the feasibility and effectiveness of this method.

5 Conclusions

Based on the ray theory, this paper conducts forward calculation of the travel time of ray in the shortest path ray tracing method and, through the comparative study of various inversion algorithms, chooses the SIRT algorithm for rapid inversion imaging and gets the comparative results of the simulation calculation. Therefore, we prove that the shortest path ray tracing method and the SIRT algorithm adopted in this paper can effectively reflect the media distribution inside the test body and then determine the location of the inside anomalous body. Consequently, this method can be applied in data processing and interpretation in areas like engineering geophysics.

References

1. Guohua, L., Wang, Z., Sun, J.: Elastic wave tomography and its application in civil engineering. Civ. Eng. J. **36**(5), 76–81 (2003)
2. He, R., Jiansi, Y., Zhang, Y.: The seismic tomography method review. CT Theory Appl. **16**(1), 35–48 (2007)
3. Isaacaon, A.R.: A 500 kHz–5 MHz CW stepped frequency borehole tomographic imaging system. Cape Town, Dec 2002
4. Giroux, B., Gloaguen, E., Chouteau, M.: bh_tomo-a Matlab borehole georadar 2D tomography package. Comput. Geosci. **33**(1), 126–137 (2007)
5. Wang, Y.: Elastic wave CT key technology and application examples. Eng. Invest. 66–68 (2005)
6. Shengdong, L.: Li, C.: Seismic travel time tomography algorithm. J China Univ. Min. Technol. **29**(2), 211–214 (2000)

Protein Secondary Structure Prediction Based on Improved C-SVM for Unbalanced Datasets

Ao Pei

Abstract When protein secondary structure is predicted, the number of helix, sheet, and coiled coil is quite different in datasets. A multiple classifier based on improved clustering support vector machine (C-SVM) algorithm is proposed to predict protein secondary structure for unbalanced training datasets. Firstly, different weights are used for different types of samples to improve classification accuracy of traditional C-SVM on unbalanced samples. Secondly, the multiple classification strategy of one-versus-one (OVO) is used to build three binary classifiers. These binary classifiers are H/E, H/C, and E/C. The Majority-Voting law is used to integrate the results of three binary classifiers. Finally, sevenfold cross-validation based on grid method is used to optimize the parameters of classifiers. Simulation results show that, compared with the other prediction methods, the classification method proposed in this paper can obtain better classification accuracy on unbalanced datasets.

Keywords Protein secondary structure prediction · C-SVM · Unbalanced datasets · One-versus-one · The majority-voting law

1 Introduction

Protein secondary structure prediction [1–3] is an important problem in molecular biology. All proteins are connected by different amino acids. Primary structure of protein is composed of the sequence of amino acids. Protein secondary structure is determined by primary structure of protein. Thus, the spatial configuration of the polypeptide chain is determined by protein secondary structure.

A. Pei (✉)
College of Computer and Information Engineering, Henan Normal University,
No. 46, Eastern Construction Road, Xinxiang, Henan, China
e-mail: aopei16@sina.cn

S. Patnaik and X. Li (eds.), *Proceedings of International Conference on Soft Computing Techniques and Engineering Application*, Advances in Intelligent Systems and Computing 250, DOI: 10.1007/978-81-322-1695-7_58, © Springer India 2014

Predicting protein secondary structure and the number of helix (H), sheet (E), and coiled coil (C) is quite different in datasets. Ideal classification results cannot be obtained by using traditional clustering support vector machine (C-SVM) [4, 5] because the same penalty parameter C is given to point sets of the positive class and negative class. In order to improve the classification accuracy, a multiple classifier based on improved C-SVM algorithm is proposed in this paper. Firstly, different weights for different types of samples are used to improve classification accuracy of the traditional C-SVM on unbalanced samples. Secondly, the multiple classification strategy of one-versus-one (OVO) [6–10] is used to build three binary classifiers. These binary classifiers are H/E, H/C, and E/C. The Majority-Voting law is used to integrate the results of individual binary classifiers. Finally, sevenfold cross-validation [12–14] based on grid method [15, 16] is used to optimize the parameters of classifiers. Simulation results show that, compared with the other prediction methods, the classification method proposed in this paper can obtain better classification accuracy on unbalanced datasets.

2 Theory and Algorithm of C-SVM

C-SVM is a kind of nonlinear soft margin support vector machine. For the problem of nonlinear classification, slack variable ξ is introduced to relax constraints on the one hand. Transformation ϕ is introduced to map the data sample sets (x_i, y_i) from the input space R^n to the Hilbert space H on the other hand. The original problem is obtained as follows:

$$
\begin{cases}
\min\limits_{\omega \in H, b \in R, \xi \in R^l} \frac{1}{2}\|\omega\|^2 + C \sum\limits_{i=1}^{l} \xi_i \\
\text{s.t.} \quad y_i(\omega \cdot x_i + b) \geq 1 - \xi_i, \xi_i \geq 0, \quad i = 1, \ldots l
\end{cases}
\tag{1}
$$

In Eq. (1), classification plane equation is $\omega \cdot x + b = 0 (\omega \in R^d, b \in R)$, and C is the penalty parameter greater than zero. If the kernel function $K(x_i, x_j)$ corresponds to the transformation ϕ, the dual problem of the original problem is expressed as follows:

$$
\begin{cases}
\min\limits_{\alpha} \frac{1}{2} \sum\limits_{i=1}^{n} \sum\limits_{j=1}^{n} y_i y_j \alpha_i \alpha_j K(x_i, x_j) - \sum\limits_{j=1}^{l} \alpha_j \\
\text{s.t.} \quad \sum\limits_{i=1}^{n} y_i \alpha_i = 0, 0 \leq \alpha_i \leq C, \quad i = 1, \ldots l
\end{cases}
\tag{2}
$$

In Eq. (2), α_i is Lagrange multiplier. While solving optimal α_i, only the part of the training samples that meet $\alpha_i \neq 0$ are called support vector. Support vector that meets $0 < \alpha_i < C$ is called boundary support vector. Support vector that meets $\alpha_i = C$ is called non-boundary support vector. It is actually training samples that are classified wrongly.

If K is positive definite kernel, the dual problem must have a solution, and its solution is $\alpha^* = \left(\alpha_1^*, \ldots, \alpha_l^*\right)^T$ in Eq. (3).

$$
\begin{cases}
\omega^* = \displaystyle\sum_{i=1}^{l} \alpha_i^* y_i x_i \\[4mm]
b^* = y_j - \displaystyle\sum_{i=1}^{l} y_i \alpha_i^* K\left(x_i, x_j\right)
\end{cases}
\tag{3}
$$

In Eq. (3), only when the data sample point (x_i, y_i) makes the sign of equality true in the constraint equality, coefficient α_i can be non-zero. (ω^*, b^*) is the solution about (ω, b) in Eq. (1) of the original problem. Then, the decision function can be calculated in Eq. (4).

$$
f(x) = \text{sign}\left(\sum_{i=1}^{n} y_i \alpha_i K\left(x_i, x_j\right) + b\right)
\tag{4}
$$

3 Improved C-SVM Based on Unbalanced Datasets

It is pointed out that the same penalty parameter C is using to the point sets of positive class and negative class that means the higher numbers of sample points are more important in the literature [17]. Therefore, the C-SVM algorithm is not suitable for the classification problem with unbalanced training sets. In order to overcome this defect, a penalty-coefficient-weighted support vector machine algorithm is proposed. The basic idea is that different weights are introduced according to the different categories and different penalties are imposed on error classification. As long as the weights of penalty coefficients of the minority class are increased and the weights of penalty coefficients of the majority class are decreased, the misclassification costs of two types of samples may be in equilibrium roughly. Therefore, there are Eqs. (5) and (6) according to the selected parameters of C properly.

$$
C_+ = w_+ C = \frac{N_-}{N_+ + N_-} C
\tag{5}
$$

$$
C_- = w_- C = \frac{N_+}{N_+ + N_-} C
\tag{6}
$$

N_+ is the number of the positive training points. N_- is the number of the negative training points. w_+ is the weight of the positive training points. w_- is the weight of the negative training points. C_+ is the penalty parameter of the positive training points. C_- is the penalty parameter of the negative training points. Then, the original problem is described as Eq. (7).

$$\begin{cases} \min_{\omega,b,\xi^l} \frac{1}{2}\|\omega\|^2 + \sum_{y_i=1} C_+\xi_i + \sum_{y_i=-1} C_-\xi_i \\ \text{s.t.} \quad \begin{array}{l} y_i(\omega \cdot x_i + b) - 1 + \xi_i \geq 0 \\ \xi_i \geq 0, \quad i = 1,\ldots l \end{array} \end{cases} \tag{7}$$

The dual problem is described in Eq. (8).

$$\begin{cases} \min_{\alpha} \frac{1}{2} \sum_{i=1}^{n} \sum_{j=1}^{n} y_i y_j \alpha_i \alpha_j K\left(x_i, x_j\right) - \sum_{j=1}^{l} \alpha_j \\ \sum_{i=1}^{n} y_i \alpha_i = 0 \\ \text{s.t.} \quad \begin{array}{l} 0 \leq \alpha_i \leq C_+, y_i = +1 \\ 0 \leq \alpha_i \leq C_-, y_i = -1 \end{array} \end{cases} \tag{8}$$

Training datasets at every time, the penalty parameters C_+ and C_- are selected by using this method. And the decision function is unchanged.

Here, $N_{BSV+}, N_{BSV-}, N_{SV+}$ and N_{SV-} express the number of positive class boundary support vector, the number of negative class boundary support vector, the total number of positive class support vector, and total number of negative class support vector, respectively. N_+ and N_- express the number of positive class sample and negative class sample. That is, $N = N_+ + N_-$. Equation (9) is obtained from Eq. (8).

$$\sum_{i=1}^{N} \alpha_i y_i = \sum_{y_i=+1} \alpha_i - \sum_{y_i=-1} \alpha_i = 0 \tag{9}$$

Then, Eq. (10) is obtained by Eq. (9).

$$\sum_{y_i=+1} \alpha_i = \sum_{y_i=-1} \alpha_i \tag{10}$$

For the positive class sample, formula $\alpha_i = C_+$ is true for the boundary support vector, so the sum of N_{BSV+} boundary support vectors is always less than $\sum_{y_i=+1} \alpha_i$. It is expressed by Eq. (11).

$$N_{BSV_+} \cdot C_+ \leq \sum_{y_i=+1} \alpha_i \tag{11}$$

In addition, C_+ is the maximum value of α_i in supported vector, so the sum of N_{SV+} maximum positive class support vector is greater than $\sum_{y_i=+1} \alpha_i$. It is expressed by Eq. (12).

$$N_{SV_+} \cdot C_+ \geq \sum_{y_i=+1} \alpha_i \tag{12}$$

Equation (13) is obtained by integrating Eqs. (12) and (11).

$$N_{\mathrm{BSV}_+} \cdot C_+ \leq \sum_{y_i=+1} \alpha_i \leq N_{\mathrm{SV}+} \cdot C_+ \tag{13}$$

Similarly, Eq. (14) is obtained.

$$N_{\mathrm{BSV}-} \cdot C_- \leq \sum_{y_i=-1} \alpha_i \leq N_{\mathrm{SV}-} \cdot C_- \tag{14}$$

Equations (13) and (14) are divided by N_+ and N_-, respectively, and $A = \sum_{y_i=+1} \alpha_i = \sum_{y_i=-1} \alpha_i$ is assumed. Equations (15) and (16) can be obtained.

$$\frac{N_{\mathrm{BSV}+}}{N_+} \leq \frac{A}{C_+ \cdot N_+} \leq \frac{N_{\mathrm{SV}+}}{N_+} \tag{15}$$

$$\frac{N_{\mathrm{BSV}-}}{N_-} \leq \frac{A}{C_- \cdot N_-} \leq \frac{N_{\mathrm{SV}-}}{N_-} \tag{16}$$

Equations (17) and (18) are obtained by taking Eqs. (5) and (6) into Eqs. (15) and (16), respectively.

$$\frac{N_{\mathrm{BSV}+}}{N_+} \leq \frac{A}{C \cdot \frac{N_-}{N_++N_-} \cdot N_+} \leq \frac{N_{\mathrm{SV}+}}{N_+} \tag{17}$$

$$\frac{N_{\mathrm{BSV}-}}{N_-} \leq \frac{A}{C \cdot \frac{N_+}{N_++N_-} N_-} \leq \frac{N_{\mathrm{SV}-}}{N_-} \tag{18}$$

Equation (19) can be obtained from Eqs. (17) and (18).

$$\frac{A}{C \cdot \frac{N_-}{N_++N_-} \cdot N_+} = \frac{A}{C \cdot \frac{N_+}{N_++N_-} \cdot N_-} \tag{19}$$

It means that two classes are the same upper bound of error classification rate. So error rate between positive class and negative class is more balanced, which means that the importance is increased in the class with small number of samples. However, classification accuracy of the class with small number of samples is improved on reducing classification accuracy of the class with large number of samples.

4 Protein Secondary Structure Prediction Based on Improved C-SVM Algorithm

Usually protein secondary structure is classified into three kinds: helix (*H*), sheet (*E*), and coiled coil (*C*). However, there is a great deal of difference in the number of the three structures. Especially for sheet, the number of sheet is twice the number of helix and coiled coil. In order to overcome this problem, improved C-SVM algorithm should be used to predict protein secondary structure in the third section of this paper. Concrete steps are described as follows.

Step 1: Datasets are selected. CB396 datasets presented by cuff are used in this paper. It is a non-redundant datasets obtained through sequence alignment and cluster analysis. The similarity between sequences is less than 30 %. The datasets include 2,184 helix (H), 1,397 sheet (E), and 2,565 coiled coil (C). Especially, the number of E is nearly twice C.

Step 2: Feature information is extracted. Feature information is composed by position-specific scoring matrix (PSSM) with evolutionary information and four amino acid physical–chemical properties (hydrophobic, charged, hydrogen bonding, and huge residue content). PSSM is obtained by making sequence alignment with CB396 datasets in non-redundant database of NCBI.

Step 3: Encoding. Similar to PHD encoding method, sliding window method is used to encode. The results are taken as the inputs of C-SVM classifier. PSSM and four amino acid properties are included in feature information, so each input vector has $25 * \omega$ elements, which ω represents the size of the sliding window.

Step 4: The classifier is constructed based on improved C-SVM algorithm. OVO classification method is used. First of all, three binary classifiers are constructed to train and test based on the improved C-SVM algorithm in the third section of this paper. Three binary classifiers are H/E, H/C, and E/C. RBF function is selected as kernel function. Then, the final decision is made on the discrimination results of each classifier by using the Majority-Voting law.

Step 5: Parameters of the classifier are optimized. The optimum parameters of classifier are obtained by sevenfold cross-validation based on grid method. The parameters include penalty parameters $C_{H/E}$, $C_{H/C}$, $C_{E/C}$ and the widths of RBF kernel function $\gamma_{H/E}$, $\gamma_{H/C}$, $\gamma_{E/C}$ of each classifier.

Step 6: The prediction results are evaluated. $Q3$ prediction accuracy $Q_3 = \frac{p_H + p_E + p_C}{N_H + N_E + N_C} \times 100\%$ and each kind of residue prediction accuracy $Q_I = \frac{P_I}{N_I} \times 100\%$ are used to evaluate the prediction results. N_H, N_E, and N_C are taken as total number of H, E and, C residues, respectively. p_H, p_E, and p_C are taken as the number of correctly predicted as H, E, and C residues, respectively. P_I is the number of each kind of correctly predicted residue. N_I is the total number each kind of residue.

Table 1 Effect comparison of prediction before and after using the improvement algorithm

	Q_3	Q_H	Q_E	Q_C
Before balance	78.4	77.4	71.7	88.5
After balance	81.3	74.6	78.7	94.2

Table 2 Effect of prediction by using seven kinds of algorithm based on CB396 datasets

Algorithm	PHD	NNSSP	DSC	PREDATOR
Q_3	72.0	71.4	68.4	68.6
Algorithm	ZPRED	Consensus	Method in this paper	
Q_3	59.6	72.7	81.3	

5 Simulation Results and Analysis

Simulation is carried out according to the steps of protein secondary structure prediction based on the improved C-SVM algorithm in the fourth section of this paper. The size of the sliding window ω is equal to 13. The optimum parameters of each classifier are obtained by sevenfold cross-validation based on grid method. That is, $C_{H/E} = 1$, $\gamma_{H/E} = 0.02$, $C_{H/C} = 3$, $\gamma_{H/C} = 0.1$, $C_{E/C} = 2$, and $\gamma_{E/C} = 0.08$.

In order to compare the effect prediction, it is shown in Table 1 that the effect of prediction before and after using the improvement algorithm. From Table 1, Q_3 reaches 81.3 % by using algorithm in this paper. It is increased by 2.9 % than before using the improvement algorithm. The other Q_I are all increased except Q_H. This also shows that this classification accuracy of the class with small number of samples is improved on reducing the accuracy classification of the class with large number of samples.

In order to compare the effect prediction between the other prediction algorithms and algorithm of this paper, PHD, NNSSP, DSC, PREDATOR, ZPRED, and consensus algorithm are used to predict based on CB396 datasets. The results of comparison are shown in Table 2. Prediction accuracy by using algorithm in this paper is higher than the other algorithm.

6 Conclusions

In order to overcome the defects of the traditional C-SVM algorithm, an improved C-SVM algorithm is proposed to predict protein secondary structure based on unbalanced datasets. Firstly, different weights are used for different types of samples to improve classification accuracy of the traditional C-SVM on unbalanced samples. Secondly, the multiple classification strategy of OVO is used to build three binary classifiers. These binary classifiers are H/E, H/C, and E/C. The

Majority-Voting law is used to integrate the results of individual binary classifiers. Finally, sevenfold cross-validation based on grid method is used to optimize the parameters of classifiers. Simulation results show that, compared with the other prediction methods, the classification method proposed in this paper can obtain better classification accuracy on unbalanced datasets.

References

1. Dai, Q., Li, Y., Liu, X., et al.: Comparison study on statistical features of predicted secondary structures for protein structural class prediction: from content to position. BMC Bioinf. **14**, 152 (2013)
2. Zangooei, M.N., Jalili, S.: Protein secondary structure prediction using DWKF based on SVR-NSGAII. Neurocomputing **94**, 87–101 (2012)
3. Kountouris, P., Agathocleous, M., Promponas, V.J.: A comparative study on filtering protein secondary structure prediction. IEEE/ACM Trans. Comput. Biol. Bioinf. **9**(3), 731–739 (2012)
4. Li, Y., Huang, Q., Xu, J., et al.: Research on prediction of streamflow based on C-SVM. Shuili Fadian Xuebao/J. Hydroelectr. Eng. **27**(6), 42–47 (2008)
5. Zhang, T., Xu, X.: Fault diagnosis based on integrated navigation system using C-SVM technology. Zhongguo Guanxing Jishu Xuebao/J. Chin. Inertial Technol. **19**(2), 239–242 (2011)
6. Wu, D.: Multi-class SVM based on improved voting strategy and its application in fault diagnosis. Xi Tong Gong Cheng Yu Dian Zi Ji Shu/Syst. Eng. Electron. **31**(4), 982–987 (2009)
7. Hu, H., Li, Y., Liu M., et al.: Classification of defects in steel strip surface based on multiclass support vector machine. Multimedia Tools Appl. 1–18 (2012)
8. Bryan, J.D., Kwon, J., Lee, N., et al.: Application of ultra-wide band radar for classification of human activities. IET Radar Sonar Navig. **6**(3), 172–179 (2012)
9. Waegeman, W., Verwaeren, J., Slabbinck, B., et al.: Supervised learning algorithms for multi-class classification problems with partial class memberships. Fuzzy Sets Syst. **184**(1), 106–125 (2011)
10. Amini, S., Razzazi, F., Nayebi, K.: A multi-class SVM based phonemes classifier based on a trainable confidence measure. In: IEEE International Symposium on Signal Processing and Information Technology, pp. 49–54 (2009)
11. Xie, X., Yang, B., Chen, Y.: Prediction of secondary structure of protein using neural network. J. Univ. Jinan **22**(2), 111–115 (2008)
12. Dai, Q.: A competitive ensemble pruning approach based on cross-validation technique. Knowl.-Based Syst. **37**, 394–414 (2013)
13. Bahar, S.F., Clarke, S.: Cross-validation of an employee safety climate model in Malaysia. J. Saf. Res. **45**, 1–6 (2013)
14. Liu, W.Y., Han, J.G.: The optimal Mexican hat wavelet filter de-noising method based on cross-validation method. Neurocomputing **108**, 31–35 (2013)
15. Feng, G.: Parameter optimizing for support vector machines classification. Comput. Eng. Appl. **47**(3), 123–124 (2011)
16. Xu, W., Xing, Z., Li, F.: Research on SVM algorithm based on parameters selection and optimization. J. Shandong Jiaotong Univ. **18**(2), 79–82 (2010)
17. Liu, S., Jia, C., Chen, P.: A weighted support vector machines with automatic parameters selection. Comput. Eng. Appl. **42**(2), 64–66 (2006)

An Algorithm for Speckle Noise Based on SVD and QSF

Weizhou Zhao, Hui Zhang, Baozhen Yang and Huili Jing

Abstract The complexity of speckle noise and the fact that original image is unknowable bring filtering algorithm design much difficulty. Traditional algorithms use the contaminated pixels to estimate the original ones and omit the influence from their location information. Based on singular value decomposition (SVD) and quadric surface fitting (QSF), a denoising algorithm is given in this paper. Firstly, SVD is operated on local window in the noise image and median filtering on unitary matrix. Secondly, a parameter is introduced to revise the singular values and to replace traditional threshold method. Finally, QSF is used to reconstruct the original image. Experiments show that, comparing with traditional mean filtering and median filtering, the proposed algorithm can get a more satisfactory result including a higher peak signal to noise ratio (PSNR) and edge preserved index (EPI).

Keywords Speckle noise · Singular value decomposition · Quadric surface fitting · PSNR · EPI

W. Zhao (✉) · H. Zhang · B. Yang · H. Jing
Xi'an Research Inst. of High Tech.,, Xi'an, China
e-mail: happywz_zhao@163.com

H. Zhang
e-mail: h_zhang@sina.com

B. Yang
e-mail: bz_yang@yahoo.cn

H. Jing
e-mail: hl_jing@sina.com

S. Patnaik and X. Li (eds.), *Proceedings of International Conference on Soft Computing Techniques and Engineering Application*, Advances in Intelligent Systems and Computing 250, DOI: 10.1007/978-81-322-1695-7_59, © Springer India 2014

1 Introduction

It is difficult to remove speckle noise for complexity of noise model and uncertainty of much knowledge in image processing. So, many researches focus on speckle noise for images, especially for remote sense images. Traditional filtering methods, such as mean filtering, median filtering, and other some new methods, for examples, nonlinear median filtering [1] and Savitzky–Golay filtering [2], are all based on information in pixel's neighborhood and polynomial fitting to recover its gray. Singular value decomposition (SVD) is a technology that is to decompose an object as two sub-objects and operate on them. SVD is often applied into 1-D signal processing, and filtering operation is run in the sub-spaces, which can remove noise in the whole signal space [3, 4]. Unfortunately, SVD is seldom applied into 2-D image filtering. A sequence collected data in 1-D signal can make polynomial fitting be easily used. But for 2-D image, it must consider both gray information and location information for every pixel. Another technology, surface fitting is often used in industry measurement and surface design [5, 6], but its application into image filtering has not been formally reported.

Speckle noise is different with other noises in that every pixel is contaminated in varying degrees. The traditional methods mentioned above use a sliding window and estimate by noise pixels, which is difficult to get a genuine gray. So, a revising processing of the gray in the local window before the estimation of the centered pixel may improve filtering results. The proposed algorithm in this paper combines local information with SVD to reduce speckle noise and applies QSF to reconstruct the original image. Different from other existed methods, it can reduce much speckle noise and preserve much edge information.

2 Algorithm

Assuming that an image with speckle noise I, its size is $m \times n$ and $I(i, j)$ is the gray value of pixel (i, j). The noise image can be described as:

$$I(i, j) = \alpha(i, j) \cdot I_c(i, j) \quad (i = 1, 2, \ldots, m; j = 1, 2, \ldots, n) \tag{1}$$

where $I_c(i, j)$ is the genuine gray of the pixel (i, j) and the coefficient $\alpha(i, j)$ shows the contamination degree of pixel (i, j). As usual, every pixel's genuine gray and its precise contamination degree are unknown but can be estimated. In order to research speckle noise, a log algorithm can be used to convert multiplicative noise such as speckle noise to additive one.

Considering the correlation between gray and distance, an eight-neighborhood can be selected to remove noise and preserve edge as much as possible. SVD can be operated on a $r \times r$ window centered on pixel (i, j), namely

$$I_w = U_w S_w V_w^T \tag{2}$$

where $U_w = (u_1, u_2, \ldots, u_r)$ and $V_w = (v_1, v_2, \ldots, v_r)$ are two orthogonal matrices. To get a satisfactory result, we can filter the decomposition results obtained in SVD by traditional median method to preserve more detailed information as follows:

$$U'_w = \text{Median}\,(u_i), V'_w = \text{Median}\,(v_i) \quad (i = 1, 2, \ldots, r) \tag{3}$$

Reference [3] having filtered the time domain signal by SVD shows that singular values are often small in noise space, and a good result can be got by let them be zero. But many experiments show that the smaller singular values can not be omitted in 2-D image filtering, or some useful information will be lost. So in this paper, the smaller singular values can be revised before reconstructing the original image, moreover, the revising is based on the gray values in local window.

Let $\gamma(\gamma < 1)$ be the contamination degree of the noise image, which is often known in simulation and can be measured in application by the noise image. An important index, $\beta_w = \text{std}(w)/\text{mean}(I)$, where $\text{std}(w)$ is the standard deviation in the window w and $\text{mean}(I)$ is the gray mean of the image I, shows a possibility that the pixel exists in edge region or texture region. The more possibility, the more gray information should be preserved. However, if the pixel is badly contaminated, it is very possible to judge it wrongly. So another parameter, $d = \min\{\beta_w, 1 - \gamma\}$, is introduced and revised as S_w by the following:

$$S'_w = c_w \cdot s_i \quad (i = 1, 2, \ldots, r) \tag{4}$$

where c_w is

$$c_w = \begin{cases} 0.95, & \text{if } d \geq 0.95 \\ 1 - 0.5 \times (1 - d)^2 & \text{if } 0.5 \leq d < 0.95 \\ 1 - 0.25 \times (1 - d)^2, & \text{else} \end{cases} \tag{5}$$

Now, the denoised results of every sub-space can be obtained, and the reconstruction of the local image I'_w can be got. Then, view the local image just reconstructed as a gray surface, and apply QSF into gray recovery of the central pixel in the window. Finally, the whole denoised image can be obtained.

3 Experiment

To testify the effectivity of this algorithm, some standard images are shown in Fig. 1 and a 5×5 sliding window is used. Considering the limitation of the length of the paper, we only list comparison between tradition methods and the proposed algorithm. Two important indexes, peak signal to noise ratio (PSNR) and edge preserved index (EPI) are used in quantitative comparison. The two indexes are, respectively, defined as:

Fig. 1 Standard images. **a** Couple. **b** Cameraman. **c** Lake. **d** Baboon. **e** Barbara. **f** Lena

$$\text{PSNR} = 10 \times \log\left(\frac{255^2}{\text{MSE}}\right) \tag{6}$$

$$\text{EPI} = \frac{\sum_{i,j} |P(i,j) - P(i+1,j)| + |P(i,j) - P(i,j+1)|}{\sum_{i,j} |I(i,j) - I(i+1,j)| + |I(i,j) - I(i,j+1)|} \tag{7}$$

The quantitative comparison with different densities are shown in Table 1. From it, mean filtering and median filtering have a lower PSNR and EPI, which shows that some information is lost in them, especially much edge information is lost in mean filtering, and block effect existing in median filtering. Compared with the two traditional methods, a higher PSNR and EPI can be obtained in the proposed algorithm, which can preserve more edge information and texture information. But, it should be noted that the higher density of speckle noise is, the lower PSNR and EPI will be, namely, there is a reverse relation between the performance of the proposed algorithm and the density of speckle noise.

Table 1 Comparison of different methods

Image	Method	0.04		0.10		0.20	
		PSNR	EPI	PSNR	EPI	PSNR	EPI
a	MEAF	78.7881	0.2040	75.4016	0.1594	72.5871	0.1360
	MEDF	79.0279	0.2391	75.3173	0.1972	72.3980	0.1797
	PROF	80.3115	0.3466	76.3199	0.2966	73.2282	0.2839
b	MEAF	76.3182	0.1503	72.9078	0.1273	70.3239	0.1148
	MEDF	76.4264	0.1903	72.7898	0.1735	70.1010	0.1682
	PROF	77.3169	0.2821	73.5838	0.2655	70.8396	0.2729
c	MEAF	77.7119	0.1612	74.5786	0.1326	72.0263	0.1174
	MEDF	77.5712	0.1955	74.3241	0.1735	71.6652	0.1675
	PROF	78.7345	0.3060	75.2549	0.2758	72.5164	0.2742
d	MEAF	75.4325	0.1347	72.4452	0.1152	69.9695	0.1046
	MEDF	75.3282	0.1670	72.2635	0.1549	69.7574	0.1531
	PROF	76.3194	0.2668	73.0879	0.2508	70.4745	0.2568
e	MEAF	77.1009	0.1323	73.8474	0.1138	71.1631	0.1044
	MEDF	76.8872	0.1698	73.6217	0.1574	70.9265	0.1535
	PROF	77.8399	0.2479	74.3879	0.2423	71.6077	0.2514
f	MEAF	77.8486	0.1339	73.8631	0.1134	70.8357	0.1038
	MEDF	77.7415	0.1780	73.6673	0.1624	70.5674	0.1607
	PROF	78.5371	0.2626	74.3832	0.2510	71.2552	0.2599

MEAF mean filtering; *MEDF* median filtering; *PROF* proposed filtering

4 Conclusions

It is difficult to get a precise model for speckle noise, so a very satisfactory result cannot been easily obtained. For the unreliability of estimation only by noise pixels, considering both the relationship in pixel's neighborhood and advantage of denoising sub-space, some technologies including SVD, local reconstruction, and QSF are used to estimate the gray value of every pixel, and then to obtain the final denoised result. Compared with the traditional mean filtering and median filtering, the parameters in this paper are self-adaptive, moreover, the proposed algorithm is fit for images with different noise densities. However, a sliding window may decline the speed of image processing. These will be our future work.

References

1. Windyga, P.S.: Fast impulsive noise removal. IEEE Trans. Image Proc. **10**(1), (2001)
2. Chinrungrueng, C., Suvichakorn, A.: Fast edge-preserving noise reduction for ultrasound images. IEEE Trans. Nucl. Sci. **48**(3), (2000)
3. Zehtabian, A., Hassanpour, H.: A non-destructive approach for noise reduction in time domain. World Appl. Sci. J. **6**(1), (2009)

4. Hermus, K., Wambacq, P.: Assessment of signal subspace based speech enhancement for noise robust speech recognition. IEEE International Conference on Acoustics, Speech and Signal Processing (2004)
5. Yuanpeng, L., Hui, Z., Liangji, C.: Algorithm of quadric surface fitting from oriented point cloud. Mechine Tool Hydraulics **36**(8), (2008)
6. Jiexian, W.: A method for fitting of conicoid in industrial measurement. Geomatics Inf. Sci. Wuhan Univ. **32**(1), (2007)

A Generation Model of Function Call Based on the Control Flow Graph

Weizhen Sun and Xiangyan Du

Abstract An approach of generating function call graph based on the intermediate results of the GCC compiler is described. The relational files applied to generate function call graph by graph visualization (Graphviz) is provided, based on the compiling and storing its results, using Graphviz to generate the corresponding function call diagram SVG. From users' expectation, it can generate function call graph in nodes and also can generate function call graph in nodes of files or directories from source code. It is appeared that the SVG can convenient to understand the Linux OS kernel's architecture.

Keywords GCC's compiling · Intermediate data · Graphviz · Function call graph

1 Introduction

GCC (the GNU Compiler Collection, GNU compiler suite) is C, object-oriented C (object C), C++, and other programming language compiler. It can translate source code into object code. In the GCC compiler implementation, compilation is generally divided into two steps [1]. The first step is that compiler's front-end accept input source code, getting some sort of intermediate representation of source code. The second step is that compiler back-end will do some optimization of the intermediate representation from the front-end and eventually generate the code that can be run on the target machine.

W. Sun (✉) · X. Du
Collage of Information Engineering, Capital Normal University,
105 West Third Ring Road, Beijing, China
e-mail: sunweizhen@cnu.edu.cn

X. Du
e-mail: duxiangyan20063@163.com

S. Patnaik and X. Li (eds.), *Proceedings of International Conference on Soft Computing Techniques and Engineering Application*, Advances in Intelligent Systems and Computing 250, DOI: 10.1007/978-81-322-1695-7_60, © Springer India 2014

In the GCC compilation, it uses GENERIC, GIMPLE, and RTL three kinds of intermediate language to represent program [1]. While the compiler is producing target program, it can output these intermediate languages in the form of the intermediate results at the same time [2]. The intermediate results using the compiler's options to convertibly store the message which can find how grammar compiler understand the program, which greatly reduce the difficulty of analysis than the direct analysis of the source code [3, 4].

Control flow analysis is mapping the body of the function as the control flow graph and analyzes the control flow graph. Nodes of the control flow graph are the basic block, which is a set of sequential execution sequence of instructions, and it is only a single out/into the point. The edges between basic blocks represent the execution path between basic functions [5]. From the intermediate results that produced during GCC compilation, we can extract the function definition information and the function calls information, thereby to analyze the relationship between function calls.

Existing function call graph generation tools, such as Codeviz [6], Source Insight [7], and Egypt [8], can generate function call relations between functions. Codeviz is based on compilation process information. Source Insight is based on the source information. Both can generate function call graph in the node of functions. Egypt also can generate function call graph in the node of functions, but the call graph corresponding to the first two function analysis tool and it does not depend on a specific version of the GCC compiler. On the basis, we can generate function call graph, which is satisfied with user requirement.

This article hopes to raise a generation module according to demand generating function call relationships between modules or between the functions, to analyze the function call relationships between modules or between files.

2 Model Building

Generating function call graph contains two main sections, to find the corresponding relations and to finish generating the corresponding function call graph.

Egypt [8] extracts function call relationships from such documents, that is the intermediate results of RTL from the GCC compiler, which is more convenient than that extracting function call relationships from the C source file. And then, Egypt puts it into formats used for graph visualization (Graphviz). Finally, Graphviz generates function call graph. Egypt itself as a Perl script puts these two tools together. Scripts can determine which kind of function call graph can be generated, in nodes of functions or files. This will not only make use of the advantages of the two tools, but also they can achieve their objective analysis of function call relationships.

According to their own needs between different levels of function calls, it can use different compiler options, such as Egypt with "-fdump-rtl-expand." In older

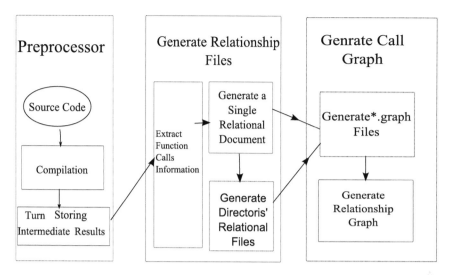

Fig. 1 Basic model structure

versions of GCC, this option is equivalent to "-dr," but GCC 4.4.0 and later versions accept "-fdump-rtl-expand" form.

According to a file or a directory or even a large source code to store the intermediate results while compiling, you can get different sizes of function call graph. As a result of the difference of the Perl script, the function call graph can be in nodes of function or in nodes of a directory.

Graphviz [9] developed by AT&T is an open-source graph visualization tool. Graphviz's layout program which generates graphics is described by a simple text language, and Graphviz can generate a variety of graphics formats, such as images, SVG Web pages, PDF, or other documents contained in the Postscript, and even displayed in an interactive graphical view device. Graphviz support rich icon display attributes, such as may be specified colors, fonts, tables, node layouts, line styles, custom options. It is convenient for us in the function call graph using different colors on lines representing different relationships between modules and within the modules. Due to the format of data requirements for Graphviz, in order to analyze the function call relationships in this paper, we deal with the intermediate results to turn them into the formation required to meet the form tool.

It presents a model, which is based on the intermediate results of the control flow model and based on users' demand, to generate function call graph. Basic model structure is shown in Fig. 1.

Firstly, compile the source code, and ultimately, we approach the front and rear according to the conditions, to carry out the program using Graphviz [10], it can easily draw structured graphic network. Through the influence of analysis implementation on the before and after conditions, we gradually generate the description file of the Graphviz graph and finally being the resulting state graph. The whole process is basically a fully automated process, without human intervention.

Model structure is mainly divided into three parts: the source code compiler module, relational file generation module, and function call graph generation module.

The source code compiler module includes the source code compilation and turning to store intermediate results.

Relational file generation module includes such steps as extracting function definition information and function call information from the intermediate results and is compiled function definitions and call information, and obtaining a series of function call relationships document through the handling of scripting language.

Function call graph generation module includes, on the basis of function call documents, in nodes of files or modules, in edges of the number of function calls between each other, generating function call graph between files or modules.

3 Implementation Methods

The intermediate results text file is obtained using the compiler options while compiling. In the process of cross-compiler to compile the Linux operating system kernel, we use an intermediate option and do analysis of the results generated and find that when doing with the option *-fdump-rtl-sched2*, the intermediate results contain the required function definitions and function call information.

For the function

```
void read_persistent_clock(struct timespec *ts)
{
        __read_persistent_clock(ts);
}
```

In use of -fdump-rtl-sched2 option, the generated *.????? R.s ched2 file has the following information,

```
;; Function read_persistent_clock
(read_persistent_clock)
***
**
***
  (insn/f:TI 15 3 16 2 (set (reg:SI 12 ip)
        (reg/f:SI 13 sp)) arch/arm/kernel/time.c:123
179 {*arm_movsi_insn}
      (nil))
```

On behalf of the function name information and the location of the function definition,

```
(call_insn:TI 10 8 23 2 (parallel [
            (call (mem:SI (reg/f:SI 3 r3 [orig:137
__read_persistent_clock ] [137]) [0 S4 A32])
                 (const_int 0 [0]))
            (use (const_int 0 [0]))
            (clobber (reg:SI 14 lr))
        ]) arch/arm/kernel/time.c:124 236
{*call_reg_armv5}
    (expr_list:REG_DEAD (reg/f:SI 3 r3 [orig:137
__read_persistent_clock ] [137])
        (expr_list:REG_DEAD (reg:SI 0 r0)
            (nil)))
    (expr_list:REG_DEP_TRUE (use (reg:SI 0 r0))
        (nil)))
```

This represents *read_persistent_clock (read_persistent_clock)* calls functions' information, and it calls function *__read_persistent_clock* at the location of *arch/arm/kernel/time.c:124*.

For the above intermediate results file, which has been generated and contains function definitions information and function calls information, we use a string processing language Ruby to analyze, and to obtain function definitions information and function call information within a single source file. Then, we distinguish internal and external function call to a single source file and try to draw the function call relations between different directory files and directories. In order to extract function definition and reference information, we need to understand some conventions of the RTL [11]. In sched2 files, firstly, it declares the function definition, after the first one *insn* or *insn/f*, it shows the specific information notes of the function, and after *call_insn* is the specific information of the function call.

4 Implementation Results

Based on the above model, this paper implements a function call graph generation tool, not only can generate function call graph, but can also generate corresponding function calls list according to user needs on the basis of generated relational files. For members who program analysis, it is more easy and intuitive to understand the source code.

Next, this paper uses the implemented tool to do an analysis of *driver* toward *network* module in Linux OS kernel. While painting net module function call graph, it then gets the graphics of function calls between the net with other modules.

From Fig. 2, the relationships between *drivers* toward *network* module total are divided into five parts. For example, *drivers—> net/rfkill*, the number on the line represents that *drivers* for *net/rfkill* directory the functions are called 16 times in total. Then, we can go from here to the corresponding function call list.

Fig. 2 Net module function
call graph_partly

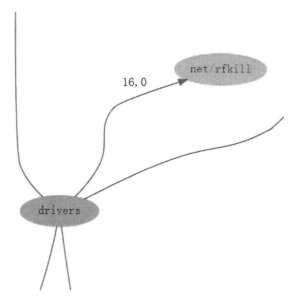

Fig. 3 From the function
call graph to function call list

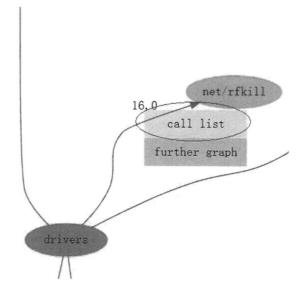

As shown in Figs. 3 and 4, the call relationship from *drivers* toward *net/rfkill* shows that *drivers* call *net/rfkill* six functions, the total number of called times is 16. The called functions are *rfkill_destroy, rfkill_set_sw_state, rfkill_init_sw_-state, rfkill_alloc, rfkill_register*, and *rfkill_unregister*, their respective defined location is referenced to *the Path*, and their respective simple notes are referenced to *the function note*. However, in the second table, it shows in detail the function

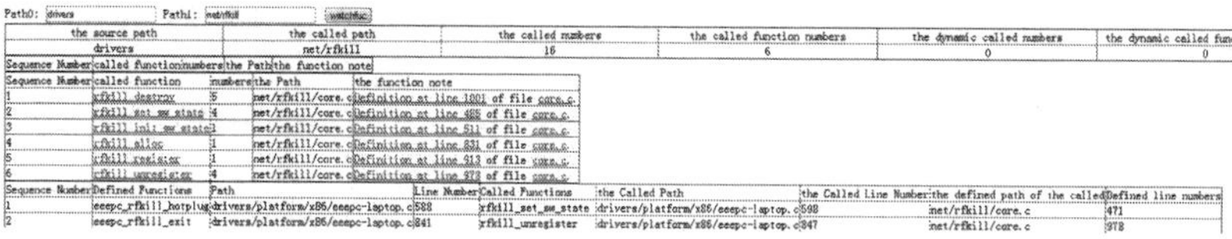

Fig. 4 The function call list from *drivers* toward *net/rfkill*

call relationships from *drivers* toward *net/rfikill*. *Defined Functions* is the calling functions in drivers, and *Path* and *Line Number* represent the calling functions' detailed definition information. *Called Functions* means the called functions in *net/rfkill*, and *the Called Path* and *the Called Line Number* are the location information where the functions are called. Furthermore, *the defined path of the called* and *Defined line numbers* are the called functions' detailed definition information.

Through Figs. 2, 3, and 4, we can intuitively and in detail understand the relationship from *drivers* toward *net/rfkill*. The process is convenient and concise.

5 Conclusions

This paper presents an approach generating static function call graph based on the intermediate results generated during the GCC compiling, making the use of the control flow analysis. The advantage of the designed model is that it uses the existing tools to easily achieve their goals of analyzing function call relationships to different levels. On the basis of relational files, this model can not only generate function call graph in nodes of files or directories or functions, but also can generate function call list. This is more intuitive and clear.

References

1. Wang, Y.: GCC's compile process and the initial exploration of intermediate representation layer of RTL. http://www.ibm.com/developerworks/cn/linux/l-gccrtl/index.html (2013). Accessed 25 June 2013
2. Stallman, R.M.: GNU compiler collection internals. Free Software Foundation (2002)
3. Griffith, A., En-Hua, H.: GCC: The Complete Reference. Tsinghua University Press, Beijing (2004)
4. Yang, X.H., Zhu, R., Zhang H.: The C model analysis in GCC compiler generated code GCC. Electron. World (20) (2012)
5. Xiong, Y.: Reverse analysis of Microsoft intermediate language. Comput. Appl. **26**(7), 1523–1525 (2006)
6. Karen, M.: CodeViz: A call graph visualiser. http://www.csn.ul.ie/~mel/projects/codeviz/.2013,06,15 (2013). Accessed 25 June 2013

7. SOURCE INSIGHT: Need to understand code?. http://www.sourceinsight.com/support.HTML#Docs.2013,06,15 (2013). Accessed 25 June 2013
8. Gustafsson, A.: Egypt: Create call graph from Gcc RTL Dump. http://www.gson.org/egypt/egypt.html.2013,06,15 (2013). Accessed 25 June 2013
9. Gansner, E.R., North, S.C.: An open graph visualization system and its applications to software engineering. Softw.: Pract. Experience **30**, 1203–1233 (2000)
10. Huang, Z., Peng, X., Zhao, W.: Behavior protocols recovery based on dependency analysis. Comput. Sci. **35**(8), 265–268 (2008)
11. Dietmeyer, D.L., Duley J.R.: Register transfer languages and their translation. In: Digital System Design Automation: Languages, Simulation and Data Base, pp. 117–218. Computer Science Press, Woodland Hills (1975)

A Novel Community-Based Trust Model for P2P Networks

Songxin Wang

Abstract In a heterogeneous peer-to-peer network, different peers may provide different qualities of service, so it is very important and helpful to identify those peers that can provide better services than others. A trust model is thus needed to provide a way for building trust through social control without trusted third parties. In this paper, we propose a new community-based trust model for P2P networks. There are three main features of this model. First, referral trust information is given to communities for each requestor, and in this sense, the prior knowledge of requestors about the environment is used to promote the collaboration. Second, the method to update referral trust value of each community is given, and thus, peers can learn from experiences about the environment. Third, the time aging factors of the both direct experience and indirect experience are taken into account. In order to justify the proposed model, we compare it with the existing models. The experiment results show that the use of community structure indeed improves the performance of the P2P networks.

Keywords P2P networks · Trust model · P2P community

1 Introduction

The emergence of decentralized and dynamic applications draws a lot of attention to P2P system. In a heterogeneous ubiquitous peer-to-peer network, different peers may provide different qualities of service, so it is very important and helpful to identify those peers that can provide better services than others. A trust model is

S. Wang (✉)
Department of Computer Science and Technology, Shanghai University of Finance and Economics, Shanghai, China
e-mail: sxwang@mail.shufe.edu.cn

S. Patnaik and X. Li (eds.), *Proceedings of International Conference on Soft Computing Techniques and Engineering Application*, Advances in Intelligent Systems and Computing 250, DOI: 10.1007/978-81-322-1695-7_61, © Springer India 2014

thus needed to provide a way for building trust through social control without trusted third parties.

In many applications, peers sharing same interests construct communities. These communities are implicitly formed, self-organizing structures that depend on the declared interests of peers. The different communities within which a peer can participate due to its claimed interest constitute the roles of the peer. Every peer belongs to at least one predetermined community, and in many cases, it belongs to more than one possibly overlapping community. Rather than being a single, homogeneous community, they become a collection of communities of users as P2P system becomes more and more heterogeneous as their scale increases. For every peer, there are many communities that it is familiar with, and when searching for a service, it can ask for those community members about their opinion about the performance of service provider. How to utilize this community structure, especially how the prior knowledge of requestor about the environment evolved, is then a changeling question for P2P networks.

Referral trust [1, 2] is trust in recommendatory, which is user's belief about the trustworthiness of other users' referral knowledge, while functional trust is trust in target agent about its ability for providing good service. In real applications, peers usually have prior knowledge of communities about how reliable the referral trust coming from community is and it can then place different trust values to different communities. Many works have been present to build trust model in P2P networks, and in these works, function trust information is often used to assist peer to find a suitable service provider, and however, referral trust information that comes from different communities is seldom considered.

This paper extends our previous work [3] and presents a new community-based trust model for P2P networks. In this work, referral trust information is given to communities for each requestor, and in this sense, the prior knowledge of requestors about the environment is used to promote the collaboration. The method to update referral trust value of each community is then given, and thus, peers can learn from experiences about the environment. Furthermore, the time aging factors of the both direct and indirect experience are taken into account. We finally compare it with the existing models. The experiment results show that the use of community structure indeed improves the performance of the P2P networks.

The remainder of the paper is organized as follows: In the following section, we compare our work with related works. Then, in Sect. 3, the community-based trust model is given, and this is followed by a simulation and analysis of the proposed model in Sect. 4. And finally, the conclusion and future work are discussed in Sect. 5.

2 Related Works

In our previous work (Wang [3]), we give a model of community-based trust model for P2P networks; however, the time aging factor was not considered in that paper. Furthermore, a new Euclidean distance, which is a more accurate metric than the one used there, is used to update the referral trust to communities in this work.

Sabater [4] proposes a reputation system that takes advantage of social relations between agents to overcome the problem that direct interactions are not always available, and however, the cooperation relation of witness and the service provider must be provided to obtain a reliable degree, and there are many cases that this information is not available either.

Khambatti et al. [5] present an optimistic-role-based model for trust among peers and show that it is scalable, dynamic, revocable, secure, and transitive. A new metric is defined to calculate peer links among communities, which is then used as trust value of a peer; however, the referral trust value of the agent to a community is not considered at all in this work, so the model cannot deal with the prior knowledge of requestors about the environment.

Singh [6] proposes a method to rate authentication information by a level of trust, which describes the strength of an authentication method, and gives a mathematical model to calculate the trust level when combining two authentication methods. However, the trust coming from the community cannot be used to help a peer to make the right decision.

Tian et al. [7] give a group-based reputation system GroupRep to establish the trust relationship in large-scale P2P networks. The trust relationship in GroupRep can be viewed as many tiers and can be used to assist the collaboration. This work has similarity with ours in that the trust relationship between group and peer is considered, but only the members of the group where the requestor belonging to are used to calculate the overall evaluation, and in our model, however, any group that the requestor is familiar can be an information source to find a service provider.

3 Community-Based Trust Model

In this section, we give a novel community-based trust model in which direct trust value and indirect trust value are combined to get overall evaluation. Community structure is used when calculating indirect trust value. The method to update referral trust value of communities after each interaction is given and the time aging factor are considered as well.

3.1 Direct Trust Calculation

Experience is one of the most important information that could influence the choice of the service requestor. In our model, an aggregation of a peer's experience to a special service provider is defined as its trust to that provider.

A peer r gets its direct trust value from its direct experience, it come from interaction history: With service provider p, let $S_{r,p}$ be the satisfaction of r to p, then direct trust value is calculated as follows:

$$\mathrm{DT}'_{r,p} = \sum_{i \in k} S_{r,p} \times f_w(i) \tag{1}$$

Note that $\mathrm{DT}_{r,p}$ is equal to zero, if there is no interaction between r and p at all.

Without further interaction, direct trust decreases over time according to a time-based aging factor Y. The closer the Y is to 1, the lower the value of Y, and the value of previous direct trust decreases with the decrease in Y. For static environments, where very few number of interactions take place within a particular time period, the value of Y might be chosen close to 0. Then, we get

$$\mathrm{DT}_{r,p} = \mathrm{DT}'_{r,p} \times \left(1 - \frac{(t - t_0)\Upsilon}{t}\right) \tag{2}$$

3.2 Indirect Trust Calculation Using Community Structure

It is less likely that repeat interactions will occur between same peers for the asymmetric interests between them. In many cases, it is difficult to establish the direct trust relationship between peers. Although interactions with someone in the past are of course the most reliable source of information about the agent's trustworthiness, however, relying only on direct experience is inefficient.

In community-based model, indirect trust value is used to get the overall evaluation. A requestor places different referral trust to different communities in prior, and this referral trust information is combined with functional trust of members of communities to get the overall evaluation; in this sense, the prior knowledge of requestor about the environment is used to promote the collaboration.

For an agent r, the set of all communities that it can interact with is denoted as N_r. The trust value to different communities is also different. We associate every $G_r \in N_r$ with a $h_r \in [0, 1]$, which is the referral trust value of r to community G_r.

The evaluation of service provider is the aggregation of all evaluations of the different communities. Note that the weight to each community is just the referral trust value of the community.

$$\mathrm{IT}'_{r,p} = \sum_{G_i \in N_r} \omega_i \times T_{G_i,p} \tag{3}$$

where $\omega_i = h_i$.

Just like direct trust, indirect trust value decreases over time according to a time-based aging factor Y also. Then, we get

$$IT_{r,p} = IT'_{r,p} \times (1 - \frac{(t - t_0)\Upsilon}{t}) \tag{4}$$

In the above formula, $T_{G_i,p}$ is used to denote the trust value to service provider by each community, and in the next part of this section, the method to calculate $T_{G_i,p}$ is given.

Let G_i be a community and $a_1^i \ldots a_j^i$ be all members of G_i, the trust value of G_i to p, which is denoted as $T_{G_i,p}$, is

$$T_{G_i,p} = \mathrm{aggr}(T_{a_1^i,p}, \ldots T_{a_j^i,p}) \tag{5}$$

Much kind of operators can be used as *aggr* operator, and in this study, we choose OWA operator. OWA operators were originally introduced by Yager [8] to provide a means for aggregating scores associated with the satisfaction of multiple criteria, which unifies in one operator the conjunctive and disjunctive behaviors:

$$OWA(x_1, x_2, \ldots, x_n) = \sum_{j=1}^{n} w_j * x_{\sigma(j)} \tag{6}$$

where σ is a permutation that orders the elements under consideration: $x_{\sigma 1} \leq x_{\sigma 2} \leq \ldots \leq x_{\sigma n}$. While ω_j is given as

$$w_j = Q\left(\frac{j}{n}\right) - Q\left(\frac{j-1}{n}\right) = \frac{j}{n} - \frac{j-1}{n} \tag{7}$$

3.3 Putting it Together: The Overall Evaluation

Using just direct trust value or just indirect trust value for a given requestor is only useful in extreme situations, and in most situations, it is preferable to consider both direct experience and indirect experience. Therefore, in our trust model, we adopt the combined degree as follows:

$$T_{r,p} = \kappa T_{r,p} + (1 - \kappa)IT_{r,p} \tag{8}$$

where κ is a parameter reflecting the weight given to DT and IT, respectively.

3.4 Updating the Referral Trust to Communities

It is necessary to feedback the experience of following a particular recommendation into the trust relationship in order to enable the agent to learn from their experience. This is done as follows: After each interaction, a peer r who has acted on the advice of its neighbor groups updates its referral trust value to each community.

Note that each neighbor community as a whole instead of individual member is updated.

Let $T'_{r,p}$ be the real satisfaction degree of peer r to service provider p, and let

$$r_k = \frac{1}{|G_i|} \sqrt{\sum_{a_j \in G_i} \left| T_{a_j} - T'_{r,p} \right|^2} \tag{9}$$

Note that Euclidean distance is used here, which is a more accurate metric than the metric used in Wang [3].

It is convenient to define the update of $h_i(T+1)$ in terms of an intermediate variable $\tilde{h}_i(T+1)$:

$$\tilde{h}_i(T+1) = \begin{cases} r_k h_i(T) + (1-\gamma)r_k & \text{for } r_k \geq 0 \\ (1-\gamma)h_i(T) + \gamma r_k & \text{for } r_k < 0 \end{cases} \tag{10}$$

We then map it back to the interval [0,1]:

$$h_i(T+1) = \frac{1 + \tilde{h}_i(T+1)}{2} \tag{11}$$

4 Evaluations

The method given in this study is implemented based on Query-Cycle Simulator [9]. At the same time, we also implement RMS-PDN [10] and a conventional reputation system with shared information noted as RSSI [7]. The data are collected after 100th query cycle, and the results are averaged 5 runs.

The efficiency of the networks describes how good peers can efficiently get reliable files. They are as follows:

Satisfaction level (SAT): If the size of authentic contents downloaded by i is authentic$_i$, and the size of inauthentic contents downloaded by i is inauthentic$_i$, let

$$\text{Sat}_i = \frac{\text{authentic}_i - \text{inauthentic}}{\text{authentic}_i + \text{inauthentic}} \tag{12}$$

And then Sat is defined as

$$\frac{\sum_{i \in V_g} \text{Sat}_i}{|V_g|} \tag{13}$$

In order to compare with RMS-PDN, the simulated network is partially decentralized with 700 peers. Peers are constructed as 10 groups.

As shown in the below figure, SAT always changes with the same trend as RSQ does, so we can concluded that our method is also more efficient in this situation.

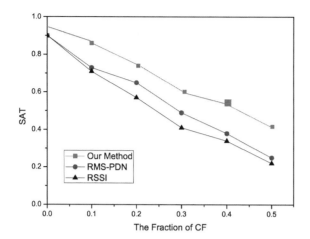

5 Summary and Outlook

In large-scale P2P networks, it is less likely that repeat interactions will occur between same peers for the asymmetric interests between them. So it is difficult to establish the direct trust relationship between peers.

This paper proposes a new community-based trust model for P2P networks. There are three main features of our model. First, referral trust information is given to communities for each requestor, and in this sense, the prior knowledge of requestors about the environment is used to promote the collaboration. Second, the method to update referral trust value of each community is given, and thus, peers can learn from experiences about the environment. Third, the time aging factors of the both direct experience and indirect experience are taken into account. Experiment results show that the use of community structure indeed improves the performance of the P2P networks.

In the future, we plan to add voting mechanism to the model so that the trust value given by the community can be more accurate.

Acknowledgments This research work is funded by 211 Project of Shanghai University of Finance and Economics in its fourth stage.

References

1. Li, Y.-M., Kao, C.-P.: TREPPS: A trust-based recommender system for peer production services. Expert Syst. Appl. **36**(2), 3263–3277 (2009)
2. Thirunarayan, K., Althur, D.K., Henson, C.A., Sheth, A.P.: A local qualitative approach to referral and functional trust. In: Proceedings of 2009 Indian International Conference on Artificial Intelligence, pp. 574–588. (2009)

3. Wang, S.: A community-based trust model for P2P networks. In: Proceedings of 2nd IITA International Conference on Artificial Intelligence, pp. 141–147. IEEE Press, New York (2010)
4. Sabater, J.: Evaluating the regret system. Appl. Artif. Intell. **18**(9–10), 797–813 (2006)
5. Khambatti, M., Dasgupta, P., Ryu, K.D.: A role-based trust model for peer-to-peer communities and dynamic coalitions. In: Proceedings of 2008 IEEE International Conference on Mechatronics and Automation, pp. 141–154. IEEE Press, New York (2008)
6. Singh, A.: Reputation based distributed trust model for P2P networks. Int. J. Eng. Res. Appl. (IJERA) **2**, 122–189 (2012)
7. Tian, H., Zou, S., Wang, W., Cheng, S.: A group based reputation system for p2p networks. In: Proceedings of Autonomic and Trusted Computing, pp. 342–351. IEEE Press, New York (2006)
8. Yager, R.: On ordered weighted averaging aggregation operators in multicriteria decisionmaking. IEEE Trans. Syst. Man Cybern. **18**(1), 183–190 (1998)
9. Schlosser, M., Condie, T., Kamvar, S.: Simulating a file-sharing p2p network. Technical report. Stanford InfoLab.Standford University (2003)
10. Mekouar, L., Iraqi, Y., Boutaba, R.: A contribution-based service differentiation scheme for peer-to-peer systems. Peer-to-Peer Networking Appl. **2**(2), 146–163 (2009)

The Algorithm of Mining Frequent Itemsets Based on MapReduce

Bo He

Abstract Cloud computing is large scale and highly scalable. The data mining based on cloud computing was a very important field. The paper proposed the algorithm of mining frequent itemsets based on mapReduce, namely MFIM algorithm. MFIM algorithm distributed data according horizontal projection method. MFIM algorithm made nodes compute local frequent itemsets with by FP-tree and mapReduce, then the center node exchanged data with other nodes and combined; finally, global frequent itemsets were gained by mapReduce. Theoretical analysis and experimental results suggest that MFIM algorithm is fast and effective.

Keywords FP-tree · MapReduce · Frequent itemsets · Cloud computing

1 Introduction

The key for mining association rules is finding frequent itemsets [1]. There are various serial algorithms for mining association rules, such as Apriori [2]. However, the database for mining association rules is generally large, traditional serial algorithms cost much time. In order to improve efficiency, some parallel mining algorithms were proposed, which include PDM [3], CD [4], FDM [5]. Most of them

B. He (✉)
School of Computer Science and Engineering, ChongQing University of Technology,
Chongqing 400054, China
e-mail: heboswnu@sina.com

B. He
State Key Laboratory for Novel Software Technology, Nanjing University,
Nanjing 210093, China

B. He
Shenzhen Key Laboratory of High-Performance Data Mining, Shenzhen 518055, China

S. Patnaik and X. Li (eds.), *Proceedings of International Conference on Soft Computing Techniques and Engineering Application*, Advances in Intelligent Systems and Computing 250, DOI: 10.1007/978-81-322-1695-7_62, © Springer India 2014

divide global transaction database into equal n fractions according to horizontal method. In addition, most parallel mining algorithms adopt Apriori-like algorithm, so that a lot of candidate itemsets are generated and database is scanned frequently. Cloud computing is large scale and highly scalable. The data mining based on cloud computing was a very important field. Then, the paper proposed the algorithm of mining frequent itemsets based on mapReduce, namely MFIM algorithm.

2 Related Description

The global transaction database is DB, and the total number of tuples is M. Suppose, $P_1, P_2, ..., P_n$ are n nodes, node for short, there are M_i tuples in DB_i, if DB_i ($i = 1, 2, ..., n$) is a part of DB and stores in P_i, then $DB = \bigcup_{i=1}^{n} DB_i$, $M = \sum_{i=1}^{n} M_i$ mining association rules can be described as follows: Each node P_i deals with local database DB_i and communicates with other nodes; finally, global frequent itemsets of global transaction database are gained by mapReduce.

Definition 1 For itemsets X, the number of tuples that contain X in local database DB_i ($i = 1, 2, ..., n$) is defined as local frequency of X, symbolized as $X.si$.

Definition 2 For itemsets X, the number of tuples that contain X in global database is global frequency of X, symbolized as $X.s$.

Definition 3 For itemsets X, if $X.si \geq$ min_sup*M_i ($i = 1, 2, ..., n$), then X is defined as local frequent itemsets of DB_i, symbolized as F_i. min_sup is the minimum support threshold.

Definition 4 For itemsets X, if $X.s \geq$ min_sup*M, then X is defined as global frequent itemsets, symbolized as F. If $|X| = k$, then X is symbolized as F_k.

Theorem 1 *If itemsets X are local frequent itemsets of DB_i, then any nonempty subset of X is also local frequent itemsets of DB_i.*

Theorem 2 *If itemsets X are global frequent itemsets, then X and all nonempty subset of X are at least local frequent itemsets of a certain local database.*

Theorem 3 *If itemsets X are global frequent itemsets, then any nonempty subset of X is also global frequent itemsets.*

3 MFIM Algorithm

MFIM distributes data according to horizontal projection method that divides M tuples in global transaction database into $M_1, M_2, ..., M_n$ ($\sum_{i=1}^{n} M_i = M$). The aggregation including M_i tuples in the ith node represents $\{T_i^j | T_i^j = O_q$ and $q = n \times (j - 1) + i\}$, T_i^j represents the jth tuple of the ith node, O_q represents the

qth tuple of global transaction database DB. DB is divided into n local databases $DB_1, DB_2, ..., DB_n$ as large as $\lfloor \frac{M}{n} \rfloor$, namely $DB = \bigcup_{i=1}^{n} DB_i$. Because, DB_i gets the tuples of DB via regular separation distance, and global transaction database is divided into n local database evenly, MFIM reduces data deviation.

MFIM sets one node P_0 as the center node, other nodes P_i send local frequent itemsets F_i to the center node P_0. P_0 gets local frequent itemsets $F'(F' = \bigcup_{i=1}^{n} F_i)$ which are pruned by the strategy of top–down. P_0 sends the remaining of F' to other nodes. For local frequent itemsets $d \in$ the remaining of F', P_0 collects local frequency $d.s_i$ of d from each node and gets global frequency $d.s$ of d. Global frequent itemsets are gained by mapReduce.

F' are pruned by the strategy of top–down. Pruning lessens communication traffic.

The strategy of top–down is described as follow.

(1) Confirming the largest size k of itemsets in F'.
(2) Collecting global frequency of all local frequent k-itemsets in F' from other nodes P_i.
(3) Judging all local frequent k-itemsets in F', if local frequent k-itemsets Q are not global frequent itemsets, then Q are deleted from F', else turn to (4).
(4) Adding Q and any nonempty subset of Q to global frequent itemsets F according to Theorem 3 and Deleting Q and any nonempty subset of Q from F'.

The pseudo code of MFIM is described as follows:

Algorithm MFIM Input: The local transaction database DB_i that has M_i tuples and $M = \sum_{i=1}^{n} M_i$, n nodes P_i ($i = 1, 2, ..., n$), the center node P_0, the minimum support threshold min_sup.

Output: The global frequent itemsets F.

Methods: According to the following steps:

Step 1: /* distributing data according to horizontal projection method*/

```
for(q=1;q<=M;q++)
{ if (q mod n= =i)
  { if (i= =0)
      i=n;  /*the qth tuple is in the nth node*/
P0 makes Oq insert to DBi ;  /*Oq represents the qth tuple in
DB */
    }
}
for(i=1;i<=n;i++)
P0 transmits DBi to Pi ;
```

Step 2: /*each node adopts FP-growth algorithm to produce local frequent itemsets by FP-tree and mapReduce*/

```
for(i=1;i<=n;i++) /*gaining global frequent items*/
  {Scanning DBᵢ once;
    computing local frequency of local items Eᵢ;
      Pᵢ sends Eᵢ and local frequency of Eᵢ to P₀;
  }
```

P_0 collects global frequent items E from E_i;

E is sorted in the order of descending support count;

P_0 sends E to other nodes P_i;   /*transmit global frequent items to other nodes P_i */

```
for(i=1;i<=n;i++)
{creating the FP-treeⁱ; /*FP-treeⁱ represent FP-tree of DBᵢ
*/
```

$$F_i = \text{FP-growth}(FP\text{-}tree^i, \ null);$$

/* node adopts FP-growth algorithm to produce local frequent

itemsets F_i */

```
}
```

Step 3: /* P_0 gets the union of all local frequent itemsets and prunes*/

```
for(i=1;i<=n;i++)
```

P_i sends F_i to P_0;   /* F_i represent local frequent itemsets of P_i */

$$F' = \bigcup_{i=1}^{n} F_i$$

P_0 combines F_i and produces F';   /* $\quad$ */

Pruning F' according to the strategy of top-down;

P_0 broadcasts the remain of F';

Step 4: /*computing global frequency of itemsets*/

```
for(i=1;i<=n;i++)
{ for each items d∈the remain of F'
    Pᵢ sends d.si to P₀; /*computing d.si aiming at FP-treeⁱ
*/
}
```

for each items $d \in$ the remain of F'

$$d.s = \sum_{i=1}^{n} d.si$$;   /* $d.s$ represents global frequency of itemsets d */

Step 5: /*getting global frequent itemsets by mapReduce*/

for each items $d \in$ the remain of F'

```
if (d.s>=min_sup*M)
```

$F = F \cup d$;

Fig. 1 Comparison of communication traffic

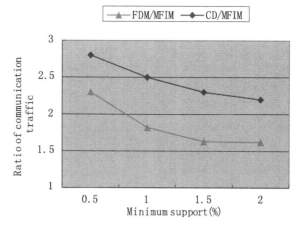

Fig. 2 Comparison of runtime

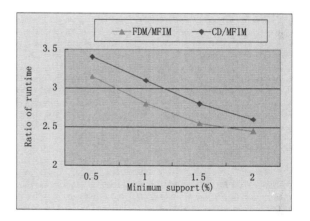

4 Experiments of MFIM

This paper compares MFIM to classical parallel algorithm CD and FDM, takes advantage of VC++6.0 to realize CD and FDM. MFIM compares to CD and FDM in terms of communication traffic and runtime. In the experiments, the number of tested nodes is five except center node. The experimental data comes from the sales data in June 2012 from a supermarket. The results are reported in Figs. 1 and 2.

The comparison experiment results indicate that under the same minimum support threshold, the communication traffic and runtime of MFIM decrease while comparing with CD and FDM.

5 Conclusions

MFIM makes nodes calculate local frequent itemsets independently by FP-growth algorithm and mapReduce, then the center node exchanges data with other nodes and combines by the strategy of top–down. It can promote highly the efficiency of data mining.

Acknowledgments This research is supported by the fundamental and advanced research projects of Chongqing under grant No. CSTC2013JCYJA40039 and the science and technology research projects of Chongqing Board of Education under grant No. KJ130825. This research is also supported by the Nanjing university state key laboratory for novel Software technology fund under grant No. KFKT2013B23 and the Shenzhen key laboratory for high-performance data mining with Shenzhen new industry development fund under grant No. CXB201005250021A.

References

1. Chen, Z.B., Han, H., Wang, J.X.: Data Warehouse and Data Mining. Tsinghua University Press, Beijing (2009)
2. Agrawal, R., Srikant, R.: Fast algorithms for mining frequent itemsets. In: Proceedings of the 20th International Conference Very Large Data Base, Santiago, pp. 487–499 (1994)
3. Park, J.S., Chen, M.S., Yu, P.S.: Efficient distributed data mining for frequent itemsets. In: Proceedings of the 4th International Conference on Information and Knowledge Management, Baltimore, pp. 31–36 (1995)
4. Agrawal, R., Shafer, J.C.: Distributed mining of frequent itemsets. IEEE Trans. Knowl. Data Eng. **8**(6), 962–969 (1996)
5. Cheung, D.W., Han, J.W., Ng, W.T., Tu, Y.J.: A fast distributed algorithm for mining association rules. In: Proceedings of IEEE 4th International Conference on Management of Data, Miami Beach, pp. 31–34 (1996)
6. He, B.: Fast mining of global maximum frequent itemsets in distributed database. Control Decis. **26**(8), 1214–1218 (2011). (in Chinese with English abstract)

Multi-Feature Metric-Guided Mesh Simplification

Hailing Wang, Fu Qiao and Bo Zhou

Abstract This paper proposes a multi-feature metric-guided mesh simplification algorithm, to preserve the crucial visual feature information for simplified model, to solve some mesh simplification algorithms generate poor quality at low levels of detail. Our algorithm is an extended quadric error metrics based on geometric and visual characteristics. The proposed method does not increase the computing time associated with original data and model shape analysis. Moreover, it can preserve the shape features of the models in the processing of simplification. The results of our algorithm have been compared execution time and visual quality with other mesh simplification algorithms, and the results show that our algorithm improved visual shape features for simplified model and got a good quality without additional computing time.

Keywords Mesh simplification · Geometric feature metric · Visual feature metric · Half-edge collapse

1 Introduction

Mesh simplification is a very important research in many domains, such as virtual reality, computer graphics, medical simulation, and scientific visualization. It reduces the number of mesh and allows fast rendering of large-scale models,

H. Wang (✉) · F. Qiao · B. Zhou
College of Computer and Information Engineering, Heilongjiang University
of Science and Technology, Harbin 150027, China
e-mail: whl020519@163.com

F. Qiao
e-mail: 463667708@qq.com

B. Zhou
e-mail: 807571875@163.com

S. Patnaik and X. Li (eds.), *Proceedings of International Conference on Soft
Computing Techniques and Engineering Application*, Advances in Intelligent Systems
and Computing 250, DOI: 10.1007/978-81-322-1695-7_63, © Springer India 2014

which approximate the original ones, to make the compromise between realism and rendering speed. Mesh simplification is a fundamental method for the large amount of triangle surface models.

Many impressive mesh simplification algorithms have been developed for mesh simplification. All the algorithms can approximately address the different aspects of large amount of triangles and get different objectives based on vertex merging [1], edge collapse [2], and triangle decimation [3], and the key issue of them is simplification metric [4, 5], such as distance metric [1, 6], area metric [7, 8], volume metric [9, 10], normal vector [11], and discrete curvature [12], but these metrics always a single one, not retaining the important feature information after simplification. In recent years, multi-feature metric has received increasing attention as an alternative feature-preserving metric for mesh simplification [13]. To determine the performance of simplification algorithm and the quality of simplification result is following fundamental criteria: visual quality, running time, and memory requirement. Different prior algorithms have different advantage and disadvantage in terms of the three factors. Some algorithms can simply estimate error metric by computing the distance of the vertex to the average plane, or the normal angle metric of vertex, so fast to simplify the original model in running times, but produce poor approximated models [14]; some methods pursue better quality simplification results to improve precise error control and generate good quality approximated models, but perform poorly in execution times [15]; many of the algorithms are fast and generate good quality simplification result, but need larger memory overhead [16]. According to our analysis, most of the above algorithms mainly aim to reduce the quantity of polygons, get low level of detail simplification model while maintaining the shape characteristics of the original model with fast running time.

In this paper, we present a novel approach to preserve shape feature of the model. Our algorithm combines geometric error and visual characteristic based on QSlim, which is driven by half-edge collapse. The error metrics include quadric error metrics, weighted by normal information, and the visual feature of torsion metric, weighted by shape of triangle, to retain the shape and detail characteristics. To verify the conclusion, we make comparison with the several published algorithms on visual image result and error detection; experimental results show our algorithm can effectively get high-quality simplified models and reduce error in the simplification process.

2 Multi-Feature Metric Algorithm

2.1 Geometrical Feature Metric

Quadric error metric [6] associates an error to a vertex according to its distance to a quadratic surface. Quadric error metric is an area-weighted metric, that is, the simplification algorithm based on quadric error metric easily preserves large triangles and eliminates small triangles. However, the features of the model generally require many small triangles to represent them. Therefore, if one vertex pair with some large triangles is present on a flat surface and another with some small triangles is present on the feature parts of the model, the QSlim is likely to choose the vertex pair on the feature parts to contract and preserve the vertex pair on the flat surface.

Normal field of a surface model plays fundamental role in its appearance, and it has been used for the metric of the geometric distortion in some geometry analysis and simplification processing [17]. Normal field variation truly represents the importance of a vertex. In view of this evidence in support of the strength of normal field, we use the normal field variation across one-ring neighborhood of a vertex to define its importance. So we make use of normal information to describe the features of the mesh surface:

$$C_{\text{norm}}(v) = \frac{\sum_{i=1}^{k} (1 - \mathbf{n} \cdot \mathbf{n_i})}{k} \tag{1}$$

where $\mathbf{n_0}$ is the normal vector of vertex, v, $\{\mathbf{n_1}, \mathbf{n_2}, \ldots, \mathbf{n_k}\}$ is the set of one-ring neighborhood of vertex v, the included angle between vertex v, and $\{\mathbf{n_1}, \mathbf{n_2}, \ldots, \mathbf{n_k}\}$ is $\mathbf{n} \cdot \mathbf{n_1}, \mathbf{n} \cdot \mathbf{n_2}, \ldots, \mathbf{n} \cdot \mathbf{n_k}$, and they are monotonically decreasing functions for weighting the difference of two normal vectors. $C_{\text{norm}}(v)$ is a higher value in high-curvature areas and small triangles. We use normal information weight quadric error metric for geometrical feature metric as $CA(v) = C_{\text{norm}}(v) \cdot Q(v)$.

2.2 Visual Feature Metric

Visual feature is a very important characteristic for model [18–20]. Although several algorithms, curvature metric, and normal metric can describe the visual feature of the mesh model, the curvature metrics are time-consuming for simplification process, and normal field always does not get good quality approximations because of losing volume. According to the Frenet-Serret formulas, they denote that there is some fixed relationship between curvature and torsion [21]. This illustrates that the torsion is a very important visual feature if with reasonable designing form. To retain the important shape characteristics and not destroy crucial features of original model, the torsion of the model should be considered during the simplification process. Our algorithm reforms the edge contraction based on the foregoing reasons with the torsion error metric; the shape feature cost of edge collapse is defined as follows:

$$T(v) = \frac{\sum\limits_{v_i \in \text{planes}(v)} \|\mathbf{B}_i - \mathbf{B}\|}{\sum\limits_{v_i \in \text{planes}(v)} \|v_i - v\|}, \quad \mathbf{B}_v = v \sum\limits_{v_i \in \text{planes}(v)} h_i \mathbf{n}_i,$$

$$h_i = \begin{cases} (\mathbf{n} \cdot \mathbf{n}_i - \varepsilon)^2 & \text{if } \mathbf{n} \cdot \mathbf{n}_i > \varepsilon \\ 0 & \text{if } \mathbf{n} \cdot \mathbf{n}_i \le \varepsilon \end{cases}$$

(2)

The torsion error metric of vertex, edge, or plane is some approximation to curvatures. In fact, for the same torsion error metric with some big and flat triangles and another with some small and high torsion triangles, the simplification algorithm should likely choose the vertex pair of small and flat triangles to contract and to preserve the vertex pair on the large and high torsion one.

One common measure of surface area is edge length, for example, the area of one triangle is $a = \sqrt{s(s - l_1)(s - l_2)(s - l_3)}$, and $s = 1/2(l_1 + l_2 + l_3)$. However, this measure is very complex and takes more computing time. For geometry theory, given a fixed-length edge e, the larger the area of the triangle e, the better the quality is , for example, equiangular triangle is best; the obtuse triangle is poor. This method is same as with angle, e.g., equiangular triangle is best, obtuse triangle, the opposite angle of e is smaller, it is worse. To improve computing efficiency, one such approximation can be defined as length ratio in the triangles of vertex pair as follows:

$$\varsigma_{\text{length}} = \begin{cases} \dfrac{l_i}{l_{i+1} + l_{i+2} + l_{i+3} + l_{i+4}}, & l_i \text{ is inner edge} \\ \dfrac{l_i}{l_{i+1} + l_{i+2}}, & l_i \text{ is boundary edge} \end{cases}$$

(3)

l_i is one edge of triangles; $\{l_{i+1}, l_{i+2}, l_{i+3}, l_{i+4}\}$ is the other edges in the triangles, which include l_i. So, the visual feature can be obtained as $CB(v) = \varsigma_{\text{length}} \cdot T(v)$

3 Simplification Algorithm Summary

The paper develops an improved method for the surface simplification, combining two parts; one is geometrical feature metric that derives from quadric error metric, weighted by vertex normal information, and the other is visual feature metric that is torsion metric, weighted by edge length ratio of triangles. According to the foregoing analysis, the total error metric function is defined as $\text{Cost}_{\text{total}} = CA + CB$.

The proposed algorithm refers to and amends half-edge contraction methods and QSlim algorithm. Our algorithm is summarized in the five major steps described as follows:

Step 1: Compute the quadric error metric $Q(v)$, normal information $C_{norm}(v)$, torsion metric $T(v)$, and edge length ratio ς_{length} for all the initial vertices in section two.

Step 2: Select the valid vertex-pair (v_i, v_j) as QSlim.

Step 3: Calculate the total cost of the valid vertex-pair from $Cost_{total}$; construct the priority queue in increasing sequences.

Step 4: Iteratively remove the lowest-cost edge from the data sequence based on half-edge contraction, and then update all occurrences of related edges and triangles for the simplification one.

Step 5: Repeat the previous steps until there are no candidate or the error is up to threshold.

4 Experimental Results

The experimental hardware environment is Windows XP with Core E6400, 4G memory, and NVIDIA 9600GT, and the software programming development tool is Visual C++ with DirectX. The experimental models in this work are dog, ox, hand, and raptor. Their detailed information is presented in Table 1.

In the experiments, we also compared our results with those methods by Discrete Curvature Norm (DCN) [12], Visual Feature-Preserving (VFP) [7], QSlim [6], and our algorithm.

Figure 1 shows the simplification results for the dog model using DCN, VFP, QSlim, and our algorithm. The number of triangles of the simplified model is 913 (30 % triangle optimization). The results show that the DCN degenerates the dog model because its curvature characteristics is weaker to lose the details and shape features, while the VFP, QSlim, and our algorithm get a good quality simplified result.

Figure 2 shows the simplification results for the OX model using DCN, VFP, QSlim, and our algorithm. The number of triangles of the simplified model is 11,409(8 % triangle optimization). The results show that the DCM, QSlim, VFP, and our algorithm all can get a good simplification with some shape feature information because simplification ratio is small; however, QSlim and our algorithm are better than the other two methods.

Besides comparison the visual simplification results, the error assessment between the original model and the simplified one, with a percentage of the number of faces in the approximation, is calculated by maximum and mean geometrical errors of Metro tools under Metro detection. In the experiments, we compare the results with those methods by VFP, QSlim, and our algorithm, in which the running time is close. These data show that there is no significant

Table 1 Information of the original models

Model	Vertices	Triangles
Dog	1,263	1,331
Ox	6,211	12,397

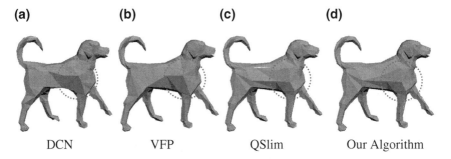

Fig. 1 Simplified model of dog model with different algorithms **a** DCN, **b** VFP, **c** QSlim, and **d** our algorithm

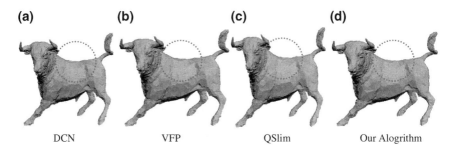

Fig. 2 Simplified model of OX model with different algorithms **a** DCN, **b** VFP, **c** QSlim, and **d** our algorithm

difference between our algorithm and the other algorithms in terms of the high-resolution models, with a small number of triangles simplified; however, these methods differ significantly when a large number of triangles are simplified. Figure 3 shows the experimental results. For the dog model, the proposed algorithm gets respectively, less 6–28 %, 6–36 % than Qslim, VFP in mean geometric error, and less 2–11 % than QSlim in maximum geometric error. For the OX model, the proposed algorithm gets respectively, less 19–25 %, 16–37 % than QSlim, VFP in mean geometric error, and less 6–20 % than QSlim in maximum geometric error.

We can obtain from the experiments that the results in terms of maximum geometrical error are not as clear as the mean geometrical error; the development

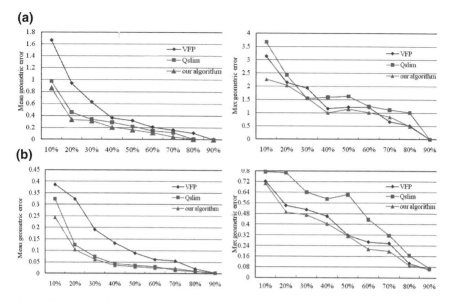

Fig. 3 Error comparison between the simplified model and the original one. **a** Dog model. **b** OX model

tendency and conclusions can be get from these data; our algorithm gives a better simplification quality than QSlim and VFP. The data show the proposed method can achieve a 2–72 % improvement in quality, or error reduction.

5 Conclusions

This paper proposed a novel mesh simplification algorithm with improvements to surface simplification for complex models. The algorithm combines geometrical error metric with quadric-error-metric-weighted normal information and visual feature metric with torsion-metric-weighted edge length ratio. With these metrics, our algorithm preserves the shape feature of its input mesh model in the simplification process. The experiment demonstrated that our algorithm can get efficient effects with the more error parameter for the mesh simplification.

References

1. González, C., Gumbau, J., Chover, M., et al.: User-assisted simplification method for triangle meshes preserving boundaries. Comput. Aided Des. **41**(12), 1095–1106 (2009)
2. Van, J., Shi, P., Zhang, D.: Mesh simplification with hierarchical shape analysis and iterative edge contraction. IEEE Trans. Visual Comput. Graphics **10**(2), 142–151 (2004)

3. Wang, X., Wu, J., Wu, E.: A novel technique based on triangle decimation for tetrahedral simplification. Acta Electronica Sinica **35**(12), 2343–2346 (2007)
4. Qu, L., Meyer, G.W.: Perceptually guided polygon reduction. IEEE Trans. Visual Comput. Graphics **14**(5), 1015–1029 (2008)
5. Menzel, N., Guthe, M.: Towards perceptual simplification of models with arbitrary materials. Comput. Graphics Forum. **29**(7), 261–270 (2010)
6. Garland, M., Heckbert, P.S.: Simplification using quadric error metrics, Proceedings of the 24th Annual Conference on Computer Graphics and Interactive Techniques 1997. pp. 209–216. ACM Press, New York (1997)
7. Hussain, M., Okada, Y., Niijima, K.: Efficient and feature-preserving triangular mesh decimation. J. WSCG **12**(1–3), 167–174 (2004)
8. Park, I., Shirani, S., Capson, D.W.: Mesh simplification using an area-based distortion measure. J. Math. Model. Algorithms **5**, 309–329 (2006)
9. Tang, H., Shu, H.Z., Dillenseger, J.L., et al.: Moment-based metrics for mesh simplification. Comput. Graphics **31**(5), 710–718 (2007)
10. Yuan-Feng, Z., Cai-Ming, Z., Ping, H.: Feature preserving mesh simplification algorithm based on square volume measure. Chin. J. Comput. **32**(2), 203–212 (2009)
11. Schroder, F., Rossbach, P.: Managing the complexity of digital terrain models. Comput. Graphics **18**(6), 775–783 (1994)
12. Kim, S.J., Jeong, W.K., and Kim, C.H. LOD generation with discrete curvature error metric. J. Kiss: Comput. Syst. Theory, 245–254 (2000)
13. Hoppe, H. New quadric metric for simplifying meshes with appearance attributes. Proceedings of the IEEE Visualization'99, Piscataway, pp. 59–66. IEEE Press, New York (1999)
14. Bao-cai, D.X.H.Y., De-hui, K.O.N.G.: Edge collapse simplification based on weighted quadric error metrics. J. Beijing Univ. Technol. **33**(7), 731–736 (2007)
15. Li, L., He, M., Wang, P. Mesh simplification algorithm based on absolute curvature-weighted quadric error metrics. Proceedings of the 5th IEEE Conference on Industrial Electronics and Applications, pp. 399–403. TaiChung (2010)
16. Yang, Jun, Zhu, Shijiao: An image retrieval method by integrating semantics and visual features. J. Comput. Inf. Syst. **7**(9), 3219–3225 (2011)
17. Wei, J., Lou, Y.: Feature preserving mesh simplification using feature sensitive metric. J Comput. Sci. Technol. **25**(3), 595–605 (2010)
18. Shi, Z., Luo, H., Niu, X.: Saliency-based structural degradation evaluation of 3D mesh simplification. IEICE Electron. Express **8**(3), 161–167 (2011)
19. Thomas, D.M., Yalavarthy, P.K., Karkala, D., et al.: Mesh simplification based on edge collapsing could improve computational efficiency in near infrared optical tomographic imaging. IEEE J. Sel. Top. Quantum Electron. **18**(4), 1493–1501 (2012)
20. Lavoue, G., Corsini, M.: A comparison of perceptually-based metrics for objective evaluation of geometry processing. IEEE Trans. Multimedia **12**(7), 636–649 (2010)
21. Bin-Shyan, J., Juin-Ling, T., Wen-Hao, Y.: An efficient and low-error mesh simplification method based on torsion detection. Visual Comput. **22**, 56–67 (2006)

Research on Medical Image Fusion Algorithms Based on Nonsubsampled Contourlet

Junyu Long, Hong Yu and Aiming Yu

Abstract Aiming at the characters of the computer tomography (CT) and Magnetic Resonance Imaging (MRI) medical images, on the basis of nonsubsampled contourlet transform (NSCT) algorithm, an image fusion algorithm on which the low-frequency coefficients are fused on the basis of the largest regional energy and the high-frequency coefficients are fused on the basis of the largest regional variance is proposed. Experiments show that this algorithm can achieve better effects than other image fusion algorithms.

Keywords NSCT · Image fusion · Regional energy · Regional variance · Medical image

1 Introduction

With the development of modern science and technology, medical image fusion has been an important technology in the field of information fusion. Medial image fusion refers to the application of image fusion technology to medical images, that is, acquiring original images from different medical imaging equipments, using some special image processing algorithms, and then fusing a new medical image that contains more disease information of the target objects. There are many kinds of medical images, such as computer tomography (CT) images, Magnetic Resonance Imaging (MRI) images, and so on. Different kinds of medical images can show different information of human organs and sickness tissues, and each kind of images has their strong points and weak points.

J. Long (✉) · H. Yu · A. Yu
Guangdong Vocational Institute of Technology, Zhuhai, 519090 Guangdong, China
e-mail: 401680464@qq.com

S. Patnaik and X. Li (eds.), *Proceedings of International Conference on Soft Computing Techniques and Engineering Application*, Advances in Intelligent Systems and Computing 250, DOI: 10.1007/978-81-322-1695-7_64, © Springer India 2014

CT images cannot display the parenchyma lesions clearly, but as they have high resolution ratio, they can display clearly bones which have large difference in density, which can help doctors to locate the lesions efficiently. The resolution ratio of the MRI images is not very high, and the MRI images can hardly image the bone tissues, but it can image the parenchyma clearly. If we synthesize the strong points of each kind of images, merging a new medical image by modern computer technology, then the medical value of each kind of images will be enhanced, thus assisting the doctors to make a right decision.

The commonly used medical image fusion algorithms at present include the weighted average, Laplace multi-scale pyramid, wavelet transform, and so on [1–8]. Among these algorithms, the weighted average algorithm is simplest, but it ignores the difference of the important messages between the target area from different sensor images, and the apparent jointing marks will be generated in the fusion image, thus not conducive to the following image processing [1]. Compared with the multi-resolution pyramid algorithm, image fusion based on wavelet transform can achieve better image fusion effects [2–7], but there are some defaults in the directional and anisotropy while applying the 2D separable wavelet basis, and only the horizontal, vertical, and diagonal directions can be used. So the wavelet transform cannot describe the image direction information perfectly. In 2002, a "really" 2D image representing algorithm contourlet transform was put forward by Do, M.N [9]. Comparing to the wavelet transform, the contourlet transform not only has the features of multi-scale and regional time–frequency, but also has the feature of multi-direction. But the image should be subsampled while transforming by this algorithm, so the contourlet transform does not have the feature of translation invariance, the pseudo-Gibbs phenomenon will occur by this algorithm, thus causing the image distortion. Aiming at these defects, nonsubsampled contourlet transform (NSCT) algorithm was put forward by Arthur L Cunha and Minh N. Do; this algorithm has the feature of translation invariance and enough redundance information; therefore, the direction information of the being fused image can be extracted effectively, and thereby getting better fusion effect [10, 11].

In this paper, aiming at the features of CT images and MRI images, NSCT algorithm is applied to fuse images and different fusion algorithms are applied to fuse the low-frequency subband and high-frequency subband separately. Experiment shows that by this algorithm, the useful information can be extracted and added to the fusion image effectively and better fusion result can be obtained.

2 NSCT Algorithm

The decomposition of the NSCT algorithm can be done by two steps. Firstly, the image is decomposed into band-pass subimages through multi-scale by non-subsampled pyramid filter bank (NSPFB), and then the band-pass subimages are

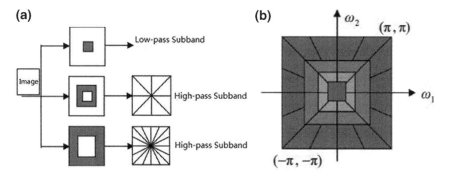

Fig. 1 The flow chart of the NSCT: **a** filter group, **b** ideal frequency distribution

decomposited through directional nonsubsampled directional filter bank (NSDFB); thus, different scales and different directional subband images can be obtained. The nonsubsampled pyramid decomposition has the feature of multi-scale, and the NSDFB has the feature of multi-directional; the combination of these two steps makes the NSCT multi-scale and multi-directional. Besides, NSCT cancels the down-sample and the up-sample step while decompositing and reconstructing an image, only sampling to the filter, the NSCT has the feature of the translation invariance. The flow chart of the NSCT decomposition is shown in Fig. 1; including, Fig. 1a shows the NSCT filter group, and Fig. 1b shows the ideal frequency domain distribution of the origin image after the NSCT decomposition.

After the image is decomposited by NSPFB, all the filters at the current level will be up-sampled to form as the filter group for the next level and each level satisfies the reconstruction condition. Let J be the decomposition scale number, then $J + 1$ band-pass subband images with the same size as the original image will be generated after NSPFB decomposition.

NSDFB module contains a dual-channel nonsubsampled filter group, which can decompose the band-pass images through directional. Using this filter group iteratively, up-sampling the filter at previous level to form as the filter group of current level, more precise directional decomposition will be obtained. Figure 2 shows a four-channel output of NSDFB structural diagram and frequency distribution diagram. Including, up-sampling matrix $Q = [1\ 1;\ 1\ -1]$ is applied to the filter; passing the first scallop filter, image is analyzed into horizontal and perpendicular two directions. Passing the second quadrant filter, image is analyzed into four pennant direction subbands as shown in Fig. 2b.

Let l_j be the directional decomposition number of the jth decomposition scale, then after NSCT decomposition, $1 + \sum_{j=1}^{j} 2^{l_j}$ subband images with the same size as the origin image will be generated.

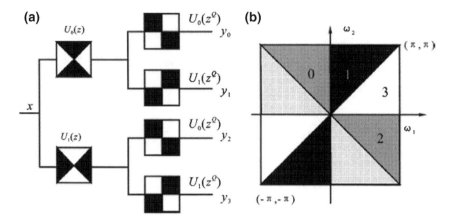

Fig. 2 Four-channel output NSDFB **a** NSDFB (Nonsubsampled Directional Filter Banks) **b** frequency decomposition diagram

3 Research on the Fusion Rules about CT Images and MRI Images Based on NSCT Algorithm

Suppose that the CT and MRI images have been matched strictly before fusing. After the NSCT transform, each image will be decomposed into two parts. The low-frequency part contains the approximate information of the image, that is, the result of lowering the original image's resolution. The fusion image will have better visual effects if the low-frequency coefficients are fused efficiently. The high-frequency part represents the edge and the contour information correspondingly. According to the above characters, in this paper, aiming at the low-frequency part and high-frequency part of the transformed image, different fusion rules are applied separately to get the NSCT coefficients of the fusion image, and then the NSCT coefficients are inversely transformed to rebuild the fusion image.

3.1 Fusion of Low-Frequency Subband

After NSCT, the low-frequency part is the approximate description of the original image, including the original image's average gray scale, texture information, etc. Fusion of the original image's low-frequency part will affect the final fusion result directly. Traditional fusion algorithm for the low-frequency part includes weighted average algorithm, although this algorithm is simple, it will cause some problems such as the fusion image's edge blur, which will interfere with the visual effect of the final fusion image. To some extent, the image's local energy can represent the amount of the information in that area, so the larger the local energy is, the larger the amount of the information will be. Therefore, a fusion algorithm based on local energy is applied to fuse the low-frequency part of the image in this paper.

Let (i, j) be the coordinate for the pixel that is to be fused, $W(m,n) = \frac{1}{16} \begin{bmatrix} 1 & 2 & 1 \\ 2 & 4 & 2 \\ 1 & 2 & 1 \end{bmatrix}$ be the weighted coefficient matrix, and $A_J^0(i,j)$, $B_J^0(i,j)$ be the low-frequency subband pixel value at (i,j) of original image A and B separately, take (i,j) as center, the regional energy $E_A(i,j)$ and $E_B(i,j)$ in the 3×3 regional window can be calculated as follows:

$$E_A(i,j) = \sum_{m=-1}^{1} \sum_{n=-1}^{1} W(m,n) \left[A_J^0(i+m,j+n) \right]^2 \tag{1}$$

$$E_B(i,j) = \sum_{m=-1}^{1} \sum_{n=-1}^{1} W(m,n) \left[B_J^0(i+m,j+n) \right]^2 \tag{2}$$

If there is not much difference between $E_A(i,j)$ and $E_B(i,j)$, we can conclude that the difference at (i,j) between the two images is not evidence, weighted algorithm can be applied to get the fusion result. Otherwise, the pixel value that has larger regional energy should be selected as the fusion result. Defining the normalized local energy $M_A(i,j)$ and $M_B(i,j)$ as follows:

$$M_A(i,j) = \frac{E_A(i,j)}{E_A(i,j) + E_B(i,j)} \tag{3}$$

$$M_B(i,j) = \frac{E_B(i,j)}{E_A(i,j) + E_B(i,j)} \tag{4}$$

Then, final fusion result can be calculated as:

$$F(i,j) = \begin{cases} A_J^0(i,j) & \text{if } M_A(i,j) > T \\ B_J^0(i,j) & \text{if } M_B(i,j) > T \\ M_A(i,j) \times A_J^0(i,j) + M_B(i,j) \times B_J^0(i,j) & \text{otherwise} \end{cases} \tag{5}$$

where T is a threshold value, and in this paper, the value of T is 0.6, and $F(i,j)$ is the fusion result.

3.2 Fusion of High-Frequency Subband

The high-frequency subband coefficients obtained from NSCT decomposition represent the geometrical features and edge features of two original images at different resolution ratios. Between the two images, the one that has larger high-frequency coefficient absolute value corresponds with some sudden change, such as the image's edge, texture, and so on. In this paper, a fusion algorithm based on maximum regional deviation is applied to fuse the high-frequency subband coefficients. Let (m, n) be the coordinate for the pixel that is to be fused at the high-frequency subband, taking (m, n) as center, to the region with size $M \times N$ (in this paper the size is 3×3), the regional deviation can be calculated as follows:

$$V_j^i(m, n) = \frac{1}{M \times N} \sum_{m=1}^{M} \sum_{n=1}^{N} \left[f(m, n) - \overline{f} \right]^2 \tag{6}$$

where $V_j^i(m, n)$ is the image's regional deviation at the jth decomposition level and the ith decomposition direction. $\overline{f}$ is the average deviation value of the region.

Let $V_{A,J}^i$ and $V_{B,J}^i$ be the high-frequency subimage's regional deviation of the source image A and B separately at the jth decomposition level and the ith decomposition direction, then the normalized regional deviation can be defined as follows:

$$\Delta_{A,J}^i = \frac{V_{A,J}^i}{V_{A,J}^i + V_{B,J}^i} \tag{7}$$

$$\Delta_{B,J}^i = \frac{V_{B,J}^i}{V_{A,J}^i + V_{B,J}^i} \tag{8}$$

If there is not much difference between $\Delta_{A,J}^i$ and $\Delta_{B,J}^i$, we can conclude that the difference in the position (i, j) between the two images is not evidence; weighted average algorithm can be used to get the fusion coefficient, otherwise the value of the pixel that has the larger regional deviation should be applied as the fusion coefficient. The final fusion algorithm is shown as follows:

$$F(m, n) = \begin{cases} A^i(m, n) & \text{if } \Delta_{A,J}^i - \Delta_{B,J}^i \geq T \\ B^i(m, n) & \text{if } \Delta_{B,J}^i - \Delta_{A,J}^i \geq T \\ \Delta_{A,J}^i A^i(m, n) + \Delta_{B,J}^i B^i(m, n) & \text{otherwise} \end{cases} \tag{9}$$

where T is the threshold (in this paper, the value of T is 0.6) and $F(m, n)$ is the fusion result.

4 Fusion Result and its Evaluation

Figure 3 shows the original CT image and MRI image applied in our experiment. The steps of our experiment are shown as follows:

(1) Firstly, the original CT and MRI images are transformed by NSCT algorithm separately to get the subband image correspondingly.

Fig. 3 original CT (**a**) and MRI (**b**) images in our experiments

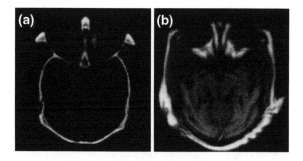

Fig. 4 Experiment results **a** result of weighted average algorithm, **b** result of the Laplace pyramid algorithm, **c** result of the NSCT custom algorithm, **d** the result of the fusion algorithm mentioned in this paper

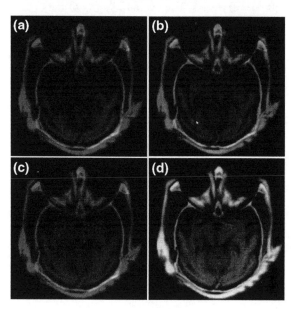

(2) Secondly, each low-frequency subband and high-frequency subband are fused by the algorithm mentioned in this paper separately.

(3) Finally, The NSCT inverse transform is done for the fusion coefficients to get the fused image.

To evaluate the effectiveness of the algorithm mentioned in this paper, we also fused the CT and MRI images by some custom fusion algorithms such as the weighted average, the Laplace Pyramid, and the NSCT custom algorithm. The fusion results are shown in Fig. 4, including, Fig. 4a is the result of weighted average algorithm, Fig. 4b is the result of the Laplace pyramid algorithm, Fig. 4c is the result of the NSCT custom algorithm, and Fig. 4d is the result of the fusion algorithm mentioned in this paper.

From Fig. 4, we can see that the image obtained from weighted average algorithm is lack of resolution, in which the edge is blurry, and some detail edge information is hard to be recognized. The result of the fusion image obtained from

Table 1 The objective evaluation to the image fusion result of the CT image and MRI image

	Information entropy	Mutual information	Average gradient
Weighted average	5.7602	1.9036	0.0286
Laplace pyramid	6.2900	2.0801	0.0439
NSCT custom algorithm	6.6142	3.3267	0.0465
Algorithm of this paper	6.7203	3.4520	0.0472

the Laplace pyramid algorithm is better than the result of the image obtained from the weighted average algorithm, but some detail information of the fusion image is still not very clear. The resolution ratio of the image obtained from NSCT custom algorithm has raised a lot, but the detail contrasts still remain some defaults in the image. The fusion result of the algorithm mentioned in this paper is better than others, in the fusion image, the detail contours are clearly visible, and the resolution ratio has raised evidently.

To evaluate the fusion effect objectively, we took information entropy, mutual information and average gradient as evaluation indexes, and the weighted average, the Laplace Pyramid, the custom NSCT algorithm and the algorithm mentioned in this paper are evaluated by these indexes. The final results are shown in Table 1. From Table 1, we can see that all the indexes obtained from the algorithm mentioned in this paper have raised evidently, which sufficiently proves the effectiveness of the algorithm mentioned in this paper.

5 Conclusions

Aiming at the features of the CT and MRI images, on the basis of NSCT algorithm, NSCT fusion algorithm is studied in this paper, and an image fusion algorithm on which low-frequency coefficients are fused on the basis of maximum local energy and high-frequency coefficients are fused on the basis of maximum region deviation is proposed. Experiments show that in this algorithm the detail information of the CT and MRI images can add to the fusion image accurately, and comparing with the custom image fusion algorithm such as the weighted average algorithm, Laplace pyramid algorithm, and NSCT custom algorithm, better fusion results can be obtained at each fusion indexes. So the algorithm mentioned in this paper is an efficient image fusion algorithm.

References

1. Wang, Z., Djemel, Z., Costas, A., et al.: A comparative analysis of image fusion methods [J]. IEEE Trans. Geosci. Remote Sens. **9**(18), 2137–2143 (2009)
2. Burt, P.J., Adelson, E.: The Laplacian pyramid as a compact image code. IEEE Trans. Commun. **31**(4), 523–540 (1983)

3. Akerman, A.: Pyramid techniques for multisensor fusion. In: SPIE Sensor Fusion, p. 124 (1992)
4. Toet, A.: Image fusion by a ratio of low-pass pyramid. Pattern Recogn. Lett. **9**(4), 245–253 (1989)
5. Park, J.H., Kim, K.O., Yang, Y.-K.: Image fusion using multiresolution analysis. In: Proceedings of the International Geoscience and Remote Sensing Symposium, vol. 2, pp. 864–866 (2001)
6. Li, S., Yang, B.: Mulitfocus image fusion using region segmentation and spatial frequency. Image Vis. Comput. **26**(7), 971–979 (2008)
7. Pajares, G.: A wavelet-based image fusion tutorial. Pattern Recogn. **9**(37), 1855–1872 (2004)
8. Shi, W., Zhu, C., Tian, Y., Nichol, J.: Wavelet-based image fusion and quality assessment. Int. J. Appl. Earth Obs. Geoinf. **6**, 241–251 (2005)
9. Do, M.N., Vetterli, M.: Pyramidal directional filter banks and curvelets. In: Proceedings of IEEE International Conference on Image Processing, vol. 3, pp. 158–161, Thessaloniki, Greece (2001)
10. Do, M.N., Vetterli, M.: The contourlet transform: an efficient directional multiresolution image representation. IEEE Trans. Image Process. **14**(12), 2091–2106 (2005)
11. Da Cunha, A.L., Zhou, J., Do, M.N.: The nonsubsampled contourlet transform: Theory, design, and applications. IEEE Trans. Image Process. **15**(10), 3089–3101 (2006)

Influence of Previous Cueing Validity on Gaze-Evoked Attention Orienting

Qian Qian, Yong Feng, Lin Shi and Feng Wang

Abstract Perception of an averted gaze can automatically shift an observer's attention toward the location gazed at. The present study intended to investigate the sequential processes between trials in gaze-cueing paradigm. The results showed that cueing effects for the group of trials that was preceded by a valid trial are larger than that for the group of trials that was preceded by an invalid trial. This sequence effect is due to the fact that repeating a trial type (valid or invalid) quickens reaction time (RT), and switching a trial type slows RT. The results also showed that similar sequence effect patterns are induced by both gaze cues and arrow cues. The present results suggest that symbolic cues, no matter biologically significant or not, can induce sequence effects, and sequential processes are common mechanisms in human attention orienting systems.

Keywords Sequence effect · Attention orienting · Gaze perception

1 Introduction

Eye gaze, as an important communication tool, has been shown to attract observers' attention [1] and to shift their attention toward the location that the eyes are looking at [2]. In order to investigate the gaze-evoked attention shift, gaze-cueing paradigm has been used by many researchers. In a typical study of this paradigm, a face stimulus looking left or right was presented to observers, and after a certain time interval [stimulus onset asynchrony (SOA)], observers responded to the appearance of a target to the left or right of the face. Although observers were instructed that the gaze direction of the face stimulus did not predict where the target would occur, reaction time (RT) was reliably faster when the face's gaze was toward the target,

Q. Qian (✉) · Y. Feng · L. Shi · F. Wang
Key Laboratory of Computer Technology Applications, Kunming University of Science and Technology, Cheng Gong Da Xue Cheng, Kunming, 650500 Yunnan, China
e-mail: qianqian1025@gmail.com

S. Patnaik and X. Li (eds.), *Proceedings of International Conference on Soft Computing Techniques and Engineering Application*, Advances in Intelligent Systems and Computing 250, DOI: 10.1007/978-81-322-1695-7_65, © Springer India 2014

rather than away from it. This facilitation of RT is referred to as the gaze-cueing effect and is considered to be an evidence of attention orienting [3].

In gaze-cueing paradigm, there were two different cue validity states for experimental trials: valid trials, in which the target occurred either on the left or right side as indicated by the gaze cue, and invalid trials, in which the target occurred at the location that was not indicated by the gaze cue. Randomly changing the cue validity states (valid and invalid) from trial to trial has been used as a common experimental setting in many previous investigations [2, 4]. However, to the best of our knowledge, whether the change of cue validity states between trials could influence gaze-cueing effects has not been investigated. Such sequence effects are important because they may reflect some memory mechanisms of attention orienting in humans and can provide better understanding of the gaze-cueing paradigm for future researches.

Similar to gaze cues, a pointing arrow also represents a direction of interest in the space and has been proved to be able to induce cueing effects reflexively. Furthermore, several recent studies have found a sequence effect between trials in arrow-cueing paradigm. By using a non-predictive central arrow cue, Qian et al. [5] reported that the cueing effect (i.e., RT of invalid trials—RT of valid trials) was larger after a valid trial than after an invalid trial. They explained this sequence effect as automatic memory processes [6] in which information of previous trials is automatically retrieved from memory to facilitate performance on current trials. For example, when the previous trial type (valid or invalid) is consistent with the current trial type, performance will be facilitated, whereas when the previous and current trial types differ, performance is slowed due to the conflict between the two trial types. This automatic retrieval hypothesis is in line with the view from exogenous cueing studies that used peripheral cues [7].

Although there are some subtle differences, gaze cues and arrow cues are similar in many respects [8, 9]. Therefore, it is very likely that the sequence effects can also be induced by gaze cues.

The goal of this study is to investigate whether the change of cue validity states between trials influences gaze-cueing effect and whether the sequence effects by gaze cues and arrow cues are different. In addition, in order to investigate the influence of cue-target SOAs, one relatively short SOA (100 ms) and one relatively long SOA (500 ms) were included in the present study.

2 Method

2.1 Participants

A total of 21 students (with a mean age of 26 years, range 20–32 years, 8 females) from Kochi University of Technology in Japan consented to participate in this experiment. All participants reported normal or corrected-to-normal vision and were naive as to the purpose of the experiment.

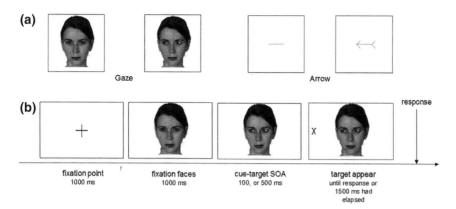

Fig. 1 *Panel A* Stimuli illustration in the present experiment. *Panel B* Illustration of the sequence of events in a valid trial. **a** Stimuli. **b** Procedure

2.2 Apparatus

The stimuli were presented on a LCD display operating at a 60 Hz frame rate that was controlled by a Dell Pentium computer. The participants were seated approximately 60 cm away from the screen.

2.3 Stimuli

A cross, subtending 1°, was placed at the center of the screen as a fixation point. Face photographs and arrows, as illustrated in Fig. 1a, were included as central cues. For real gaze cues, the central fixation stimulus was a photograph of a female face with direct gaze, about 4° wide and 7° height, displayed in eight-bit grayscale. The face photograph was manipulated to produce the left-gaze and right-gaze cues by cutting out the pupil/iris area of each eye and pasting it into the left and right corners of each eye, respectively, using Photoshop CS2 software. Thus, only the area within the eyes differed between the cue and straight-gaze stimuli. For arrow cues, the central fixation stimulus was a horizontal line centered on the screen, 3° in length. An arrow head and an arrow tail appeared at the ends of the central line, both pointing left or both pointing right. The length of an arrow, from the tip of the arrow head to the ends of the tail, was 4°. The target stimulus was a capital letter "X" measuring 1° wide, 1° high, and was presented 15° away from the fixation point on the left or right side.

2.4 Design

The cue-target SOAs were 100 and 500 ms. On each trial, cue direction, target location, and SOA duration were selected randomly and equally. There were two sessions, one for gaze cues and one for arrow cues. The order of sessions was counterbalanced across participants. Each session contained four blocks with 100 trials each. In each block, 20 trials were catch trials in which the target did not appear. The participants were instructed not to respond if the target did not appear. Including 20 training trials for each session, there were in total 840 trials for each participant. The RT of the first trial on each block and the RT of the trials followed a catch trial were excluded from analysis.

2.5 Procedure

Participants were instructed to keep fixating on the center of the screen. Figure 1b provides a representative illustration of the sequence of events on a single trial. First, a fixation cross appeared at the center of the screen for 1,000 ms, followed by a central fixation stimulus for 1,000 ms, and then, the cue stimulus appeared. After a certain cue-target SOA, a target letter "X" appeared either at left or right until participants had responded or 1,500 ms had elapsed. The cue stimulus still remained on the screen after the appearance of the target. Participants were instructed to respond when the target appeared by pressing the "SPACE" key with the index finger of their right hand as quickly and accurately as possible. Participants were also informed that the central stimuli did not predict the location in which target would appear and that they should try to ignore the central cues.

3 Results

The participants missed an average of about 0.5 % of the targets and made false-alarm errors on approximately 1.0 % of the catch trials. Anticipations (RT of less than 100 ms) and outliers (RT over 1,000 ms) were classified as errors and were excluded from analysis. After that, responses with RTs exceeding plus or minus two standard deviations of the participant's mean RT were also removed as errors. As a result, about 5.5 and 5.2 % of all trials were removed as errors in gaze-cue and arrow-cue conditions, respectively. The error rates did not vary systematically, and no signs of any speed–accuracy trade-off were observed.

The mean RTs under different conditions can be seen in Fig. 2 (gaze cue) and Fig. 3 (arrow cue). A five-way ANOVA with cue type (gaze cue and arrow cue), previous SOA (pre-100 ms and pre-500 ms), SOA (100 and 500 ms), previous cue

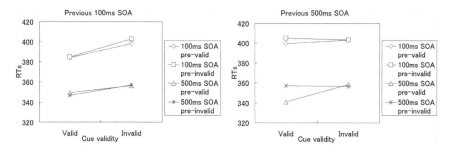

Fig. 2 Mean reaction times (*RTs*) for gaze cue condition under different previous and current cue validity, previous and current SOAs

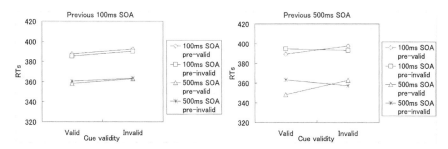

Fig. 3 Mean reaction times (*RTs*) for arrow cue condition under different previous and current cue validity, previous and current SOAs

validity (pre-valid and pre-invalid), and cue validity (valid and invalid) as within-participants factors was conducted on the RTs. There was a main effect of cue validity, $F(1, 20) = 10.607$, $p < 0.004$, indicating cueing effects, i.e., RTs were shorter in valid than in invalid trials. There was significant interaction between previous cue validity and cue validity, $F(1, 20) = 7.628$, $p < 0.012$, demonstrating that the cueing effect was stronger after a valid trial than after an invalid trial, i.e., a typical sequence effect as reported by previous studies. Importantly, the three-way interaction of previous SOA × previous cue validity × cue validity was significant, $F(1, 20) = 7.502$, $p < 0.013$, indicating that the sequence effect mainly appeared only when the SOA of previous trials was 500 ms. The three-way interaction of SOA × previous cue validity × cue validity was not significant ($p > 0.11$), indicating that the sequence effect was showed at both short and long current SOAs. Furthermore, no significant influence of cue type on the cueing effect or the sequence effect was found (ps > 0.10).

4 Discussion

The present study investigated whether the sequence effect could be found in gaze-cueing paradigm. The results showed significant sequence effects in gaze cueing, and the sequence effects by gaze cues did not significantly differ from that induced by arrow cues. Furthermore, for both gaze cueing and arrow cueing, sequence effects were significantly influenced by the SOA of previous trials: When the previous SOA was short (100 ms), no sequence effect was observed, but when the previous SOA was long (500 ms), sequence effects were shown at both short and long current SOAs.

It is now well established that when a gaze shift is observed, the observer's attention rapidly and automatically orients to the same location in space [2]. The common view about such attention shifts is that they are relatively transient and reflexive. One may expect gaze-cueing effects to be purely automatic and not influenced by high-level processes, such as memory. However, recent studies have consistently shown the influence of memory on gaze cueing. Frischen and Tipper [10] demonstrated a memory effect for gaze direction in gaze-cueing paradigm wherein perceiving the gaze cue (e.g., a left gaze) of a specific face can induce attention shift when the face with direct gaze is re-encountered some minutes later. In another study, Frischen and Tipper [11] reported a long-term inhibitory effect wherein RTs were inhibited at a long cue-target SOA (2,400 ms) [12]. These studies clearly suggested that gaze cueing is influenced by the attentional states in previous views. Consistent with this suggestion, the present results showed another form of influence from previous views on gaze cueing. Specifically, gaze-cueing effects are influenced by the memory to the attentional states between two consecutive trials. Such observation provides a more general and complete understanding of the role of memory on gaze-evoked attention orienting.

Furthermore, it was demonstrated that the sequential memory processes took effect in both gaze cueing and arrow cueing. These observations are in line with recent studies that reported similar attentional effects between gaze and arrow cues [9, 13]. The present results extend previous studies by showing that biologically relevant (gaze) and biologically irrelevant (arrow) central cues trigger very similar attention shifts even when the effects between trials are measured. In addition, for both gaze cueing and arrow cueing, the sequence effects observed in present study are influenced by the SOA latency of previous trials. This result can be explained by the different time course of two phases (i.e., initial encoding phase in previous trials and later retrieval phase in current trials) in the sequential processes. Specifically, during the encoding phase, the relationship between a cue and a target may not be encoded into memory if the perceiving time of the cue is not sufficient. On the other hand, once the relationship information is encoded, it will be retrieved rapidly in an automatic way at both short and long SOAs. Another possibility is that when the perceiving time of the cue is short, the encoded relationship information from the previous trial will not be totally updated by the new relationship in the current trial, which in turn impairs the sequence effect in the next trial.

Finally, the significant influence of previous cue validity on cueing effects also has some implications on studies that involved gaze-cueing paradigm, especially for studies that aimed to investigate predictive or counterpredictive gaze cues [8, 14]. For example, when the gaze cue predicts the target location with a rate of 80 percent, there will be more pre-cued trials than pre-uncued trials. As a result, larger average cueing effects for predictive cues than non-predictive cues are due in part to sequence effects, not only due to the voluntary control of observers. Clearly, it is important for future studies to take the influence of sequence effects into account when results are evaluated.

Acknowledgments This research is supported by NSFC (61063027, 61262042, and 31300938), Scientific Research Fund of Yunnan Provincial Department of Education (2013Z134) and Grant-in-Aid of KUT.

References

1. Birmingham, E., Bischof, W., Kingstone, A.: Gaze selection in complex social scenes. Visual Cogn. **16**, 341–355 (2008)
2. Friesen, C., Kingstone, A.: The eyes have it! Reflexive orienting is triggered by nonpredictive gaze. Psychon. Bull. Rev. **5**, 490–495 (1998)
3. Frischen, A., Bayliss, A., Tipper, S.: Gaze cueing of attention: Visual attention, social cognition, and individual differences. Psychonomic Bull. **133**, 694–724 (2007)
4. Langton, S., Bruce, V.: Reflexive visual orienting in response to the social attention of others. Visual Cogn. **6**, 541–567 (1999)
5. Qian, Q., Shinomori, K., Song, M.: Sequence effects by non-predictive arrow cues. Psychol. Res. **76**, 253–262 (2012)
6. Logan, G.: Toward an instance theory of automatization. Psychol. Rev. **95**, 492–527 (1988)
7. Dodd, M., Pratt, J.: The effect of previous trial type on inhibition of return. Psychol. Res. **71**, 411–417 (2007)
8. Friesen, C., Ristic, J., Kingstone, A.: Attentional effects of counterpredictive gaze and arrow cues. J. Exp. Psychol. Hum. Percept. Perform. **30**, 319–329 (2004)
9. Tipples, J.: Orienting to counterpredictive gaze and arrow cues. Percept. Psychophys. **70**, 77–87 (2008)
10. Frischen, A., Tipper, S.: Long-term gaze cueing effects: Evidence for retrieval of prior states of attention from memory. Vis. Cogn. **14**, 351–364 (2006)
11. Frischen, A., Tipper, S.: Orienting attention via observed gaze shift evokes longer term inhibitory effects: Implications for social interactions, attention, and memory. J. Exp. Psychol. Gen. **133**, 516–533 (2004)
12. Frischen, A., Smilek, D., Eastwood, J., Tipper, S.: Inhibition of return in response to gaze cue: Evaluating the roles of time course and fixation cue. Vis. Cognition **15**, 881–895 (2007)
13. Birmingham, E., Kingstone A. "Human social attention: A new look at past, present, and future investigations." The Year in Cognitive Neuroscience 2009: Annals of the New Work Academy of Sciences 2009, 1156, 118-140, 2009
14. Driver, J., Davis, G., Ricciardelli, P., Kidd, P., Maxwell, E., Baron-Cohen, S.: Gaze perception triggers reflexive visuospatial orienting. Vis. Cogn. **6**, 509–540 (1999)

Application of RBF Neural Network in Intelligent Fault Diagnosis System

Yingying Wang, Ming Chang, Hongwei Chen and Ming qian Wang

Abstract With the problem of large-scale system safety and stability increasingly prominently, intelligent fault diagnosis is becoming very important, and this paper puts forward the application of neural network in fault diagnosis system and analyzes in detail the principle, structure model, learning algorithm of radial basis function (RBF) neural network based on the basic principle of neural network knowledge. At last, taking the numerical control (NC) module in a system as an example, combined with the specific characteristics of NC module, this paper gives the structure of neural network diagnosis system and builds the RBF network model for simulation, training, and learning, the result of which shows that the intelligent fault diagnosis method can improve the back-propagation (BP) neural network, is feasible, and has a strong practical value.

Keywords Neural network · Fault diagnosis · RBF neural network · BP neural network

1 Introduction

With the development of computer science and control technology, the scale and complexity of all kinds of system has been increased rapidly, the safety and reliability of equipments is becoming more and more serious, some tiny fault of the system cannot be detected and excluded in time, the whole system may be failed, and even disastrous consequences may be caused. Therefore, fault warning and diagnosis is more important. The fault diagnosis system means to identify causes or properties which cause the system dysfunctional in a certain

Y. Wang (✉) · M. Chang · H. Chen · M. q. Wang
Changchun Institute of Engineering Technology, Changchun, China
e-mail: wyywyqn@163.com

S. Patnaik and X. Li (eds.), *Proceedings of International Conference on Soft Computing Techniques and Engineering Application*, Advances in Intelligent Systems and Computing 250, DOI: 10.1007/978-81-322-1695-7_66, © Springer India 2014

environment, judge parts or components that the deterioration state occurs in, and predict the development trend of the deterioration state, and so on [1]. The significance of the fault diagnosis technology is mainly in reducing the accident rate, decreasing repair costs, cutting the repair time, and increasing the running time. The existing fault diagnosis methods can be divided generally into three categories: (1) based on signal processing; (2) based on analytical model; and (3) based on knowledge. Among them, the diagnosis method based on knowledge is a new theoretical foundation that the computer artificial intelligence technique provides for the fault diagnosis. Because this method does not need precise mathematical model of the object, but has some "smart" characteristics, it is a method with great vitality. The fault diagnosis method of neural network is an important branch of the fault diagnosis method based on knowledge. The author used back-propagation (BP) neural network to construct intelligent fault diagnosis system, and the analysis of the experimental results proves that the BP neural network can play an important role in fault diagnosis, but the BP neural network is easy to fall into local minimum in the application, the radial basis function (RBF) proposed in recent years solves the above problems effectively, and the theory proposes that the RBF is more suitable for fault diagnosis, but lacks empirical. Therefore, the author established the RBF network model, which can be used in the simulation with the same data as [2] to determine its effectiveness in fault diagnosis.

2 The Basic Principle of the Neural Network Knowledge

Simply put, the neural network is an artificial system, which can simulate the structure and function of the human brain with a physical device, system or existing computer. It consists of a large number of simple neurons extensively interconnected to constitute a computational structure, which can simulate the working process of biological neural systems of the human brain in a way.

2.1 RBF Neural Network Theory

RBF neural network uses the RBF as the "base" to build the space of the hidden layer, which can map directly the input vectors to the hidden layer space. When the center point of the RBF is determined, the nonlinear mapping relationship can be defined. The output of the network is the linear weighted sum of the output of the hidden layer unit, and the weights of the network can be obtained directly from the linear equation or least mean square (LMS) method, thus greatly accelerating the speed of learning and avoiding the problem of local minimum.

2.2 RBF Neural Network Structure

The topological structure of RBF neural network is similar to that of multilayer feedforward network, and it is a kind of three-layer feedforward neural network. The nodes of RBF network's input layer transmit only the input signal to the hidden layer, the hidden layer nodes are constituted by radial functions such as Gauss function, and the output layer nodes usually are simple linear functions. The functions (kernel function) of the hidden layer nodes respond to the input signal in local, i.e., when the input signal is near the central range of the nuclear function, the hidden layer nodes will produce larger output. The nodes of the input layer, the hidden layer, and the output layer are, respectively, X, Y, and Z. The function of the hidden layer unit is to transform the input pattern, making the input data from a low-dimensional space into a high-dimensional one in order to classify and recognize in the output layer. The transformation role of hidden units actually can be regarded as extracting the features of the input data, and the transfer function of hidden units of the RBF network is the Gauss kernel.

2.3 Learning Algorithm of RBF Neural Network

The RBF network has three parameters to learn, which are as follows: the center of the basic function Ci, the variance σi, and the weight between the hidden layer and the output layer ωi. According to the selection method of the RBF center, the most common learning methods are self-organizing selection center method, orthogonal least square method, and so on. The method of organizational learning to determine Ci and σi is the clustering method. The clustering method is to cluster the samples into several classes, taking the center of the class as the center of the RBF function, and the common method is k-mean clustering method. Learning algorithm of LMS method is used to learn the weight ωi, and the pseudo-inverse method or least square method can also be used directly. The adjustment rule of LMS weight ω is as follows:

$$e(n) = d(n) - X^T(n)\,\omega(n); \quad \omega(n+1) = \omega(n) + \eta X(n)\,e(n)$$

where $X(n)$ is the output of the hidden layer; $\omega(n)$ is the weight vector; $d(n)$ is the desired output; η is the learning rate; and n is the number of iterations. The structure of the RBF neural network is simple, and the design of it needs less time than the other general feedforward network training. If the number of neurons in the hidden layer is enough, and each layer's weight and threshold are correct, then the RBF function network can accurately approximate any function [3].

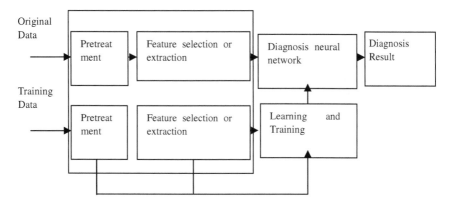

Fig. 1 Structure of the diagnosis system

3 Structure of Diagnosis System

The structure of the diagnosis system based on RBF neural network is shown in Fig. 1.

4 Application Examples and Simulation

This paper takes the numerical control (NC) module fault diagnosis of some equipment as example and applies the RBF neural network knowledge into fault diagnosis and simulation. Output of neurons corresponding to fault types is as shown in Table 1. The RBF model uses one hidden layer structure. The number of nodes in the hidden layer and the center of the hidden layer are determined by the k-means clustering method on the actual output sample data. We mainly focus on the learning accuracy of the network and the generalization ability of it, from which the performance of the network can be measured.

The RBF network is trained by adjusting three parameters, namely the center of the basic function C_i, the variance σ_i, and the weight between the hidden layer and the output layer ω_i, to make the performance tend to optimum, satisfying the requirements of the error and accuracy. This paper uses the orthogonal least square OSL to train the RBF network. Through training the network continuously, finally it is determined that the number of hidden layer neurons is one hundred.

After the RBF neural network is trained using MATLAB, the RBF neural network model containing one hundred neurons is obtained, the function between the input layer and the hidden layer of the model is the Gauss function, and the

Table 1 Output of neurons corresponding to the type of fault

Type of fault	Output of neurons expected				
	Y1	Y2	Y3	Y4	Y5
Transmission fault of spindle	1	0	0	0	0
Power fault of spindle	0	1	0	0	0
Fault of cooling motor	0	0	1	0	0
Fault of digital transfer board A, B	0	0	0	1	0
Disconnection of intermediate cables	0	0	0	0	1

Table 2 Actual simulation data of test samples

Number of the sample	Actual simulation data of the RBF network				
1	1.0000	0.0009	0.0044	0.0066	0.0000
2	0.0000	0.8945	0.0007	0.0020	0.0006
3	1.0000	0.0436	0.0008	0.0015	0.0099
4	0.0000	0.0004	0.9856	0.0090	0.0097
5	0.0000	0.0000	0.0089	0.0058	1.0004
6	0.0000	0.0660	0.0000	0.0037	0.0008
7	0.0000	0.0038	0.0025	0.0000	0.0007
8	0.0000	0.0069	0.0078	1.0000	1.0008
9	0.0000	0.06843	0.0034	1.0000	1.0000
10	0.0000	0.0061	0.9987	0.9984	1.0000

number of training steps is one hundred. The simulation results of the RBF network with MATLAB are shown in Table 2. Through the data comparison, the accurate rate of fault diagnosis based on the RBF neural network is up to 92.8 %.

5 Conclusions

This paper, based on the theory of neural network and the study of the RBF neural network, puts forward the fault diagnosis expert system model based on the RBF neural network knowledge, through simulation in the application process of concrete instance, proves that this method has the advantages of simple structure, easy method, short training time, low error, and high accuracy of the diagnostic results, and overcomes the shortcomings of the BP neural network that is easy to fall into local minimum, especially suits to the system diagnosis that the expert knowledge is difficult to organize and represent. However, RBF remains to be further improved in the treatment of multiple faults; in the future, if we can combine the RBF neural network and other fault diagnosis methods such as the expert system, the diagnosis ability of the system will improve certainly, which will be the next research.

References

1. Abou-Ali, M.G., Khamis, M.: TIREDDX: an integrated intelligent defects diagnostic system for tire production and service. Expert Syst. Appl. **24**(3), 247–259 (2003)
2. Yahia, M.E., Mahmod, R., Sulaiman, N., et al.: Rough neural expert system. Expert Syst. Appl. **18**(2), 87–99 (2000)
3. Zhanf, J., Morris, A.J., Martin, E.B.: Robust process fault detection and diagnosis using neural fuzzy networks. In: Proceeding of IFAC World Congress USA, San Francisco, 169–174 (1996)

An Analysis of the Keys to the Executable Domain-Specific Model

Qing Duan, Junhui Liu, Zhihong Liang, Hongwei Kang and Xingping Sun

Abstract Software development is switched from code-centric to model-centric with model-driven development (MDD). Model is expected to bring an essential leap of software development and drive the whole software development process. It means that model is not only an analysis and design specification, but also a software product which can be automatically transformed into the executable system. That requires the executability of model. This paper presents an executable domain-specific model named xDSM and discusses the keys to making xDSM executable. The keys are accuracy and integrality of the model and its behavior modeling. They all are built based on domain-specific meta-modeling.

Keywords Domain-specific modeling · Executable model · Model-driven development · Meta-modeling

1 Introduction

Efficiency and quality of software development is a matter of the utmost concern of the computer society. In the era of advanced language, the development method goes from structured development to object-oriented development and then to model-driven development (MDD). Model describes system and its environment

Q. Duan
Key Laboratory for Software Engineering of Yunnan Province, Kunming, Yunnan, China
e-mail: qduan@ynu.edu.cn

J. Liu (✉) · Z. Liang · H. Kang · X. Sun
School of Software, Yunnan University, Kunming 650091 Yunnan, China
e-mail: qduan@ynu.edu.cn

Z. Liang
e-mail: zhliang@ynu.edu.cn

H. Kang
e-mail: hankslau@gmail.com

S. Patnaik and X. Li (eds.), *Proceedings of International Conference on Soft Computing Techniques and Engineering Application*, Advances in Intelligent Systems and Computing 250, DOI: 10.1007/978-81-322-1695-7_67, © Springer India 2014

from a given view. It is an abstract representation of system and its environment. For a specific aim, model extracts a set of concepts relevant to the subject in order to make developers focusing on the whole system and ignoring irrelevant details [1].

The executability of model is always an underbelly of MDD for a long time. Software itself is dynamic. Static models can describe some profiles of software. But it can describe neither the entire software nor the running process of software. At the same time, the abstract of models restricts accuracy of models, which makes models lack many of the key elements that are used to construct the entire software. In MDA system, UML can be used to build models of the system from different perspectives and aspects. Model views represent a part or a profile of the system. However, there are neither positive connections nor constraints among those model views. Model views can be more or less and be concrete or abstract. The process of building a model can be ceased at any phase. It is very difficult for modellers to construct a complete software model unless they understand all the details of code generator. That makes the executable models difficult to achieve in UML system.

This paper presents an executable domain-specific model named xDSM and discusses the keys to making xDSM executable. The keys are accuracy and integrality of the model and its behavior modeling. They all are built based on domain-specific meta-modeling.

2 xDSM: Executable Domain-Specific Model

xDSM is constructed based on the domain-specific model and is technically applied to solve the software development problems existing in a certain application domain. xDSM represents the concepts and rules of the domain. The model is targeted that narrows the scope of the description effectively and is helpful to define the model accurately. xDSM modeling process is divided into two phases: the xDSM meta-modeling phase and the xDSM application modeling phase. The former is carried out by domain experts and technical experts, and the latter is carried out by end users. The duty and the role of modellers in each modeling phase are different, as shown in Fig. 1.

xDSM is required to meet Modeling Maturity Levels (MMLs) standard 5 [2]. It requires the model definition to be sufficiently precise. Accuracy here is to describe the details relevant to the modeling objectives accurately rather than to describe all aspects of modeling. The core of xDSM is behavior modeling. It is required to describe domain concepts and system behaviors unambiguously. In the meta-modeling phase, domain concepts are described unambiguously, including domain objects, relationships, constraints, and any operations embodied in the domain concept. In the application modeling phase, the target is to meet all the requirements to software systems. The accurate software behavior modeling is carried out using meta-model. The model does not care about the implementation of local software functions, but it does not ignore the necessary details of the

Fig. 1 xDSM
meta-modeling and
application modeling

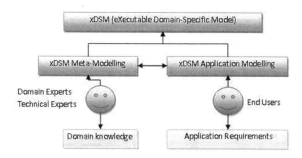

behavior execution yet—the data flow, the control flow, and the related constraints of behaviors must be described in detail.

On the one hand, the measurement of accuracy of models is taken by domain experts and technical experts through xDSM meta-modeling and the domain-specific model execution infrastructure (DSMEI). Namely, if the application model, which is built according to the definition of the meta-model, can be accurately and completely executed by DSMEI, the models can be regarded accurate enough. On the other hand, the application model, which is built in accordance with end users' requirements, can ensure the integrity of the model. Namely, if the results of the application model execution meet the system requirements completely, or the generation system realizes the functional requirements completely, the models can be regarded complete enough. Moreover, application modeling also facilitates the improvement of meta-modeling and the execution environment, to meet the requirements of application modeling better.

Furthermore, the description of the behavior details in xDSM also increases the complexity of modeling. It requires to adjust the complexity of modeling through meta-modeling and application modeling. That is guided by domain experts and developers, mainly in the meta-modeling phase. On the one hand, the behavior complexity is encapsulated in the meta-model, while the behavior details are hidden in domain objects and relationships with the different granularity; on the other hand, the complex behavior descriptions are hidden by the implementation convention of the meta-model and the execution environment. So end users can do the application modeling simply and flexibly. So it is easier for end users to build the executable model with high quality.

3 Keys to xDSM

It is difficult that a model is transformed into the concrete realization if the model is ambiguous and if it is at a high abstract level. The keys to making xDSM executable are accuracy and integrality of the model and its behavior modeling. They all are built based on domain-specific meta-modeling.

3.1 XDSM Meta-Modeling

In domain-specific software development, it is required to define the special modeling language and establish the corresponding modeling environment for the different application domain. But it costs much higher to develop special modeling tools for different modeling languages. The meta-modeling technique is a good solution to this problem [3]. The main idea is that the domain-specific meta-model is customized by domain experts according to the requirements of specific domain, and the meta-model is parsed by the corresponding tools. So a domain-specific modeling language needs to be elicited to support the meta-modeling, and the modeling tools need to be developed to support the domain-specific modeling language. A large number of engineering practices show that the efficiency of domain-specific modeling based on meta-modeling is 10 times higher than that based on UML [4]. There are two kinds of meta-modeling frameworks [5]:

- Modeling method with generic modeling as the core: A meta-model, which is used to describe a modeling language, is established by domain experts with generic modeling tools. It is configured for the generic modeling tools to make the generic modeling tools support the modeling language described by the meta-model.
  The generic modeling tools are also known as Generic Modeling Environment (GME). GME can be used not only to create meta-models (meta–meta-models are configured for the generic modeling tools), but also to build application models (meta-models are configured for the generic modeling tools) [6].
- Meta-modeling based on the modeling tool generator: The first step is to establish meta-model by the modeling tool generator to describe the modeling language. It does not produce the configuration files for the modeling tools during this process, but generate the modeling tools directly, which support that modeling language.

In this paper, the modeling method with generic modeling as the core is used to build the executable model and define xDSM meta-model and xDSM application model in a unified GME. Domain-specific meta-modeling is an approach of the systematic model abstract. The abstract is able to reduce the complexity of models and modeling language, while it is used to describe system characteristics and maintain the validity of the model. The xDSM modeling process is based on the domain-specific meta-modeling approach, which is divided into meta-modeling phase and application modeling phase, while the roles of modellers are separated at the same time, as shown in Fig. 2.

In the phase of xDSM meta-modeling, domain experts analyze the specific domain and establish xDSM meta-model. In other words, domain experts construct models of domain knowledge, extract domain-specific concepts, constraints, rules, and the form of representation, and create domain objects, relationships, and the related constraints. According to xDSM meta-model, domain-specific supporting services are developed by technical experts at the same time. Meta-modeling and

Fig. 2 xDSM
meta-modeling process

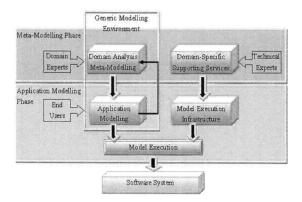

the development of domain-specific supporting services are complementary. While meta-model built in the top-down way determines the requirement specifications and organizational relationship of domain-specific supporting service, the execution of xDSM application model, which is built based on xDSM meta-model, is also supported by domain-specific supporting services. Moreover, xDSM meta-modeling and the development of domain-specific supporting services are negotiated and completed by domain experts and technical experts together.

In the phase of xDSM application modeling, corresponding to application requirements and based on xDSM meta-model, end users use domain-specific concepts to carry out the application entity modeling for the problem domain. The specifications and constraints defined by xDSM meta-model must be abided strictly in the modeling process. xDSM application model established by end users can be executed in DSMEI so as to validate users' application requirements and ensure that application modeling can meet the requirements of software system completely.

Through the separation of meta-modeling and application modeling as well as the role division of modellers, the responsibility of each role can be defined. By integrating system modeling and modellers for xDSM modeling, the maximum value of each role can be brought into play in MDD. The domain knowledge is modeled by domain experts, and the software is controlled and adjusted by end users according to software requirements. The more are controlled by xDSM, the cost of software development and maintenance will be lower, thereby the software productivity is maximized.

3.2 xDSM Behavior Modeling

Software is dynamic and composed of various behavior sets which accomplish the different system objectives. Software specification is objective-oriented because only system objectives are the most direct expression of software system [7].

Fig. 3 Behavior structure

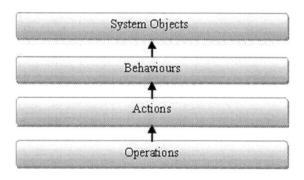

Behavior is the main expression of the system objective. A series of actions are executed in software specification to achieve the system objective. To extract the behavior model corresponding to system objectives and to describe system objectives with behaviors are the keys to the problem-oriented modeling method.

Software behavior is divided into two basic types: the state-related behavior and the state-free behavior. The state-related behavior can be expressed by finite state machine, and the state-free behavior can be expressed with operations. The most of software behaviors are state-free. To the state-related behavior, it is understood here as follows: Being given a message, the responding behavior of the state subject is decided by its current state. The state-related behavior can be also expressed with operations, which is the outcome from parameterizing the states and merging the state transition operations. Consequently, software behaviors can be expressed with operations entirely.

The behavior structure is composed of behaviors, actions, and operations, which are the keys to the domain-specific behavior modeling, as shown in Fig. 3.

Behavior is the direct result of actions of at least a domain concept. Behavior does not exist by itself. It must depend on domain concepts and actions. Action is the basic unit of behavior, which is contained in behavior. Behavior provides execution context and constraint for action and decides their coordination and the execution timing. Action is more concerned about the independence and atomicity of semantics, which is built based on the conceptions that are proved in computer science. Operation is the main representation of action and the basic unit of action specification. An operation gets a group of inputs, which is transformed into a group of output by executing actions. All the input and output can be defined and described by the value specification in detail.

Behavior modeling of xDSM is carried out according to behavior logic, not the simple expression logic and computational logic. Behavior and computation are blended with each other. For decoupling the behavior logic and the computational logic, the logical behavior can focus on describing the coordination relationship between the domain concepts, while the computational process of the implementation details can be ignored. And the computational logic of the atomic operation of domain business can be encapsulated in the services. So the behavior logic based on the above can be modeled, configured, and dynamically loaded.

3.3 Accuracy and Integrality of xDSM

The executable model that is in conformity with MMLs 5 is built based on the accurate and integrate xDSM behavior modeling. MMLs 5 requires that the model can describe the system completely, consistently, detailedly, and accurately and can be transformed into a software system completely and automatically, so as to realize the model execution in the real sense. xDSM modeling is targeted well enough to narrow down the description scope of the mode. The most important thing is that xDSM modeling process is divided into the meta-modeling phase with domain experts and technical experts as the core and the application modeling phase with end users as the core. Through the separation between meta-modeling and application modeling as well as the role division of modellers, the responsibility of each role can be defined, and accuracy and integrity of xDSM can be ensured.

Integrality of xDSM: Corresponding to application requirements and based on xDSM meta-model, end users use domain-specific concepts to carry out the application entity modeling for solving the application problems. The specifications and constraints defined by xDSM meta-model must be abided strictly in the modeling process. xDSM application model established by end users can be executed in DSMEI so as to validate users' application requirements and ensure that application modeling can meet the requirements of software system completely. At the same time, in the process of application modeling, if xDSM application model correctly constructed by end users is insufficient to achieve the domain-specific system objective, xDSM meta-model and domain-specific supporting services will continue to be improved by domain experts and technical experts. It is an iterative process. It will enhance the overall integrity of xDSM.

Accuracy of xDSM: In the process of xDSM modeling, to add the definition of action specifications besides the definition of model elements that improves accuracy of xDSM substantially. In the phase of xDSM meta-modeling, the construction of xDSM meta-model and the development of domain-specific supporting services are negotiated and completed by domain experts and technical experts together. During the process, it is involved with many implicit conventions and constraints. To measure models accurately is determined by domain experts and technical experts with xDSM meta-model and DSMEI. Namely, if the application model, which is built on the meta-model definition, can be executed by DSMEI accurately and completely, the models will be regarded as accurate enough.

4 Conclusions

The paper mainly studies the model executability based on domain-specific model. xDSM is considered as the core of MDD. xDSM emphasizes on the capacity of meta-modeling and adopts the separation of meta-modeling and domain application modeling to establish models that adapt better to specific domain. At the same time, it is able to separate users' application modeling from domain experts' meta-modeling as well as developers' creating support tools.

xDSM is customized for solving software development problems in a certain application areas. It is dedicated and problem-oriented. Although it is at the expense of commonality, it improves the accuracy of the description on domain-specific problems and its solutions and reduces the complexity of modeling.

Acknowledgments This work has been supported by the Open Foundation of Key Laboratory in Software Engineering of Yunnan Province under Grant Nos. 2012SE301, 2010KS04, and 2011SE13 and by the research funding for teacher with a doctorate under Grant No. XT412004.

References

1. Kuhne, T.: What is Model language engineering for model driven software development. Dagstuhl Seminar Proceedings, 2005
2. Davis, M.D., Sigal, R., Weyuker, E.J.: Computability, complexity, and languages, fundamentals of theoretical computer science. Academic Press, Inc, London (2008)
3. Wand, Y.: Ontology as a foundation for meta-modeling and method engineering. J. Inf. Softw. Technol. **38**(4), 281–288 (1996)
4. MetaEdit Inc.: Domain-specific modeling with MetaEdit + 10 times faster than UML, MetaCase, USA, White Paper, 2005
5. Liu, H., Ma, Z.Y., Shao, W.Z.: Progress of research on metamodeling. J. Softw. **19**(6), 1317–1327 (2008)
6. Ledeczi, A., Bakay, A., Maroti, M., Volgyesi, P.: Composing domain-specific design environments. IEEE Comput. **34**(11), 44–51 (2001)
7. Boehm, B., Egyed, A., Kwan, J., Madachy, R.: Developing multimedia applications with the winwin spiral model. Proceedings of ESEC/FSE 97 and ACM Software Engineering Notes, Nov1997

About the Editors

Prof. Srikanta Patnaik is Professor of Computer Science and Engineering at SOA University, Bhubaneswar, India, and also serves as adjunct/visiting professor to many universities inside India and abroad. He is the Chairman and Founder Director of Interscience Institute of Management and Technology, Bhubaneswar, India.

Presently, he is the Editor-in-Chief of the International Journal of Information and Communication Technology and International Journal of Computational Vision and Robotics, published by Inderscience Publishing House, England. He is also Editor-in-Chief of Book Series on "Modeling and Optimization in Science and Technology" [MOST], published by Springer, Germany, and other two series namely "Advances in Computer and Electrical Engineering" (ACEE) and "Advances in Medical Technologies and Clinical Practice" (AMTCP) published by IGI-Global, USA. He has published more than 100 research papers in International Journals and Conference Proceedings. He is now actively engaged for the promotion of creativity and innovation among the engineering and management students in India.

Dr. Xiaolong Li received his bachelor and master degrees in Electrical and Information Engineering Department at the Huazhong University of Science and Technology, Wuhan, China, in 1999 and 2002, respectively. He received his Ph.D. degree in Electrical and Computer Engineering department at the University of Cincinnati in 2006. He joined the Morehead State University in 2006, where he was an Assistant Professor in the department of Industrial and Engineering Technology. In 2008, he joined the department of Electronics and Computer Engineering Technology at the Indiana State University, where he is currently an assistant professor.

His current research interest is in the areas of wireless and mobile networking and microcontroller-based applications. Dr. Li has published more than 40 papers in International Journals and Proceedings. He is one of the editors of Springer book

S. Patnaik and X. Li (eds.), *Proceedings of International Conference on Soft Computing Techniques and Engineering Application*, Advances in Intelligent Systems and Computing 250, DOI: 10.1007/978-81-322-1695-7, © Springer India 2014

series on "Modeling and Optimization in Science and Technology" [MOST] and editor of book "Recent Development on Ad hoc and Sensor Networks" to be published by Springer in April 2014. He also serves as general chair and technical program chair of many international conferences. He is IEEE Communication Society Chapter chair at Central Indiana Section.

Author Index

A

Alamgir, Imtiaz, 457

B

Babyn, Paul, 435
Ba, Haihe, 299
Bai, Xiaoming, 345
Bao, Xingchuan, 33, 315
Bian, Yixin, 141

C

Cao, Wantian, 315
Chang, Ming, 561
Chehmasong, Sulkiplee, 131
Cheng, Xiao-yan, 97
Chen, Hongwei, 561
Chen, Li, 273
Chen, Zhiyong, 147

D

Damkliang, Kasikrit, 131
Deng, Junquan, 185
Deshmukh, A. Y., 403
Ding, Gangyi, 43
Dong, Xuan, 185
Duan, Qing, 567
Duan, Wenying, 345
Du, Xiangyan, 513

F

Fan, Baode, 213
Feng, Gu, 33
Feng, Yong, 553
Fuchao, Hu, 257
Fu, Xiaowei, 273

G

Gang, Lei, 63
Guo, Jinglei, 281
Guo, Libing, 195
Guo, Qian, 33

H

Han, Geyang, 263
Hanghu, Warakorn, 325
Han, Jinhua, 205
Han, Yan, 451
He, Bo, 529
He, Feng, 309
He, Jianwei, 195
He, Jing, 223

J

Jiang, Jiya, 81
Jiang, Rong, 419
Jie, Feng, 63
Jin, Cong, 281
Jing, Huili, 507
Junjie, Liang, 411

K

Kaeoaiad, Chawee, 131
Kang, Hongwei, 567
Kang, Qing, 223
Khandale, Gopichand D., 403
Kui, Wang, 63

L

Liang, Kaidong, 427
Liang, Zhihong, 567
Liao, Hongzhi, 419

S. Patnaik and X. Li (eds.), *Proceedings of International Conference on Soft Computing Techniques and Engineering Application*, Advances in Intelligent Systems and Computing 250, DOI: 10.1007/978-81-322-1695-7, © Springer India 2014

Li, Demin, 105
Li, Fengyin, 123
Li, Hong, 223
Li, Junqiao, 381
Li, Mengmeng, 213
Lin, Xinshi, 369
Lin, Zhongda, 345
Li, Peng, 105
Li, Shuguo, 233
Liu, Chen, 475
Liu, Jihong, 55, 467
Liu, Junhui, 419, 567
Liu, Kai, 369
Liu, Litian, 233
Liu, Lu, 345
Liu, Peiyu, 123
Liu, Tong, 81
Liu, Xiaopeng, 451
Liu, Yongkang, 43
Li, Xia, 247, 361
Li, Xiangming, 195
Li, Xichun, 17
Li, Yaping, 491
Li, Yong-fei, 97
Li, Yonggang, 195
Long, Junyu, 543
Lou, Yan, 71
Luan, Yu-huan, 435
Lu, Changhua, 81
Lv, Jianming, 427
Lv, Shaohe, 185

M

Majumder, S. P., 333
Ma, Peijun, 141, 247, 361
Ma, Yeyun, 369
Mei, Songzhu, 299
Meng, Hai-Dong, 1
Ming, Hai-Ting, 289

N

Ning, He, 257

P

Pan, Yi, 205
Pan, Yijia, 351
Pei, Ao, 499

Q

Qian, Qian, 553
Qian, Yuji, 451
Qiao, Fu, 535
Qin, Xue-Huan, 289
Qiu, Taorong, 345
Quanchao, Hou, 257

R

Rahman, Aminur, 457
Rahman, Munshi Mahbubur, 333
Ren, Jiangchun, 299
Ren, Jing-Pei, 1
Ren, Zhipeng, 71

S

Shen, Hu, 185
Shen, Li, 157
Shi, Lin, 553
Shi, Xuefei, 483
Shi, Yanqing, 81
Song, Shang-ling, 435
Song, Yu-Chen, 1
Song, Yugui, 71
Su, Donglin, 351
Sukpisit, Sukgamon, 325
Sun, Bing, 263
Sun, Feng-rong, 435
Sun, Huaiyu, 309
Sun, Weizhen, 389, 513
Sun, Xingping, 567
Su, Xiangyong, 147
Su, Xiaohong, 141, 247, 361

T

Tan, Chengxiang, 17
Tang, Jun, 381
Tan, Qing, 25
Thakare, Laxman P., 403
Tian, Jing, 273
Tong, Dan, 97
Tu, Fang, 299

W

Wang, Bing, 369
Wang, Eric Ke, 115

Wang, Feng, 553
Wang, Hailing, 535
Wang, Ming qian, 561
Wang, Mingyu, 451
Wang, Qiong, 157
Wang, Rui, 451
Wang, Songxin, 521
Wang, Tiantian, 247, 361
Wang, Wei, 97
Wang, Xiaodong, 185
Wang, Xinggang, 167
Wang, Xiujin, 389
Wang, Yi, 273
Wang, Yingying, 561
Wang, Zheng, 397
Wang, Zhiying, 157, 299
Wan, Shuang, 381
Wongsirichot, Thakerng, 325
Wu, Tao, 233
Wu, Xi, 289

X

Xiao, Qiong, 43
Xiao-ying, Chen, 11
Xia, Yamei, 475
Xie, Hong, 351
Xing, Jianqiang, 369
Xu, Jiajun, 443
Xu, Ran, 177
Xu, Wenting, 55, 467
Xu, Xin, 273
Xu, Yang, 97
Xu, Zhongfu, 351

Y

Yang, Baozhen, 507
Yang, Can, 427
Yang, Lina, 17
Yang, Luwei, 87
Yang, Ming, 419
Yang, Xiaoqiang, 205
Yao, Gui-hua, 435
Yao, Shuzhen, 443
Ye, Yunming, 115
Yong, Xian, 63
Yu, Aiming, 543
Yu, Chaofa, 147
Yue, Xiaojing, 25
Yue-zhai, Zheng, 11
Yu, Hong, 543
Yu, Suping, 491

Z

Zhang, Fanlong, 247
Zhang, Hui, 507
Zhang, Huijuan, 177
Zhang, Xiaolu, 105
Zhang, Yulin, 167
Zhan, Hongfei, 55, 467
Zhao, Weizhou, 507
Zhao, Yiwu, 71
Zhou, Bo, 535
Zhou, Jinbiao, 195
Zhou, Xingming, 185
Zhou, Zelong, 147
Zhu, Mi, 309
Zhu, Wen-Jing, 289

Printed by Publishers' Graphics LLC